Auxin and Its Role in Plant Development

Eva Zažímalová • Jan Petrášek • Eva Benková
Editors

# Auxin and Its Role in Plant Development

Springer

*Editors*
Eva Zažímalová
Jan Petrášek
Laboratory of Hormonal Regulations in Plants
Institute of Experimental Botany AS CR
Prague 6
Czech Republic

Eva Benková
Hormonal Cross-talk Group
Institute of Science and Technology (IST), Austria
Klosterneuburg
Austria

ISBN 978-3-7091-1969-3 ISBN 978-3-7091-1526-8 (eBook)
DOI 10.1007/978-3-7091-1526-8
Springer Wien Heidelberg New York Dordrecht London



Printed on acid-free paper

Springer is part of Springer Science+Business Media (www.springer.com)

# What Is Auxin? How It Operates?

Many articles dealing with plant growth and development start with the "auxin mantra", such as: Auxin is involved in control of many developmental processes in plants.

When in 1881 Charles Darwin and his son Francis examined coleoptiles exposed to unidirectional light, and proposed the existence of a signalling molecule directing their bending, they might have not been fully aware of enormous significance of their discovery for understanding the key principles governing plant growth and development. Since then the mysterious signalling molecule was identified, named auxin, and an immense number of observations confirmed a crucial importance of this tiny compound throughout life of any plant. Besides its for a long time known function in regulation of organ bending in response to light and gravity, auxin was revealed to mediate growth reactions of plants to current environmental conditions in general, and on top of that to control also genetically pre-programmed physiological processes such as embryogenesis, and initiation and formation of diverse organs including flowers, leaves, shoots, roots, and ovules. However, in spite of tremendous progress in the auxin research in last decades, it is still not fully understood how auxin operates and how it can regulate so many and so different processes. So, in spite of years of intensive research bringing much essential information, auxin still remains rather enigmatic.

In this book, respected scientists—experts in different fields of "auxinology"—summarize recent progress in understanding of how auxin operates to control and coordinate plant development. In 18 chapters various aspects of auxin biology focusing on auxin metabolism, transport, signalling, and principles of auxin-regulated plant organogenesis, tropic responses, as well as other interactions with environment are reviewed and future perspectives are outlined.

We hope this compact contemporary overview on the enigma called auxin will inspire new fresh research ideas to address remaining auxin challenges.

Prague and Klosterneuburg
May 2014

Eva Zažímalová
Jan Petrášek
Eva Benková

# Contents

# Part I
# Auxin: Definition; Metabolism, Transport and Signalling

# Chapter 1
# The Auxin Question: A Philosophical Overview

Tom Bennett and Ottoline Leyser

**Abstract** In this opening chapter, we ruminate upon "the auxin question": what is auxin? It is a seemingly simple question with no simple answer. We firstly try to provide a philosophical framework for understanding the question itself. We then discuss some possible answers to the question, and examine how these answers might help to drive the future direction of auxin research. We also offer some speculations on the evolution of auxin, and how such a simple molecule may have accrued such diverse functions.

## 1 Introduction: The Auxin Question

At a recent conference, the following conversation between two imaginary auxin researchers was overheard:

A: "Auxin does everything."
B: "Yes, but what is auxin?"
A: "I have to go now."

As an author, the simultaneous excitement and terror caused by an apparently open-ended brief is difficult to match. For this opening chapter, we were asked simply to reflect upon the question "what is auxin?" It seems innocuous as it lies there on the page, just three words, twelve characters—and yet, like a coiled serpent, it is *very* much longer than it appears and should be approached *very* carefully. The longer the question is contemplated, the harder it becomes to give a straight answer; it is the sort of metaphysical question that might drive one to

T. Bennett • O. Leyser (✉)
Sainsbury Laboratory, University of Cambridge, Bateman Street, Cambridge CB2 1LR, UK
e-mail: OL235@cam.ac.uk

E. Zažímalová et al. (eds.), *Auxin and Its Role in Plant Development*,
DOI 10.1007/978-3-7091-1526-8_1, © Springer-Verlag Wien 2014

madness. It would be pleasing if there was a simple answer at least to begin at, but even taken at its most literal, the question does not have a definitive answer. Perhaps it is not possible to give a perfectly clear, unifying answer, only dim reflections on individual facets of the question; then again, perhaps the answer *is* simple, but we do not really understand the question we are asking.

What follows is our attempt to provide some kind of answer to this question, and to the deeper question that is wrapped wolf-like within the clothing of the first; how does such a simple molecule have such incredibly various effects? We also offer some speculation as to *why* auxin is so complicated; how and why has the system evolved in this way? Forced for once to confront these questions, rather than skirt round them, our own perspectives on auxin have shifted, and as much as anything else, we are critiquing our own previous (and perhaps current) misconceptions. This is inevitably a personal reflection on auxin, and we claim no great authority to answer "the auxin question", but if it is enough to stimulate debate as to the nature of "the auxin answer", then it will have served its purpose.

## 2 A Conceptual Framework

Conceptual frameworks can be very useful in driving science forward, but equally, they can hinder progress if the framework does not correctly formulate the question being addressed. Although it is not quite the sense in which Kuhn (1962) used the term, we might reasonably describe the current "regulator of development" model of auxin action as a paradigm. Over the last 80 years, there has been a repeated shift in the paradigm of auxin, as new discoveries (in many fields) have challenged existing theories (reviewed in Abel and Theologis 2010; see also Estelle 2009). We might persuade ourselves that the recent explosion in our knowledge of auxin means that our paradigm is more robust than in past generations—that we are now nearer "*the answer*"—but such optimistic thinking pervades any paradigm (Kuhn 1962). Certainly, we know more, but the shallowness of our *understanding* is amply demonstrated by the fact that we can still legitimately ask "what is auxin?" and not give a clear answer. It is inevitable that the next generation of auxin scientists will look back on many of our theories and smile at aspects of our misguided logic, just as we view many of the ideas of previous generations.

A conceptual framework is at its most useful when it allows us to synthesise previously unconnected ideas, or to reconcile previously problematic observations. They are also undoubtedly useful in the communication of science, particularly to researchers in other fields. However, they begin to be counter-productive when they become a "truth" in their own right, to be proved and defended. Then, effort is diverted into experiments that seek to confirm rather than test a model ("confirmation bias"), or seek to smooth over its inconsistencies, instead of striving for greater understanding of the actual biological problem. To pick an example from plant science, the ABC model of flower development was, and still is, an excellent framework for understanding mutant phenotypes of *Arabidopsis* and *Antirrhinum* flowers; however, arguably for some it had a constraining effect, with too much effort spent in "perfecting" it, despite its inconsistencies and shortcomings, rather than using it as a starting point for a deeper understanding of flower development (e.g. Gutierrez-Cortines and Davies 2000).

It is inconceivable that we can answer "the auxin question" without some kind of conceptual framework, but what form should this framework take? It seems fairly clear that our current theories are insufficient to explain the bewildering mass of data that we have generated. Partly this is a function of the current funding-dissemination structure of science; we conceptualise our research as easy-to-digest, easy-to-sell, "provable" theories, which are by necessity self-limiting. Can we instead, as a community, establish a robust concept of auxin which brings together all our knowledge, which drives our research forwards, and leads to a genuinely deeper understanding of auxin and its role in plant biology? Such a framework must be only that; a flexible set of ideas, constantly modified as our knowledge increases; we must not cling on to ideas simply because they are convenient or pleasingly tidy.

So, what is auxin? Where do we begin?

## 3 Auxin Is Indole-3-Acetic Acid

Our philosophical struggle to understand auxin begins even with its chemical identity. At least we can reassure ourselves that it was ever thus; the original attempts at purifying auxin identified two auxins (auxin-A and auxin-B) that were in all probability non-existent (Wildman 1997); only somewhat later was an actual auxin identified, namely indole-3-acetic acid (IAA) (Kogl et al. 1934). Nevertheless, the idea of multiple "auxins" became firmly entrenched, no doubt helped by the fact that the second plant hormone to be identified, cytokinin, really does have multiple active forms. Furthermore, the definition of auxin was for a long time based on bioassays rather than an explicit chemical description (Abel and Theologis 2010; Simon and Petrášek 2011). This means that a large number of chemicals showing auxinic activity are described as auxins (or anti-auxins), even though most are not naturally occurring. This has led to the terms auxin and IAA being used non-synonymously. To an extent, this might not matter too much, and the continued use of vague terminology does pleasingly hark back to a golden age in plant research. However, after nearly a century, we really ought finally to define what we mean by auxin. It is remarkable that despite the revolutionary advances in our understanding, we persist with such a hazy definition of a molecule that is so important it warrants a whole book.

The first step in a strict definition must be to reject as an "auxin" any molecule that does not occur naturally in plants or green algae. Standard biological terminology would be to describe these molecules, such as naphthalene-1-acetic acid (NAA) and 2,4-dichlorophenoxyacetic acid (2,4-D) as auxin analogues (or auxin mimics, maybe even auxin agonists). Based on our current knowledge, this leaves us with four endogenous molecules that have been shown to exhibit auxinic activity to some degree: IAA itself, indole-3-butyric acid (IBA), 4-chloro-indole-3-acetic acid (4-Cl-IAA) and phenyl acetic acid (PAA). In the case of IBA, there is relatively little evidence that it has inherent activity, but it is readily converted to IAA and may therefore represent a storage form of IAA (reviewed in Simon and Petrášek 2011). Other storage/degradation forms of IAA (e.g. IAA-lysine and IAA-leucine) are not referred to as "auxins", so there is no pressing reason to describe IBA as an auxin either.

The case for 4-Cl-IAA and PAA is more complex; in both cases auxinic activity has been observed, although PAA does not have as strong effects as 4-Cl-IAA (Simon and Petrášek 2011). There is evidence of PAA binding to both ABP1 and canonical auxin signalling proteins, but it inhibits carrier-mediated auxin transport (Simon and Petrášek 2011; Strader and Nemhauser 2013). On the other hand, 4-Cl-IAA has strong auxinic effects, but has also been found to modulate processes that IAA does not (reviewed in Simon and Petrášek 2011). This suggests that PAA and 4-Cl-IAA are related but fundamentally different signals to auxin, which could perhaps be described as "auxin-like" signals. One possibility is that PAA and 4-Cl-IAA might have evolved in certain plant groups by neofunctionalisation from the core auxin synthesis and signalling pathways; certainly 4-Cl-IAA has only been identified in a few taxonomic groups. However, the genetic basis for PAA and 4-Cl-IAA synthesis and signalling is not clear at the moment, and further work will be needed to establish their exact relationship to the primary IAA signal.

Following these arguments to their logical conclusion, one option is to define auxin as IAA and only IAA. Our rapidly developing understanding of auxin perception raises the possibility that more specific assays could help provide a stricter definition of auxin. Excellent work has elucidated the structure of the auxin binding pocket in the TIR1/AFB family of auxin receptors (Tan et al. 2007). Binding of auxin in this pocket promotes association with members of the Aux/IAA transcriptional repressor family, with the auxin acting as a molecular glue between the two proteins. Different Aux/IAA-TIR1/AFB combinations have different affinities for IAA and auxin analogues, leading to the idea that these two protein families act as co-receptors (Calderón Villalobos et al. 2012; see also Chap. 6). Assessment of binding affinities of IBA, 4-Cl-IAA or PAA could allow them to be ruled in or ruled out as auxins. Similar evidence could be derived from studies of ABP1-type receptors, for which the crystal structure predicts different binding affinities for auxin and auxin analogues (Woo et al. 2002).

## 4 Auxin Is Not a Hormone

Auxin is produced in many tissues and is detected in both the same and other tissues. It is a mobile molecule that can be actively transported over long distances. These properties should be enough to convince anyone that auxin is a signal, but what kind of signal? A word very commonly used in conjunction with auxin is "hormone". In its infancy auxin was often called a "plant growth substance"—*the growth substance*, even (from the German "Wuchsstof")—but there are also early references to it as a hormone (e.g. Snow 1935), and this latter terminology has been widely used, though by no means universally accepted, for a considerable time now. A general definition of a hormone is a substance secreted by specialised cells in one tissue that affects the behaviour of cells in another tissue in a specific manner. Given what we now know about auxin—it has no specialised site of synthesis or action, nor does it have specific effects on cells—it is clear that, whatever else it might be, it does not meet this definition of a hormone.

It could be argued instead that auxin is a "phytohormone": something quite different to the classical concept of a hormone from animal biology defined above. However, given the disparate substances that are usually classed as phytohormones, and the disparate ways in which they work, how exactly can "phytohormone" as a concept be defined? Any definition that included auxin would have to be so vague as to render the concept meaningless. None of this is to say that the hormone concept cannot be applied to plants. The FLOWERING TIME (FT) protein is a much better fit to the classical hormone concept, but is referred to as hormone only infrequently. Similarly, the CLAVATA3 peptide acts in quasi-hormonal manner, albeit over very small distances. We may also eventually come up with a set of properties that defines a separate "phytohormone" concept that includes the lower molecular weight signals from plants, including for example cytokinins and brassinosteroids. Auxin, however, is really something unique, and even the properties of auxin *itself* are difficult to unify into a single concept.

In some ways, this is a rather semantic argument and a rose by any other name would smell as sweet. However, as described above, it is likely that the tacit acceptance of the hormone concept for auxin creates artificial expectations and limitations in our research. We call auxin a hormone, so we expect, sub-consciously perhaps, that auxin will *act* like a hormone. It is reflected in our experimental designs, in our interpretations of data, and in the way that we write about auxin. To give an example; when TIR1 was identified as a proto-typical auxin receptor, there was widespread surprise that the signal transduction pathway was so short (Kepinski and Leyser 2005; Dharmasiri et al. 2005; Callis 2005; Abel and Theologis 2010). We previously knew little about auxin perception, so why should we have been surprised by this particular answer? We were surprised primarily because we conceptualised auxin as a hormone, and most hormones do have long and complicated transduction pathways, full of exotic sounding kinases and GTP-binding proteins. Yet, we knew at the time that auxin was capable of generating incredibly fast transcriptional responses, indicative of a short signal transduction pathway.

However, the real problem with the hormone paradigm runs much deeper. It is absolutely implicit in the concept that hormones are an *instruction* to the target cells to *do* something in particular (e.g. store glucose! moult! flower!). Similarly, it is implicit in our view of auxin that it is an instructive factor, i.e. that it makes cells *do* things. This leads to the "complexity problem"; if auxin is indeed an instruction, how can it be taken to mean so many different things in so many places? An instruction that can be so liberally interpreted is not instruction at all, and this alone should convince us that we are looking at the problem from the wrong angle. Whichever way it is approached, the complexity problem certainly needs an explanation, but perhaps if we stop viewing auxin as an instruction, we can make life easier for ourselves and gain a deeper understanding of the auxin enigma.

It would clearly be wishful thinking to hope that the use of "hormone" to describe auxin might be phased out—in the absence of another convenient category, auxin will probably have to live with its label. Still, auxin is not a hormone. It is *far* more interesting than that.

## 5 Auxin Is Impetus

This list of things that auxin does grows longer by the month. No one working on auxin today needs to be convinced that auxin activity is complex, though some of the pioneers of auxin research might be a little terrified if they had to catch up on the literature. The "complexity problem", outlined above, is the absolute crux of auxin research; how can we understand a signal that has seemingly limitless powers?

One answer to this question is that the canonical auxin signalling system translates auxin into different responses depending on concentration and context (e.g. Kieffer et al. 2010). There is increasing evidence that there are different AFB-Aux/IAA-ARF complexes expressed in each cell type, each with a different affinity for auxin, and different sets of target genes, with the result that in each context a different set of ARFs at a different set of promoters will be activated, with different dynamics (Kieffer et al. 2010; Abel and Theologis 2010). In *Arabidopsis*, there are 6 AFBs, 29 Aux/IAAs and 23 ARFs and therefore a very great number of possible combinations that could operate in different cell types at different times to provide specificity in auxin response (e.g. Rademacher et al. 2011). It is certainly clear that these different complexes have highly quantitative effects on auxin signalling due to differences in the stability of the Aux/IAAs and differences in the affinity of the components for each other (Havens et al. 2012). Since they are the DNA-binding components of the system, the main qualitative effects of auxin signalling on transcriptional output (i.e. which set of genes is activated) are probably mediated by the combination of ARFs that are present in each cell. However, recent work suggests that there are only five major auxin-activated ARFs in *Arabidopsis* (ARFs 5, 6, 7, 8 and 19), meaning that the multiple effects of auxin cannot be explained simply as a function of which ARFs are activated. The remaining ARFs (the so-called repressive type) only interact weakly with the core auxin signalling machinery (Vernoux et al. 2011) and are therefore probably not activated by auxin in themselves; instead, it seems likely that these proteins are cell-type specific inhibitors of auxin response. If these ARFs block specific promoters, they could determine the subset of genes available for activation by positive ARFs in given cell types, thus contributing to the specificity of auxin response; however, there is currently little evidence that repressive ARFs do act in this manner. Indeed, recent work shows that ARFs probably have few intrinsic differences in their

DNA-binding specificity, although it was demonstrated that ARFs act as dimers and that ARF1 and ARF5 dimers have different tolerances for the spacing between pairs of auxin response elements (Boer et al. 2014), so some specificity may arise from the configuration of promoter elements in target genes. The overall impression is that we cannot answer the complexity problem through canonical signalling alone. Where does this leave us? There are of course other signalling systems for auxin; indeed, given the ubiquity of auxin and its very long evolutionary history, it would be rather surprising if there were not. In addition to ABP1, there are tentative suggestions in the community about other possible receptors or signalling components. We should certainly expect more receptors to emerge in the forthcoming years.

As presaged above though, the simplest solution to the complexity problem may be that auxin is not actually an *instruction* to do anything at all; in which case, the concern about how auxin causes so many responses could be reframed and perhaps better understood. However, if auxin is not an instruction, then what kind of signal is it? A previous review by Stewart and Nemhauser (2010) proposed that auxin may act as a kind of cellular currency, permitting many different "transactions" to occur, the nature of the transaction depending on exactly where the auxin is being "spent"; auxin here is seen as a "permissive" type of signal.

The key element of this approach is the switch in focus from an auxin-centric viewpoint to a process-dominated one. Perhaps auxin can be best understood if it is viewed as a signal that provides "impetus" to processes, but does not specify what those processes are. Thus, auxin might actually *do nothing*, but rather motivate everything. In this model, each cell type has a set of processes that are inherent to it (i.e. are developmentally specified), any of which might be "boosted" by auxin, but none of which is directly specified by auxin. To put this in a molecular framework, whether an auxin-influenced gene is "on" or "off" depends on the *other*, cell-type specific transcription factors bound to the promoter of that gene, while the ARFs binding to the promoter act as a kind of rheostat, specifying how much transcription occurs—but ultimately do not control specificity. To give an example, we might imagine a cell type (A) that can differentiate into a second cell type (B) under the influence of a gene (C). C is normally expressed in A, but not at high enough levels to trigger differentiation; however, when auxin is present, transcription of C is boosted, leading to differentiation into B. Thus, auxin does not specify the differentiation of B—that is implicit in the developmental context of A—but it would *appear* to the observer that auxin is the causative factor for B. Auxin does not cause differentiation to B in other contexts, because C is only ever expressed in the context of A; auxin specifies neither C activity nor B formation and can only drive those processes in the context of A.

While this might not seem very different from current models of auxin action, in many ways that is the whole point: it is a subtle shift in emphasis, but it shifts the complexity problem away from auxin. Under this lens, auxin action is not actually complex; the complexity is in the tissue systems themselves, and the ways in which they each utilise auxin as an impetus to drive different processes. The recent theoretical struggle with auxin has been to try and derive 100 qualitatively different instructions from one molecule; the resolution to this problem may simply be that there is only one auxin signal—impetus.

## 6 Auxin Is Complex

Pleasingly simple though this concept is, auxin will never be *that* straightforward. This "impetus" model may particularly useful in developmental responses to auxin (as contrasted to "simple" growth responses), especially the more subtle patterning effects such as gynoecium development, embryo patterning, etc. The impetus model may also explain the long-standing puzzle of the role of auxin in cell division. Auxin is often seen to regulate cell division (particularly in callus, for instance), and it has long been suggested that it may be a general regulator of division, directly integrated with the cell cycle machinery (e.g. den Boer and Murray 2000). While there is now some evidence for a direct cell cycle effects (Jurado et al. 2010; see also Chap. 7), it is also clear that auxin does not universally promote division. However, if the general role of auxin is to potentiate processes that are already specified, then its role in cell division would become clearer; auxin promotes division (by canonical signalling and/or other pathways) only in contexts where the potential for division is already developmentally specified; for instance, the root meristem (Sabatini et al. 1999), or cambial cells (Snow 1935).

The impetus model will not explain all aspects of auxin action. For instance, there could be some developmental processes in which auxin does act as an instruction, specifying events rather than just driving them; the specification of vasculature, which can often happen out-of-context (Sachs 1981; Sauer et al. 2006), might be an example. Moreover, the difference between instruction and impetus might be blurred in some developmental processes. Auxin has long been considered a general regulator of cell elongation—indeed, this is its proto-typical function, for which it is named—and the many "classical" growth responses to auxin are manifestations of this same phenomenon. The prominence of this response makes it tempting to conclude that a principal, and perhaps ancestral, role of auxin is as an instruction to cells to elongate, but it is nevertheless also clear that elongation is highly context dependent.

Thus, even if we can understand how complexity emerges in responses to auxin mediated by canonical signalling, it may not be possible to explain all these transcriptional effects in a single convenient concept. Add to this the poorly understood role of ABP1-dependent signalling, and the probability of other minor signalling mechanisms, and it is clear that complexity is not merely an artefact of "observer effect", but a real feature of auxin responses—at least in the flowering plants that are most commonly studied. Complexity does not lie just with signalling systems however, because when auxin transport is factored into the equation, things become even more convoluted.

## 7 Auxin Is Connectivity

It is not generally remarked upon, but the properties of the auxin transport system (see also Chap. 5) are not particularly well suited to the actual transport of auxin. Admittedly, it is better than diffusion, but transport rates in the order of 1 cm/h are scarcely impressive, and repeatedly transporting auxin into and out of cells is very inefficient. As an engineering solution, transporting auxin in the phloem or xylem (as does happen to some degree) would be much more effective for either signal transmission or bulk movement of auxin as a "commodity". We might therefore conclude that auxin transport is not really about transport per se, but rather *connectivity*. What the transport system very effectively achieves is to connect every cell in the plant together, in an intricate web of auxin, flowing slowly but inexorably from shoot to root. An important implication of the "impetus model" is that auxin has a simple, quantitative informational content, interpreted by cells to modulate cellular processes. Viewed in this perspective, the transport system can therefore be seen a system for distributing this information to cells. However, it is much more than that; it is not simply a one-dimensional flow of information; the auxin stream is constantly altered as it moves through cells and tissues, and further spatial and temporal stimuli can be integrated into the system en route. In this model, moving auxin through cells (rather than bypassing them) is a key feature of the system, because it allows all cells both to connect to the system and to modify it. Other important properties of the system in this respect are that it is hierarchical—so that, for instance, some cells/tissues have disproportionate input into the system or disproportionate exposure to the transport stream—and it is directional—so that the propagation of information is not equal between all cells or in all directions.

All cells can modify, in reflection of their own status, the amount of signal they pass on to neighbouring cells by adding or subtracting auxin, or by changing the immediate kinetics of transport. Due to the high connectivity of the system, these modulations can have both local and global effects, meaning cells can influence multiple processes on a global scale, without having to emit multiple signals. Of course, the effect of a single cell will be minimal, but coordinated action by a group of cells (a meristem, for instance) would be able to alter significantly development across the whole plant. Overall, the topology of the auxin transport system permits auxin to have a much greater influence over development than if it were merely a long-distance signal, travelling through the vascular tissues. This is not to argue that this topology is inherently advantageous per se—indeed, for other signals which mediate specific effects, direct transmission is much preferable—only that in the peculiar case of auxin this highly connective method of transport greatly increases the influence and complexity of this singular signal.

As an example, we can consider the *Arabidopsis* root meristem. At the hub of this system is the tip of the root; the quiescent centre (QC), which acts as the organising centre of the meristem, and the columella root cap, which is particularly important in detecting environmental stimuli (including gravity) and directing root growth accordingly. Auxin promotes cell division and elongation in the root meristem and also drives tropic behaviour in the growth of the root (see also Chap. 16). These effects are mediated by basipetal/shootward streams of auxin moving from the root tip through cell files in the lateral root cap and epidermal layers of the root. The columella controls allocation of auxin into these cell files, and by varying the distribution of auxin between two sides of the root can control asymmetric root growth. Auxin can also be produced in the root tip, allowing the root to control the amount of growth (as distinct from the direction of growth) in relation to local stimuli (Stepanova et al. 2008). Here, auxin is being used as an informational signal to connect the "sensory" cells in the root tip, which control the amount and distribution of auxin, to the meristematic cells where auxin provides impetus for cell division. In theory, the meristematic cells could synthesise their own auxin to drive division, but the lack of connectivity with the tip (and by extension with other meristematic cells) would prevent whole-organ spatio-temporal coordination of growth; so in this context, the key feature of auxin transport is the connectivity it allows, and not the final distribution of auxin that it produces. Furthermore, as is long established, the roots act as a sink for shoot auxin, and there is certainly bulk transport of auxin from shoot to root (e.g. Bhalerao et al. 2002). There is thus a rootward stream of auxin, linking active shoot meristems with active root meristems, that developmentally connects the shoot and root. This auxin stream is not required to drive root growth, on short timescales at least, since the roots can both synthesise their own auxin and recycle the existing pool (Stepanova et al. 2008; Grieneisen et al. 2007), but it allows global root growth to be modulated with respect to shoot growth. Again, connectivity is the key feature of the system and not the supply of auxin per se. The auxin from the shoot moves through the centrak cell files of the root, to the tip, where this global signal can be integrated with the local auxin stimuli. The root tip cells can either add auxin to the pool, or remove it, and spatially distribute the auxin as appropriate, to generate an integrated impetus signal tailored to the circumstances of that individual root meristem (see also Bennett and Scheres 2010). The directionality of the system means that roots cannot back-signal to shoots directly through transport of auxin; instead, roots produce other signals, such as strigolactones, which are transported in the xylem (Kohlen et al. 2011) and act to regulate PIN1 protein abundance in the shoot (Crawford et al. 2010; Shinohara et al. 2013), thus allowing shoot growth to be regulated with respect to root growth and establishing feedback between the two tissue systems.

We can thus observe that the connectivity of the auxin transport system allows many different stimuli, local and global, to be integrated into a single quantitative signal at the point of effect. The auxin transport system greatly increases the apparent, and indeed actual, complexity of auxin responses, since to understand how auxin "does" so many things, it is not merely enough to know how much auxin is in a given tissue and how that tissue will respond to the auxin; it is also important to know how the auxin is transported through the tissue, and how the tissue modifies the signal. However, a further dimension of complexity still is added by the remarkable emergent properties of the auxin transport system.

## 8 Auxin Is Spooky

When it comes to astonishing insights into the very nature of the universe, Albert Einstein certainly has a good publication record, including many contributions to the development of quantum mechanical theory. One of the emergent properties of this theory is "quantum entanglement", which suggests that particles that physically interact and are then separated continue to be "entangled"; they instantaneously detect and respond to changes in each other's state, violating the speed of light in the apparent transmission of information between them. This was a step too far for even Einstein, who dismissed the notion as "*spukhafte Fernwirkung*" ("spooky action at a distance"), although entanglement has since found considerable experimental support and is now a well-established part of standard models of quantum theory. Einstein made rather fewer memorable contributions to the field of auxin research—one cannot help but feel that he ducked the *really* difficult questions—but perhaps he was put off by plant science's own version of the entanglement problem and the rather spooky effects generated by the auxin transport system.

It is generally accepted that auxin transport streams are self-organising, although we still have little grasp on how such self-organisation occurs (see also Chap. 14). Many models have been proposed, and mathematically analysed, to explain the emergence of various types of auxin transport phenomena (particularly "up-the-gradient" and "with-the-flux" patterns), but those models do not capture all phenomena and also lack an explicit mechanistic basis for core aspects of their operation. It is currently a key challenge in auxin biology to unite the experimental and theoretical aspects of this problem into a cogent explanation of these self-organising behaviours. From the point of view of the "auxin question", the most interesting aspects of self-organisation in the auxin transport system are those properties that apparently allow auxin to act as a "signal" between cells or tissues without any actual movement of signal from one to the other. In essence, it seems that because cells and tissues are "entangled" through the transport system, events in one location can affect auxin transport in other (non-downstream) locations, allowing cells to detect those events and respond accordingly to them, without any actual movement of auxin (or other signals) between the two locations. This is, as Einstein obviously feared, the botanical equivalent of action at a distance; and as far as attempting to explain auxin action to a 6-year-old goes, it is probably the final nail in the coffin. It is not enough to know where auxin is, or where it is going; the connectivity of the whole system must also be understood.

The starkest example of this action-at-a-distance is seen in Sachs' classic canalisation experiments (Sachs 1981). For instance, an auxin source (A), connected to a vascular bundle with high levels of auxin transport (V), will prevent a second auxin source (B) from establishing a transport connection with V, but if A is removed then a connection between B and V is established. Both the "finding" of V by B and the inhibition of that process by A occur without any apparent signal transmission between V and B, or A and B, but as somewhat *spooky* properties of the system as a whole. Further examples of these emergent properties can be seen in the regulation of vein initiation in leaves (Sachs 1969; Rolland-Lagan and Prusinkiewicz 2005; Scarpella et al. 2006), phyllotaxis in the meristem (Smith et al. 2006; Jönsson et al. 2006; see also Chaps. 10 and 15) and in the regulation of axillary buds by auxin exported from active meristems (Prusinkiewicz et al. 2009).

## 9 Auxin Is Ancient

In the preceding sections, we have tried to provide a framework in which the intricate complexity of responses to auxin can at least be contemplated, but there is no escaping that complexity, and no set of simple rules that can predict what those responses will be. Ultimately auxin can only be understood as a function of the whole system, and reductionist approaches seeking to explain auxin responses in terms of simple molecular events will be limited in their power to do so. To the parsimonious mind, this naturally poses the question of *why* the system is so complicated—how and why did the architecture of the system evolve in this way and not as a system of many separate signals? It is possible that by studying the evolution of auxin, we can begin to unravel some of that complexity and ask whether auxin was ever simple, and if so, what its proto-typical functions were. To misquote Dobzhansky: "nothing to do with auxin makes sense except in the light of evolution".

At least part of the problem in understanding auxin is that there are no real precedents on which to base an explanation; such a generalist signal does not, to our knowledge, exist in other systems and even in plants, auxin is a unique molecule. We must therefore try and understand which particular aspects of plant biology might have promoted the evolution of a signal with these properties that are not present in other systems; the quest to understand auxin thus draws us into questioning the very nature of plants themselves. Since there is convincing evidence for auxin synthesis, response and transport in streptophyte algae, the sister taxa to land plants—indeed, it seems that all the main elements of auxin biology were in place before the evolution of land plants (De Smet et al. 2011; see also Chap. 13)—the answers must at least partially lie underwater and in the past.

Photosynthesis is undoubtedly the quintessence of plants, and at heart a plant is a machine for optimal extraction of light and other key nutrients from the environment. We should therefore expect that early in algal/plant evolution the major factors driving (and constraining) development were environmental cues. A second defining characteristic of plants is the cellulosic cell wall, which also plays a key role in determining the way plants develop. While the cell wall absolutely does not preclude complex cell shapes/differentiation pathways, it does promote the use of relatively undifferentiated cells with simple morphology in development unless there is very good reason to use something more complex. Development in plants therefore tends to be more a consequence of the ways cells are arranged at a tissue level, rather than due to major morphological changes in the cells themselves (though of course there are counter-examples). The plant cell wall also allows two fundamentally different mechanisms of "growth" to occur: cell division, preceded by cell growth in an energy-intensive process, and cell expansion, driven primarily by water-driven expansion of the vacuole, which is in turn made possible by the rigidity of the cell wall. Together, photosynthesis and the cell wall seem like good candidates for plant-specific processes that might have promoted the evolution of auxin, especially in the algal context.

Although many green algae do have complex morphology and differentiation, these are clearly derived forms, and most green algae assume simple morphologies such as filaments or laminae. We can use Coleochaete as an example; these freshwater algae have simple development, producing planar discs of tissue or branching filamentous structures that expand in size with relatively little further differentiation. Like all plants, the ultimate selective force underlying development in Coleochaete must be to optimise resource acquisition to support reproductive effort, and an important determinant of the development of Coleochaete is therefore light harvesting capacity. We would expect the main drivers for growth in these organisms to be the availability of light, together with $CO_2$, nitrogen, phosphorous and other minerals. The inputs into growth are therefore quite complex, but the low developmental potential of the system means that the output is very simple; in discoidal Coleochaete, the disc must expand in a more or less even-fashion, through a combination of cell expansion/elongation and division. Indeed, the cell divisions in this system follow as a natural consequence of cell expansion, and their orientation is apparently determined by the dimensions of the cell (Dupuy et al. 2010), so the primary developmental output is actually coordinated cell expansion. From a parsimony perspective, it would make some sense to integrate the various inputs into a single signal that controls this primary developmental output; and of course, that signal would closely match the properties of auxin. It is currently unknown whether development in Coleochaete is controlled in this manner, or whether auxin plays any role, but given the central role of auxin in cell expansion in land plants, it seems like a reasonable hypothesis.

Regardless of how development is actually regulated in Coleochaete, we can at least rationalise the evolution of an auxin-like signal in an organism that is environmentally sensitive and has low developmental potential. This hypothetical auxin-like signal would also have two further important properties. Firstly, it is a widely distributed signal containing integrated information regarding the environment that could be harnessed to regulate any other process that also required environmental input. Secondly, although it is acting as an "instruction" for growth, the signal is highly generic in its effect, and is not tied to any particular developmental or differentiation pathway; it affects primarily growth rather than development. Thus, utilising this signal for other purposes does not mean that those processes would only happen in a certain context; the signal is contextually neutral. We could therefore imagine that for a new developmental process that was coordinated with the environment—the production of rhizoid-like cells for instance—our hypothetical signal could be used to give "impetus" to the process, but would not directly specify the differentiation of these cells. This would not interfere with the expansion-promoting role of the signal, nor would the expansion role interfere with the production of rhizoids. Any number of processes can subsequently "piggyback" onto the signal in the same way, because the signal is still not tied to any particular developmental context. If the hypothetical signal had a simple transcriptional output (like auxin), then piggybacking onto the signal could simply be a case of introducing appropriate binding sites into the promoters of relevant differentiation genes—a very simple process that can evolve quickly by single base-pair modifications.

We can thus see how an auxin-like system might suit simple algal developmental systems. We can also see that, if such a simple generic signal evolved, it would subsequently provide a very cost-effective way of providing the same information to other processes—because it encodes powerful informational content, and yet has little instructional value. Attuning many processes to the same signal would also allow the efficient co-regulation of many processes in line with the general status of the plant. It is clear that in *Arabidopsis*, a huge proportion of the genome is regulated by auxin, which makes little sense as a response to a specific hormonal signal, but much sense in the context of a universal co-regulator. The power of such a system lies in the initial integration a range of environmental (and perhaps other) inputs into a single signal that controls growth in a generic way; once such a signal existed, it would perhaps be inevitable that this rich vein of information would be tapped over and over again. Of course, there is currently little evidence to suggest that auxin did initially evolve in such a manner, but it is at least a plausible and testable hypothesis. Ultimately, it is clear that auxin exists in and may well regulate growth in green algae; and therefore that as far as understanding the origins of complexity goes, a key future direction for auxin research lies in the past.

In higher plants the homeostasis, transport and signalling of auxin, coupled with the extensive feedback regulation of all these processes by auxin itself, adds up to a staggeringly complex network, usually characterised in terms of loops and more loops (Benjamins and Scheres 2008; Leyser 2010). However, auxin must have started as a simple algal signal, and even alongside its complexity, auxin in modern plants still possesses the kind of simple, generic and universal characteristics that we have discussed in this section. Although it has existed for a very long time, auxin has not undergone the type of sub- or neofunctionalisation that typically occurs in peptide signals (for instance, insulin-like growth factors from animals, or CLE peptides from plants) that would have allowed repeated separation of accumulated functions into new signals, and the subsequent streamlining of each new signal. We must therefore assume that the value of auxin as a universal signal has been sufficiently high throughout plant evolution to retain it as a single signal and to warrant the evolution of this tangled network of regulatory proteins, which allow the information encoded in the signal to be deciphered and utilised in an ever-increasing number of ways. Thus, the ultimate paradox of auxin might be that its complexity is simply an emergent property of its elegant simplicity.

# 10 Conclusion

What is auxin? It is a question that operates on several philosophical levels, from the categorical—*of what type is it?*—through the metaphysical—*what is its essence?*—to the logical—*how does it work?* The urge to categorise is one of philosophy's oldest motivations, but auxin is difficult to classify in a meaningful biological way. As far as we can tell, it is a unique kind of signal, both in its myriad functions and in the complexity of responses to it; to try and put a label on auxin is to obscure its singular nature. Thus, at the simplest level, the somewhat unsatisfactory answer to the auxin question is that auxin is *auxin*, no more and no less.

However, another major motivation of philosophy is to explain the enigmatic—auxin certainly qualifies in that respect—and it is possible to give rather more satisfactory answers to the deeper aspects of the auxin question. In this chapter, we have attempted both to explore the meaning of the question and to formulate some answers to it: not definitive ones, and indeed mostly speculative ones, but hopefully at least inspiring and testable ones. Ultimately, the auxin question is answerable—although definitely serpentine, it is not venomous—but the answer is still very long. It is already as long as this book, and it is likely to continue growing for some time; after all, auxin has been growing plants for perhaps a billion years, and we have only barely just noticed.

# References


Abel S, Theologis A (2010) Odyssey of auxin. Cold Spring Harb Perspect Biol 2:a004572

Benjamins R, Scheres B (2008) Auxin: the looping star in plant development. Annu Rev Plant Biol 59:443–465

Bennett T, Scheres B (2010) Root development-two meristems for the price of one? Curr Top Dev Biol 91:67–102

Bhalerao RP, Eklöf J, Ljung K, Marchant A, Bennett M, Sandberg G (2002) Shoot-derived auxin is essential for early lateral root emergence in Arabidopsis seedlings. Plant J 29:325–332

Boer DR, Freire-Rios A, van den Berg WA, Saaki T, Manfield IW, Kepinski S, López-Vidrieo I, Franco-Zorrilla JM, de Vries SC, Solano R, Weijers D, Coll M (2014) Structural basis for DNA binding specificity by the auxin-dependent ARF transcription factors. Cell 156:577–589

Calderón Villalobos LI, Lee S, De Oliveira C, Ivetac A, Brandt W, Armitage L, Sheard LB, Tan X, Parry G, Mao H, Zheng N, Napier R, Kepinski S, Estelle M (2012) A combinatorial TIR1/AFB-Aux/IAA co-receptor system for differential sensing of auxin. Nat Chem Biol 8:477–485

Callis J (2005) Plant biology: auxin action. Nature 435:436–437

Crawford S, Shinohara N, Sieberer T, Williamson L, George G, Hepworth J, Müller D, Domagalska MA, Leyser O (2010) Strigolactones enhance competition between shoot branches by dampening auxin transport. Development 137:2905–2913

de Smet I, Voss U, Lau S, Wilson M, Shao N, Timme RE, Swarup R, Kerr I, Hodgman C, Bock R, Bennett M, Jürgens G, Beeckman T (2011) Unraveling the evolution of auxin signaling. Plant Physiol 155:209–221

den Boer BG, Murray JA (2000) Control of plant growth and development through manipulation of cell-cycle genes. Curr Opin Biotechnol 11:138–145

Dharmasiri N, Dharmasiri S, Estelle M (2005) The F-box protein TIR1 is an auxin receptor. Nature 435:441–445
Dupuy L, Mackenzie J, Haseloff J (2010) Coordination of plant cell division and expansion in a simple morphogenetic system. Proc Natl Acad Sci U S A 107:2711–2716
Estelle M (2009) Journal club: growth versus development. Nat Rev Mol Cell Biol 10:813
Grieneisen VA, Xu J, Marée AF, Hogeweg P, Scheres B (2007) Auxin transport is sufficient to generate a maximum and gradient guiding root growth. Nature 449:1008–1013
Gutierrez-Cortines ME, Davies B (2000) Beyond the ABCs: ternary complex formation in the control of floral organ identity. Trends Plant Sci 5:471–476
Havens KA, Guseman JM, Jang SS, Pierre-Jerome E, Bolten N, Klavins E, Nemhauser JL (2012) A synthetic approach reveals extensive tunability of auxin signalling. Plant Physiol 160:135–142
Jönsson H, Heisler MG, Shapiro BE, Meyerowitz EM, Mjolsness E (2006) An auxin-driven polarized transport model for phyllotaxis. Proc Natl Acad Sci U S A 103:1633–1638
Jurado S, Abraham Z, Manzano C, López-Torrejón G, Pacios LF, Del Pozo JC (2010) The Arabidopsis cell cycle F-box protein SKP2A binds to auxin. Plant Cell 22:3891–3904
Kepinski S, Leyser O (2005) The Arabidopsis F-box protein TIR1 is an auxin receptor. Nature 435:446–451
Kieffer M, Neve J, Kepinski S (2010) Defining auxin response contexts in plant development. Curr Opin Plant Biol 13:12–20
Kogl F, Haagen-Smit AJ, Erxleben H (1934) Uber die neues auxin (heteroauxin) aus hain, XI Mitteilung. Zeitschr Physiol Chem 228:90sses
Kohlen W, Charnikhova T, Liu Q, Bours R, Domagalska MA, Beguerie S, Verstappen F, Leyser O, Bouwmeester H, Ruyter-Spira C (2011) Strigolactones are transported through the xylem and play a key role in shoot architectural response to phosphate deficiency in nonarbuscular mycorrhizal host Arabidopsis. Plant Physiol 155:974–987
Kuhn TS (1962) The structure of scientific revolutions. University of Chicago Press, Chicago, IL. ISBN 9780226458113
Leyser O (2010) The power of auxin in plants. Plant Physiol 154:501–505
Prusinkiewicz P, Crawford S, Smith RS, Ljung K, Bennett T, Ongaro V, Leyser O (2009) Control of bud activation by an auxin transport switch. Proc Natl Acad Sci U S A 106:17431–17436
Rademacher EH, Möller B, Lokerse AS, Llavata-Peris CI, van den Berg W, Weijers D (2011) A cellular expression map of the Arabidopsis AUXIN RESPONSE FACTOR gene family. Plant J 68:597–606
Rolland-Lagan AG, Prusinkiewicz P (2005) Reviewing models of auxin canalization in the context of leaf vein pattern formation in Arabidopsis. Plant J 44:854–865
Sabatini S, Beis D, Wolkenfelt H, Murfett J, Guilfoyle T, Malamy J, Benfey P, Leyser O, Bechtold N, Weisbeek P, Scheres B (1999) An auxin-dependent distal organizer of pattern and polarity in the Arabidopsis root. Cell 99:463–472
Sachs T (1969) Polarity and the induction of organized vascular issues. Ann Bot 33:263–275
Sachs T (1981) The control of the patterned differentiation of vascular tissues. Adv Bot Res 9:151–162
Sauer M, Balla J, Luschnig C, Wisniewska J, Reinöhl V, Friml J, Benková E (2006) Canalization of auxin flow by Aux/IAA-ARF-dependent feedback regulation of PIN polarity. Genes Dev 20:2902–2911
Scarpella E, Marcos D, Friml J, Berleth T (2006) Control of leaf vascular patterning by polar auxin transport. Genes Dev 20:1015–1027
Shinohara N, Taylor C, Leyser O (2013) Strigolactone can promote or inhibit shoot branching by triggering rapid depletion of the auxin efflux protein PIN1 from the plasma membrane. PLoS Biol 11:e1001474
Simon S, Petrášek J (2011) Why plants need more than one type of auxin. Plant Sci 180:454–460
Smith RS, Guyomarc'h S, Mandel T, Reinhardt D, Kuhlemeier C, Prusinkiewicz P (2006) A plausible model of phyllotaxis. Proc Natl Acad Sci U S A 103:1301–1306

Snow R (1935) Activation of cambial growth by pure hormones. New Phytol 34:347–360
Stepanova AN, Robertson-Hoyt J, Yun J, Benavente LM, Xie DY, Dolezal K, Schlereth A, Jürgens G, Alonso JM (2008) TAA1-mediated auxin biosynthesis is essential for hormone crosstalk and plant development. Cell 133:177–191
Stewart JL, Nemhauser JL (2010) Do trees grow on money? Auxin as the currency of the cellular economy. Cold Spring Harb Perspect Biol 2(2):a001420
Strader LC, Nemhauser JL (2013) Auxin 2012: a rich mea ho'oulu. Development 140(6):1153–1157
Tan X, Calderon-Villalobos LI, Sharon M, Zheng C, Robinson CV, Estelle M, Zheng N (2007) Mechanism of auxin perception by the TIR1 ubiquitin ligase. Nature 446(7136):640–645
Vernoux T, Brunoud G, Farcot E, Morin V, Van den Daele H, Legrand J, Oliva M, Das P, Larrieu A, Wells D, Guédon Y, Armitage L, Picard F, Guyomarc'h S, Cellier C, Parry G, Koumproglou R, Doonan JH, Estelle M, Godin C, Kepinski S, Bennett M, De Veylder L, Traas J (2011) The auxin signalling network translates dynamic input into robust patterning at the shoot apex. Mol Syst Biol 7:508
Wildman SG (1997) The auxin-A, B enigma: scientific fraud or scientific ineptitude? Plant Growth Regul 22:37–68
Woo EJ, Marshall J, Bauly J, Chen JG, Venis M, Napier RM, Pickersgill RW (2002) Crystal structure of auxin-binding protein 1 in complex with auxin. EMBO J 21(12):2877–2885

# Chapter 2
# Auxin Biosynthesis and Catabolism

**Yangbin Gao and Yunde Zhao**

**Abstract** Auxin concentrations in plants are tightly regulated through both biosynthesis and degradation. In the past few years, much progress was made in the area of auxin metabolism. Genetic and biochemical studies in *Arabidopsis* unequivocally established a complete tryptophan (Trp)-dependent two-step auxin biosynthesis pathway in which Trp is first converted into indole-3-pyruvate (IPA) by the TAA family of aminotransferases and subsequently indole-3-acetic acid (IAA) is produced from IPA by the YUC family of flavin monooxygenases. The TAA/YUC pathway is highly conserved in the plant kingdom and is probably the main auxin biosynthesis pathway in plants. Recent work also demonstrated that oxidative degradation of auxin plays an essential role in maintaining auxin homeostasis and in regulating plant development. In this chapter, we discuss the recent advancements in auxin biosynthesis and catabolism.

## 1 Introduction

Auxin is an essential hormone for many aspects of plant growth and development (Zhao 2010). Plants have evolved a sophisticated network to control auxin levels with spatial and temporal precision in response to environmental cues and developmental signals. Indole-3-acetic acid (IAA), the main natural auxin in plants, can be produced from de novo biosynthesis. Free IAA, which is the presumed active form of auxin, can also be released from IAA conjugates including IAA esters, IAA-saccharides, and IAA-amino acids. A third probable route for producing IAA is to convert indole-3-butyric acid (IBA) to IAA using enzymes similar to those

Y. Gao • Y. Zhao (✉)
Section of Cell and Developmental Biology, University of California San Diego, 9500 Gilman Drive, La Jolla, CA 92093-0116, USA
e-mail: y3zhao@ucsd.edu

E. Zažímalová et al. (eds.), *Auxin and Its Role in Plant Development*,
DOI 10.1007/978-3-7091-1526-8_2, © Springer-Verlag Wien 2014

used in β-oxidation of fatty acids. When auxin levels need to be lowered, plants employ several mechanisms to deactivate IAA. IAA can be quickly converted into the presumed inactive forms by reaction of the carboxyl group of IAA with amino acids, sugars, and other small molecules. The IAA conjugates may serve as a first step for the eventual complete degradation of IAA. IAA is also inactivated by oxidation of the indole ring of IAA. For example, IAA can be converted to 2-oxindole-3-acetic acid (OxIAA). In this chapter, we discuss the progress made in the area of auxin biosynthesis and metabolism in the past few years.

## 2 De Novo Auxin Biosynthesis

De novo auxin biosynthesis is broadly divided into two categories: Tryptophan (Trp) dependent and Trp independent. Trp-independent auxin biosynthesis pathway was proposed two decades ago based on results from feeding plants with labeled Trp and Trp biosynthetic intermediates and from studies on Trp-deficient mutants (Wright et al. 1991; Normanly et al. 1993). However, the molecular mechanisms and genes for the Trp-independent pathway are not known. Therefore, the Trp-independent pathway will not be discussed further in this chapter.

Trp has long been known as a precursor for the production of IAA in plants. Feeding plants with labeled Trp yields labeled IAA, indicating that plants have the enzymes to convert Trp to IAA (Wright et al. 1991; Normanly et al. 1993). Many biosynthetic pathways have been elucidated using analytic biochemistry techniques in combination with labeled precursors and intermediates. For example, the biosynthetic routes for brassinolide and ethylene have been established long before the biosynthetic genes have been identified (Yang and Hoffmann 1984; Sakurai and Fujioka 1993). However, the classic feeding and analytic biochemical approaches failed to identify the key components for Trp-dependent plant auxin biosynthesis pathways. There are several reasons for this apparent failure. First, Trp is a precursor for many metabolites (Fig. 2.1). Trp is a precursor for indole-3-pyruvate (IPA), tryptamine (TAM), indole-3-acetaldoxime (IAOx), indole-3-acetamide (IAM), indole-3-acetonitrile (IAN), and indole-3-acetaldehyde (IAAld) (Fig. 2.1). *Arabidopsis* and many other plants have the capacity to produce all of the above-mentioned intermediates (Fig. 2.1) at a given developmental stage (Ouyang et al. 1999; Sugawara et al. 2009). Some of the intermediates such as IAN exist in very high concentrations (Fig. 2.1) (Sugawara et al. 2009). Such a complex profile of Trp metabolism makes it difficult to identify Trp-dependent IAA synthesis intermediates. Second, some of the intermediates are intrinsically unstable in vitro and can be nonenzymatically converted to other compounds during the experimental process, therefore complicating the analysis of metabolic profiling. For example, IPA is readily converted nonenzymatically into IAA in vitro (Bentley et al. 1956). Third, most of the Trp metabolic intermediates display auxin activities during in vitro bioassays (Fig. 2.2). In the presence of IAM in growth media, light-grown *Arabidopsis* seedlings have long hypocotyls and epinastic cotyledons

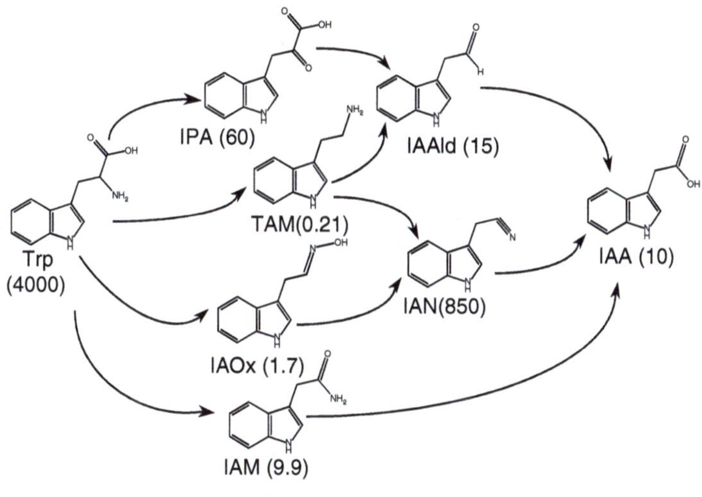


**Fig. 2.1** Selected tryptophan metabolic intermediates. *Arabidopsis* plants produce all of the intermediates shown in the figure. The numbers in parenthesis refer to the actual concentrations in ng/g fresh weight. *IAA* indole-3-acetic acid, *IAAld* indole-3-acetaldehyde, *IAOx* indole-3-acetaldoxime, *IAM* indole-3-acetamide, *IAN* indole-3-acetonitrile, *IPA* indole-3-pyruvate, *TAM* tryptamine, and *Trp* tryptophan

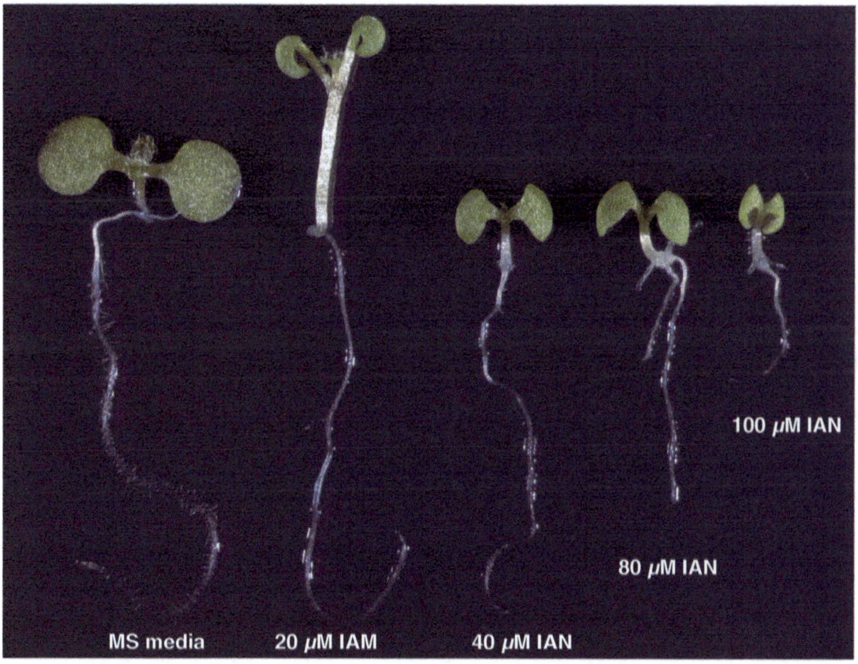


**Fig. 2.2** Some tryptophan metabolites display auxin activities. Indole-3-acetamide (IAM) stimulates hypocotyl elongation and causes epinastic cotyledons. Indole-3-acetonitrile (IAN) inhibits primary root elongation and stimulates adventitious root initiation

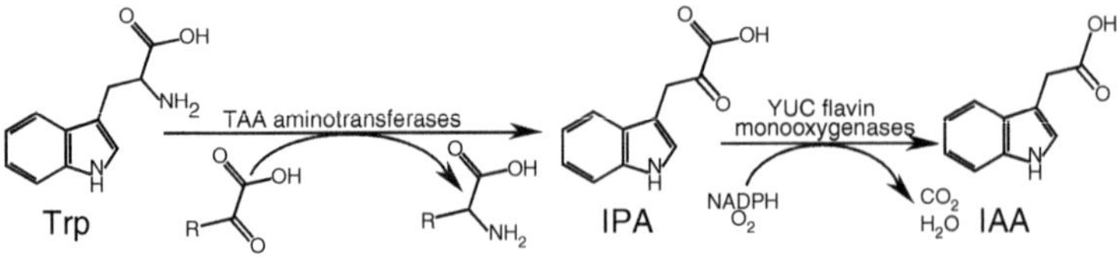


**Fig. 2.3** A tryptophan-dependent auxin biosynthesis pathway in plants. The TAA family of aminotransferases produces indole-3-pyruvate (IPA) from tryptophan (Trp) and the YUC flavin-containing monooxygenases catalyze the conversion of IPA into indole-3-acetic acid (IAA)

(Fig. 2.2). The IAM-induced phenotypes are identical to those observed in auxin overproduction mutants (Boerjan et al. 1995; Romano et al. 1995; Zhao et al. 2001). Therefore, the phenotypes of plants grown on IAM media are likely caused by overaccumulation of IAA in plants. *Arabidopsis* seedlings grown on IAN-containing media produce more adventitious roots and have short primary roots (Normanly et al. 1997) (Fig. 2.2). The IAN-induced phenotypes are very similar to those observed in plants grown on IAA-containing media, suggesting that IAN is probably converted to IAA in plants (Fig. 2.2). Indeed, a genetic screen for mutants insensitive to IAN identified *Arabidopsis* nitrilase genes that encode enzymes for the hydrolysis IAN to IAA (Bartel and Fink 1994; Normanly et al. 1997). Interestingly, treatments with IAN or IAM cause auxin overaccumulation in plants and high-auxin phenotypes. However, the IAM-induced phenotypes are dramatically different from those caused by IAN (Fig. 2.2). It is speculated that both IAM and IAN need to be metabolized into IAA to show auxin activities as the observed phenotypic differences may be simply caused by different tissue specificities of the hydrolytic enzymes for IAM and IAN. Although it is very clear that Trp metabolic intermediates can be converted to IAA in plants, it is difficult to determine how important their contribution to the total IAA pool under natural conditions is. The fact that plants produce a large number of Trp metabolic intermediates (Fig. 2.1) and that some of the Trp metabolites have auxin activities when added to growth media (Fig. 2.2) made it very difficult to dissect Trp-dependent auxin biosynthesis pathways using classic analytic biochemistry techniques alone.

The main criterion for determining whether a Trp metabolite is important for de novo auxin biosynthesis is to use the "deletion test." If the intermediate is important for auxin biosynthesis, we expect that plants show dramatic developmental defects similar to those observed in mutants defective in auxin transport or signaling if the plants lose the ability to make the intermediate. Recent results from a combination of analytic biochemical studies and *Arabidopsis* genetics research have established that the main auxin biosynthesis pathway in *Arabidopsis* is a simple two-step pathway that converts Trp to IAA (Fig. 2.3). The pathway is highly conserved throughout the plant kingdom.

## 2.1 The TAA/YUC Pathway as the Main Auxin Biosynthesis Pathway

*The YUCCA flavin-containing monooxygenases catalyze the rate-limiting step.* The *YUCCA* (*YUC*) *flavin-containing monooxygenase* (*FMO*) gene was identified as a key auxin biosynthesis gene a decade ago from an activation-tagging screen for long hypocotyl mutants in *Arabidopsis* (Zhao et al. 2001). The dominant *yucca* (later renamed as *yuc1D*) mutant was caused by the insertion of four copies of the CaMV 35S transcriptional enhancer downstream of the *YUC* gene (Zhao et al. 2001). The enhancers greatly increase the *YUC* expression levels, resulting in dramatic developmental defects. Physiological and molecular studies demonstrated that *yuc1D* is an auxin overproduction mutant (Zhao et al. 2001). Direct auxin measurements show that *yuc1D* contains 50 % more free IAA than wild-type *Arabidopsis* plants. Moreover, the auxin reporter DR5-GUS is greatly upregulated in *yuc1D* further supporting that *yuc1D* is an auxin overproduction mutant. It was suggested that YUC flavin-containing monooxygenases catalyze a rate-limiting step in auxin biosynthesis (Zhao et al. 2001).

*YUC* was later found to be a member of a gene family with 11 genes in the *Arabidopsis* genome. The founding member was renamed as *YUC1*. Overexpression of any of the *YUC* family members leads to auxin overproduction phenotypes in *Arabidopsis*, suggesting that all of the *YUC* genes participate in auxin biosynthesis (Cheng et al. 2006, 2007). The *YUC* genes have overlapping functions and inactivation of a single *YUC* gene does not cause any obvious developmental defects (Cheng et al. 2006, 2007). The observed genetic redundancy among *YUC* genes may provide an explanation for why *YUC* genes had not been discovered previously by forward loss-of-function genetic screens. Detailed analyses of various *yuc* mutant combinations have demonstrated that *YUC* genes are essential for almost all of the major developmental processes including embryogenesis, seedling growth, vascular initiation and patterning, flower development, and plant architecture (Cheng et al. 2006, 2007). For example, the *yuc1 yuc4* double mutants do not make tertiary veins in rosette leaves and fail to make continuous vascular boundless in flowers. Overall *yuc1 yuc4* flowers contain fewer floral organs and are completely sterile. A key piece of evidence that demonstrates the roles of *YUC* genes in auxin biosynthesis is the genetic rescue of *yuc1 yuc4* mutants with the bacterial auxin biosynthesis gene iaaM under the control of the *YUC1* promoter (Cheng et al. 2006, 2007).

The biochemical mechanisms of YUC-mediated auxin biosynthesis have been solved recently (Mashiguchi et al. 2011; Dai et al. 2013). YUC enzymes use NADPH and molecular oxygen to catalyze the oxidative decarboxylation of IPA to generate IAA (Fig. 2.3). On the basis of sequence homology to the mammalian microsome FMOs, it is expected that YUCs use a flavin (FAD or FMN) as a cofactor. Expressed in and purified from *E. coli*, the *Arabidopsis* YUC6 displayed a bright yellowish colour, suggesting that YUC6 contains a flavin cofactor. HPLC and other experiments demonstrate that the cofactor in YUC6 is FAD, not FMN

(Dai et al. 2013). The YUC6-catalyzed conversion of IPA to IAA can be divided into three consecutive chemical steps: (1) reduction of FAD to $FADH_2$ using electrons from NADPH; (2) binding of molecular oxygen to $FADH_2$ to form the C4a-(hydro)peroxyl flavin intermediate; (3) the reaction of the C4a intermediate with IPA to produce IAA from decarboxylation of IPA (Fig. 2.3) (Dai et al. 2013). Interestingly, the reduction of YUC6 by NADPH takes place regardless of the presence of IPA. IPA also does not affect the rate of YUC6 reduction. The kinetic pattern and rate of the formation of the C4a intermediate is also not affected by IPA (Dai et al. 2013). However, the decomposition of C4a intermediate is greatly accelerated by IPA (Dai et al. 2013). The oxidized YUC6, reduced YUC6, and the C4a intermediate display distinct spectroscopic properties and can be monitored spectroscopically. The oxidized YUC6 shows two peaks at 376 and 448 nm in the UV-visible spectrum, while reduction of YUC6 causes the disappearance of the 448 nm peak. The YUC6 C4a-(hydro)peroxyl flavin intermediate has a maximum absorbance at 381 nm in a UV-visible spectrum (Dai et al. 2013). The FAD cofactor in YUC6 provides a convenient handle to follow the progression of the YUC-catalyzed reactions.

Besides IPA as a substrate, YUC6 can also catalyze the decarboxylation of phenyl-pyruvate (PPA) to produce phenyl-acetic acid (PAA), suggesting that YUC enzymes do not have strict substrate specificities (Dai et al. 2013). It is not known whether the YUC6-catalyzed conversion of PPA to PAA has any physiological significance. However, it is known that PAA displays auxin activities when added into growth media. Both YUC enzymes and mammalian FMOs share sequence homologies and form the C4a-(hydro)peroxyl flavin intermediate. Mammalian FMOs are mainly known for their ability to oxygenate soft nucleophiles such as nitrogen- or sulfur-containing molecules, whereas YUCs such as YUC6 oxygenate electrophilic substrates such as IPA and PPA (Ziegler 1988, 2002; Dai et al. 2013). However, mammalian FMOs recently have been shown to use electrophilic substrates as well and YUCs were previously shown to oxygenate soft nucleophiles in vitro (Zhao et al. 2001; Kim et al. 2007; Lai et al. 2010). The stability of the C4a intermediate is also quite different for YUCs and mammalian FMOs. The YUC6 intermediate has a half-life of about 20 s, whereas that of some FMOs from mammalian cells is more than 30 min (Ziegler 1988, 2002; Dai et al. 2013). It is important to use both in vitro enzymatic assays and in vivo genetic evidence to determine the physiological functions of flavin-containing monooxygenases.

In the presence of excess PPA or IPA, some uncoupled YUC6 reactions still take place and produce hydrogen peroxide. The uncouple ratio is about 4 % (Dai et al. 2013). It is not clear whether the uncoupled reaction plays any physiological role. It is conceivable that $H_2O_2$ produced from the uncoupled reaction may participate in deactivating YUC enzymes, providing an intrinsic mechanism for turning off auxin biosynthesis.

Genetic, physiological, and biochemical studies have unambiguously demonstrated that the YUC family of flavin-containing monooxygenases plays a key role in auxin biosynthesis. Genetic evidence suggests that the conversion of IPA to IAA is the rate-limiting and the committed step for IAA biosynthesis.

*Tryptophan Aminotransferase of* Arabidopsis *(TAA) family of aminotransferases plays a key role in auxin biosynthesis.* Three groups independently discovered that TAA1, the founding member of a large family of aminotransferases, is an important auxin biosynthesis enzyme (Stepanova et al. 2008; Tao et al. 2008; Yamada et al. 2009). Mutations in *TAA1*, which is also called *SAV3*, *WEI8*, and *TIR2*, alter shade-avoidance responses, cause resistance to ethylene and to the auxin transport inhibitor NPA (Stepanova et al. 2008; Tao et al. 2008; Yamada et al. 2009). Although inactivation of *TAA1* along does not cause dramatic developmental phenotypes, simultaneously disruption of *TAA1* and its close homolog *TAR2* leads to defects in vascular pattern formation and in flower development in *Arabidopsis*. The *taa* mutants produce less free IAA compared to wild-type plants (Stepanova et al. 2008; Tao et al. 2008; Yamada et al. 2009).

TAA1 and its related proteins catalyze the transfer of the amino group from Trp to pyruvate or to α-ketoglutarate to produce IPA and Ala or Glu (Fig. 2.3) in vitro. Therefore, it is important to keep in mind that *TAA* genes not only produce IPA but also affect the homeostasis of other α-keto acids and other amino acids. It is not clear which α-keto acid is the preferred in vivo acceptor of the amino group from Trp.

TAAs and YUCs were previously placed in two separate pathways (Zhao et al. 2001; Stepanova et al. 2008; Tao et al. 2008). But several recent genetic studies have demonstrated that YUCs and TAAs participate in the same pathway (Mashiguchi et al. 2011; Stepanova et al. 2011; Won et al. 2011). The *yuc* mutants and *taa* mutants share many similarities. For example, *yuc1 yuc2 yuc4 yuc6* quadruple mutants have dramatic vascular and floral defects, which are also observed in *taa1 tar2* double mutants (Cheng et al. 2006; Stepanova et al. 2008). In fact, all of the characteristics of *taa* mutants can be phenocopied by inactivating certain combinations of *YUC* genes (Won et al. 2011). Overexpression of *YUC* genes leads to auxin overproduction phenotypes, which are dependent on the presence of functional *TAA* genes (Won et al. 2011). Furthermore, *taa* mutants are partially IPA deficient, whereas *yuc* mutants accumulate IPA, suggesting that *TAA* genes participate in IPA production and that YUCs use IPA as a substrate (Mashiguchi et al. 2011; Won et al. 2011). Finally, recent biochemical studies on the catalytic mechanisms of YUC flavin monooxygenases provide the final proof of the TAA/YUC two-step pathway as the main auxin biosynthesis pathway (Mashiguchi et al. 2011; Dai et al. 2013; Zhao 2013).

The TAA/YUC pathway is widely distributed throughout the plant kingdom. *YUC* genes from maize (Gallavotti et al. 2008), rice (Woo et al. 2007; Yamamoto et al. 2007; Fujino et al. 2008; Abu-Zaitoon et al. 2012), tomato (Exposito-Rodriguez et al. 2011), petunia (Tobena-Santamaria et al. 2002), strawberry (Liu et al. 2012), and other species (Kim et al. 2012; Cheol Park et al. 2013) have been functionally characterized and they all participate in auxin biosynthesis. The *TAA* genes in maize have also been shown to participate in auxin biosynthesis (Phillips et al. 2011). The committed step for auxin biosynthesis is catalyzed by the YUC flavoproteins. Thus the YUC-catalyzed reaction has to be tightly controlled. It has been shown that *YUC* genes are only expressed in discrete groups of cells (Cheng et al. 2006, 2007). Such tight control of *YUC* transcription provides a mechanism for temporal and spatial regulation of auxin production.

## 2.2 *Other Trp-Dependent Auxin Biosynthesis Pathways*

Trp is metabolized into several other indolic compounds (Fig. 2.1), some of which show auxin activities when applied to plants (Fig. 2.2). The physiological roles of the indolic compounds other than IPA in auxin biosynthesis are still ambiguous. That a compound can be metabolized into IAA both in vitro and in vivo does not mean that the compound is actually an important contributor to auxin biosynthesis in plants. Further genetic analysis of the genes responsible for generating the Trp metabolic intermediates (Fig. 2.1) is needed to assess the roles of the compounds in auxin biosynthesis.

*IAM pathway*. *Arabidopsis* and maize have detectable amount of IAM (Sugawara et al. 2009), which is the key intermediate in the bacterial auxin biosynthesis pathway characterized in *Agrobacterium* and *Pseudomonas* two decades ago (Yamada et al. 1985; Romano et al. 1995). In plant pathogenic bacteria, Trp is oxidized by the iaaM Trp-2-monooxygenase to IAM that is subsequently hydrolyzed by iaaH to produce IAA. Unlike the bacterial IAM pathway, the genes and enzymes responsible for producing IAM in plants have not been identified. It appears that plants do not have genes with high sequence homology to the bacterial *iaaM* gene. Therefore, IAM may be synthesized using a different mechanism. It is possible that IAM may be synthesized from IAA as a way to control free IAA levels. Conversion of IAA to IAM may be accomplished using mechanisms similar to glutamine biosynthesis.

Hydrolysis of IAM occurs in plants as feeding plants with IAM leads to elevated auxin levels and "high-auxin" phenotypes (Fig. 2.2). It is proposed that a group of hydrolases, which are homologous to the bacterial hydrolase iaaH, plays a role in converting IAM to IAA (Pollmann et al. 2006; Hoffmann et al. 2010). It is still inconclusive whether IAM is an important auxin biosynthesis intermediate in plants because IAM-deficient mutants have not been identified.

*TAM pathway*. Tryptamine is presumably produced by Trp decarboxylase, but the enzymes responsible for the reaction in *Arabidopsis* have not been characterized. Sequence homology-based prediction may not lead to the correct identification of the genes. TAM was a proposed substrate for the YUC flavin monooxygenases (Zhao et al. 2001; Kim et al. 2012), which have now been shown to catalyze the conversion of IPA to IAA in vitro and in vivo. However, all of the flavin-containing monooxygenases form the C4a-(hydro)peroxyl flavin intermediates, which are the catalytically active intermediates. The C4a intermediate can do both nucleophilic and electrophilic reactions, depending on the reaction conditions. For example, mammalian FMOs have long been recognized for their roles in xenobiotic metabolism by reacting with soft nucleophiles such as nitrogen-containing compounds (Ziegler 2002). It has also been shown that Human FMOs can catalyze a Baeyer–Villiger type reaction, in which the C4a intermediate reacts with an electrophilic carbonyl carbon (Lai et al. 2010). To date, it has not been ruled out that TAM is an important intermediate in auxin biosynthesis; however, biosynthesis and metabolism of TAM are not well understood.

*IAN pathway.* IAN is very abundant compared to other Trp metabolites (Fig. 2.1). IAN stimulates adventitious root development and inhibits primary root elongation (Fig. 2.2). The conversion of IAN to IAA is catalyzed by nitrilases. Inactivation of nitrilase genes leads to resistance to exogenous IAN, but the nitrilase mutants do not display obvious developmental defects observed in known auxin signaling and transport mutants (Bartel and Fink 1994; Normanly et al. 1997). *Arabidopsis* genome contains four copies of the nitrilase gene. The developmental consequences of disrupting all four nitrilase genes have not been investigated, partially due to the fact that two of the copies are immediately adjacent to each other on the same chromosome. Therefore, it is still an open question whether IAN plays a significant role in auxin biosynthesis.

The routes for IAN production are not well understood either. It has been reported that metabolism of indolic glucosinolate yields IAN (de Vos et al. 2008). However, maize does not produce glucosinolates, but still produces IAN, suggesting that other routes can produce IAN. It has been suggested that IAN may also be produced from other indolic compounds such as IAOx (Sugawara et al. 2009).

*IAAld pathway.* IAAld was previously proposed as an intermediate in the IPA pathway for auxin biosynthesis (Zhao 2010). In plants, it is now known that IAAld is not an intermediate in the IPA pathway (Mashiguchi et al. 2011; Won et al. 2011) as IPA is converted to IAA by the YUC flavin-containing monooxygenases without producing IAAld (Dai et al. 2013). In some IAA-producing bacteria, IAAld is produced from IPA by IPA decarboxylases (Carreno-Lopez et al. 2000). IAAld can be further oxidized into IAA by aldehyde oxidases. In *Arabidopsis*, genes homologous to the bacterial IPA decarboxylases appear not to play a role in auxin biosynthesis. Inactivation of *Arabidopsis* aldehyde oxidases does not disturb auxin homeostasis, suggesting that it is very likely that IAAld does not contribute significantly to de novo auxin biosynthesis (Mashiguchi et al. 2011). However, IAAld can also be oxidized by aldehyde dehydrogenases, which have not been characterized in *Arabidopsis*.

*IAOx pathway.* IAOx has only been detected in *Arabidopsis* and related species (Mashiguchi et al. 2011). Monocots such as rice and maize do not have detectable levels of IAOx (Mashiguchi et al. 2011). In *Arabidopsis*, CYP79B2 and CYP79B3 convert Trp directly to IAOx (Hull et al. 2000; Zhao et al. 2002). Overexpression of *CYP79B2* in *Arabidopsis* leads to long hypocotyl and epinastic cotyledons, a phenotype that is also observed in *YUC* overexpression lines, suggesting that IAOx can be a precursor for IAA biosynthesis (Zhao et al. 2002). IAOx is also a precursor for indolic glucosinolate biosynthesis (Boerjan et al. 1995; Bak and Feyereisen 2001). When the genes encoding glucosinolate biosynthesis enzymes are mutated, more IAOx is fluxed into IAA biosynthesis, causing auxin overproduction phenotypes (Boerjan et al. 1995; Bak and Feyereisen 2001). For example, the *sur1* and *sur2* mutants that are defective in glucosinolate biosynthesis overproduce auxin, which leads to the development of long hypocotyls and numerous adventitious roots.

It appears that *CYP79B2* and *B3* are the only genes responsible for producing IAOx in *Arabidopsis*. The *cyp79b2 cyp79b3* double mutants appear to completely abolish the biosynthesis of IAOx and the double mutants have no detectable amount of IAOx (Sugawara et al. 2009). The double mutants have subtle growth defects when grown at high temperature, but have no obvious phenotypes under normal growth conditions (Zhao et al. 2002). Therefore, it is believed that IAOx is not an essential intermediate for auxin biosynthesis. Nor is IAOx a universal intermediate for auxin biosynthesis.

In summary, after three decades molecular genetics studies in *Arabidopsis*, the picture of de novo auxin biosynthesis has become clearer. The two-step Trp-dependent pathway catalyzed by TAAs and YUCs is the main auxin biosynthesis pathway that plays essential roles in almost all of the main developmental processes. In retrospect, *Arabidopsis* probably is not the best model for auxin biosynthesis studies for two main reasons. First, the *Arabidopsis* glucosinolate biosynthesis pathway really complicated the analyses of IAA biosynthesis because the glucosinolate biosynthesis intermediate IAOx can be converted into IAA. The aforementioned glucosinolate mutants such as *sur1* and *sur2* had dramatic auxin overproduction phenotypes (Boerjan et al. 1995; Bak and Feyereisen 2001). Second, the genetic redundancy within *YUC*s and *TAA*s in *Arabidopsis* made it difficult for loss-of-function studies. The single *Arabidopsis yuc* or *taa* mutants do not show dramatic auxin phenotypes. Only the multiple mutants of *taa* or *yuc* display dramatic developmental defects (Cheng et al. 2006, 2007; Stepanova et al. 2008). In contrast, some monocots such as maize offer a relatively simpler system for analyzing auxin biosynthesis. Maize does not produce indolic glucosinolate (Sugawara et al. 2009). Furthermore, inactivation of a single *YUC* gene or *TAA* gene in maize leads to dramatic developmental phenotypes, whereas inactivation of at least two *YUC* genes or *TAA* genes in *Arabidopsis* is needed to cause main developmental defects (Gallavotti et al. 2008; Phillips et al. 2011).

## 3 IAA Production from Non-De Novo Pathways

Besides de novo in loco synthesis and transportation from neighboring cells, free IAA can also be made available by releasing IAA from its conjugated forms or from indole butyric acid (IBA) (Woodward and Bartel 2005). In fact, the majority of IAA in plants exists in the conjugated forms, which are proposed to serve as a storage pool. It is known that IAA can be conjugated via ester bonds with simple alcohols and with sugars such as glucose and *myo*-inositol or via amide bonds with amino acids, peptides, or proteins. Free IAA can be produced when the conjugates are hydrolyzed. Hydrolysis of conjugates provides plants with a potentially faster way to modulate free IAA levels than de novo biosynthesis. For example, in the germinating seeds of maize, large amount of IAA is released from the endosperm from its ester form to support the growth of developing seedlings (Bialek et al. 1992). Some of the enzymes responsible for hydrolyzing IAA-sugar or IAA-amino acids have been characterized and they show different substrate specificities and are developmentally regulated (Bartel and Fink 1995; Davies et al. 1999; Lasswell et al. 2000; LeClere et al. 2002; Rampey et al. 2004).

## 3.1 Conversion of IBA to IAA

IBA has long been used in agriculture for promoting root initiation/growth from plant cuttings. *Arabidopsis* plants accumulate detectable amount of IBA. However, it is not understood how IBA is synthesized in plants. IBA is known to inhibit primary root elongation and stimulate lateral root formation. Genetic screens for *Arabidopsis* mutants resistant to exogenous IBA have identified many loci (*ibr*, IBA resistant). The majority of the *ibr* loci encode enzymes related to β-oxidation of fatty acids or biogenesis of peroxisome, where β-oxidation takes place. The genetic data suggest that the observed auxin activities of IBA depend on the conversion of IBA to IAA (Zolman et al. 2000; Strader et al. 2010). However, it has not been completely ruled out that IBA itself has some biological activities (Simon et al. 2013).

The physiological roles of IBA-derived IAA are difficult to determine because the enzymes responsible for IBA to IAA conversion may also participate in other pathways such as fatty acid metabolism. Recent characterization of mutations resistant to IBA leads to the discovery that disruption of *ENOYL-COA HYDRATASE2* (*ECH2*) gene causes defects in IBA responsiveness, but appears not to affect sugar and fatty acid metabolism. Further analysis of *ech2* and other *ibr* mutants demonstrated that IBA-derived IAA plays important roles in root hair development and cotyledon cell expansion (Strader et al. 2010, 2011).

## 3.2 Release of Free IAA from IAA Conjugates

IAA conjugates with ester-link to sugars and small alcohols or amide-link to amino acids and peptides have been identified in plants. The various conjugates may serve as a storage form of IAA and can release free IAA when needed. The most studied case of releasing free IAA from conjugates is the hydrolysis of IAA-amino acid conjugates. Among the 20 potential amino acid conjugates, 19 (except IAA-Arg) were tested for their ability to release free IAA in a bioassay based on root elongation in *Arabidopsis* (LeClere et al. 2002). It was shown that IAA-Ala, -Leu, -Phe, -Asn, -Gln, -Glu, -Gly, -Met, -Ser, -Thr, and -Tyr inhibited root elongation by more than 50 % at 40 μM, suggesting that these amino acid conjugates can be hydrolyzed to release free IAA. In contrast, IAA-Asp, -Cys, -His, -Ile, -Lys, -Pro, -Trp, and -Val appeared not a source for free IAA (LeClere et al. 2002). Genetic screens for *Arabidopsis* mutants resistant to IAA-Leu and IAA-Ala identified a family of hydrolases including *IAA–Leu Resistant 1* (*ILR1*), *IAA–Ala resistant* (*IAR3*), and the *ILR1-like protein* (*ILL2*) responsible for releasing free IAA from the IAA-amino acid conjugates (Davies et al. 1999; Lasswell et al. 2000; LeClere et al. 2002; Rampey et al. 2004). The *ilr1 iar3 ill2* triple mutants are resistant to several IAA-amino acid conjugates and have shorter hypocotyl and fewer lateral roots than wild-type plants, suggesting that releasing free IAA from conjugates plays important roles in IAA homeostasis and plant development (Rampey et al. 2004).

# 4 Deactivation of IAA

The active form of IAA is believed to be free IAA. The carboxyl group in IAA is essential for its auxin activities. IAA is inactivated by complete oxidation, a process that is still not well understood. IAA can also be taken out of action by forming various conjugates with alcohols, sugars, and amino acids (Woodward and Bartel 2005).

## *4.1 Synthesis of IAA Conjugates*

Great progresses have been made in recent years towards understanding the enzymes responsible for synthesizing IAA esters and amide conjugates. In maize, synthesis of IAA-ester with sugar starts with the formation of IAA-glucose that is preceded by activation of glucose by the formation of glucose-UDP that is then joined with IAA. IAA-glucose is further converted to other IAA-sugar ester conjugates that are mostly believed to be a storage form of IAA (Michalczuk and Bandurski 1982; Leznicki and Bandurski 1988a, b). The formation of methyl IAA by the IAMT1 methyl transferase has been implicated in regulating leaf development in *Arabidopsis* (Qin et al. 2005).

In *Arabidopsis*, 20 amidosynthases encoded by the large *Gretchen Hagen 3* (*GH3*) family of genes conjugate IAA as well as some other plant hormones such as jasmonic acid and salicylic acid with amino acids to form amide conjugates (Hagen et al. 1991; Liu et al. 1994; Staswick et al. 2005). *GH3* genes are among the early-induced genes by auxin treatments (Hagen et al. 1991). Originally discovered as being able to adenylate IAA in vitro, GH3 amidosynthases are later shown to be responsible for synthesizing IAA-amino acid conjugates. The adenylyl-IAA serves as the activated intermediate and readily reacts with some amino acids (Staswick et al. 2005). Some of the IAA-amino acid conjugates can be hydrolyzed to release free IAA, while some of the conjugates appear non-hydrolyzable in vivo (LeClere et al. 2002). The latter group of IAA-conjugates may serve as a way to inactivate IAA. For example, once IAA-Asp is formed, it would not be hydrolyzed and the conjugated IAA is consequently permanently deactivated. IAA-Asp is also known as a target for oxidative degradation. GH3 proteins have also been shown to play roles in response to environmental stimuli such as light and wounding processes, possibly through the regulation the formation of IAA, jasmonic acid, and/or salicylic acid conjugates (Woodward and Bartel 2005). Interestingly, some of the IAA conjugates possesses antagonist effects against IAA. Externally applied IAA-Trp effectively antagonizes the inhibitory effects of IAA treatment in *Arabidopsis* roots (Staswick 2009a, b). IAA-peptide and IAA-protein conjugates have also been discovered (Walz et al. 2002), indicating that IAA may serve as a small molecular tag but their functions are still unclear.

## *4.2 IAA Degradation via Oxidation*

IAA starts the oxidative degradation either with decarboxylation on the side chain or with oxidation of the indole ring. Very little is known regarding oxidative degradation of IAA. It has been reported that peroxidase may be involved in the oxidative decarboxylation of IAA (Normanly 2010). Oxidative intermediates including OxIAA have been discovered in plants (Reinecke and Bandurski 1983; Ostin et al. 1998; Kai et al. 2007; Peer et al. 2013). In *Arabidopsis*, other IAA metabolites such as *N*-(6-hydroxyindol-3-ylacetyl)-phenylalanine (6-OH-IAA-Phe), *N*-(6-hydroxyindol-3-ylacetyl)-valine (6-OH-IAA-Val), and 1-*O*-(2-oxoindol-3-ylacetyl)-beta-*d*-glucopyranose (OxIAA-Glc) have been observed with OxIAA-Glc being the main oxidative product. Recently, it was reported that in *Arabidopsis* roots, OxIAA is the major catabolic product of IAA (Pencik et al. 2013). Because OxIAA has little auxinic effects, irreversible oxidation of IAA into OxIAA effectively removes the IAA from the auxin pool. Another recent discovery in rice shed light on the genes underlying the conversion of IAA to OxIAA (Zhao et al. 2013). Rice plants with a mutation in the *Dioxygenase for Auxin Oxidation* (*DAO*) gene have elevated free IAA levels in anthers and ovaries and are defective in anther dehiscence, pollen fertility, and seed development (Zhao et al. 2013). The *dao* mutants also do not have detectable level of oxIAA, and the purified DAO protein expressed in *E. coli* could convert IAA to oxIAA in vitro (Zhao et al. 2013). The new findings mark the beginning of understanding the molecular and genetic mechanisms underlying IAA oxidation and the roles of oxidative degradation of IAA in auxin homeostasis.

# 5 Conclusions

Plants employ many ways to control auxin levels thus to ensure proper growth and development. The major advancement in our understanding of auxin homeostasis in the past few years is the elucidation of a complete two-step Trp-dependent auxin biosynthesis pathway where Trp is converted to IPA by the TAA family of amino transferases and the YUC flavin monooxygenases catalyzes the production of IAA using IPA as a substrate. However, there are still some gaps in our understanding of both auxin biosynthesis and degradation. The identification of a complete network of auxin metabolic pathways would allow us to effectively modulate auxin levels in plants with temporal and spatial control and thus greatly facilitate the dissection of the molecular mechanisms by which auxin regulates various aspects of plant growth and development.

**Acknowledgment** Our research was supported by the NSF Plant Genome grant (DBI-0820729 to YZ) and the NIH (R01GM68631 to YZ).

# References

Abu-Zaitoon YM, Bennett K, Normanly J, Nonhebel HM (2012) A large increase in IAA during development of rice grains correlates with the expression of tryptophan aminotransferase OsTAR1 and a grain-specific YUCCA. Physiol Plant 146:487–499

Bak S, Feyereisen R (2001) The involvement of two p450 enzymes, CYP83B1 and CYP83A1, in auxin homeostasis and glucosinolate biosynthesis. Plant Physiol 127:108–118

Bartel B, Fink GR (1994) Differential regulation of an auxin-producing nitrilase gene family in Arabidopsis thaliana. Proc Natl Acad Sci U S A 91:6649–6653

Bartel B, Fink GR (1995) ILR1, an amidohydrolase that releases active indole-3-acetic acid from conjugates. Science 268:1745–1748

Bentley JA, Farrar KR, Housley S, Smith GF, Taylor WC (1956) Some chemical and physiological properties of 3-indolylpyruvic acid. Biochem J 64:44–49

Bialek K, Michalczuk L, Cohen JD (1992) Auxin biosynthesis during seed germination in Phaseolus vulgaris. Plant Physiol 100:509–517

Boerjan W, Cervera MT, Delarue M, Beeckman T, Dewitte W, Bellini C, Caboche M, Van Onckelen H, Van Montagu M, Inze D (1995) Superroot, a recessive mutation in Arabidopsis, confers auxin overproduction. Plant Cell 7:1405–1419

Carreno-Lopez R, Campos-Reales N, Elmerich C, Baca BE (2000) Physiological evidence for differently regulated tryptophan-dependent pathways for indole-3-acetic acid synthesis in Azospirillum brasilense. Mol Gen Genet 264:521–530

Cheng Y, Dai X, Zhao Y (2006) Auxin biosynthesis by the YUCCA flavin monooxygenases controls the formation of floral organs and vascular tissues in Arabidopsis. Genes Dev 20:1790–1799

Cheng Y, Dai X, Zhao Y (2007) Auxin synthesized by the YUCCA flavin monooxygenases is essential for embryogenesis and leaf formation in Arabidopsis. Plant Cell 19:2430–2439

Cheol Park H, Cha JY, Yun DJ (2013) Roles of YUCCAs in auxin biosynthesis and drought stress responses in plants. Plant Signal Behav 8(6):e24495

Dai X, Mashiguchi K, Chen Q, Kasahara H, Kamiya Y, Ojha S, DuBois J, Ballou D, Zhao Y (2013) The biochemical mechanism of auxin biosynthesis by an arabidopsis YUCCA flavin-containing monooxygenase. J Biol Chem 288:1448–1457

Davies RT, Goetz DH, Lasswell J, Anderson MN, Bartel B (1999) IAR3 encodes an auxin conjugate hydrolase from Arabidopsis. Plant Cell 11:365–376

de Vos M, Kriksunov KL, Jander G (2008) Indole-3-acetonitrile production from indole glucosinolates deters oviposition by Pieris rapae. Plant Physiol 146:916–926

Exposito-Rodriguez M, Borges AA, Borges-Perez A, Perez JA (2011) Gene structure and spatio-temporal expression profile of tomato genes encoding YUCCA-like flavin monooxygenases: the ToFZY gene family. Plant Physiol Biochem 49:782–791

Fujino K, Matsuda Y, Ozawa K, Nishimura T, Koshiba T, Fraaije MW, Sekiguchi H (2008) NARROW LEAF 7 controls leaf shape mediated by auxin in rice. Mol Genet Genomics 279:499–507

Gallavotti A, Barazesh S, Malcomber S, Hall D, Jackson D, Schmidt RJ, McSteen P (2008) sparse inflorescence1 encodes a monocot-specific YUCCA-like gene required for vegetative and reproductive development in maize. Proc Natl Acad Sci U S A 105:15196–15201

Hagen G, Martin G, Li Y, Guilfoyle TJ (1991) Auxin-induced expression of the soybean GH3 promoter in transgenic tobacco plants. Plant Mol Biol 17:567–579

Hoffmann M, Lehmann T, Neu D, Hentrich M, Pollmann S (2010) Expression of AMIDASE1 (AMI1) is suppressed during the first two days after germination. Plant Signal Behav 5:1642–1644

Hull AK, Vij R, Celenza JL (2000) Arabidopsis cytochrome P450s that catalyze the first step of tryptophan-dependent indole-3-acetic acid biosynthesis. Proc Natl Acad Sci U S A 97:2379–2384

Kai K, Horita J, Wakasa K, Miyagawa H (2007) Three oxidative metabolites of indole-3-acetic acid from Arabidopsis thaliana. Phytochemistry 68:1651–1663

Kim JI, Sharkhuu A, Jin JB, Li P, Jeong JC, Baek D, Lee SY, Blakeslee JJ, Murphy AS, Bohnert HJ, Hasegawa PM, Yun DJ, Bressan RA (2007) yucca6, a dominant mutation in Arabidopsis, affects auxin accumulation and auxin-related phenotypes. Plant Physiol 145:722–735

Kim JI, Baek D, Park HC, Chun HJ, Oh DH, Lee MK, Cha JY, Kim WY, Kim MC, Chung WS, Bohnert HJ, Lee SY, Bressan RA, Lee SW, Yun DJ (2012) Overexpression of Arabidopsis YUCCA6 in potato results in high-auxin developmental phenotypes and enhanced resistance to water deficit. Mol Plant 6:337–349

Lai WG, Farah N, Moniz GA, Wong YN (2010) A Baeyer-Villiger oxidation specifically catalyzed by human flavin-containing monooxygenase 5. Drug Metab Dispos 39:61–70

Lasswell J, Rogg LE, Nelson DC, Rongey C, Bartel B (2000) Cloning and characterization of IAR1, a gene required for auxin conjugate sensitivity in Arabidopsis. Plant Cell 12:2395–2408

LeClere S, Tellez R, Rampey RA, Matsuda SP, Bartel B (2002) Characterization of a family of IAA-amino acid conjugate hydrolases from Arabidopsis. J Biol Chem 277:20446–20452

Leznicki AJ, Bandurski RS (1988a) Enzymic synthesis of indole-3-acetyl-1-O-beta-d-glucose. II. Metabolic characteristics of the enzyme. Plant Physiol 88:1481–1485

Leznicki AJ, Bandurski RS (1988b) Enzymic synthesis of indole-3-acetyl-1-O-beta-d-glucose. I. Partial purification and characterization of the enzyme from Zea mays. Plant Physiol 88:1474–1480

Liu ZB, Ulmasov T, Shi X, Hagen G, Guilfoyle TJ (1994) Soybean GH3 promoter contains multiple auxin-inducible elements. Plant Cell 6:645–657

Liu H, Ying YY, Zhang L, Gao QH, Li J, Zhang Z, Fang JG, Duan K (2012) Isolation and characterization of two YUCCA flavin monooxygenase genes from cultivated strawberry (Fragaria x ananassa Duch.). Plant Cell Rep 31:1425–1435

Mashiguchi K, Tanaka K, Sakai T, Sugawara S, Kawaide H, Natsume M, Hanada A, Yaeno T, Shirasu K, Yao H, McSteen P, Zhao Y, Hayashi K, Kamiya Y, Kasahara H (2011) The main auxin biosynthesis pathway in Arabidopsis. Proc Natl Acad Sci U S A 108:18512–18517

Michalczuk L, Bandurski RS (1982) Enzymic synthesis of 1-O-indol-3-ylacetyl-beta-D-glucose and indol-3-ylacetyl-myo-inositol. Biochem J 207:273–281

Normanly J (2010) Approaching cellular and molecular resolution of auxin biosynthesis and metabolism. Cold Spring Harb Perspect Biol 2:a001594

Normanly J, Cohen JD, Fink GR (1993) Arabidopsis thaliana auxotrophs reveal a tryptophan-independent biosynthetic pathway for indole-3-acetic acid. Proc Natl Acad Sci U S A 90:10355–10359

Normanly J, Grisafi P, Fink GR, Bartel B (1997) Arabidopsis mutants resistant to the auxin effects of indole-3-acetonitrile are defective in the nitrilase encoded by the NIT1 gene. Plant Cell 9:1781–1790

Ostin A, Kowalyczk M, Bhalerao RP, Sandberg G (1998) Metabolism of indole-3-acetic acid in Arabidopsis. Plant Physiol 118:285–296

Ouyang J, Chen M, Li J (1999) Measurement of soluble tryptophan and total indole-3-acetic acid in Arabidopsis by capillary electrophoresis. Anal Biochem 271:100–102

Peer WA, Cheng Y, Murphy AS (2013) Evidence of oxidative attenuation of auxin signalling. J Exp Bot 64:2629–2639

Pencik A, Simonovik B, Petersson SV, Henykova E, Simon S, Greenham K, Zhang Y, Kowalczyk M, Estelle M, Zazimalova E, Novak O, Sandberg G, Ljung K (2013) Regulation of auxin homeostasis and gradients in Arabidopsis roots through the formation of the indole-3-acetic acid catabolite 2-oxindole-3-acetic acid. Plant Cell 25:3858–3870

Phillips KA, Skirpan AL, Liu X, Christensen A, Slewinski TL, Hudson C, Barazesh S, Cohen JD, Malcomber S, McSteen P (2011) vanishing tassel2 encodes a grass-specific tryptophan aminotransferase required for vegetative and reproductive development in maize. Plant Cell 23:550–566

Pollmann S, Neu D, Lehmann T, Berkowitz O, Schafer T, Weiler EW (2006) Subcellular localization and tissue specific expression of amidase 1 from Arabidopsis thaliana. Planta 224:1241–1253

Qin G, Gu H, Zhao Y, Ma Z, Shi G, Yang Y, Pichersky E, Chen H, Liu M, Chen Z, Qu LJ (2005) An indole-3-acetic acid carboxyl methyltransferase regulates Arabidopsis leaf development. Plant Cell 17:2693–2704

Rampey RA, LeClere S, Kowalczyk M, Ljung K, Sandberg G, Bartel B (2004) A family of auxin-conjugate hydrolases that contributes to free indole-3-acetic acid levels during Arabidopsis germination. Plant Physiol 135:978–988

Reinecke DM, Bandurski RS (1983) Oxindole-3-acetic acid, an indole-3-acetic acid catabolite in Zea mays. Plant Physiol 71:211–213

Romano CP, Robson PR, Smith H, Estelle M, Klee H (1995) Transgene-mediated auxin overproduction in Arabidopsis: hypocotyl elongation phenotype and interactions with the hy6-1 hypocotyl elongation and axr1 auxin-resistant mutants. Plant Mol Biol 27:1071–1083

Sakurai A, Fujioka S (1993) The current status of physiology and biochemistry of brassinosteroids. Plant Growth Regul 13:147–159

Simon S, Kubes M, Baster P, Robert S, Dobrev PI, Friml J, Petrasek J, Zazimalova E (2013) Defining the selectivity of processes along the auxin response chain: a study using auxin analogues. New Phytol 200:1034–1048

Staswick P (2009a) Plant hormone conjugation: a signal decision. Plant Signal Behav 4:757–759

Staswick PE (2009b) The tryptophan conjugates of jasmonic and indole-3-acetic acids are endogenous auxin inhibitors. Plant Physiol 150:1310–1321

Staswick PE, Serban B, Rowe M, Tiryaki I, Maldonado MT, Maldonado MC, Suza W (2005) Characterization of an Arabidopsis enzyme family that conjugates amino acids to indole-3-acetic acid. Plant Cell 17:616–627

Stepanova AN, Robertson-Hoyt J, Yun J, Benavente LM, Xie DY, Dolezal K, Schlereth A, Jurgens G, Alonso JM (2008) TAA1-mediated auxin biosynthesis is essential for hormone crosstalk and plant development. Cell 133:177–191

Stepanova AN, Yun J, Robles LM, Novak O, He W, Guo H, Ljung K, Alonso JM (2011) The Arabidopsis YUCCA1 flavin monooxygenase functions in the indole-3-pyruvic acid branch of auxin biosynthesis. Plant Cell 23:3961–3973

Strader LC, Culler AH, Cohen JD, Bartel B (2010) Conversion of endogenous indole-3-butyric acid to indole-3-acetic acid drives cell expansion in Arabidopsis seedlings. Plant Physiol 153:1577–1586

Strader LC, Wheeler DL, Christensen SE, Berens JC, Cohen JD, Rampey RA, Bartel B (2011) Multiple facets of Arabidopsis seedling development require indole-3-butyric acid-derived auxin. Plant Cell 23:984–999

Sugawara S, Hishiyama S, Jikumaru Y, Hanada A, Nishimura T, Koshiba T, Zhao Y, Kamiya Y, Kasahara H (2009) Biochemical analyses of indole-3-acetaldoxime-dependent auxin biosynthesis in Arabidopsis. Proc Natl Acad Sci U S A 106:5430–5435

Tao Y, Ferrer JL, Ljung K, Pojer F, Hong F, Long JA, Li L, Moreno JE, Bowman ME, Ivans LJ, Cheng Y, Lim J, Zhao Y, Ballare CL, Sandberg G, Noel JP, Chory J (2008) Rapid synthesis of auxin via a new tryptophan-dependent pathway is required for shade avoidance in plants. Cell 133:164–176

Tobena-Santamaria R, Bliek M, Ljung K, Sandberg G, Mol JN, Souer E, Koes R (2002) FLOOZY of petunia is a flavin mono-oxygenase-like protein required for the specification of leaf and flower architecture. Genes Dev 16:753–763

Walz A, Park S, Slovin JP, Ludwig-Muller J, Momonoki YS, Cohen JD (2002) A gene encoding a protein modified by the phytohormone indoleacetic acid. Proc Natl Acad Sci U S A 99:1718–1723

Won C, Shen X, Mashiguchi K, Zheng Z, Dai X, Cheng Y, Kasahara H, Kamiya Y, Chory J, Zhao Y (2011) Conversion of tryptophan to indole-3-acetic acid by TRYPTOPHAN AMINOTRANSFERASES OF ARABIDOPSIS and YUCCAs in Arabidopsis. Proc Natl Acad Sci U S A 108:18518–18523

Woo YM, Park HJ, Su'udi M, Yang JI, Park JJ, Back K, Park YM, An G (2007) Constitutively wilted 1, a member of the rice YUCCA gene family, is required for maintaining water homeostasis and an appropriate root to shoot ratio. Plant Mol Biol 65:125–136

Woodward AW, Bartel B (2005) Auxin: regulation, action, and interaction. Ann Bot 95:707–735

Wright AD, Sampson MB, Neuffer MG, Michalczuk L, Slovin JP, Cohen JD (1991) Indole-3-acetic acid biosynthesis in the mutant maize orange pericarp, a tryptophan auxotroph. Science 254:998–1000

Yamada T, Palm CJ, Brooks B, Kosuge T (1985) Nucleotide sequences of the Pseudomonas savastanoi indoleacetic acid genes show homology with Agrobacterium tumefaciens T-DNA. Proc Natl Acad Sci U S A 82:6522–6526

Yamada M, Greenham K, Prigge MJ, Jensen PJ, Estelle M (2009) The TRANSPORT INHIBITOR RESPONSE2 gene is required for auxin synthesis and diverse aspects of plant development. Plant Physiol 151:168–179

Yamamoto Y, Kamiya N, Morinaka Y, Matsuoka M, Sazuka T (2007) Auxin biosynthesis by the YUCCA genes in rice. Plant Physiol 143:1362–1371

Yang SF, Hoffmann NE (1984) Ethylene biosynthesis and its regulation in higher plants. Annu Rev Plant Biol 35:155–189

Zhao Y (2010) Auxin biosynthesis and its role in plant development. Annu Rev Plant Biol 61:49–64

Zhao Y (2013) Auxin biosynthesis: a simple two-step pathway converts tryptophan to indole-3-acetic acid in plants. Mol Plant 5:334–338

Zhao Y, Christensen SK, Fankhauser C, Cashman JR, Cohen JD, Weigel D, Chory J (2001) A role for flavin monooxygenase-like enzymes in auxin biosynthesis. Science 291:306–309

Zhao Y, Hull AK, Gupta NR, Goss KA, Alonso J, Ecker JR, Normanly J, Chory J, Celenza JL (2002) Trp-dependent auxin biosynthesis in Arabidopsis: involvement of cytochrome P450s CYP79B2 and CYP79B3. Genes Dev 16:3100–3112

Zhao Z, Zhang Y, Liu X, Zhang X, Liu S, Yu X, Ren Y, Zheng X, Zhou K, Jiang L, Guo X, Gai Y, Wu C, Zhai H, Wang H, Wan J (2013) A role for a dioxygenase in auxin metabolism and reproductive development in rice. Dev Cell 27:113–122

Ziegler DM (1988) Flavin-containing monooxygenases: catalytic mechanism and substrate specificities. Drug Metab Rev 19:1–32

Ziegler DM (2002) An overview of the mechanism, substrate specificities, and structure of FMOs. Drug Metab Rev 34:503–511

Zolman BK, Yoder A, Bartel B (2000) Genetic analysis of indole-3-butyric acid responses in Arabidopsis thaliana reveals four mutant classes. Genetics 156:1323–1337

# Chapter 3
# Identification and Profiling of Auxin and Auxin Metabolites

**Ondřej Novák, Aleš Pěnčík, and Karin Ljung**

**Abstract** During the last 10 years, the analytical techniques used in different areas of "life science" have improved tremendously. Mass spectrometry (MS) has become the most versatile and sensitive technique available for identifying and quantifying organic molecules, and liquid chromatography-mass spectrometry is the modern analytical tool of choice for analyzing samples of plant, animal and human origin. Both the sensitivity and the selectivity of the available techniques have increased immensely; modern instruments are much smaller, more user-friendly and more versatile than before, and the overall cost of the method has been greatly reduced. However, the required equipment is not available to most plant research laboratories, and most researchers in biology have limited experience with MS techniques. In this chapter, we aim to explain the advantages and limitations of these techniques, and how they can be used in plant research today. More specifically, we demonstrate how different MS techniques can be used for auxin metabolite identification, quantification and profiling. Efficient sample extraction and purification is essential for highly sensitive and selective analyses. We therefore describe selected novel approaches that have been developed to increase the sensitivity of these analyses and make them applicable at the tissue and cellular levels. We also discuss how these techniques can be combined with isotope labelling and mutant analyses to get a better understanding of the metabolic pathways involved in auxin biosynthesis and degradation. Finally, we examine

O. Novák
Umeå Plant Science Centre, Department of Forest Genetics and Plant Physiology, Swedish University of Agricultural Sciences, 901 83, Umeå, Sweden

Laboratory of Growth Regulators, Centre of Region Haná for Biotechnological and Agricultural Research, Institute of Experimental Botany ASCR and Palacký University, Šlechtitelů 11, 783 71, Olomouc, Czech Republic

A. Pěnčík • K. Ljung (✉)
Umeå Plant Science Centre, Department of Forest Genetics and Plant Physiology, Swedish University of Agricultural Sciences, 901 83, Umeå, Sweden
e-mail: Karin.Ljung@slu.se

E. Zažímalová et al. (eds.), *Auxin and Its Role in Plant Development*,
DOI 10.1007/978-3-7091-1526-8_3, © Springer-Verlag Wien 2014 

the future prospects for the use of MS and other analytical techniques in auxin research as well as the potential for combining these techniques to obtain more information from single samples, and perhaps even from single cells.

## 1 Extraction of IAA Metabolites from Plant Tissues

In general, the isolation and/or pre-concentration of natural bioactive compounds, such as phytohormones, involves multiple critical and often time-consuming steps based on extracting and purifying analytes from a complex plant matrix. Conventional sample preparation methods involve the use of protein precipitation (PP), liquid/liquid extraction (LLE) or solid phase extraction (SPE) to facilitate the subsequent chromatographic separation and detection of the targeted analytes (Nováková and Vlčková 2009). Modern approaches to sample preparation often involve parallel sample processing and/or the use of new sample preparation techniques (which may permit the use of smaller sample sizes) designed to minimize analyte losses and reduce the levels of interfering substances such as proteins, carbohydrates, pigments and lipids (Núñez et al. 2012).

High-efficiency homogenization is important for accurate IAA and IAA metabolite quantification. In classical methods, homogenization is achieved by manually grinding frozen or dry plant tissue in extraction buffer with a mortar and pestle or using knife homogenizers. Recent advances in mass spectrometry and improvements in analytical sensitivity have made it possible to use vibration mills/ball grinders for the convenient homogenization of small amounts of plant material with cold extraction media directly in plastic vials (Barkawi et al. 2008; Novák et al. 2012). Miniaturized homogenization is readily adapted for use with high-throughput methods because it can be performed on large numbers of samples simultaneously.

The choice of a specific extraction method depends on both the compounds to be analyzed and the type of tissue. During the last decade, many different combinations of organic solvents have been used to extract free IAA: methanol/water (Kowalczyk and Sandberg 2001; Durgbanshi et al. 2005; Sugawara et al. 2009), isopropanol/acetic acid (Chiwocha et al. 2003), 1-propanol/water/concentrated HCl (Schmelz et al. 2003; Pan et al. 2008), methanol/water/formic acid (modified Bieleski's solvent) (Dobrev and Kamínek 2002; Kojima et al. 2009), isopropanol/imidazole buffer (pH 7) (Barkawi et al. 2008; Liu et al. 2012) and methanol/isopropanol/acetic acid (Müller and Munné-Bosch 2011). Similarly, a mixture of buffers and various concentrations of organic solvents, such as methanol, acetone or isopropanol, have been used as extraction solvents for IAA metabolites (Kowalczyk and Sandberg 2001; Barkawi et al. 2008; Sugawara et al. 2009). However, IAA is also readily soluble in water at neutral pH and therefore sodium phosphate buffer is an attractive alternative to organic solvents for this purpose because aqueous extracts tend to be less heavily contaminated with non-polar substances (Sundberg

1990). A study of IAA metabolite stability in different solvents showed that sodium phosphate buffer at pH 7.0 was the most efficient and least destructive extraction solvent for the majority of the tested IAA metabolites (Novák et al. 2012).

In order to obtain the greatest possible sensitivity and selectivity in the final MS analysis, it is necessary to combine a very efficient extraction method with a purification protocol that yields very high analyte recovery while simultaneously producing low background noise levels. Moreover, hydrolysis and/or chemical degradation during the extraction process must also be considered to avoid under- or overestimating analyte levels. The addition of antioxidants during the extraction process protects auxins in the sample against oxidation and degradation. For example, Ernstsen et al. (1986) reported the beneficial effects of sodium diethyldithiocarbamate on the breakdown of various IAA precursors to IAA during the purification and analysis of extracts from *Pinus sylvestris* L. needles. Butylhydroxytoluene is another useful antioxidant and exhibits good solubility in organic solvents such as methanol (Huang et al. 1992). On the other hand, mild alkaline conditions (65 % isopropanol and 35 % 0.2 M imidazole, pH 7.0) have been shown to allow optimal recovery of IAA without inadvertent hydrolysis of its conjugated forms (Baldi et al. 1989). Auxins can also be protected against degradation by performing the extraction process in a low-oxygen environment created by imposing a partial vacuum. This was demonstrated by Hu et al. (2011), who found that efficient vacuum microwave-assisted extraction (VMAE) of IAA and related compounds was significantly faster than conventional solvent extraction. The degradation of IAA and its metabolites can also be minimized by reducing the time spent on the extraction process or by performing the extraction at a low temperature, e.g. +4 °C.

The isotope dilution method (Rittenberg and Foster 1940), which involves determining the concentration of an unlabelled (endogenous) compound and comparing it to that of a labelled internal standard, can be used to correct for losses of IAA and its metabolites during extraction without knowing the extraction yield. However, in order to obtain an accurate quantitative estimate of these losses, the isotopically labelled standard must be equilibrated with the endogenous pool of the compound at an early stage in the extraction process (Sundberg 1990).

## 2 Methods for Sample Purification

Liquid/liquid extraction (LLE) is the most widely used traditional preparative technique for isolating analytes from plant tissues. However, in many recent applications, it has been replaced by solid phase extraction (SPE). Several conventional LLE methods based on the differential solubility of IAA in two liquid phases have been described. Solvent pairs used in these processes include sodium phosphate buffer/ethyl acetate, aqueous 1-propanol/methylene chloride, acidified 80 % methanol/diethyl ether and potassium sulphate buffer/chloroform (Liu et al. 2002; Schmelz et al. 2003; Durgbanshi et al. 2005; Quittenden et al. 2009). More recently, two novel methods, dispersive and hollow fibre liquid–liquid microextraction (LLME), have been used to pre-concentrate acidic plant hormones (including IAA) from, for example crude plant extracts (Lu et al. 2010) and natural coconut juice (Wu and Hu 2009). Ideally, LLME techniques should use negligible volumes of toxic, volatile and flammable organic solvents and involve as few steps as possible in order to minimize extraction times.

As mentioned above, the most common method for purifying IAA and its metabolites is SPE, using either silica or polymer-based SPE columns. SPE is rapid and reproducible, and protocols can easily be modified for use with a wide range of compounds and to accommodate different types and quantities of plant material. Ideally, a one-step SPE method with good recovery would minimize losses during purification and prevent the degradation of sensitive metabolites. Various low-specificity, compound-specific and class-specific sorbents are commonly used to isolate auxins from crude plant extracts on the basis of diverse interactions, including adsorption, hydrogen bonding, polar and non-polar interactions (Van der Waals and dipole–dipole forces), ionic interactions (via cation–anion exchange mechanisms) and/or size exclusion. For many applications, low-specificity silica and polymer-based materials can be used interchangeably, although the chemical backgrounds of the purified extracts might be different. Compound-specific ion exchange sorbents with fixed ionic sites that are complementary to the analytes of interest can be used to improve the selectivity of the analysis by isolating specific groups of ions from the sample solution. Several SPE methods rely on silica-based, siloxane-bonded sorbents with long alkyl chains such as $C_{18}$ or $C_8$ (Kowalczyk and Sandberg 2001; Pěnčík et al. 2009; Petersson et al. 2009). However, modern porous co-polymers of divinylbenzene and *N*-vinylpyrrolidone (such as Oasis HLB) have become the preferred sorbents for one-step SPE in more recently developed methods (Izumi et al. 2009; Sugawara et al. 2009; Liu et al. 2012; Novák et al. 2012). In addition, a two-step SPE method that uses amino SPE followed by poly(methyl methacrylate) epoxide resin SPE has been developed for purifying extracts from small amounts of tissue (Barkawi et al. 2008). Multi-step purification methods often involve an ion exchange or mixed-mode SPE stage in order to more thoroughly purify the targeted plant hormones. Mixed-mode SPE processes that have both reverse-phase and ion-exchange characteristics can retain different IAA metabolites (precursors, catabolites and conjugates) depending on the chemical properties of the sorbents used. For example, a process involving $C_{18}$ SPE followed by SPE with a mixed cation-exchange cartridge (Oasis MCX) was used to separate cytokinins from IAA and abscisic acid, resulting in high overall extraction and concentration efficiencies (Dobrev and Kamínek 2002). While the basic analytes were effectively retained by the cation-exchange mechanism, acidic and neutral hydrophobic compounds were only retained by the reversed phase mechanism and eluted when the column was washed with pure methanol. To address this problem, Dobrev et al. (2005) introduced an additional purification step that uses a mixed-mode reversed phase anion-exchange polymeric sorbent (Oasis MAX). This made it possible to separate the acidic analytes from the neutral compounds. Mixed-mode sorbents have been used to purify IAA and its metabolites from samples originating from diverse plant species including *Arabidopsis*, radish, rice, tobacco and wheat (Dobrev and Kamínek 2002; Dobrev et al. 2005; Izumi et al. 2009; Kojima et al. 2009; Sugawara et al. 2009; Farrow and Emery 2012). However, ion-exchange mechanisms based on pH gradients would be unsuitable for isolating labile IAA precursors such as indole-3-acetaldoxime (Novák et al. 2012). Ideally, a general purification method should be able to separate a broad range of IAA metabolites and would require optimized conditions that use mixtures of water and organic solvents as eluents.

Traditional SPE aside, a range of alternative high-affinity and highly selective sample preparation techniques have been used for the isolation, concentration and clean-up of IAA and related compounds from complex plant matrices. In 1986, an immunoaffinity sorbent prepared by covalently binding polyclonal rabbit antibodies to activated silica was used to purify IAA in extracts from the cambial zone and shoots of *Pinus sylvestris* (Sundberg et al. 1986). Pěnčík et al. (2009) subsequently introduced antibodies that were capable of interacting specifically not only with IAA but also with other 3-substituted indoles. These antibodies were used to develop a complex analytical protocol based on an immunosorbent with relatively high cross-reactivity that recognizes a whole class of structurally related compounds. The new protocol was used to isolate IAA and its conjugates in samples of the pericarps of immature seeds of the Christmas rose (*Helleborus niger* L.). Another class-specific method is based on molecularly imprinted polymers (MIPs), synthetic analogues of immunosorbents that bind specifically to a target molecule even in the presence of closely related compounds. MIPs exhibit good thermal stability, are physically robust and are easy and inexpensive to prepare. They have recently been used to isolate phytohormones by SPE (Du et al. 2012a). Other reports have discussed the use of IAA-imprinted polymers based on *N*,*N*-dimethylaminoethyl methacrylate, methacrylic acid and 9-vinyladenine as functional monomers (Kugimiya and Takeuchi 1999a, b; Chen et al. 2006). More recently still, magnetic MIP beads that have molecular recognition elements and can be used in magnetic separation have been synthesized (Zhang et al. 2010). These particles can be dispersed in the plant extract and then easily separated from the liquid phase using a magnetic field, which is more convenient than conventional centrifugation or filtration steps. The practical utility of this technique was investigated by using it to purify IAA in tissue samples from three different plant species (pea, rice and wheat) that had been extracted using LLE or VMAE and were then subjected to MIP-based clean-up procedures (Zhang et al. 2010; Hu et al. 2011).

## 3 Identification of Auxin Metabolites

When studying auxin metabolism, it is essential to combine data gained from genetic investigations with the identification and chemical characterization of individual metabolites. Due to the very low concentrations of phytohormone metabolites and the large number of potentially interfering compounds present in the complex matrices of plant tissue samples, the identification of these metabolites is very challenging. In early studies, very large amounts of tissue had to be extracted and purified by extensive, multistep procedures in order to isolate these substances (Reinecke and Bandurski 1983). However, the ongoing development of analytical instrumentation over the last two decades has greatly simplified the analysis of organic compounds in plant material. The most useful and widely used tool for identifying individual metabolites is mass spectrometry, especially tandem mass spectrometry (MS/MS). Low-resolution MS/MS can be performed with triple quadrupole or linear ion trap instruments and is capable of providing sufficient selectivity and sensitivity for the analysis of complex plant samples. High-resolution mass spectrometers (HRMS), such as the new time-of-flight (TOF) instruments or machines based on Orbitrap technology, are also being developed and improved on an ongoing basis and are powerful tools, particularly for determining the structures of unknown compounds. Before MS detection, the individual components of samples are usually separated using either gas chromatography (GC) or liquid chromatography (LC). While GC-MS offers very good sensitivity, LC-MS is more versatile and permits the analysis of a wider range of compounds without requiring prior derivatization.

When a sample is introduced into a mass spectrometer, its components are initially converted into positively or negatively charged ions in the ion source. These ions are then detected, producing a mass spectrum that shows the mass-to-charge ratios (*m*/*z* values) of the individual ions. Individual molecules can be identified by analyzing the ions' fragmentation patterns, isotopic distributions and accurate mass values. If an appropriate standard is available, the proposed identity of a compound can be verified by comparing its mass spectrum and chromatographic behaviour to that of the standard. This provides the most straightforward and reliable confirmation of compound identity and is therefore the most frequently used method for this purpose (Kai et al. 2007a; Pěnčík et al. 2009). Identifications that are based only on analyses of spectra can be unreliable, as demonstrated by the case of a putative IAA precursor that was initially identified as *N*-hydroxytryptamine (Zhao et al. 2001) based on evidence that was subsequently shown to be inconsistent (Tivendale et al. 2010).

The occurrence of IAA–Ala and IAA–Leu as endogenous IAA metabolites in Arabidopsis was demonstrated by performing a screen for IAA conjugates in Arabidopsis samples (Kowalczyk and Sandberg 2001). In this work, plant extracts were fractionated by preparative high-performance liquid chromatography (HPLC), and each individual fraction was derivatized and subjected to GC-MS analysis. Novel IAA conjugates were then identified based on the occurrence of characteristic indole fragments such as the quinolinium ion (*m*/*z* 130) in the resulting mass spectra. A similar strategy based on MS/MS screening for characteristic fragments was subsequently used to identify oxidative metabolites of IAA conjugates such as 6-OH-IAA-Phe, 6-OH-IAA-Val and oxIAA-Glc in Arabidopsis, and IAA-*N*-Glc and its conjugates with Asp and Glu in rice (Kai et al. 2007a, b). In the latter two studies, the identities of the detected compounds were confirmed

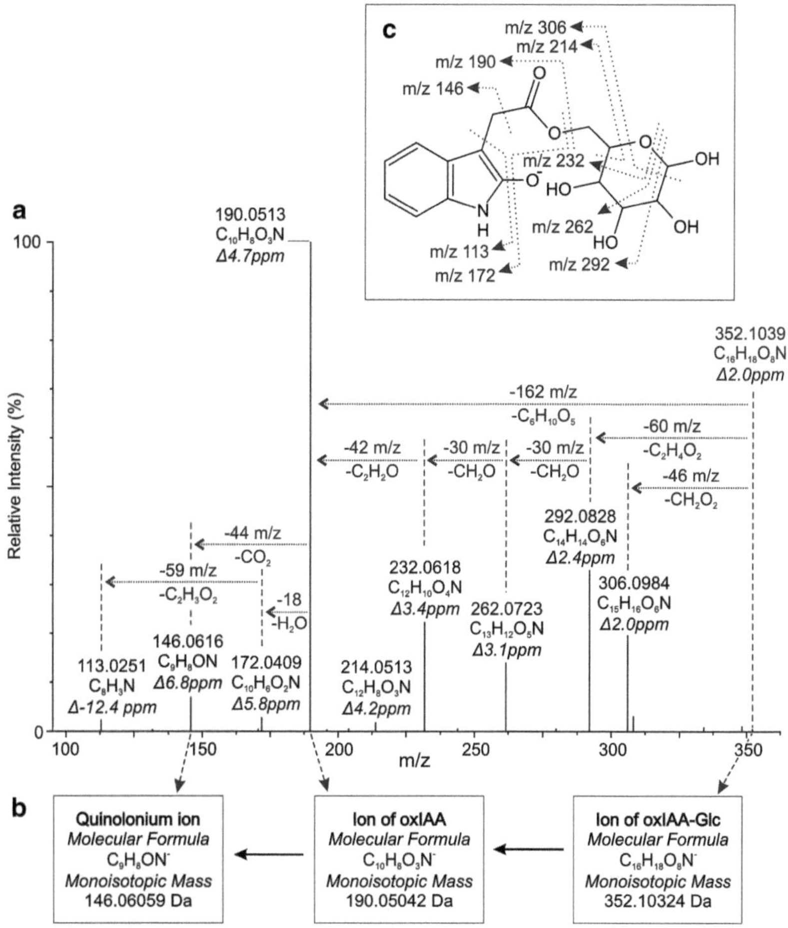


**Fig. 3.1** Identification of 1-*O*-(2-oxoindol-3-ylacetyl)-β-D-glucopyranose (oxIAA-Glc) from *Arabidopsis thaliana*. To elucidate the structure of oxIAA-Glc, the purified extract was separated using a Thermo Accela LC system and analyzed by tandem mass spectrometry using an LTQ/Orbitrap mass spectrometer (Thermo Fisher Scientific, Bremen, Germany). Centroid mass spectra of negative ions were collected in the Orbitrap mass analyzer, with a target mass resolution of 30,000 (full width at half maximum peak height, defined at *m/z* 400), after collision-induced dissociation (CID) in the LTQ cell, using a normalized collision energy of 20 %. The accurate masses of the precursor ion (*m/z* 352) and its fragments were calculated and used to determine the elementary composition and structure of the analyte, with fidelity within 12.5 ppm. (**a**) Full $MS^3$ spectrum of oxIAA-Glc. The ions at *m/z* 352 and *m/z* 190 were selected for fragmentation, using CID energies of 45 eV and 35 eV, respectively. (**b**) Proposed MS fragmentation of oxIAA-Glc according to Kai et al. (2007a). (**c**) OxIAA-Glc structure with proposed bond breakages during fragmentation

using synthetic standards. The use of highly selective sample preparation techniques before the final analysis can greatly increase the scope for identifying individual compounds with a given analytical method. For example, the use of immunoaffinity extraction followed by LC-MS/MS analysis made it possible to isolate and identify the IAA conjugates IAA-Gly, IAA-Val and IAA-Phe in samples of the Christmas rose (Pěnčík et al. 2009). In addition, various IAA conjugates and metabolites were detected within the hormone metabolite profile of tomato samples using a high-resolution MS Orbitrap instrument (Van Meulebroek et al. 2012). Figure 3.1 shows the mass spectrum of a compound found in an Arabidopsis extract that was identified as oxIAA-Glc using a Fourier Transform Orbitrap MS instrument.

Another technique that is widely used for metabolite identification is nuclear magnetic resonance (NMR). While NMR is a very powerful tool for determining the structures of chemical compounds, it is less sensitive than MS and is therefore not suitable for analyzing small biological samples. However, it can be a very useful tool for obtaining detailed structural information about a specific metabolite (Östin et al. 1995).

## 4 Quantification of Auxin Metabolites

At present, endogenous IAA and IAA metabolites (precursors, catabolites and conjugates) are generally quantified by mass spectrometry using the isotope dilution technique (Rittenberg and Foster 1940). As mentioned above, internal standards labelled with stable isotopes such as $^{2}H$ (deuterium), $^{13}C$, $^{15}N$ and/or $^{18}O$ atoms are added to the samples at an early stage in the sample preparation process, prior to homogenization and extraction. Isotope-labelled standards are easily distinguished during MS analysis due to their unique masses. Stable isotope dilution assays can be very accurate and precise because they correct for losses or inefficiencies in the sample preparation process as well as ion suppression effects during the MS analysis.

Complete chromatographic separation is generally required during the analysis of complex plant samples using ultraviolet and/or fluorescence detection. In contrast, baseline separation is not essential if the peaks have independent MS signals. GC-MS is an important tool for the quantitative analysis of auxins, and a number of recent reports have also demonstrated that LC-MS can also be very useful for IAA and IAA metabolite quantification.

The rapid development of chromatographic techniques, such as ultra-high-performance liquid chromatography using sub-2-μm particle columns, has greatly improved the speed, separation, resolution and sensitivity of LC-based analyses relative to those achieved using conventional HPLC. LC separations of IAA and its metabolites are typically performed with silica or polymer-based columns. Additives, such as e.g. formic acid (Pěnčík et al. 2009), acetic acid (Novák et al. 2012) or ammonium acetate (Prinsen et al. 1998), are often added to the mobile phase to obtain good chromatographic separation and high ionization efficiency.

GC-MS is a very robust technique that is suitable for high-throughput analysis, providing excellent separation and high sensitivity. However, GC-MS methods also have several disadvantages compared to LC-MS, especially LC-MS instruments equipped with an electrospray interface. Electrospray ionization (ESI) is a soft ionization technique that produces very intense precursor ions. This improves the yield of the product ions and thus increases the sensitivity of the subsequent MS/MS analyses. In contrast, GC-MS instruments typically use electron impact ionization, which causes extensive fragmentation of the precursor ions. In addition, it is necessary to derivatize IAA and most IAA metabolites prior to GC-MS analysis in order to increase their volatility. The carboxyl groups of IAA and its acidic metabolites are usually methylated using ethereal diazomethane, after which any free hydroxyl groups are trimethylsilylated (Edlund et al. 1995; Kowalczyk and Sandberg 2001; Pěnčík et al. 2009). There are also a few methods that are used specifically to derivatize IAA precursors for GC-MS analysis, such as the acylation of tryptamines, the trimethylsilylation of indole-3-ethanol and methyl chloroformate derivatization of tryptophan (Quittenden et al. 2009; Liu et al. 2012). Derivatization prior to LC-MS can improve the selectivity and sensitivity of the analysis as well as the chromatographic behaviour and resolution of the targeted compounds. However, all IAA metabolites except IPyA and IAAld can be analyzed by LC-MS in positive or negative ESI mode without prior derivatization (Kai et al. 2007a, b; Sugawara et al. 2009; Novák et al. 2012).

The development of improved MS instruments and innovative technologies, such as tandem mass spectrometry (MS/MS), has greatly increased the sensitivity and selectivity of quantitative MS analyses. The main limiting factors that affect the utility of MS when studying IAA and its metabolites are the complexity of the plant matrices and the low limits of detection required for effective analysis (around 1 ppt). In conventional MS analyses, these factors necessitate the use of laborious and time-consuming sample purification protocols. This can be avoided by combining the high selectivity of MS/MS with fast chromatography, enabling rapid high-throughput MS analysis. Because of its many advantages, LC-MS/MS became widely accepted as a tool for plant hormone analysis within a few years of its introduction, and several sensitive methods for quantifying IAA and its metabolites have been reported (Kai et al. 2007a, b; Pěnčík et al. 2009; Sugawara et al. 2009; Mashiguchi et al. 2011; Novák et al. 2012). In addition, GC-MS/MS has been used to quantify most of the putative intermediates in the tryptamine pathway in peas (Quittenden et al. 2009). Liu et al. (2012) have reported that replacing an older GC-MS protocol (Barkawi et al. 2008) with a modern GC-MS/MS method and a high-throughput pipette tip SPE protocol improved the sensitivity and the accuracy of auxin analyses using very small plant tissue samples.

The use of miniaturized purification methods in conjunction with the high separation efficiency and selectivity of LC-MS/MS analysis has proven to be very useful for quantifying plant hormones in very small samples (<5 mg FW) of plant material (Svačinová et al. 2012). As a case in point, a recent publication described an ultra-sensitive GC-MS/MS method for cell-specific IAA analysis (Petersson et al. 2009). Similarly, miniaturized tip-based SPE has been used for tissue-specific quantification of IAA and its metabolites (Pěnčík and Novák, unpublished). Several methods that use LC chromatography and capillary electrophoresis in conjunction with MS using quadrupole TOF instruments to quantify auxins have been reported in the last few years (Sugawara et al. 2009; Chen et al. 2011; Mashiguchi et al. 2011). Although high-resolution MS (HRMS) is not yet widely used for quantitative analysis of plant hormones, it is likely that this technology will soon be employed for routine quantification due to the lower price of the next-generation HRMS instruments. HRMS makes it possible to perform very selective quantitative analyses due to its very narrow mass windows. Its usefulness in IAA analysis was demonstrated in 2001 during a study on Arabidopsis seedlings (Ljung et al. 2001b). At present, the detection specificity, assay precision, and sensitivity for IAA achieved using the full scan approach of the Orbitrap mass spectrometer are comparable to those achieved using conventional SRM methods that rely on triple quadrupole or linear ion trap instruments (Van Meulebroek et al. 2012).

# 5 Metabolite Profiling

## *5.1 Auxin Metabolite Profiling*

In order to understand auxin metabolism, it is important to be able to quantify not only free IAA but also different IAA precursors, conjugates and catabolites. IAA metabolites have very different chemical properties and can be acidic, neutral or basic in nature. They occur at a wide range of concentrations, and some of them are very unstable in solution. Methods for IAA metabolite profiling must therefore be able to deal with all of these factors in order to produce quantitative data of high quality. The methods used must also be thoroughly validated so that potential problems with extraction (poor extraction efficiency), purification (low recovery of analytes), matrix effects (ion suppression and high chemical background) and chromatography (co-elution of structurally related compounds) can be identified and avoided where possible (Taylor 2005; Novák et al. 2012).

Early methods for IAA metabolite profiling were primarily developed to quantify free IAA and some acidic IAA metabolites. These analyses were mainly performed by GC-MS (Tam et al. 2000; Barkawi et al. 2008), although LC-MS-based methods for the analysis of these metabolites have also been described (Kowalczyk and Sandberg 2001). GC-MS has the advantage of being a very robust analytical technique with a high sample throughput, but it is not suitable for compounds that are thermolabile, and most IAA metabolites also have to be derivatized in order to make them volatile enough for GC analysis. This limits the range of IAA metabolites that can be quantified using GC-MS, and derivatization might have negative effects on the sensitivity of the method towards some IAA metabolites and their recovery. In contrast, LC-MS has proven to be a very versatile and sensitive method for IAA metabolite profiling and is suitable for the analysis of a wide range of IAA metabolites (Sugawara et al. 2009; Mashiguchi et al. 2011; Novák et al. 2012).

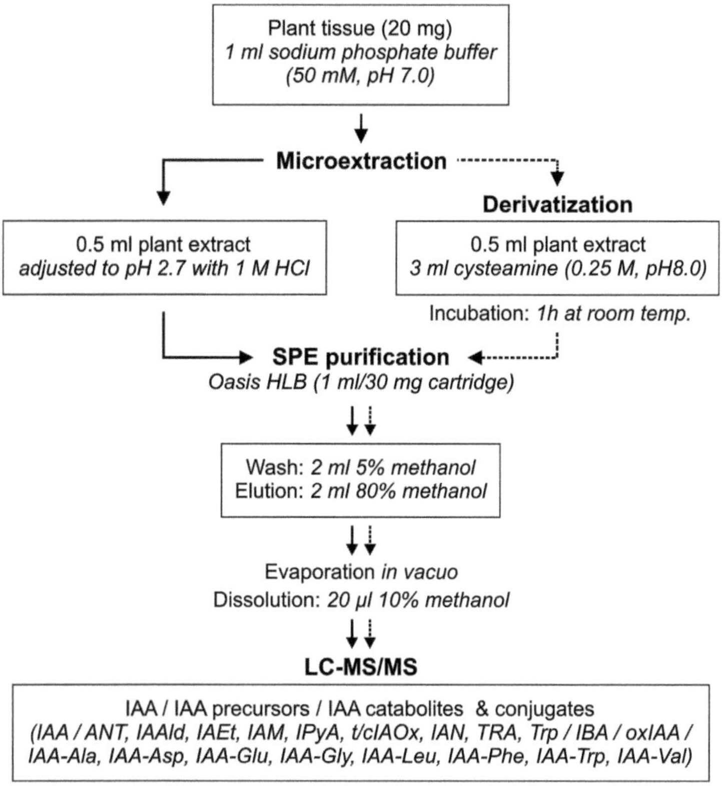


**Fig. 3.2** Protocol for extracting and purifying IAA metabolites for LC-MS/MS profiling of small amounts of plant material. The plant material (<20 mg) is homogenized and extracted in sodium-phosphate buffer (1 ml) containing labelled internal standards and then divided into two sub-samples. One sub-sample is acidified and applied to a pre-conditioned HLB column (30 mg), which is then washed and eluted with 80 % methanol. The second sub-sample is derivatized with cysteamine to produce the thiazolidine derivatives of IPyA and IAAld and then purified using the same SPE protocol. Both eluates are evaporated to dryness, reconstituted and analyzed by LC-MS/MS (for more details see Novák et al. 2012)

Figure 3.2 illustrates the workflow of a method for IAA metabolite profiling using small amounts of plant tissue. This method is based on microextraction using sodium phosphate buffer, purification on a polymer-based SPE column and LC-MS/MS analysis. To facilitate the analysis of labile IAA metabolites such as IPyA and IAAld, the extract is separated into two sub-samples, one of which is derivatized with cysteamine to convert these labile compounds into their thiazolidine (TAZ) derivatives (Novák et al. 2012, Fig. 3.3). Other methods for derivatizing these compounds have also been published (Tam and Normanly 1998; Quittenden et al. 2009; Mashiguchi et al. 2011; Liu et al. 2012). Derivatization is not only necessary to prevent the degradation of these metabolites during extraction and purification but can also be used to increase the sensitivity of GC-MS and

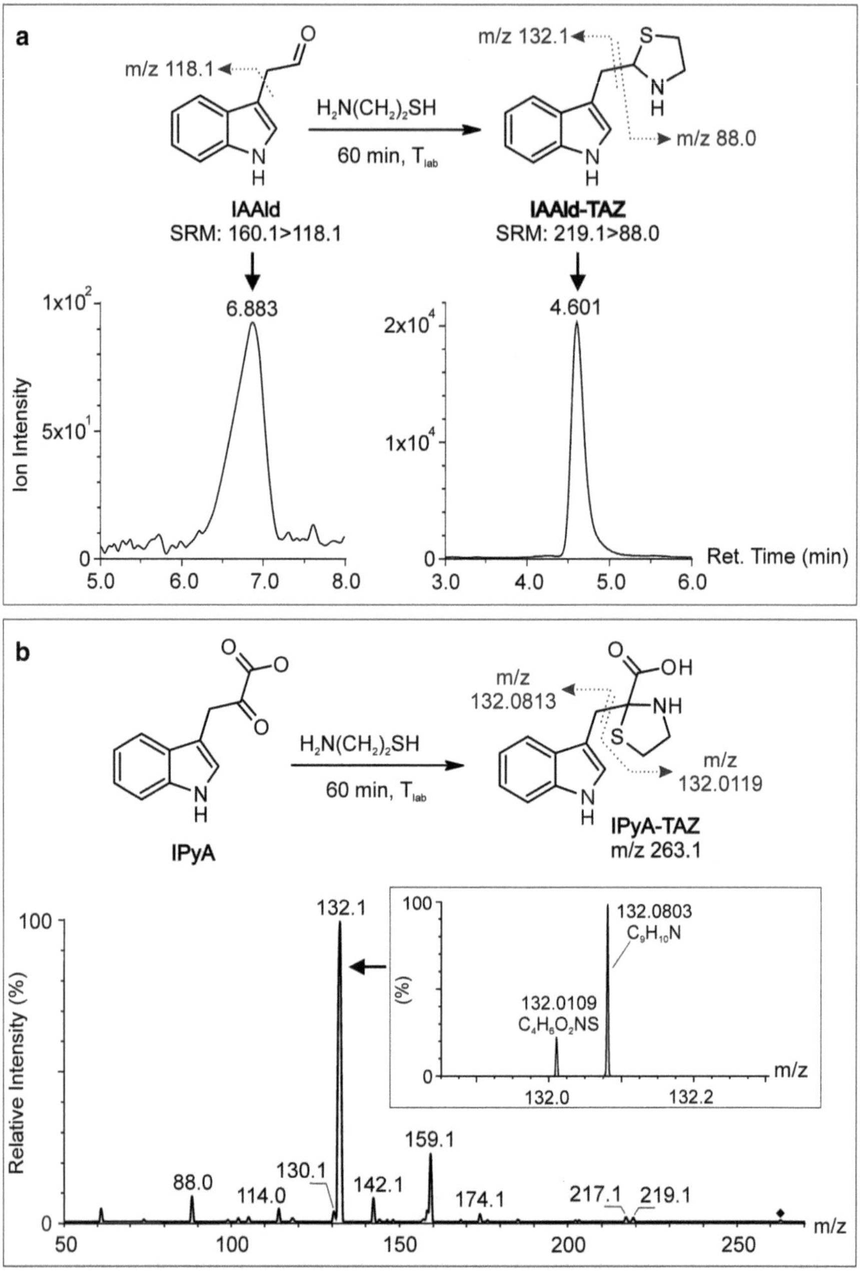


**Fig. 3.3** Derivatization of IAAld and IPyA. IAAld and IPyA were derivatized with cysteamine to produce the corresponding thiazolidines (TAZ). (**a**) The sensitivity and the peak symmetry of IAAld was greatly improved after derivatization with cysteamine. 1 pmol IAAld and IAAld-TAZ were injected onto a reverse-phase LC column and analyzed by MS/MS. (**b**) MS/MS spectrum of IPyA-TAZ formed by derivatizing IPyA with cysteamine. The product ion scan was obtained by low-resolution MS analysis (using a triple quadrupole instrument) under optimized MS conditions, at a collision energy of 20 eV. *Inset*: Ultra high-resolution MS analysis (Orbitrap technology) was used to separate and identify the two main product ions, whose masses differed by only 0.07 Da

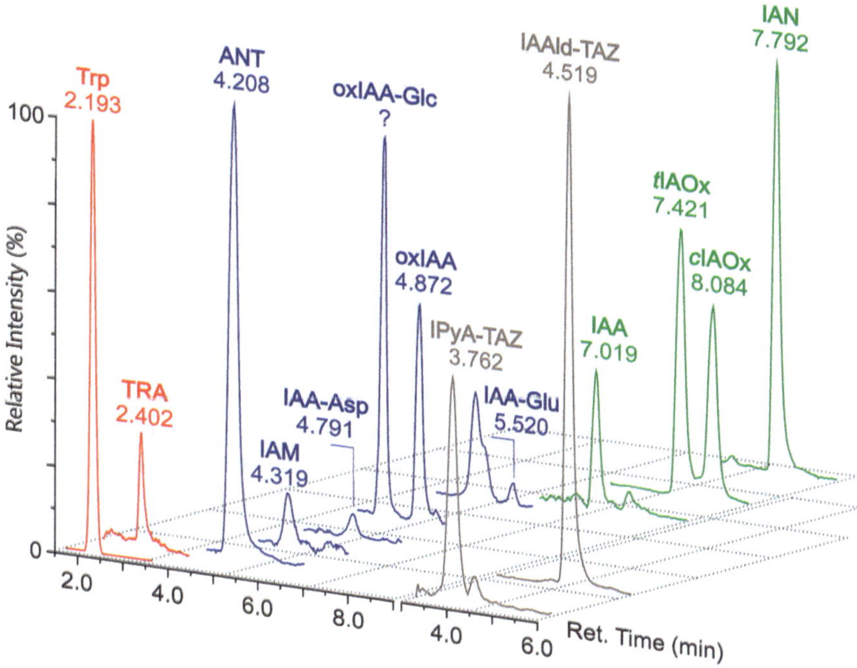


**Fig. 3.4** LC-MS/MS chromatograms of IAA metabolites. IAA metabolites were purified from the primary root tip (1 mm) of 6-day-old Arabidopsis wild-type seedlings, and analyzed by LC-MS/MS. The underivatized and derivatized compounds were injected onto a reversed-phase column, and the chromatographic analysis was split into four scan segments (for more experimental details see Novák et al. 2012)

LC-MS analyses and to improve peak shapes (Edlund et al. 1995; Kojima et al. 2009, Fig. 3.3a).

In order to obtain a better understanding of developmental processes in plants, it will be necessary to perform IAA and IAA metabolite analyses on more well-defined parts of the plants, such as specific tissues and cell types in, for example the root and shoot apex. This was earlier very difficult, but has become feasible due to the high sensitivity and selectivity of modern LC-MS and GC-MS instruments, and methods for tissue- and cell-type specific analysis of IAA and IAA metabolites have recently been published (Petersson et al. 2009; Novák et al. 2012). Figure 3.4 shows chromatograms obtained during IAA metabolite profiling of the Arabidopsis root apex.

## 5.2 *Profiling of Different Classes of Hormones in the Same Sample*

The idea of analyzing different classes of plant hormones in a single sample is not new, and several attempts have been made to develop methods for the simultaneous analysis of plant hormones such as auxins, cytokinins (CKs), abscisic acid (ABA) and gibberellins (GAs) (Du et al. 2012b). Because many of these compounds are acidic in nature, several authors have proposed that it should be possible to extract and analyze them together (Müller et al. 2002; Schmelz et al. 2003; Durgbanshi et al. 2005; Chen et al. 2011). However, no single method has yet been developed that is equally capable of extracting all known classes of plant hormones due to their diverse chemical structures and physicochemical properties. In addition, there is little available information that can be used to compare or cross-validate the extraction methods that have been published and are in widespread use (Pan and Wang 2009). A wide range of purification strategies based on concepts described in the preceding sections, including LLE and/or low-specific, compound-specific and class-specific SPE have been developed.

One of the more recently developed methods for the simultaneous analysis of multiple plant hormones involves using LC-MS/MS to profile CKs, IAA and ABA in 100–300 mg Arabidopsis tissue samples after purification on a combination of mixed-mode (MCX and MAX) and polymer-based (HLB) sorbents (Farrow and Emery 2012). Other methods for the simultaneous analysis of CKs and acidic plant hormones have also been developed (Chiwocha et al. 2003; Izumi et al. 2009). Notably, Izumi et al. (2009) showed that coupling nanoflow-LC to a tandem mass spectrometer improved the sensitivity of a method for quantifying several hormones in samples extracted from minute amounts of tobacco tissue (1 mg DW). Kojima et al. (2009) created a fully automated SPE purification and LC-MS/MS method for the simultaneous analysis of six major phytohormone groups (including 23 CKs, 7 auxins, ABA and 12 GAs). This method uses bromocholine to esterify the free carboxylic acid groups of the plant hormones in order to permit their detection using positive ionization mode. Underivatized GAs are usually negatively charged, but positive ionization is generally more sensitive than negative ionization. In another recent study, up to 18 different plant hormones were profiled simultaneously by LC-MS/MS after the extraction of 50–100 mg of plant material without any further purification or derivatization (Pan et al. 2008; Müller and Munné-Bosch 2011).

As mentioned previously, HRMS analyses focus on narrower mass windows than unit-mass-based scan modes and provide accurate mass data, which can be a lot more useful and informative than simple unit masses. Specifically, the use of ultra-high-mass resolutions makes it possible to determine the exact masses of the detected ions, so it is simple to distinguish between analytes and interfering matrix compounds (Kaufmann 2012). This property was exploited in a recent study that used a very simple extraction and purification procedure combined with high-resolution mass spectrometry to identify and quantify a large number of endogenous phytohormones in tomato samples (Van Meulebroek et al. 2012).

## 5.3 Metabolic Studies

In order to study the pathways of auxin biosynthesis and degradation, sensitive methods for analyzing steady state levels of IAA and its metabolites have recently been developed (Sugawara et al. 2009; Mashiguchi et al. 2011; Novák et al. 2012). The use of heavy isotope labelling of intermediates in conjunction with MS analysis has proven to be a powerful tool for monitoring fluxes through IAA metabolic pathways. Notably, such methods for in vivo labelling have been used to identify IAA biosynthetic intermediates and IAA degradation products. They can also be combined with mutant analyses to provide additional insights into the relevant metabolic pathways and to estimate the rates of IAA biosynthesis and the half-lives of IAA metabolites.

Labelling has been performed using both general precursors such as deuterated water (deuterium oxide, $^2H_2O$) and various IAA precursors such as heavy isotope labelled tryptophan. $^2H_2O$ has the advantage that it is easily taken up by the plant root and transported to all tissues and cells. Deuterium atoms are initially incorporated into intermediates in the shikimate pathway, and later into IAA precursors and IAA itself (Normanly 2010). Feeding experiments with $^2H_2O$ in plant species, such as *Zea mays* (Pengelly and Bandurski 1983; Jensen and Bandurski 1995), *Lycopersicon esculentum* (Cooney and Nonhebel 1991), *Arabidopsis thaliana* (Ljung et al. 2001b, 2005; Sairanen et al. 2012) and *Physcomitrella patens* (Eklund et al. 2010), have provided valuable insights into IAA metabolism.

Feeding with heavy labelled IAA precursors, such as [$^{13}C_{11}$,$^{15}N_2$]-Trp (Mashiguchi et al. 2011), [$^{13}C_6$]-IAOx, [$^{15}N_2$]-TAM and [$^2H_2$]-IAN (Sugawara et al. 2009) has likewise been very important for our understanding of IAA biosynthesis in Arabidopsis. Similar studies in *Pisum sativum*, using [$^2H_5$]-Trp, [$^2H_5$]-TAM, [$^2H_5$]-IAN and [$^2H_5$]-IAOx (Quittenden et al. 2009; Tivendale et al. 2010) and *Lemna gibba*, using [$^{15}N_1$]-ANT and [$^2H_5$]-Trp (Rapparini et al. 2002) have also provided important information on IAA biosynthesis in these species.

Figure 3.5 shows the labelling of IAA metabolites after feeding with stable isotope-labelled IAA precursors ($^2H_2O$ and $^{13}C^6$-ANT). The precursors were taken up by the seedlings and the heavy atoms were incorporated into newly synthesized Trp, IPyA and IAA.

Feeding experiments using stable labelled IAA and various IAA metabolites have been performed in *Arabidopsis thaliana* (Östin et al. 1998; Kai et al. 2007a), *Oryza sativa* (Kai et al. 2007b) and *Pinus sylvestris* (Ljung et al. 2001a) to investigate IAA catabolism and conjugation. The results obtained made it possible to identify major IAA catabolites/conjugates in different plant species and to study the processes of IAA conjugate formation and hydrolysis.

It is important to be aware of the problems that may be encountered when using heavy isotope labelling. Different compounds can have very different uptake rates, and there might also be differences in their sub-cellular compartmentation. This can affect IAA metabolism and is very difficult to measure. Feeding with high,

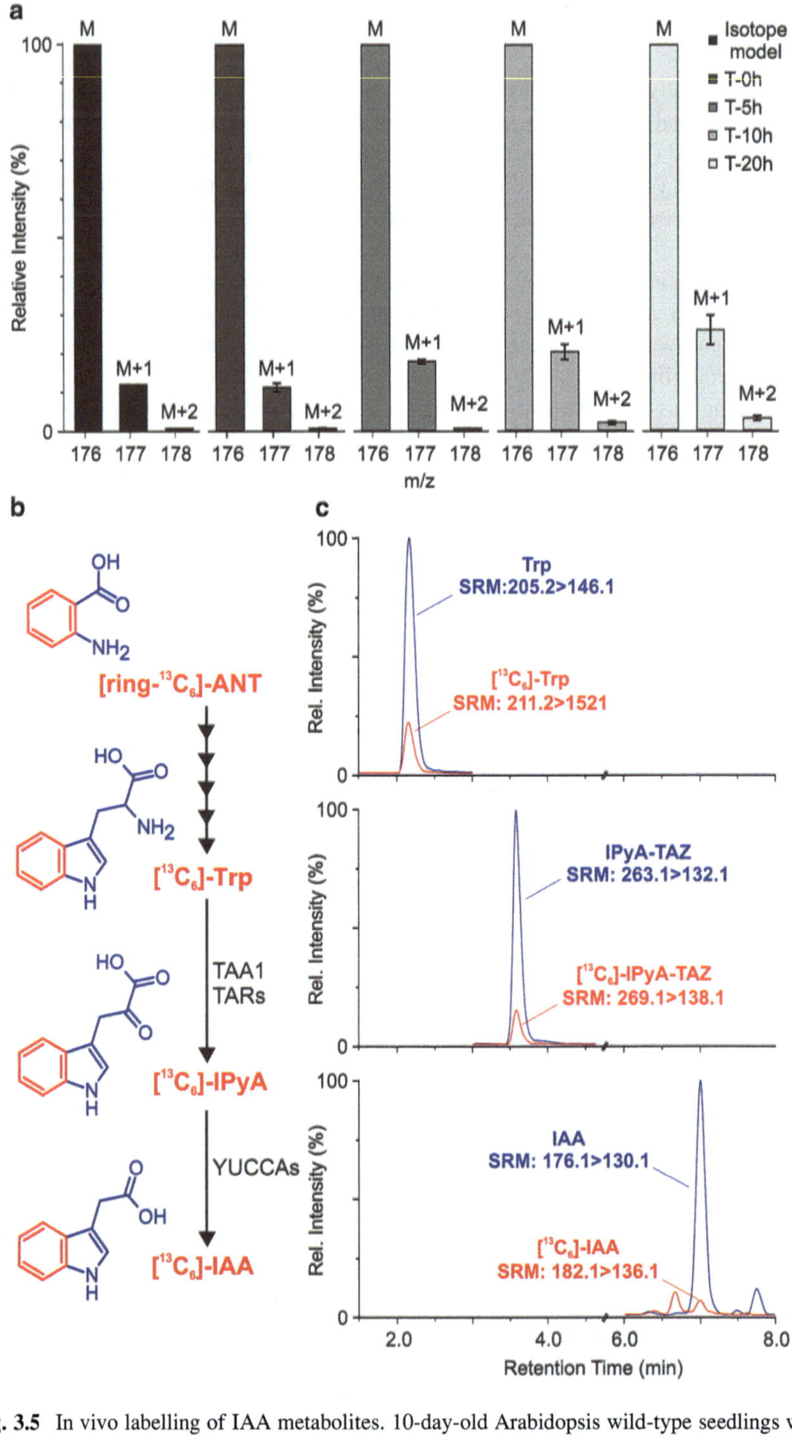


**Fig. 3.5** In vivo labelling of IAA metabolites. 10-day-old Arabidopsis wild-type seedlings were incubated with liquid medium containing 5 % glucose and 30 % deuterated water or 100 μM [$^{13}C_6$]-anthranilate (ANT). IAA and its precursors tryptophan (Trp) and indole-3-pyruvic acid

unphysiological concentrations of IAA and its precursors can cause changes in IAA metabolism due to factors such as feedback inhibition of specific enzymatic steps (Östin et al. 1998). Finally, feeding with high concentrations of deuterated water for extended periods of time has been shown to retard plant growth in some cases (Ljung et al. 2005).

# 6 Future Prospects

The analytical techniques used for plant hormone analysis have improved tremendously over the last decade, and it is likely that this trend will continue in the future (Nováková and Vlčková 2009; Du et al. 2012b; Núñez et al. 2012). The sensitivity and selectivity of mass spectrometric techniques both continue to improve, making it possible to perform tissue- and cell-specific quantification of IAA and different IAA metabolites (Petersson et al. 2009; Novák et al. 2012). The sensitivity of these analyses can be further increased by miniaturizing the preceding extraction and purification steps, since this can minimize analyte losses due to adsorption on surfaces and increase analyte recovery in the SPE step (Svačinová et al. 2012; Liu et al. 2012). The use of new SPE technologies, such as combined silica/polymeric sorbents (Svačinová et al. 2012), mixed mode sorbents (Dobrev and Kamínek 2002) and molecularly imprinted polymers (MIPs) (Zhang et al. 2010; Hu et al. 2011; Du et al. 2012a), can also increase analyte recovery during the SPE step while minimizing matrix effects during the MS analysis. It is likely that the introduction of new GC and LC methods, such as two-dimensional chromatography or separation using nanoflow-LC, together with the rapid ongoing development of MS techniques such as ultra-high-resolution MS and ion mobility mass spectrometry, will greatly increase the scope for applying these technologies in many different fields of life science (Holčapek et al. 2012). The recent improvements in MS sensitivity and selectivity have made it possible to analyze the levels of IAA and other plant hormones within single cells and sub-cellular compartments, although many technical difficulties associated with these techniques remain to be addressed (Oikawa and Saito 2012).

On the other hand, there is also a need for methods for the high-throughput analysis of different plant hormones, where ultra-high sensitivity might not be the

**Fig. 3.5** (continued) (IPyA) were purified from shoot tissue and analysed by LC-ESI-MS/MS. (**a**) Mass isotopomer profiles for the molecular ion (*m/z* 176) of IAA after incubation with deuterated water. Incorporation of one or two deuterium atoms was observed up to 20 h after the beginning of treatment (for more experimental details see Sairanen et al. 2012). (**b**) Scheme showing the in vivo labelling of IAA metabolites after feeding with [ring-$^{13}C_6$]-ANT. (**c**) The MS ion chromatograms for tryptophan (Trp), the thiazolidine derivative of indole-3-pyruvic acid (IPyA-TAZ) and IAA in [$^{13}C_6$]-ANT-fed Arabidopsis seedlings. The chromatograms for the SRM transitions for [$^{13}C_6$]-labelled and unlabelled IAA metabolites are shown in red and blue, respectively

most important issue. A range of methods for the high-throughput analysis of auxins (Barkawi et al. 2010), different classes of plant hormones (Kojima and Sakakibara 2012), and IAA and its biosynthetic precursors (Liu et al. 2012) have been reported. In a similar vein, the QuEChERS ("Quick, Easy, Cheap, Effective, Rugged, and Safe") method has been used to extract and purify multiple plant hormones from vegetable and fruit tissues prior to LC-MS/MS analysis (Flores et al. 2011; Shi et al. 2012). These high-throughput methods often use robotics to automate sample handling during extraction and purification in conjunction with fast liquid chromatography during the final MS step to facilitate the rapid analysis of a large number of samples (Nováková and Vlčková 2009; Núñez et al. 2012).

LC- and GC-MS/MS analyses have become the methods of choice for quantifying many different classes of plant hormones and their metabolites in diverse plant species. Further increases in the number of compounds that can be analyzed will require more efficient and automated data processing, as well as powerful methods for the statistical analysis of the large data sets that will be produced. Multivariate data analysis has proven to be a very valuable tool for visualizing and interpreting MS data (Wiklund et al. 2008; Kirwan et al. 2012) and has recently been used to evaluate data from IAA metabolite profiling experiments (Novák et al. 2012). It would be very appealing to combine targeted analyses of plant hormones with metabolomics, genomics and proteomics data in future plant systems biology studies, but such an approach would be very challenging from a data handling perspective.

It is our belief that plant hormone analysis will continue to be a valuable tool in plant research for the foreseeable future, and it is to be hoped that the art of "grinding and finding" will become increasingly accessible to research groups around the world. Close collaborations between analytical chemists and experts in plant biology, genetics and biochemistry will be essential in these attempts to improve our understanding of the function of these important plant growth regulators.

## References

Baldi BG, Maher BR, Cohen JD (1989) Hydrolysis of indole-3-acetic acid esters exposed to mild alkaline conditions. Plant Physiol 91:9–12

Barkawi LS, Tam YY, Tillman JA et al (2008) A high-throughput method for the quantitative analysis of indole-3-acetic acid and other auxins from plant tissue. Anal Biochem 372:177–188

Barkawi LS, Tam YY, Tillman JA et al (2010) A high-throughput method for the quantitative analysis of auxins. Nat Protoc 5:1609–1618

Chen CB, Chen YJ, Zhou J et al (2006) A 9-vinyladenine-based molecularly imprinted polymeric membrane for the efficient recognition of plant hormone 1H-indole-3-acetic acid. Anal Chim Acta 569:58–65

Chen ML, Huang YQ, Liu JQ et al (2011) Highly sensitive profiling assay of acidic plant hormones using a novel mass probe by capillary electrophoresis-time of flight-mass spectrometry. J Chromatogr B 879:938–944

Chiwocha SD, Abrams SR, Ambrose SJ et al (2003) A method for profiling classes of plant hormones and their metabolites using liquid chromatography-electrospray ionization tandem mass spectrometry: an analysis of hormone regulation of thermodormancy of lettuce (Lactuca sativa L.) seeds. Plant J 35:405–417

Cooney TP, Nonhebel HM (1991) Biosynthesis of indole-3-acetic acid in tomato shoots: measurement, mass spectral identification and incorporation of 2H from 2H2O into indole-3-acetic acid, D- and L-tryptophan, indole-3-pyruvate and tryptamine. Planta 184:368–376

Dobrev PI, Kamínek M (2002) Fast and efficient separation of cytokinins from auxin and abscisic acid and their purification using mixed-mode solid-phase extraction. J Chromatogr A 950:21–29

Dobrev PI, Havlíček L, Vágner M et al (2005) Purification and determination of plant hormones auxin and abscisic acid using solid phase extraction and two-dimensional high performance liquid chromatography. J Chromatogr A 1075:159–166

Du F, Ruan G, Liang S et al (2012a) Monolithic molecularly imprinted solid-phase extraction for the selective determination of trace cytokinins in plant samples with liquid chromatography-electrospray tandem mass spectrometry. Anal Bioanal Chem 404:489–501

Du F, Ruan G, Liu H (2012b) Analytical methods for tracing plant hormones. Anal Bioanal Chem 403:55–74

Durgbanshi A, Arbona V, Pozo O et al (2005) Simultaneous determination of multiple phytohormones in plant extracts by liquid chromatography-electrospray tandem mass spectrometry. J Agric Food Chem 53:8437–8442

Edlund A, Eklöf S, Sundberg B et al (1995) A microscale technique for gas chromatography-mass spectrometry measurements of picogram amounts of indole-3-acetic acid in plant tissues. Plant Physiol 108:1043–1047

Eklund DM, Thelander M, Landberg K et al (2010) Homologues of the Arabidopsis thaliana SHI/STY/LRP1 genes control auxin biosynthesis and affect growth and development in the moss Physcomitrella patens. Development 137:1275–1284

Ernstsen A, Sandberg G, Crozier A (1986) Effects of sodium diethyldithiocarbamate, solvent, temperature and plant extracts on the stability of indoles. Physiol Plant 68:519–522

Farrow SC, Emery RN (2012) Concurrent profiling of indole-3-acetic acid, abscisic acid, and cytokinins and structurally related purines by high-performance-liquid-chromatography tandem electrospray mass spectrometry. Plant Methods 8:42

Flores MI, Romero-González R, Frenich AG et al (2011) QuEChERS-based extraction procedure for multifamily analysis of phytohormones in vegetables by UHPLC-MS/MS. J Sep Sci 34:1517–1524

Holčapek M, Jirásko R, Lísa M (2012) Recent developments in liquid chromatography-mass spectrometry and related techniques. J Chromatogr A 1259:3–15

Hu Y, Li Y, Zhang Y et al (2011) Development of sample preparation method for auxin analysis in plants by vacuum microwave-assisted extraction combined with molecularly imprinted clean-up procedure. Anal Bioanal Chem 399:3367–3374

Huang MT, Ho CT, Lee CY (eds) (1992) Phenolic compounds in food and their effects on health I and II. American Chemical Society, Washington, DC

Izumi Y, Okazawa A, Bamba T et al (2009) Development of a method for comprehensive and quantitative analysis of plant hormones by highly sensitive nanoflow liquid chromatography-electrospray ionization-ion trap mass spectrometry. Anal Chim Acta 648:215–225

Jensen PJ, Bandurski RS (1995) Incorporation of deuterium into indole-3-acetic acid and tryptophan in Zea mays seedlings grown on 30 % deuterium oxide. Plant Physiol 147:697–702

Kai K, Horita J, Wakasa K et al (2007a) Three oxidative metabolites of indole-3-acetic acid from Arabidopsis thaliana. Phytochemistry 68:1651–1663

Kai K, Wakasa K, Miyagawa H (2007b) Metabolism of indole-3-acetic acid in rice: identification and characterization of N-beta-D-glucopyranosyl indole-3-acetic acid and its conjugates. Phytochemistry 68:2512–2522

Kaufmann A (2012) The current role of high-resolution mass spectrometry in food analysis. Anal Bioanal Chem 403:1233–1249

Kirwan GM, Johansson E, Kleemann R et al (2012) Building multivariate systems biology models. Anal Chem 84:7064–7071

Kojima M, Sakakibara H (2012) Highly sensitive high-throughput profiling of six phytohormones using MS-probe modification and liquid chromatography-tandem mass spectrometry. Methods Mol Biol 918:151–164

Kojima M, Kamada-Nobusada T, Komatsu H et al (2009) Highly sensitive and high-throughput analysis of plant hormones using MS-probe modification and liquid chromatography-tandem mass spectrometry: an application for hormone profiling in Oryza sativa. Plant Cell Physiol 50:1201–1214

Kowalczyk M, Sandberg G (2001) Quantitative analysis of indole-3-acetic acid metabolites in Arabidopsis. Plant Physiol 127:1845–1853

Kugimiya A, Takeuchi T (1999a) Application of indoleacetic acid-imprinted polymer to solid phase extraction. Anal Chim Acta 395:251–255

Kugimiya A, Takeuchi T (1999b) Effects of 2-hydroxyethyl methacrylate on polymer network and interaction in hydrophilic molecularly imprinted polymers. Anal Sci 15:29–33

Liu BF, Zhong XH, Lu YT (2002) Analysis of plant hormones in tobacco flowers by micellar electrokinetic capillary chromatography coupled with on-line large volume sample stacking. J Chromatogr A 945:257–265

Liu X, Hegeman AD, Gardner G et al (2012) Protocol: high-throughput and quantitative assays of auxin and auxin precursors from minute tissue samples. Plant Methods 8:31

Ljung K, Östin A, Lioussanne L et al (2001a) Developmental regulation of indole-3-acetic acid turnover in Scots pine seedlings. Plant Physiol 125:464–475

Ljung K, Bhalerao RP, Sandberg G (2001b) Sites and homeostatic control of auxin biosynthesis in Arabidopsis during vegetative growth. Plant J 28:465–474

Ljung K, Hull AK, Celenza J et al (2005) Sites and regulation of auxin biosynthesis in Arabidopsis roots. Plant Cell 17:1090–1094

Lu QM, Chen LH, Lu MH et al (2010) Extraction and analysis of auxins in plants using dispersive liquid-liquid microextraction followed by high-performance liquid chromatography with fluorescence detection. J Agric Food Chem 58:2763–2770

Mashiguchi K, Tanaka K, Sakai T et al (2011) The main auxin biosynthesis pathway in Arabidopsis. Proc Natl Acad Sci USA 108:18512–18517

Müller M, Munné-Bosch S (2011) Rapid and sensitive hormonal profiling of complex plant samples by liquid chromatography coupled to electrospray ionization tandem mass spectrometry. Plant Methods 7:37

Müller A, Düchting P, Weiler EW (2002) A multiplex GC-MS/MS technique for the sensitive and quantitative single-run analysis of acidic phytohormones and related compounds, and its application to Arabidopsis thaliana. Planta 216:44–56

Normanly J (2010) Approaching cellular and molecular resolution of auxin biosynthesis and metabolism. Cold Spring Harb Perspect Biol 2:a001594

Novák O, Hényková E, Sairanen I et al (2012) Tissue specific profiling of the Arabidopsis thaliana auxin metabolome. Plant J 72:523–536

Nováková L, Vlčková H (2009) A review of current trends and advances in modern bio-analytical methods: chromatography and sample preparation. Anal Chim Acta 656:8–35

Núñez O, Gallart-Ayala H, Martins CP et al (2012) New trends in fast liquid chromatography for food and environmental analysis. J Chromatogr A 1228:298–323

Oikawa A, Saito K (2012) Metabolite analyses of single cells. Plant J 70:30–38

Östin A, Catalá C, Chamarro J et al (1995) Identification of glucopyranosyl-β-1,4-glucopyranosyl-β-1-N-oxindole-3-acetyl-N-aspartic acid, a new IAA catabolite, by liquid chromatography/tandem mass spectrometry. J Mass Spectrom 30:1007–1017

Östin A, Kowalyczk M, Bhalerao RP et al (1998) Metabolism of indole-3-acetic acid in Arabidopsis. Plant Physiol 118:285–296

Pan X, Wang W (2009) Profiling of plant hormones by mass spectrometry. J Chromatogr B 877:2806–2813

Pan X, Welti R, Wang X (2008) Simultaneous quantification of major phytohormones and related compounds in crude plant extracts by liquid chromatography-electrospray tandem mass spectrometry. Phytochemistry 69:1773–1781

Pěnčík A, Rolčík J, Novák O et al (2009) Isolation of novel indole-3-acetic acid conjugates by immunoaffinity extraction. Talanta 80:651–655

Pengelly WL, Bandurski RS (1983) Analysis of indole-3-acetic acid metabolism in Zea mays using deuterium oxide as a tracer. Plant Physiol 73:445–449

Petersson SV, Johansson AI, Kowalczyk M et al (2009) An auxin gradient and maximum in the Arabidopsis root apex show by high-resolution cell-specific analysis of IAA distribution and synthesis. Plant Cell 21:1659–1668

Prinsen E, Van Dongen W, Esmans EL et al (1998) Micro and capillary liquid chromatography tandem mass spectrometry: a new dimension in phytohormone research. J Chromatogr A 826:25–37

Quittenden LJ, Davies NW, Smith JA et al (2009) Auxin biosynthesis in pea: characterization of the tryptamine pathway. Plant Physiol 151:1130–1138

Rapparini F, Tam YY, Cohen JD et al (2002) Indole-3-acetic acid metabolism in Lemna gibba undergoes dynamic changes in response to growth temperature. Plant Physiol 128:1410–1416

Reinecke DM, Bandurski RS (1983) Oxindole-3-acetic acid, an indole-3-acetic acid catabolite in Zea mays. Plant Physiol 71:211–213

Rittenberg D, Foster L (1940) A new procedure for quantitative analysis by isotope dilution, with application to the determination of amino acids and fatty acids. J Biol Chem 133:727–744

Sairanen I, Novák O, Pěnčík A et al (2012) Soluble carbohydrates regulate auxin biosynthesis via PIF proteins in Arabidopsis. Plant Cell 24:4907–4916

Schmelz EA, Engelberth J, Alborn HT et al (2003) Simultaneous analysis of phytohormones, phytotoxins, and volatile organic compounds in plants. Proc Natl Acad Sci USA 100:10552–10557

Shi X, Jin F, Huang Y et al (2012) Simultaneous determination of five plant growth regulators in fruits by modified quick, easy, cheap, effective, rugged, and safe (QuEChERS) extraction and liquid chromatography-tandem mass spectrometry. Agric Food Chem 60:60–65

Sugawara S, Hishiyama S, Jikumaru Y et al (2009) Biochemical analyses of indole-3-acetaldoxime-dependent auxin biosynthesis in Arabidopsis. Proc Natl Acad Sci USA 106:5430–5435

Sundberg B (1990) Influence of extraction solvent (buffer, methanol, acetone) and time on the quantification of indole-3-acetic acid in plants. Physiol Plant 78:293–297

Sundberg B, Sandberg G, Crozier A (1986) Purification of indole-3-acetic acid in plant extracts by immunoaffinity chromatography. Phytochemistry 25:295–298

Svačinová J, Novák O, Plačková L et al (2012) A new approach for cytokinin isolation from Arabidopsis tissues using miniaturized purification: pipette tip solid-phase extraction. Plant Methods 8:17

Tam YY, Normanly J (1998) Determination of indole-3-pyruvic acid levels in Arabidopsis thaliana by gas chromatography-selected ion monitoring-mass spectrometry. J Chromatogr A 800:101–108

Tam YY, Epstein E, Normanly J (2000) Characterization of auxin conjugates in Arabidopsis. Low steady-state levels of indole-3-acetyl-aspartate, indole-3-acetyl-glutamate, and indole-3-acetyl-glucose. Plant Physiol 123:589–596

Taylor PJ (2005) Matrix effects: the Achilles heel of quantitative high-performance liquid chromatography-electrospray-tandem mass spectrometry. Clin Biochem 38:328–334

Tivendale ND, Davies NW, Molesworth PP et al (2010) Reassessing the role of N-hydroxytryptamine in auxin biosynthesis. Plant Physiol 154:1957–1965

Van Meulebroek L, Vanden Bussche J, Steppe K et al (2012) Ultra-high performance liquid chromatography coupled to high resolution Orbitrap mass spectrometry for metabolomic profiling of the endogenous phytohormonal status of the tomato plant. J Chromatogr A 1260:67–80

Wiklund S, Johansson E, Sjöström L et al (2008) Visualization of GC/TOF-MS-based metabolomics data for identification of biochemically interesting compounds using OPLS class models. Anal Chem 80:115–122

Wu YL, Hu B (2009) Simultaneous determination of several phytohormones in natural coconut juice by hollow fiber-based liquid-liquid-liquid microextraction-high performance liquid chromatography. J Chromatogr A 1216:7657–7663

Zhang Y, Li Y, Hu Y, Li G et al (2010) Preparation of magnetic indole-3-acetic acid imprinted polymer beads with 4-vinylpyridine and β-cyclodextrin as binary monomer via microwave heating initiated polymerization and their application to trace analysis of auxins in plant tissues. J Chromatogr A 1217:7337–7344

Zhao YD, Christensen SK, Fankhauser C et al (2001) A role for flavin monooxygenase-like enzymes in auxin biosynthesis. Science 291:306–309

# Chapter 4
# Intracellular Auxin Transport

**David Scheuring and Jürgen Kleine-Vehn**

**Abstract** The phytohormone auxin is of fundamental importance in plant development. Since the identification of auxin as a plant growth substance, auxin transport has drawn considerable research attention. Intercellular (polar) auxin transport contributes to its graded distribution in cell files or entire organs and allows for dynamic, environmentally controlled rearrangements in auxin accumulation. Insights into polar auxin transport mechanisms have broadened our understanding of the phenotypic flexibility of plants. Besides intercellular auxin transport, auxin is also transported intracellularly across organelle membranes, but its importance in plant development remains nascent. The intracellular sequestration of auxin into cellular compartments, such as peroxisomes and the endoplasmic reticulum (ER), plays important roles in auxin metabolism and could furthermore, in the case of the ER, have a direct impact on auxin signaling events. In this chapter, we review the most recent insights into intracellular auxin transport and its role in cellular auxin homeostasis.

## 1 Introduction

The phytohormone auxin is an important regulator of plant growth and patterning (reviewed in Leyser 2011). Auxin controls diverse developmental aspects, such as apical–basal axis formation during embryogenesis, de novo postembryonic organogenesis, or pathogen interaction (reviewed in De Smet and Jurgens 2007; Péret et al. 2009; Kazan and Manners 2009). At a cellular level, auxin determines the rate of cell division, elongation, and differentiation (reviewed in Perrot-Rechenmann

D. Scheuring • J. Kleine-Vehn (✉)
Department of Applied Genetics and Cell Biology, University of Natural Resources and Life Sciences (BOKU), Vienna, Austria
e-mail: juergen.kleine-vehn@boku.ac.at; jkleinevehn@gmail.com

E. Zažímalová et al. (eds.), *Auxin and Its Role in Plant Development*,
DOI 10.1007/978-3-7091-1526-8_4, 

2010). Given its manifold functions, it is not too surprising that auxin levels and the cellular responsiveness to auxin are tightly regulated. It now appears that auxin metabolism and transport jointly control the spatiotemporal activity of auxin (reviewed in Rosquete et al. 2012), which is a key component in regulating plant development.

Four distinct endogenous auxin molecules, namely, indole-3-acetic acid (IAA), indole-3-butyric acid (IBA), 4-chloroindole-3-acetic acid (4-CI-IAA), and phenylacetic acid (PAA), have been identified (reviewed in Simon and Petrášek 2011), from which IAA appears to be the most abundant (reviewed in Strader and Bartel 2011). IAA is mainly synthesized in the young aerial parts of the plant (Ljung et al. 2001) and is subsequently transported in a root-ward direction (reviewed in Petrášek and Friml 2009). Moreover, local auxin biosynthesis throughout the entire plant has been shown to be developmentally important (Zhao 2012; Ljung 2013). The passage of auxin from cell to cell is carried out by multiple auxin carriers at the plasma membrane. The most prominent are the influx carrier AUXIN RESISTANT 1 (AUX1), a subfamily of ATP-BINDING CASSETTE (ABC) transporters, and the PIN-FORMED (PIN) efflux carrier family (reviewed in Grunewald and Friml 2010; Kramer and Bennett 2006; Zažímalová et al. 2010) (Fig. 4.1). Carrier-dependent polar auxin transport contributes to the enormously flexible regulation of auxin distribution within tissues (see Chap. 5).

Intracellular auxin must be perceived and converted into a signal to induce specific cellular auxin responses (see Chap. 6). In *Arabidopsis*, three independent auxin receptors have been proposed (Fig. 4.1): (1) the F-box protein TRANSPORT INHIBITOR RESISTANT1/AUXIN SIGNALING F-BOX (TIR1/AFB) (Dharmasiri et al. 2005; Kepinski and Leyser 2005), (2) the F-box protein S-PHASE KINASE-ASSOCIATED PROTEIN 2A (SKP2A) (Jurado et al. 2010), and (3) the AUXIN-BINDING PROTEIN (ABP1) (Hertel et al. 1972; Tromas et al. 2010).

Most auxin responses have been associated with nuclear TIR1/AFB action (Tan et al. 2007; Dharmasiri et al. 2005). In the absence of auxin, AUXIN/INDOLE ACETIC ACID (Aux/IAA) proteins bind and inhibit AUXIN RESPONSE FACTOR (ARF) transcription factors. Auxin binding to the co-receptor complex of TIR1/AFB and Aux/IAA induces ubiquitination and degradation of Aux/IAA and ultimately the release of ARFs, leading to transcriptional reprogramming (reviewed in Chapman and Estelle 2009).

Nuclear SKP2A is involved in cell cycle regulation and furthermore has been shown to directly bind auxin (Jurado et al. 2008; del Pozo et al. 2006). Mutations in the proposed auxin-binding pocket abolish auxin binding and SKP2A activity (Jurado et al. 2010). Further insight is needed to link SKP2A action and auxin involvement in the cell cycle.

ABP1 was the first auxin-binding protein described in the literature and has a long-standing history as a putative auxin receptor (reviewed in Sauer and Kleine-Vehn 2011; Shi and Yang 2011; Scherer 2011). ABP1 localizes mainly to the ER, but a small portion is secreted to the cell wall where it is assumed to be active (Sauer and Kleine-Vehn 2011). ABP1 responds rapidly and non-genomically to

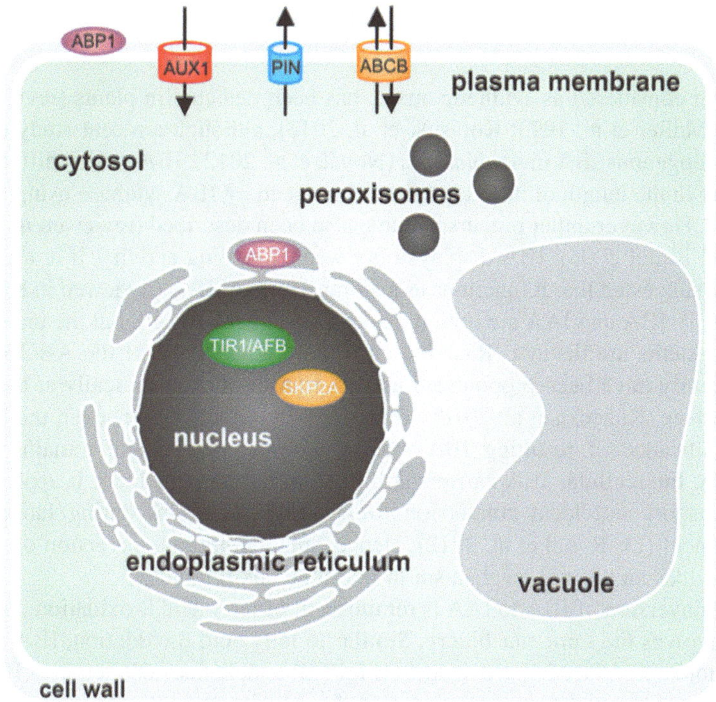


**Fig. 4.1** Intercellular auxin carriers and auxin receptors in plants. In *Arabidopsis*, three auxin receptors and at least three different types of auxin carriers at the plasma membrane have been suggested. *Abbreviations*: ABP1 = AUXIN-BINDING PROTEIN1, AUX1 = AUXIN RESISTANT1, PIN = PIN-FORMED, ABCB = ATP-BINDING CASETTE B, TIR1/AFB = TRANSPORT INHIBITOR RESISTANT1/AUXIN SIGNALING F-BOX, and SKP2A = S-PHASE KINASE-ASSOCIATED PROTEIN 2A

auxin by regulating endocytosis or organizing the cytoskeleton (Robert et al. 2010; Xu et al. 2010). These functions occur proximate to the plasma membrane; ABP1 additionally affects the cell cycle and gene expression (David et al. 2007; Braun et al. 2008; Paque et al. 2014).

Both intracellular and extracellular auxin perception is required for integrative auxin signaling. Carrier activity at the plasma membrane might therefore play a decisive role in regulating the balance between extra- and intracellular signaling. Compared with auxin perception and signaling events, auxin metabolism appears at least in part compartmentalized. The precise subcellular distribution of auxinic compounds requires specific auxin transport across organelle membranes, which adds further complexity to the mechanisms of auxin perception and metabolism. However, in contrast with cell-to-cell auxin transport, intracellular auxin transport is only poorly understood. Here, we review our current knowledge of intracellular auxin compartmentalization and transport.

## 2 Intracellular Auxin Transport at Peroxisomes

IBA, long considered as synthetic auxin, has been detected in plants (reviewed in Ludwig-Müller et al. 1993; Korasick et al. 2013), although a recent study did not detect endogenous IBA in *Arabidopsis* (Novák et al. 2012). IBA, which differs from IAA only in the length of its side chain, is produced by IBA synthase using IAA as substrate. However, other precursors have also been described (reviewed in Woodward and Bartel 2005). IBA displays only weak signaling activity, if at all, and it has been suggested that it functions as a storage form of IAA (reviewed in Korasick et al. 2013). IBA and IAA are both transported from cell to cell, but the underlying carrier systems are distinct (Rashotte et al. 2003). Members of the ABCG transporter family have been propounded as IBA transporters, specifically at the root–soil interface (Růžčcka et al. 2010; Strader and Bartel 2011); however, the biological significance of resulting IBA release into the rhizosphere remains so far unknown. Intercellular IBA carriers still need to be unearthed, but it appears that IBA transport and local conversion to IAA are important during lateral root development (De Rybel et al. 2012). Hence, spatiotemporal conversion of IBA to IAA provides an elegant mechanism to fine-tune auxin biology.

The conversion of IBA to IAA is reminiscent of fatty acid ß-oxidation and most likely involves the same machinery. Similar to fatty acid ß-oxidation, IBA to IAA conversion takes place in peroxisomes (Fig. 4.2) and, hence, also requires IBA and IAA to be imported into and exported from this organelle (reviewed in Hu et al. 2012; Fawcett et al. 1960).

The ATPase PXA1/ABCD1, initially described as transporting fatty acids at peroxisomes (reviewed in Linka and Weber 2010), might be involved in pumping IBA from the cytosol into the peroxisome (Fig. 4.2). Loss-of-function mutants of PXA1/ABCD1 are IBA resistant (Zolman et al. 2001) and show reduced IBA to IAA conversion (Strader et al. 2010). Notably, defects of *abcd1* mutants, such as a reduced number of lateral roots (Zolman et al. 2001) and a delayed stamen filament elongation (Footitt et al. 2007), could be partially rescued by exogenous auxin application.

These findings suggest that IBA transport into peroxisomes and subsequent conversion to IAA is of importance for plant growth and development. However, the identity of transporters mediating the export of IBA-derived IAA from peroxisomes into the cytosol remains unfortunately elusive (Fig. 4.2).

## 3 Compartmentalization of Auxin in the Endoplasmic Reticulum

The endoplasmic reticulum has a prominent role in the synthesis and folding of proteins and is also involved in multiple metabolism pathways. Recent evidence suggests that the ER is essential for auxin metabolism (reviewed in Barbez and

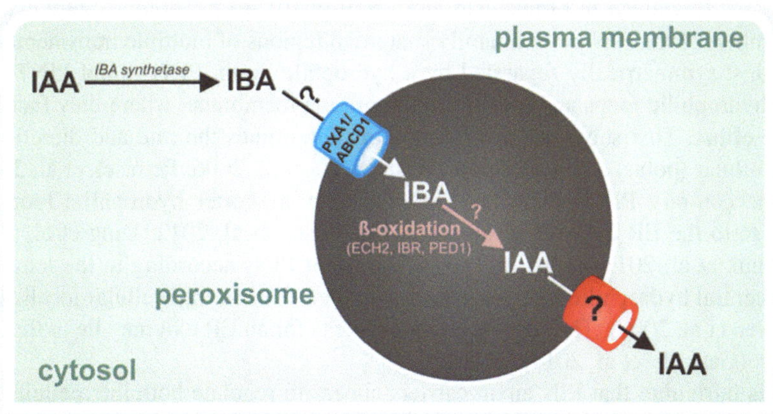


**Fig. 4.2** Auxin transport at peroxisomes. Conversion from IAA to IBA is carried out in the cytosol by the IBA synthetase. The reverse reaction, IBA to IAA takes place in peroxisomes, but the exact mechanism remains unclear. However, enzymes acting in the ß-oxidation pathway have been suggested as playing an important role in IBA to IAA conversion. ABCD1 is likely to transport IBA into peroxisomes. The export carrier for IAA is still unknown. *Abbreviations*: IAA = Indole-3-acetic acid, IBA = Indole-3-butyric, and PXA1/ABCD1 = ABC transporter. ECH2, IBR, and PED1 are candidates for IBA ß-oxidation

Kleine-Vehn 2013). For example, several auxin amid-conjugate hydrolases show an ER-retention motif (Campanella et al. 2003; Bitto et al. 2009), suggesting that conjugation-based regulation of auxin activity might be linked to the ER. A splice variant of the IAA biosynthesis gene YUCCA4 also localizes to the cytosolic face of the ER (Kriechbaumer et al. 2012). The functional importance of this localization is currently unknown, but could supply auxin for ER-directed transport. Furthermore, the auxin receptor ABP1 is localized within the ER lumen and, hence, auxin signaling could also take place in the ER (Barbez and Kleine-Vehn 2013).

While all these findings indicate that the ER might be important for auxin metabolism and signaling, it is also clear that such compartmentalized mechanisms would require auxins or auxinic compounds to be transported across the ER membrane. Therefore, it was not too surprising that putative auxin transporters were indeed discovered at the ER. However, the actual molecular player was unexpected, because the transporter family has been already identified as being involved in intercellular auxin transport.

### 3.1 *PIN Subfamily of Auxin Carriers Function at the ER*

PIN proteins are auxin efflux carriers (see Chap. 5), originally said to function in intercellular auxin transport, modulating an extensive range of developmental processes (reviewed in Grunewald and Friml 2010). In *Arabidopsis thaliana*, all 8 members of the PIN protein family share two regions of multiple transmembrane domains, symmetrically separated by a hydrophilic loop. PIN1-4 and PIN7 have long hydrophilic loops and localize to the plasma membrane, where they facilitate auxin efflux. This subgroup of PIN proteins determines the rate and direction of intercellular (polar) auxin transport (Wisniewska et al. 2006; Petrášek et al. 2006). On the contrary PIN5, PIN6, and PIN8 display a shorter hydrophilic loop and localize to the ER (Mravec et al. 2009; Dal Bosco et al. 2012; Ding et al. 2012; Sawchuk et al. 2013). Thus, the classification of PINs according to the length of their central hydrophilic loop correlates with their distinct intracellular localization (Mravec et al. 2009). The sequence requirements for an ER exit may lie in the loop region (Ganguly et al. 2014).

It is intriguing that PIN auxin carriers appear to regulate both intercellular and intracellular auxin transport and that ER-localized PINs seemingly predate the evolution of plasma membrane localized PINs (Viaene et al. 2013). ER-localized PIN5 increases cellular auxin retention, presumably by transporting auxin from the cytosol into the ER lumen (Mravec et al. 2009). The availability of cytosolic/nuclear IAA would be reduced as a result of such auxin compartmentalization in the ER, which in turn might impact on nuclear auxin signaling events (Mravec et al. 2009).

Apart from the transport of free IAA, PIN activity at the ER has been linked to auxin conjugation-based metabolism (Mravec et al. 2009; Dal Bosco et al. 2012; Ding et al. 2012). Only a small fraction of auxin is in its free and active form—the majority is conjugated to amino acids, peptides, and sugars (see Chaps. 2 and 3). Conjugation enables cells to reversibly inactivate and store auxin. In *Arabidopsis*, IAA is mainly conjugated to amino acids and the most abundant IAA amid conjugates are IA-alanine (Ala), IA-leucine (Leu), IA-aspartate (Asp), and IA-glutamate (Glu) (Tam et al. 2000; Kowalczyk and Sandberg 2001). The auxin conjugation to amino acids is catalyzed by members of the auxin-inducible GRETCHEN HAGEN3 (GH3) gene family (Staswick et al. 2005; Hagen and Guilfoyle 1985), while hydrolysis of amino acids is achieved by the IAA-LEUCINE RESISTANT 1 (ILR1)-like family (Bartel and Fink 1995; Rampey et al. 2004; Davies et al. 1999; LeClere et al. 2002). PIN5 activity correlates with higher levels of IAA amino acid and sugar conjugates. Hence, it has been hypothesized that PIN5 leads to auxin accumulation in the ER, where conjugation-based metabolism might take place (Mravec et al. 2009) (Fig. 4.3). Similar to PIN5, PIN8 localizes to the ER (Dal Bosco et al. 2012; Ding et al. 2012). Ectopic PIN8 expression leads to aberrant hypocotyl and root length as well as changes in flowering time (Ding et al. 2012; Dal Bosco et al. 2012). Surprisingly, a *pin5* loss-of-function mutation can rescue the *pin8* defects in pollen, whereas *PIN5* overexpression attenuates phenotypes caused by ectopic PIN8 expression (Ding et al. 2012). These findings indicate antagonistic roles for PIN5 and PIN8, possibly by counteracting the sequestration of auxin and auxin conjugates in the ER lumen (Ding et al. 2012). Similar to the scenario in pollen, *pin6* and *pin8* show antagonistic genetic interaction with *pin5* in the control of leaf vein patterning (Sawchuk

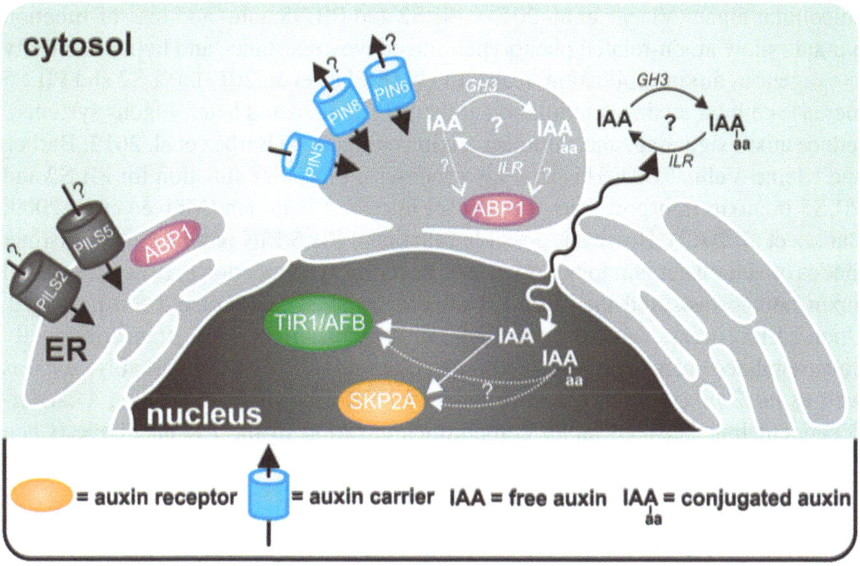


**Fig. 4.3** Simplified model of auxin transport at the ER. Distinct auxin pools in the cytosol, the nucleus, and the ER depend on different carriers (as depicted in the *cartoon*). *Abbreviations*: ER = endoplasmic reticulum, IAA = Indole-3-acetic acid, PIN = PIN-FORMED, PILS = PIN-LIKES, ABP1 = AUXIN-BINDING PROTEIN1, GH3 = GRETCHEN HAGEN 3, and ILR = IAA-LEUCINE RESISTANT1-like

et al. 2013) (Fig. 4.3). This could be explained by a scenario in which PIN5 transports auxin from the cytoplasm into the ER (Mravec et al. 2009), and PIN6/PIN8 transport auxin from the ER to the cytoplasm (Sawchuk et al. 2013), but may also depend on PIN protein translocation to the plasma membrane (Ganguly et al. 2014). Notably, intercellular (PIN1) and intracellular (PIN5, PIN6, and PIN8) auxin transport seem to converge to fine-tune auxin-dependent vein networks (Sawchuk et al. 2013), suggesting complex interplay between intercellular and ER auxin transport.

## *3.2 PILS Proteins: A New Family of Putative Auxin Transporters*

Besides PIN proteins, another family of putative auxin transporters at the ER has been identified, possibly highlighting the importance of auxin compartmentalization in the ER (Barbez et al. 2012). PIN-LIKES (PILS) proteins display only low sequence similarity and seem evolutionarily distinct from PINs (Feraru et al. 2012; Viaene et al. 2013), but show similarities to PINs in the predicted protein topology (Barbez et al. 2012). PILS proteins are conserved throughout plant evolution (Barbez et al. 2012; Feraru et al. 2012), whereas PIN proteins are absent in unicellular algae (Viaene et al. 2013). PILS2 and PILS5 gain- and loss-of-function mutants show auxin-related phenotypes and convey resistance and hypersensitivity to exogenous auxin application, respectively (Barbez et al. 2012). PILS2 and PILS5 increase cellular auxin accumulation at the ER (in plant and heterologous systems), reduce auxin signaling, and stimulate auxin conjugation (Barbez et al. 2012; Barbez and Kleine-Vehn 2013). This analogy proposes a PIN5-like function for PILS2 and PILS5 in auxin transport from the cytosol into the ER lumen (Mravec et al. 2009; Barbez et al. 2012). However, possible functional PILS/PIN redundancy or distinct modes of actions remain to be addressed. Recently, a connection between ER-based auxin homeostasis and the unfolded protein response (UPR) has been proposed. Upon ER stress, auxin receptors and transporters are transcriptionally downregulated in *Arabidopsis* seedlings (Chen et al. 2013). Remarkably, *pin5* as well as *pils2 pils5* mutants show a reduced UPR activation, suggesting a carrier-dependent link between auxin compartmentalization in the ER and UPR (Chen et al. 2013).

## 4 Auxin and Other Cellular Compartments

Auxin metabolism and/or compartmentalization in peroxisomes and ER lumen is currently gaining research attention, raising questions about the involvement of additional organelles contributing to auxin homeostasis. Recently, the vacuole and mitochondria have been advanced as playing a role in cellular auxin homeostasis.

The *Arabidopsis* transporter protein WALLS ARE THIN1 (WAT1) is ubiquitously expressed throughout the plant, but it is preferentially associated with vascular tissues. Within the cell, WAT1-GFP localizes to the tonoplast (Ranocha et al. 2010). In *wat1-1* mutants, the cell wall is defective (Denance et al. 2012), but local auxin application rescues this mutant phenotype (Ranocha et al. 2013). Auxin transport assays, using isolated *Arabidopsis* vacuoles and heterologous systems, such as yeast and *Xenopus* oocytes, revealed that WAT1 is transporting auxin, indicating that WAT1 facilitates auxin export from the vacuole (Ranocha et al. 2013). In agreement, IAA and related compounds could be isolated from *Arabidopsis* vacuoles (Ranocha et al. 2013). This finding adds another layer of complexity to auxin homeostasis, but the possible role of auxin in the vacuole remains to be demonstrated. Another link has been established between auxin homeostasis and IAA-ALANINE RESISTANT4 (IAR4) that localizes to mitochondria (Quint et al. 2009). *iar4* mutants show auxin-related phenotypes that can be rescued by exogenous auxin application. While free IAA levels are not affected, *iar4* mutants show higher levels of auxin conjugates, suggesting that IAR4 could affect auxin homeostasis.

It remains to be seen whether vacuoles or mitochondria directly or indirectly affect auxin metabolism, but it certainly illustrates the complexity of cellular auxin homeostasis.

# 5 Concluding Remarks

Auxin transport has important impacts on plant development. While we currently have a good insight into intercellular transport and its importance in plant development, we are still in the early stages of understanding intracellular auxin transport. Compartmentalization of a signaling molecule allows fine-tuning of its cellular activity and/or metabolism. IBA, for example, appears to be a largely inactive auxin molecule that becomes redistributed in plant tissues and spatially converted to an active signal. It is tempting to speculate that IBA converting cells might differ from other tissues in their activity of IBA and/or auxin carriers at peroxisomes. Compartmentalization of IBA metabolism seems to be essential for lateral rooting (De Rybel et al. 2012) and is a stunning example of how environmental signals control postembryonic organ formation (reviewed in Korasick et al. 2013).

IBA metabolism and reversible IAA conjugation to different moieties might appear mechanistically similar, with both leading to IAA inactivation. However, while IBA is actively transported from cell to cell, it remains to be seen whether IAA conjugates are readily transported within plant tissues. Therefore, it is conceivable that IAA conversion to IBA or IAA conjugation exert distinct spatial consequences.

The regulation of auxin conjugation might be linked to the ER (reviewed in Barbez and Kleine-Vehn 2013), but the localization of the molecular players is poorly defined. Apart from its probable role in auxin metabolism, the ER might also be directly involved in auxin signaling. The auxin receptor ABP1 is largely retained in the ER, based on its KDEL retention motif, and only a tiny part of ABP1, constituting its extracellular activity, escapes this regulation (reviewed in Sauer and Kleine-Vehn 2011). It thus seems possible that the ER-localized pool of ABP1 participates in auxin signaling as well (Barbez and Kleine-Vehn 2013). ABP1 in the ER could either directly affect prominent ER functions or provide a signal toward the nucleus via the nuclear envelope. In such a scenario, intracellular auxin transport at the ER would not only affect the rate of compartmentalized auxin metabolism, but might also be a decisive factor for specific auxin signaling events in the ER. Such an intracellular transport mechanism would even allow single cells within a tissue to acquire individual and distinct auxin responses.

**Acknowledgments** We are grateful to E. Barbez, M. Sauer, and C. Luschnig for helpful discussion and D. Whittaker for help in preparing the manuscript. This work was supported by the Vienna Science and Technology Fund (WWTF) (to J.K.-V.), Austrian Science Fund (FWF) (Projects: P26568-B16 and P26591-B16 to J.K.-V.), and the Deutsche Forschungsgemeinschaft (DFG) (personal postdoctoral research grant to D.S.).

## References

Barbez E, Kleine-Vehn J (2013) Divide Et Impera-cellular auxin compartmentalization. Curr Opin Plant Biol 16(1):78–84. doi:10.1016/j.pbi.2012.10.005

Barbez E, Kubeš M, Rolčík J, Beziat C, Pěnčík A, Wang B, Rosquete MR, Zhu J, Dobrev PI, Lee Y, Zažímalová E, Petrášek J, Geisler M, Friml J, Kleine-Vehn J (2012) A novel putative auxin carrier family regulates intracellular auxin homeostasis in plants. Nature 485 (7396):119–122. doi:10.1038/nature11001

Bartel B, Fink GR (1995) ILR1, an amidohydrolase that releases active indole-3-acetic acid from conjugates. Science 268(5218):1745–1748

Bitto E, Bingman CA, Bittova L, Houston NL, Boston RS, Fox BG, Phillips GN Jr (2009) X-ray structure of ILL2, an auxin-conjugate amidohydrolase from Arabidopsis thaliana. Proteins 74 (1):61–71. doi:10.1002/prot.22124

Braun N, Wyrzykowska J, Muller P, David K, Couch D, Perrot-Rechenmann C, Fleming AJ (2008) Conditional repression of AUXIN BINDING PROTEIN1 reveals that it coordinates cell division and cell expansion during postembryonic shoot development in Arabidopsis and tobacco. Plant Cell 20(10):2746–2762. doi:10.1105/tpc.108.059048

Campanella JJ, Larko D, Smalley J (2003) A molecular phylogenomic analysis of the ILR1-like family of IAA amidohydrolase genes. Comp Funct Genomics 4(6):584–600. doi:10.1002/cfg.340

Chapman EJ, Estelle M (2009) Mechanism of auxin-regulated gene expression in plants. Annu Rev Genet 43:265–285. doi:10.1146/annurev-genet-102108-134148

Chen Y, Aung K, Rolčík J, Walicki K, Friml J, Brandizzi F (2013) Inter-regulation of the unfolded protein response and auxin signaling. Plant J. doi:10.1111/tpj.12373

Dal Bosco C, Dovzhenko A, Liu X, Woerner N, Rensch T, Eismann M, Eimer S, Hegermann J, Paponov IA, Ruperti B, Heberle-Bors E, Touraev A, Cohen JD, Palme K (2012) The endoplasmic reticulum localized PIN8 is a pollen-specific auxin carrier involved in intracellular auxin homeostasis. Plant J 71(5):860–870. doi:10.1111/j.1365-313X.2012.05037.x

David KM, Couch D, Braun N, Brown S, Grosclaude J, Perrot-Rechenmann C (2007) The auxin-binding protein 1 is essential for the control of cell cycle. Plant J 50(2):197–206. doi:10.1111/j.1365-313X.2007.03038.x

Davies RT, Goetz DH, Lasswell J, Anderson MN, Bartel B (1999) IAR3 encodes an auxin conjugate hydrolase from Arabidopsis. Plant Cell 11(3):365–376

De Rybel B, Audenaert D, Xuan W, Overvoorde P, Strader LC, Kepinski S, Hoye R, Brisbois R, Parizot B, Vanneste S, Liu X, Gilday A, Graham IA, Nguyen L, Jansen L, Njo MF, Inze D, Bartel B, Beeckman T (2012) A role for the root cap in root branching revealed by the non-auxin probe naxillin. Nat Chem Biol 8(9):798–805. doi:10.1038/nchembio.1044

De Smet I, Jurgens G (2007) Patterning the axis in plants–auxin in control. Curr Opin Genet Dev 17(4):337–343. doi:10.1016/j.gde.2007.04.012

del Pozo JC, Diaz-Trivino S, Cisneros N, Gutierrez C (2006) The balance between cell division and endoreplication depends on E2FC-DPB, transcription factors regulated by the ubiquitin-SCFSKP2A pathway in Arabidopsis. Plant Cell 18(9):2224–2235. doi:10.1105/tpc.105.039651

Denance N, Ranocha P, Oria N, Barlet X, Riviere MP, Yadeta KA, Hoffmann L, Perreau F, Clement G, Maia-Grondard A, van den Berg GC, Savelli B, Fournier S, Aubert Y, Pelletier S, Thomma BP, Molina A, Jouanin L, Marco Y, Goffner D (2012) Arabidopsis wat1 (walls are thin1)-mediated resistance to the bacterial vascular pathogen, Ralstonia solanacearum, is accompanied by cross-regulation of salicylic acid and tryptophan metabolism. Plant J. doi:10.1111/tpj.12027

Dharmasiri N, Dharmasiri S, Estelle M (2005) The F-box protein TIR1 is an auxin receptor. Nature 435(7041):441–445. doi:10.1038/nature03543

Ding Z, Wang B, Moreno I, Ďupľáková N, Simon S, Carraro N, Reemmer J, Pěnčík A, Chen X, Tejos R, Skůpa P, Pollmann S, Mravec J, Petrášek J, Zažímalová E, Honys D, Rolčík J, Murphy A, Orellana A, Geisler M, Friml J (2012) ER-localized auxin transporter PIN8

regulates auxin homeostasis and male gametophyte development in Arabidopsis. Nat Commun 3:941. doi:10.1038/ncomms1941
Fawcett CH, Wain RL, Wightman F (1960) The metabolism of 3-indolylalkanecarboxylic acids, and their amides, nitriles and methyl esters in plant tissues. Proc R Soc Lond B Biol Sci 152:231–254
Feraru E, Vosolsobě S, Feraru MI, Petrášek J, Kleine-Vehn J (2012) Evolution and structural diversification of PILS putative auxin carriers in plants. Front Plant Sci 3:227. doi:10.3389/fpls.2012.00227
Footitt S, Dietrich D, Fait A, Fernie AR, Holdsworth MJ, Baker A, Theodoulou FL (2007) The COMATOSE ATP-binding cassette transporter is required for full fertility in Arabidopsis. Plant Physiol 144(3):1467–1480. doi:10.1104/pp. 107.099903
Ganguly A, Park M, Kesewat MS, Cho HT (2014). Functional analysis of the hydrophilic loop in intracellular trafficking of arabidopsis PIN-FORMED proteins. Plant Cell 2014. doi:10.1105/tpc.113.118422
Grunewald W, Friml J (2010) The march of the PINs: developmental plasticity by dynamic polar targeting in plant cells. EMBO J 29(16):2700–2714. doi:10.1038/emboj.2010.181
Hagen G, Guilfoyle TJ (1985) Rapid induction of selective transcription by auxins. Mol Cell Biol 5 (6):1197–1203
Hertel R, Thomson KS, Russo VEA (1972) *In-vitro* auxin binding to particulate cell fractions from corn coleoptiles. Planta 107:325–340
Hu J, Baker A, Bartel B, Linka N, Mullen RT, Reumann S, Zolman BK (2012) Plant peroxisomes: biogenesis and function. Plant Cell 24(6):2279–2303. doi:10.1105/tpc.112.096586
Jurado S, Diaz-Trivino S, Abraham Z, Manzano C, Gutierrez C, del Pozo C (2008) SKP2A, an F-box protein that regulates cell division, is degraded via the ubiquitin pathway. Plant J 53 (5):828–841. doi:10.1111/j.1365-313X.2007.03378.x
Jurado S, Abraham Z, Manzano C, Lopez-Torrejon G, Pacios LF, Del Pozo JC (2010) The Arabidopsis cell cycle F-box protein SKP2A binds to auxin. Plant Cell 22(12):3891–3904. doi:10.1105/tpc.110.078972
Kazan K, Manners JM (2009) Linking development to defense: auxin in plant-pathogen interactions. Trends Plant Sci 14(7):373–382. doi:10.1016/j.tplants.2009.04.005
Kepinski S, Leyser O (2005) The Arabidopsis F-box protein TIR1 is an auxin receptor. Nature 435 (7041):446–451. doi:10.1038/nature03542
Korasick DA, Enders TA, Strader LC (2013) Auxin biosynthesis and storage forms. J Exp Bot 64 (9):2541–2555. doi:10.1093/jxb/ert080
Kowalczyk M, Sandberg G (2001) Quantitative analysis of indole-3-acetic acid metabolites in Arabidopsis. Plant Physiol 127(4):1845–1853
Kramer EM, Bennett MJ (2006) Auxin transport: a field in flux. Trends Plant Sci 11(8):382–386. doi:10.1016/j.tplants.2006.06.002
Kriechbaumer V, Wang P, Hawes C, Abell BM (2012) Alternative splicing of the auxin biosynthesis gene YUCCA4 determines its subcellular compartmentation. Plant J 70(2):292–302. doi:10.1111/j.1365-313X.2011.04866.x
LeClere S, Tellez R, Rampey RA, Matsuda SP, Bartel B (2002) Characterization of a family of IAA-amino acid conjugate hydrolases from Arabidopsis. J Biol Chem 277(23):20446–20452. doi:10.1074/jbc.M111955200
Leyser O (2011) Auxin, self-organisation, and the colonial nature of plants. Curr Biol 21(9):R331–R337. doi:10.1016/j.cub.2011.02.031
Linka N, Weber AP (2010) Intracellular metabolite transporters in plants. Mol Plant 3(1):21–53. doi:10.1093/mp/ssp108
Ljung K (2013) Auxin metabolism and homeostasis during plant development. Development 140 (5):943–950. doi:10.1242/dev.086363
Ljung K, Bhalerao RP, Sandberg G (2001) Sites and homeostatic control of auxin biosynthesis in Arabidopsis during vegetative growth. Plant J 28(4):465–474

Ludwig-Müller J, Sass S, Sutter EG, Wondner M, Epstein E (1993) Indole-3-butyric acid in *Arabidopsis thaliana*. I. Identification and quantification. Plant Growth Regul 13:179–187
Mravec J, Skůpa P, Bailly A, Hoyerová K, Křeček P, Bielach A, Petrášek J, Zhang J, Gaykova V, Stierhof YD, Dobrev PI, Schwarzerová K, Rolčík J, Seifertová D, Luschnig C, Benková E, Zažímalová E, Geisler M, Friml J (2009) Subcellular homeostasis of phytohormone auxin is mediated by the ER-localized PIN5 transporter. Nature 459(7250):1136–1140. doi:10.1038/nature08066
Novák O, Henykova E, Sairanen I, Kowalczyk M, Pospíšil T, Ljung K (2012) Tissue-specific profiling of the Arabidopsis thaliana auxin metabolome. Plant J 72(3):523–536. doi:10.1111/j.1365-313X.2012.05085.x
Paque S, Mouille G, Grandont L, Alabadi D, Gaertner C, Goyallon A, Muller P, Primard-Brisset C, Sormani R, Blazquez MA, Perrot-Rechenmann C (2014) AUXIN BINDING PROTEIN1 links cell wall remodeling, auxin signaling, and cell expansion in Arabidopsis. Plant Cell 26 (1):280–295. doi:10.1105/tpc.113.120048
Peret B, De Rybel B, Casimiro I, Benková E, Swarup R, Laplaze L, Beeckman T, Bennett MJ (2009) Arabidopsis lateral root development: an emerging story. Trends Plant Sci 14 (7):399–408. doi:10.1016/j.tplants.2009.05.002
Perrot-Rechenmann C (2010) Cellular responses to auxin: division versus expansion. Cold Spring Harb Perspect Biol 2(5):a001446. doi:10.1101/cshperspect.a001446
Petrášek J, Friml J (2009) Auxin transport routes in plant development. Development 136 (16):2675–2688. doi:10.1242/dev.030353
Petrášek J, Mravec J, Bouchard R, Blakeslee JJ, Abas M, Seifertová D, Wisniewska J, Tadele Z, Kubeš M, Čovanová M, Dhonukshe P, Skůpa P, Benková E, Perry L, Křeček P, Lee OR, Fink GR, Geisler M, Murphy AS, Luschnig C, Zažímalová E, Friml J (2006) PIN proteins perform a rate-limiting function in cellular auxin efflux. Science 312(5775):914–918. doi:10.1126/science.1123542
Quint M, Barkawi LS, Fan KT, Cohen JD, Gray WM (2009) Arabidopsis IAR4 modulates auxin response by regulating auxin homeostasis. Plant Physiol 150(2):748–758. doi:10.1104/pp. 109.136671
Rampey RA, LeClere S, Kowalczyk M, Ljung K, Sandberg G, Bartel B (2004) A family of auxin-conjugate hydrolases that contributes to free indole-3-acetic acid levels during Arabidopsis germination. Plant Physiol 135(2):978–988. doi:10.1104/pp. 104.039677
Ranocha P, Denance N, Vanholme R, Freydier A, Martinez Y, Hoffmann L, Kohler L, Pouzet C, Renou JP, Sundberg B, Boerjan W, Goffner D (2010) Walls are thin 1 (WAT1), an Arabidopsis homolog of Medicago truncatula NODULIN21, is a tonoplast-localized protein required for secondary wall formation in fibers. Plant J. doi:10.1111/j.1365-313X.2010.04256.x
Ranocha P, Dima O, Nagy R, Felten J, Corratge-Faillie C, Novák O, Morreel K, Lacombe B, Martinez Y, Pfrunder S, Jin X, Renou JP, Thibaud JB, Ljung K, Fischer U, Martinoia E, Boerjan W, Goffner D (2013) Arabidopsis WAT1 is a vacuolar auxin transport facilitator required for auxin homoeostasis. Nat Commun 4:2625. doi:10.1038/ncomms3625
Rashotte AM, Poupart J, Waddell CS, Muday GK (2003) Transport of the two natural auxins, indole-3-butyric acid and indole-3-acetic acid, in Arabidopsis. Plant Physiol 133(2):761–772. doi:10.1104/pp. 103.022582
Robert S, Kleine-Vehn J, Barbez E, Sauer M, Paciorek T, Baster P, Vanneste S, Zhang J, Simon S, Čovanová M, Hayashi K, Dhonukshe P, Yang Z, Bednarek SY, Jones AM, Luschnig C, Aniento F, Zažímalová E, Friml J (2010) ABP1 mediates auxin inhibition of clathrin-dependent endocytosis in Arabidopsis. Cell 143(1):111–121. doi:10.1016/j.cell.2010.09.027
Rosquete MR, Barbez E, Kleine-Vehn J (2012) Cellular auxin homeostasis: gatekeeping is housekeeping. Mol Plant 5(4):772–786. doi:10.1093/mp/ssr109
Růžička K, Strader LC, Bailly A, Yang H, Blakeslee J, Langowski L, Nejedla E, Fujita H, Itoh H, Syono K, Hejátko J, Gray WM, Martinoia E, Geisler M, Bartel B, Murphy AS, Friml J (2010) Arabidopsis PIS1 encodes the ABCG37 transporter of auxinic compounds including the auxin

precursor indole-3-butyric acid. Proc Natl Acad Sci USA 107(23):10749–10753. doi:10.1073/pnas.1005878107

Sauer M, Kleine-Vehn J (2011) AUXIN BINDING PROTEIN1: the outsider. Plant Cell 23 (6):2033–2043. doi:10.1105/tpc.111.087064

Sawchuk MG, Edgar A, Scarpella E (2013) Patterning of leaf vein networks by convergent auxin transport pathways. PLoS Genet 9(2):e1003294. doi:10.1371/journal.pgen.1003294

Scherer GF (2011) AUXIN-BINDING-PROTEIN1, the second auxin receptor: what is the significance of a two-receptor concept in plant signal transduction? J Exp Bot 62(10):3339–3357. doi:10.1093/jxb/err033

Shi JH, Yang ZB (2011) Is ABP1 an auxin receptor yet? Mol Plant 4(4):635–640. doi:10.1093/mp/ssr050

Simon S, Petrášek J (2011) Why plants need more than one type of auxin. Plant Sci 180 (3):454–460. doi:10.1016/j.plantsci.2010.12.007

Staswick PE, Serban B, Rowe M, Tiryaki I, Maldonado MT, Maldonado MC, Suza W (2005) Characterization of an Arabidopsis enzyme family that conjugates amino acids to indole-3-acetic acid. Plant Cell 17(2):616–627. doi:10.1105/tpc.104.026690

Strader LC, Bartel B (2011) Transport and metabolism of the endogenous auxin precursor indole-3-butyric acid. Mol Plant 4(3):477–486. doi:10.1093/mp/ssr006

Strader LC, Culler AH, Cohen JD, Bartel B (2010) Conversion of endogenous indole-3-butyric acid to indole-3-acetic acid drives cell expansion in Arabidopsis seedlings. Plant Physiol 153 (4):1577–1586. doi:10.1104/pp. 110.157461

Tam YY, Epstein E, Normanly J (2000) Characterization of auxin conjugates in Arabidopsis. Low steady-state levels of indole-3-acetyl-aspartate, indole-3-acetyl-glutamate, and indole-3-acetyl-glucose. Plant Physiol 123(2):589–596

Tan X, Calderon-Villalobos LI, Sharon M, Zheng C, Robinson CV, Estelle M, Zheng N (2007) Mechanism of auxin perception by the TIR1 ubiquitin ligase. Nature 446(7136):640–645. doi:10.1038/nature05731

Tromas A, Paponov I, Perrot-Rechenmann C (2010) AUXIN BINDING PROTEIN 1: functional and evolutionary aspects. Trends Plant Sci 15(8):436–446. doi:10.1016/j.tplants.2010.05.001

Viaene T, Delwiche CF, Rensing SA, Friml J (2013) Origin and evolution of PIN auxin transporters in the green lineage. Trends Plant Sci 18(1):5–10. doi:10.1016/j.tplants.2012.08.009

Wisniewska J, Xu J, Seifertová D, Brewer PB, Růžička K, Blilou I, Rouquie D, Benková E, Scheres B, Friml J (2006) Polar PIN localization directs auxin flow in plants. Science 312 (5775):883. doi:10.1126/science.1121356

Woodward AW, Bartel B (2005) Auxin: regulation, action, and interaction. Ann Bot 95 (5):707–735. doi:10.1093/aob/mci083

Xu T, Wen M, Nagawa S, Fu Y, Chen JG, Wu MJ, Perrot-Rechenmann C, Friml J, Jones AM, Yang Z (2010) Cell surface- and rho GTPase-based auxin signaling controls cellular interdigitation in Arabidopsis. Cell 143(1):99–110. doi:10.1016/j.cell.2010.09.003

Zažímalová E, Murphy AS, Yang H, Hoyerová K, Hošek P (2010) Auxin transporters–why so many? Cold Spring Harb Perspect Biol 2(3):a001552. doi:10.1101/cshperspect.a001552

Zhao Y (2012) Auxin biosynthesis: a simple two-step pathway converts tryptophan to indole-3-acetic acid in plants. Mol Plant 5(2):334–338. doi:10.1093/mp/ssr104

Zolman BK, Silva ID, Bartel B (2001) The Arabidopsis pxa1 mutant is defective in an ATP-binding cassette transporter-like protein required for peroxisomal fatty acid beta-oxidation. Plant Physiol 127(3):1266–1278

# Chapter 5
# Intercellular Transport of Auxin

**Jesica Reemmer and Angus Murphy**

**Abstract** Auxin is an essential hormone that regulates both programmed and plastic plant development. The mechanisms that regulate auxin metabolism, transport and signal transduction are well characterized, although important unresolved questions remain. A unique feature of auxin-regulated plant development is that it involves a combination of cellular perception with polarized auxin gradients across groups of cells, tissues, and organs. Plants achieve these polarized auxin gradients via site-specific synthesis followed by directed and polar patterning of transport components in individual cells. These streams are primarily mediated by three functionally distinct plasma membrane transporter families. Apical–basal and organogenic patterning is largely defined by the polar efflux activities of full-length PIN-FORMED (PIN) facilitators. Dynamic auxin uptake into directed streams is mediated by the AUXIN RESISTANT 1 (AUX1) and LIKE AUX1 (LAX) symporters. Finally, long-distance transport streams are motivated by the ATP-BINDING CASSETTE subfamily B (ABCB) active transporters that continually pump across the plasma membrane and prevent reuptake of exported auxin. Multiple accessory proteins regulate auxin transporter activity and interactions with subcellular environments. The current understanding of cellular transport of auxin will be reviewed in this chapter.

J. Reemmer
Department of Horticulture and Landscape Architecture, Purdue University, West Lafayette, IN 47907, USA

A. Murphy (✉)
Department of Horticulture and Landscape Architecture, Purdue University, West Lafayette, IN 47907, USA

Department of Plant Science and Landscape Architecture, University of Maryland, College Park, MD 20742, USA
e-mail: asmurphy@umd.edu

E. Zažímalová et al. (eds.), *Auxin and Its Role in Plant Development*,
DOI 10.1007/978-3-7091-1526-8_5, © Springer-Verlag Wien 2014

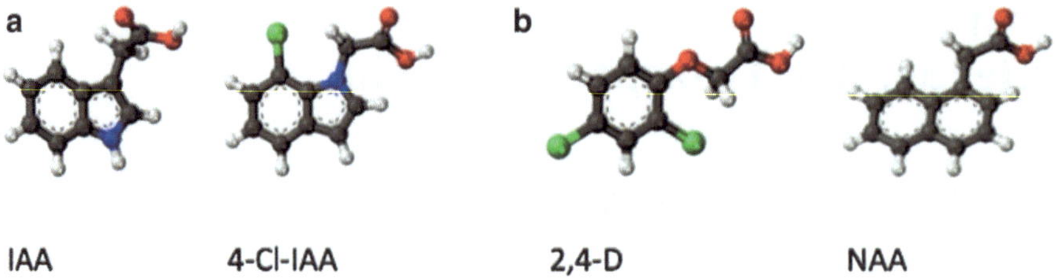


**Fig. 5.1** Molecular structures of auxin. IAA and 4-Cl-IAA are endogenous auxins (**a**), whereas 2,4-D (2,4-dichlorophenoxyacetic acid) and naphtalene-1-acetic acid (NAA) are synthetically produced (**b**). The synthetic auxins 2,4-D and NAA exhibit chemical traits, such as greater resistance to metabolism or higher rates of diffusion into cells, that make them useful tools for auxin transport studies

# 1 Introduction

Auxins are indolic plant hormones that function in regulation of cell division and elongation, polar growth, organogenesis, determination of shoot and root architecture, and tropic responses to gravity and light (reviewed in Teale et al. 2006; Benjamins and Scheres 2008). The word auxin is derived from the Greek word "auxein," meaning "to grow/increase" (Kögl and Haagen-Smith 1931). Naturally occurring auxins contain an indole and carboxylic acid group. The most abundant and developmentally important natural auxin is indole-3-acetic acid (IAA), although 4-chloro-indole acetic acid (4-Cl-IAA) is also found in some species (Fig. 5.1a). Synthetic auxins mimic IAA in structure and spacing of the carboxylic acid moiety (Fig. 5.1b).

IAA is produced primarily by multistep conversion of the aminoacyl precursor tryptophan synthesized from chorismate in the chloroplast (Radwanski and Last 1995). In *Arabidopsis*, tryptophan is converted to indole-3-pyruvic acid (IPA) by TRYPTOPHAN AMINOTRANSFERASE of ARABIDOPSIS 1 (TAA1), also known as TRANSPORT INHIBITOR RESPONSE 2 (TIR2) (Yamada et al. 2009; Tao et al. 2008; Stepanova et al. 2008; see Chap. 2). The second, and rate limiting, step of this pathway is the conversion of IPA to IAA by YUCCA, which catalyzes the oxidative decarboxylation of α-keto acids, including IPA and phenyl pyruvate (PPA) in *Arabidopsis* (Dai et al. 2013). In the model plant *Arabidopsis*, the biosynthetic pathway that generates indole glucosinolate as defense compounds also contributes to IAA pools (Sugawara et al. 2009).

Auxins are produced in the shoot apex and other sites of cell division and differentiation (Ljung et al. 2001, 2005; Bhalerao et al. 2002; Cheng et al. 2006, 2007; Stepanova et al. 2008; Tao et al. 2008; Petersson et al. 2009). Emergent evidence of localized auxin production stands in contrast to traditional models in which auxin was thought to be synthesized almost exclusively in the shoot apex and transported to sites of action (Ljung et al. 2001, 2005; Bhalerao et al. 2002; Petersson et al. 2009). Simultaneous elucidation of auxin transport mechanisms over the past 15 years has largely supported models of more distributed synthesis at the expense of shoot apical source models (reviewed in Peer et al. 2011). In particular, closely linked localized synthesis and polar auxin gradients involved in organogenesis have emerged as largely distinct from long-distance auxin transport streams.

When free auxin reaches a cell, two temporally distinct sets of responses may be observed (see Chap. 6). The earliest cell elongation responses to auxin may involve non-transcriptional events. AUXIN-BINDING PROTEIN 1 (ABP1) is a putative auxin receptor localized to the ER and cell periphery and may perceive auxin levels outside of the cell (Löbler and Klämbt 1985; Peer et al. 2013). However, transcriptional activation of genes containing auxin response promoter elements requires derepression of trans-acting Auxin Response Factor (ARF) proteins by the TIR1/AFB-AUX/IAA co-receptor system (Guilfoyle et al. 1986; Theologis et al. 1985). Auxin binding to the TIR1/AFB F-box ubiquitin ligase subunit and an AUX/IAA co-receptor promotes AUX/IAA degradation to derepress ARF activation of auxin responsive genes expression (Dharmasiri et al. 2005; Kepinski and Leyser 2005). The S-Phase Kinase-Associated Protein 2A (SKP2A) is a similar F-box protein whose function has been found to be additive to the effects of *tir1-1* (del Pozo et al. 2002; Jurado et al. 2010). SKP2A has a different target, however, and its rapid degradation of key regulators of cell cycle control suggests SKP2A mediates auxin-responsive cell cycle control (del Pozo et al. 2006; Jurado et al. 2008).

A unique aspect of auxin-dependent growth regulation is that the polarity and concentration of the auxin transport stream imparts vital information that the system can recognize. As early as 1880, Darwin described the unknown transmission of "some influence" as the agent of seedling bending in response to light (Darwin 1880). Though delayed by nearly 50 years, the theory of lateral auxin relocalization as a mechanism for bending was almost immediately proposed by both Cholodny and Went once the phytohormone auxin was discovered (Cholodny 1927; Went and Thimann 1937). The asymmetric distribution of auxin in this relocalization is responsible for differential cell elongation and the reorientation of growth evidenced in both photo- and gravitropic responses (see Chap. 16).

Another result of lateral relocalization is uneven accumulation of auxin into local maxima and minima. It is through these concentration differences that auxin sets the blueprint for plant development. These gradients are found to be essential for both the embryonic development of apical–basal polarity (see Chap. 9) and the continued patterning of organogenesis (Reinhardt et al. 2003).

## 2 Cellular Auxin Transport

### 2.1 *Cellular Auxin Import Is Motivated by Chemiosmotic Gradients*

Cellular uptake and efflux of auxin is motivated by a combination of chemiosmotic forces and ATP hydrolysis. IAA is a weak acid with a pKa of approximately 4.85.

In the acidic (pH 5.5) conditions of the apoplast, only a small fraction (a calculated 17 %) of auxin molecules are proton associated (Rubery and Sheldrake 1974; Raven 1975). While protonated auxin preferentially diffuses into the cell membrane, 83 % of the auxin pool remains unavailable to lipophilic diffusion in its dissociated form ($IAA^-$). Additionally, cells must be able to selectively take up auxin in competition with other organic acids and adjust for incorporation of the already limited pool of auxin into other tissues. Thus, in order to meet developmental demands, there is a distinct need for protein importers to actively recruit the traveling auxin signal. In *Arabidopsis*, this transport is carried out by the high-affinity auxin influx transporter AUX1 and its LAX protein family members (Goto et al. 1987; Parry et al. 2001). These permease-like proteins function by harnessing the potential of the proton gradient to drive passive anionic symport of deprotonated IAA at the plasma membrane (Yang et al. 2006).

The importance of symport-driven uptake in enhancing transport streams is observed in *aux1* mutants or transformants wherein *AUX/LAX* genes are uniformly expressed in all cells of a tissue. Increased uptake activity increases the total auxin found in transport streams, and decreased activity results in decreased concentration of auxin in provascular and vascular tissues compared to wild-type plants (Marchant et al. 2002; Kramer 2004). AUX1 activity in lateral root cap cells has been shown to be essential to uptake of auxin from the root apex into directed transport streams in the root epidermis that direct gravitropic responses (Swarup et al. 2001; Kleine-Vehn et al. 2006).

In *Arabidopsis*, the three closely related LAX proteins also function in auxin uptake. Comparison of gene structure revealed well-conserved exon/intron boundaries indicative of origination from a common ancestor through gene duplication, but regulation of subcellular trafficking and spatial expression patterns of the LAXs differ considerably from AUX1 (Parry et al. 2001; Bainbridge et al. 2008; Swarup et al. 2008; Jones et al. 2009; Péret et al. 2012). For instance, AUX1 intracellular targeting is regulated by AXR4, which encodes a putative endoplasmic reticulum (ER) chaperone thought to facilitate the correct folding of AUX1 and its export from ER to Golgi (Dharmasiri et al. 2006). However, LAX2 and LAX3 fail to target to the plasma membrane in AUX1-expressing cells, suggesting they may need their own specific ER chaperones (Péret et al. 2012). Mutations have member-specific effects on auxin-related phenotypes as well. Both mutant *aux1* and *lax3* plants show comparable reduced lateral root emergence. However, *aux1* shows a reduced level of lateral root primordia, whereas *lax3* actually has a threefold increase in primordia compared to the wild type (Swarup et al. 2008). This suggests distinct functional roles for the different family members. However, all three LAX proteins have been shown to retain an auxin influx carrier function, albeit with varying transport specificities, that strongly suggests subfunctionalization (Yang et al. 2006; Swarup et al. 2008; Péret et al. 2012).

## 2.2 *Polar Auxin Transport Defines Local Concentration Gradients*

Once inside the neutral conditions of the cytosol, auxin is deprotonated to its polar anionic form ($IAA^-$). This precludes auxin from diffusing back through the lipophilic cell membrane and, unless aided by exporters, auxin remains trapped in the cytosol. In a unique system, auxin can be transported cell to cell in a polar fashion. This short-range directional transport is not only a method of export, but the key to building the patterned auxin gradients crucial in developing tissues (Rubery and Sheldrake 1974; Raven 1975; Zažímalová et al. 2010).

PIN-FORMED (PIN) efflux facilitators derive their name from the striking phenotype of *pin1* mutants in which the inflorescences do not form floral organs and remain bare pin-like stems (Goto et al. 1987). In addition to PIN1, seven other PIN genes are present in the genome of the model species *Arabidopsis*. Studies of plasma membrane-localized, "long" PINs (PIN1, 2, 3, 4, and 7) have shown distinct roles for individual members of the family. PIN1, 4, and 7 are vital in maintaining the polar auxin streams necessary for organogenesis and development along with AUX1/LAX influx carriers (Reinhardt et al. 2003; Bainbridge et al. 2008). PIN2 activity is crucial to gravitropism and the reflux of auxin at the root tip (Chen et al. 1998; Müller et al. 1998; Friml et al. 2004; Rahman et al. 2010). PIN3 has a restrictive effect on auxin streams important for directional growth (Friml et al. 2002). However, mutational studies reveal that despite their apparent specialties, PINs largely have redundant functions; the loss of a single PIN protein can be compensated for by the ectopic activities of the other PIN family members (Blilou et al. 2005; Vieten et al. 2005). By mutating multiple PIN gene family members, greater phenotypic and developmental defects can be induced in systems such as embryonic development, root patterning, and lateral root initiation (Benkova et al. 2003; Friml et al. 2003; Blilou et al. 2005). Interestingly, application of the auxin efflux inhibitor naphthylphthalamic acid (NPA) can mimic these effects. The fact that NPA produces these mutant phenotypes indicates a function in the proximity of PINs and has long made NPA a useful tool in studying altered development (Katekar and Geissler 1977).

One of the key purposes of directing local auxin concentrations is the creation of the auxin maxima necessary for organogenesis (see Chaps. 10–12). The auxin streams created by joint AUX/LAX uptake and PIN efflux form the architectural patterns of new organs at the shoot apical meristem (Vernoux et al 2010). Interestingly, a plant lacking the function of all four members of the AUX/LAX auxin influx/transporters is still viable and moderately fertile, although its architecture is significantly altered (Bainbridge et al. 2008). Similarly, while auxin efflux transport proteins ABCB1 and ABCB19 can be visualized in developing embryos, knocking out their function did not result in observable defect to development (Mravec et al. 2008). Treating a double mutant of *pin1* and *aux1* with auxin discovered one clue as to how fertility is maintained in the face of such mutations. Auxin treatment resulted in very large, fused organs at the apex rather than single flowers being produced (Reinhardt et al. 2003). This suggests that AUX1 may be involved in the positioning of organs and restricting their boundaries, thus ensuring that sufficient auxin remains in the necessary layers of the SAM (Reinhardt et al. 2003). Furthermore, these observations imply that the function of concentrating auxin at a maximum is shared, with input from ABCB efflux transporters acting redundantly (Noh et al. 2001).

## 2.3 *ABCB Efflux Transporters Maintain Long-Distance Streams of Auxin*

Long-range transport of IAA is needed to generate auxin pools in sink tissues, which are important for such developmental processes as stimulating lateral roots and shoot branching. This transport can be accomplished via the phloem vasculature, as is the common route for metabolites.

The best-known member of the B subfamily of ATP-Binding Cassette (ABC) transporters is human ABCB1 (MULTIDRUG RESISTANT 1/PHOSPHOGLYCOPROTEIN 1), which has been extensively studied for its role in increased resistance to chemotherapeutic agents in breast, brain, and colon cancer cells (Luckie et al. 2003). However, it was apparent that plant homologues of human ABCB1 are not promiscuous drug transporters, and, while applicable in mammalian systems, the use of the multidrug resistance (MDR) nomenclature for this subgroup of proteins has been discontinued (Sidler et al. 1998). The *Arabidopsis thaliana* ABC subfamily B comprises 21 members; four proteins to date clearly mediate high-specificity auxin transport, and a number of highly homologous proteins are thought to mediate auxin transport to some degree (Kamimoto et al. 2012). While both the PIN and AUX/LAX families have been extensively studied in terms of gene expression and protein localization, ABCB proteins are not as well characterized. Phylogenetic and structural analyses indicate that the subclass of ABCB transporters function in auxin transport across plant species, and many studies have focused on the representative ABCB1 and 19 proteins of *Arabidopsis* (reviewed in Blakeslee et al. 2005; Verrier et al. 2008).

ABCB1 was discovered in the attempt to identify proteins conferring broad-spectrum herbicide resistance. Mammalian cells with increased expression of ATP-driven efflux pumps can gain resistance to a wide variety of cytotoxic drugs, and it had been proposed that a similar system might exist in plants (Dudler and Hertig 1992). Instead, it was found that overexpression of AtABCB1 resulted in elongation of seedling hypocotyls when grown under dim light, whereas antisense lines exhibited reduced elongation of the hypocotyls (Sidler et al. 1998). These phenotypes are similar to those witnessed after treating wild-type plants with low concentrations of auxin or an auxin transport inhibitor, for the overexpressor and antisense line, respectively. Further studies revealed that AtABCB1 is localized at the plasma membrane in nonpolar distributions at the shoot and root apices and is predominantly found with polar localization above the root apex. Its expression in both yeast and mammalian systems displays increased efflux of IAA and active synthetic 1-NAA, and *in planta* oxidative breakdown products of IAA are effluxed as well (Geisler et al. 2005). ABCB1 genes also have auxin transport function in other plant species. In maize, the P-glycoprotein *brachytic2* (*br2*) mutation shares 67 % identity with AtABCB1 and results in dwarfed plants with shortened lower stalk internodes (Leng and Vineyard 1951; Stein 1955; Noh et al. 2001). The mutant gene *dwarf3* similarly results in dwarfed sorghum, a close relation to maize in both genomic organization and plant form (Mullet et al. 2002). While specific phenotypes vary between species, it has been confirmed that both *brachytic2* and *dwarf3* mutant phenotypes result from loss-of-function mutations to ABCB1 genes and display reductions in long-distance transport of auxin (Multani et al. 2003; Bailly et al. 2012). The collective evidence suggests that ABCB1 functions primarily in regions of high auxin concentration to accelerate vectorial transport and maintain long-distance auxin transport streams in combination with PIN and other ABCB family members (Bandyopadhyay et al. 2007).

*Arabidopsis* ABCB19 was quickly linked to its closest homologue *Arabidopsis* ABCB1 (Noh et al. 2001). Both AtABCB1 and AtABCB19 exhibit remarkable structural similarity to the mammalian ABCB1 multidrug resistance transporter known for very broad substrate specificity. However, both AtABCB1 and AtABCB19 exhibit relatively high specificity for auxin as a transport substrate (Titapiwatanakun et al. 2009; Yang and Murphy 2009). Phenotypic analyses of *abcb19* showed epinastic cotyledons, abnormally wrinkled leaves, reduced apical dominance, partial dwarfism, and reduced basipetal polar auxin transport in hypocotyls and inflorescences, all of which are phenotypes consistent with altered auxin response and/or transport. These defects were synergistically enhanced in the double mutant *abcb1abcb19*, suggesting some functional redundancy between these efflux transporters (Noh et al. 2001; Geisler et al. 2003, 2005). Comparison of expression domains revealed that ABCB19 maintains whole-plant auxin flow from the shoot to root apices, whereas ABCB1 function is more restricted to the shoot apex (Geisler et al. 2005). In addition, ABCB19 appears to function in the regulation of differential growth in response to light and gravity stimulation, and it is the substrate target for photoreceptor kinase PHOTOTROPIN1 (PHOT1) (Liscum and Briggs 1995; Noh et al. 2001, 2003; Lin and Wang 2005; Lewis et al. 2007; Wu et al. 2007; Nagashima et al. 2008). Phosphorylation of ABCB19 halts auxin efflux activity, which increases auxin levels at and above the site of inhibition. This action is an early step in the eventual unilateral growth that causes the bend seen in phototropism (Christie et al. 2011).

### 2.4 Canalization and the Amplification of Streams by PIN Proteins

The notion that auxin and polarity are linked dates back to classic histological studies on developing vascular cells (Sachs 1969). The canalization hypothesis was

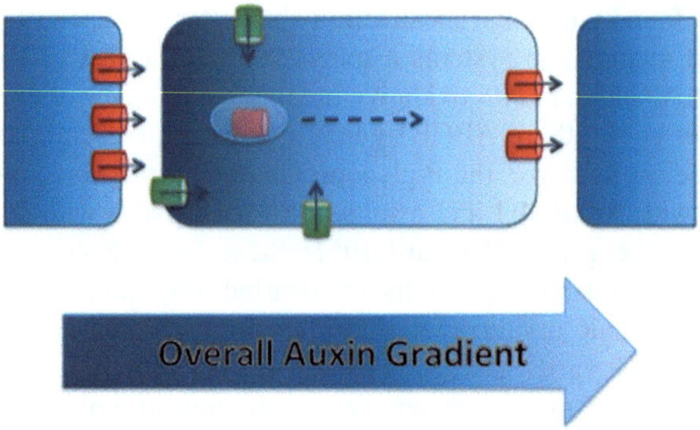


**Fig. 5.2** Micropolar auxin gradients. In addition to the auxin gradient at the tissue level, auxin gradients can be thought of to exist within individual cells. This gradient (higher auxin concentration in *darker blue*) may interact with vesicular trafficking as a way for individual cells to perceive this auxin gradient. *Red cylinders* denote PIN exporters, *green cylinders* AUX/LAX importers. *Solid black arrows* denote auxin transport, *broken arrow* possible vesicular movement

put forward by Sachs, and later mathematically formulated by Mitchison, as a proposal that a positive feedback exists between the flow of the signal molecule and the capacity for its flow (Sachs 1969; Mitchison 1980). Further experimentation employing an antibody against export protein PIN1 revealed the occurrence of upregulation and relocalization away from the site of exogenous auxin application (Sauer et al. 2006). Studies of such experimental responses strongly suggest that auxin, or a secondary signal produced in response to auxin, is transported into the cell and directs canalization by regulating the polarity of PIN positioning in some cell types. This idea was expanded to include the hypothesis that cells could effectively monitor the auxin concentration in their surrounding environment and respond by pumping auxin toward neighboring cells against a concentration gradient (Fig. 5.2). To test this hypothesis, computer simulation was employed to handle the hundreds of complex interactions that would occur in such a scenario (Jönsson et al. 2006; Smith et al. 2006). Using theoretical parameters, the canalization hypothesis was shown to amplify small auxin fluxes and generate physiologically plausible results in phyllotactic patterning and vein formation in some tissues (Scarpella et al. 2006; Sauer et al. 2006).

Interesting observations arise from computational modeling. In the attempt to describe PIN allocation and function, two radically different, but conditionally functional, models have been defined that yield discrepant functional conclusions in different locations. In meristematic tissues, PINs act to sense local auxin concentrations, yet in inner tissues a mechanism of flux sensing is called for (Jönsson et al. 2006; Smith et al. 2006; Scarpella et al. 2006; Sauer et al. 2006). Therefore, simplistic models of transport are not sufficient to fully describe the complexities that arise in the actual *in planta* scenario. Boundaries such as cell walls can pose issues that have significant effects on the expected design of a system (Stoma et al. 2008; Bayer et al. 2009). The topologic effect of networks of intracellular compartments can have an appreciable effect on cytoplasmic auxin concentration through sequestration; merely averaging over a cell volume would result in an inappropriate rate constant for describing auxin flux (Merks et al. 2007; Hošek et al. 2012). While computational models provide powerful insights and direction for further research, it is necessary to consider that some of their conclusions may diverge from *in planta* auxin patterns.

## 2.5 PIN Polarity Is Regulated by Phosphorylation

PIN function is influenced by the phosphorylation of kinases in the AGC family (named for homology to mammalian c*A*MP-dependent protein kinase A, c*G*MP-dependent protein kinase G, and phospholipid-dependent protein kinase C). This family includes members such as PINOID (PID) kinase, D6 PROTEIN KINASE (D6PK), WAVY ROOT GROWTH 1 (WAG1), WAG2, and PHOTOTROPIN 1 (PHOT1) and PHOT2 (Sakai et al. 2001; Dhonukshe et al. 2010; Huang et al. 2010). Although the mechanisms for the PHOT blue light receptors and the WAG root growth regulators have not been fully elucidated, their corresponding tropisms may be the result of changes in auxin response or transport (Harper et al. 2000; Esmon et al. 2006; Santner and Watson 2006).

PID was previously proposed to be responsible for the phosphorylation of PIN proteins, as it was observed that *pid* mutants phenocopy the *pin1* mutant phenotype. From overexpression studies, it was further seen that phosphorylation by PID leads to a basal to apical localization shift of at least PIN1, PIN2, and PIN4 in root cortex and lateral root cap cells (Friml et al. 2003). Shifts triggered by this PID kinase can be reverted by increasing the expression of *PP2A*, a gene whose product is a subunit of a compound phosphatase. This suggests that PIN polarity is at least in part controlled by PID-dependent phosphorylation (Michniewicz et al. 2007). Another group of kinases, D6PK and its three D6PK-LIKE homologues, have more recently been shown to phosphorylate and directly activate PIN proteins (Zourelidou et al. 2009; Willige et al. 2013). D6PKs colocalize at the basal ends of cells with PINs that mediate rootward auxin transport. However, D6PK does not colocalize with PIN2 in epidermal root cells and, thus, does not appear to regulate PIN2 activity. This is consistent with auxin efflux activity exhibited by PIN2, but not PIN1, 3, or 7 when heterologously expressed in *Saccharomyces cerevisiae* (Yang and Murphy 2009), where a D6PK ortholog has not been identified.

Expression, abundance, localization, and biochemical activity of D6PK are insensitive to auxin and NPA, although the genes are expressed strongly at the sites of lateral root initiation (Zourelidou et al. 2009). Consistent with direct regulation of PIN transport activity, neither loss of function nor overexpression of D6PK causes alteration in PIN polarity. Seedlings of overexpression studies show other differences: D6PK seedlings having defects in lateral root formation, while

PID seedlings exhibit agravitropic growth and meristem collapse (Benjamins et al. 2001; Friml et al. 2004).

## 2.6 PIN Proteins Interact with ABP1

Of the long PINs, all are trafficked by dynamic cellular mechanisms (reviewed in Grunewald and Friml 2010). The clathrin-mediated endocytosis of PIN proteins is positively regulated by ABP1. ABP1 normally functions in the recruitment of clathrin to the plasma membrane. However, when ABP1 is bound by auxin, its signaling is blocked. This leads to a reduced internalization of PINs by clathrin-mediated endocytosis. The effect of auxin binding is thus the enhancement of auxin efflux transport (Robert et al. 2010). ABP1 also activates the Rho GTPase ROP6 and its effector RIC1. RIC1 promotes cytoskeletal organization by physically interacting with the microtubules-severing protein KATANIN1 (KTN1) (Fu et al. 2005, 2009; Lin et al. 2013). In this way, the auxin signaling pathway is linked to the regulation of microtubule organization and physically promotes cell elongation.

## 2.7 In Silico Modeling of ABCB Proteins Suggests Exclusion

The aforementioned auxin transport mechanisms mainly address the shuttling of auxin discretely to and from the apoplast and cytoplasm of the cell. As efflux transporters, ABCB proteins have two well-studied binding sites in the central pocket through which cytoplasmic auxin can be exported from the cell. Early models of plant ABCBs were designed by threading their sequences on the crystal structure of the Sav1866 bacterial ABC transporter in the closed conformation (Dawson and Locher 2006; Yang and Murphy 2009). Further insight came from the publication and validation of the crystal structure of murine ABCB1 (MmABCB1) in the open conformation (Aller et al. 2009). New structural models identified kingdom-specific candidate substrate-binding regions and suggested an early evolutionary divergence of plant and mammalian ABCBs. While the two experimentally validated IAA substrate-binding sites identified in models based on the closed Sav1866 structure are present in open configuration models, an additional binding site within the outer leaflet was also uncovered (Bailly et al. 2012) (Fig. 5.3a).

This finding led to an elegant development in the conceptualization of auxin transport. Auxin is an amphipathic molecule, and a significant amount of anionic auxin is found partially inserted in the lipid bilayer. A mechanism for the removal of this auxin is thus necessary, particularly in cells such as those adjoining vascular tissues where the apoplastic concentration of auxin is high, and reuptake must be prevented to maintain transport flow (Mravec et al. 2008; Titapiwatanakun

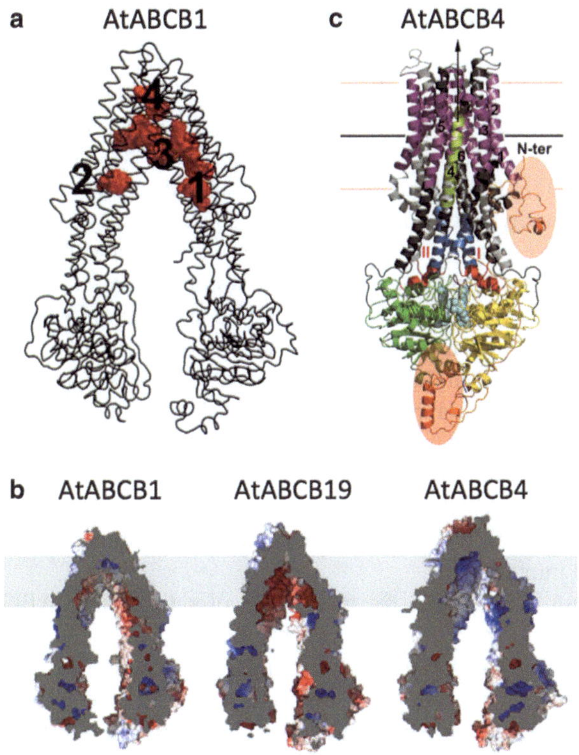


**Fig. 5.3** ABCB transporter *in silico* models. (**a**) *Red mesh* indicates IAA docking poses for AtABCB1 threaded on MmABCB1 in the open conformation. Site 3 reveals a new region of auxin binding potential that may account for exporter membrane exclusion ability (Bailly et al. 2012). (**b**) Electrostatic potentials between *Arabidopsis* ABCB proteins in cut view. The predominantly positive surface of the transport chamber of AtABCB4 suggests an evolution of transport specialization (Yang and Murphy 2009). (**c**) AtABCB4 has additional unique domains as compared to AtABCB exporters. N-terminal coiled-coil domain and linker domain between NBD1 and TMD2 are highlighted in *red*. TMH4 highlighted in *green* is shifted down off the membrane plane. These adjustments could be sufficient to alter the regulation and direction of auxin transport (Bailly et al. 2012)

et al. 2009; Matsuda et al. 2011). AtABCB1 and AtABCB19 have been shown to function primarily in such exclusion of IAA from cellular membranes (Blakeslee et al. 2007; Wu et al. 2007; Mravec et al. 2008; Bailly et al. 2012). With the rise of vascular plants, PINs appear to have emerged to provide an additional vectorial factor for the control of organogenesis and tropic responses while ABCBs maintained long-distance transport in increasingly longer and complex shoots and roots (Blakeslee et al. 2007; Titapiwatanakun et al. 2009).

## 2.8 ABCB Trafficking and Maturation

Similar to what is seen with AXR4 and AUX1, folding, trafficking, and activation of ABCB1, ABCB4, and ABCB19 is dependent on the co-chaperone immunophilin-like FK506-BINDING PROTEIN 42/TWISTED DWARF 1 (FKBP42/TWD1) (Bouchard et al. 2006; Bailly et al. 2008; Wu et al. 2010). TWD1 was originally biochemically identified in plasma membrane (PM) fractions and has been shown to be distributed to the endoplasmic reticulum (ER) and tonoplast as well (Murphy et al. 2002; Kamphausen et al. 2002; Geisler et al. 2003). TWD1 acts at the ER surface to fold and activate ABCBs, but appears to function at the PM to maintain ABCB activity as well (Bailly et al. 2008; Wu et al. 2010; Henrichs et al. 2012; Wang et al. 2013). Although ABCB1, ABCB4, and ABCB19 are largely trapped at the ER in the absence of TWD1, a percentage of all three transporters still reside at the PM (Titapiwatanakun et al. 2009; Wu et al. 2010; Wang et al. 2013). However, TWD1 also colocalizes widely with the lateral marker PEN3/ABCG36 and partially with nonpolar PM proteins and BRI1-GFP (Langowski et al. 2010; Růžička et al. 2010; Wang et al. 2013) suggesting other potential interactions.

In attempting to determine the mechanism of activation of ABCB activity by TWD1, TWD1 was found to interact with the PINOID AGC kinase that alters ABCB1 activity by protein phosphorylation (Henrichs et al. 2012). In addition, the plasma membrane localization of TWD1 provides a mechanism to minimize apoplastic reflux in tissues where high auxin contents exist, thus addressing the need to separate shoot- and rootward auxin streams in opposing root tissues and leaf epidermal cells (Geisler et al. 2005; Matsuda et al. 2011). This idea is further in agreement with the acid growth theory prediction of auxin-stimulated lateral proton extrusion for axial cell expansion (Hager 2003). Finally, the severity of the *twd1* phenotype in comparison to the *abcb1abcb19* double mutant argues that additional transporters might also be regulated by TWD1 activity.

## 2.9 Membrane Lipids Define Functional Environments

Regardless of the site of TWD1 activation, ABCB transporters must be trafficked to the plasma membrane to function properly. Sphingolipids are essential to establishing a rigid membrane environment to maintain native structure necessary for protein validation and vesicular packaging. Fluorescently labeled ABCB19, for instance, is impaired in its ability to reach the plasma membrane in *tsc10a* mutant plants (Yang et al. 2013). The *tsc10a* mutants are deficient in a key enzyme in sphingolipid biosynthesis, and a loss of this function results in epinastic cotyledon development, altered flowering patterns, and reduced hypocotyl elongation phenotypes reminiscent of *abcb19* mutants (Chao et al. 2011). In particular, very long chain fatty acid sphingolipids (VLCFA-SL) are essential for proper development (Markham et al. 2011).

Co-localization experiments showed that inhibiting the synthesis of sphingolipids, either because of the *tsc10a* mutation or treatment with fumonisin B1 (FB1), results in ABCB19 retention in the ER and the Golgi apparatus and failure to reach the PM. Additionally, fluorescently labeled ABCB19 already present on the membrane accumulates intracellularly after treatment with 1-phenyl-2- hexadecanoylamino-3-morpholino-1-propanol (PPMP) or fumonisin B1 (FB1), both inhibitors of the synthesis of the sphingolipid ceramide (Yang et al. 2013). These observations suggest sphingolipids are of particular importance at multiple points of ABCB trafficking and maintenance at the plasma membrane.

Once positioned at the plasma membrane, sterols are required to create the correct lipid composition of the plasma membrane environment to allow for the conformational changes associated with transport action. Cell membranes are necessary not only as the boundaries of living units, but also as the critical sites for interactions. The removal of sterols by methyl-beta-cyclodextrin (MβCD) induces the removal of ABCB19 from the plasma membrane. The removal is only partial, although it seems to have a distinct effect on ABCB19, as marker protein PLASMA MEMBRANE INTRINSIC PROTEIN 2A (PIP2-GFP) has no loss of signal at the plasma membrane under the same treatment conditions (Yang et al. 2013). Addition of cholesterol to ABCB19, however, enhances transport activity (Titapiwatanakun et al. 2009). This suggests that ABCB19 recruits sterols to its environment to increase its stability and functioning. It has already been shown that ABC family members are capable of transferring these sterols as well (Tarling and Edwards 2011). With the knowledge that Type 4 P-type ATPases catalyze the translocation of phospholipids between the cytosolic and apoplastic sides of the plasma membrane, it is conceivable that ABCB19 may catalyze a similar flipping action (Tanaka et al. 2011). In keeping with the idea that the third modeled auxin-binding site of ABCB transporters is less specific, this site has been suggested to flip substrates wrapped by lipids to the outer leaflet during the change to the outward-facing conformation (Aller et al. 2009; Bailly et al. 2012).

In *Arabidopsis*, sphingolipids and sterols have been shown to contribute to trafficking of PIN1 and AUX1 in their respective membrane domains as well (Carland et al. 2002; Willemsen et al. 2003; Men et al. 2008; Pan et al. 2009; Roudier et al. 2010; Markham et al. 2011). It is well documented that ABCB19 has a profound effect on the stabilization of PIN1 at the PM; PIN and ABCB proteins function together, and ABCB19 is actually required for PIN1 retention in those membranes (Blakeslee et al. 2007; Mravec et al. 2008; Titapiwatanakun et al. 2009; Yang et al. 2013). Observations of this web of interactions point to the existence of an interactive and dynamic environment that allows multiple facets of regulation to exert control over particular stimuli.

# 3 Homeostasis

As further components of fine-tuning auxin transport, homeostatic transport and subcellular compartmentalization have developed. Among the ABCB and PIN efflux transporters, there are proteins that diverged in function from their family members to play more conditional refining roles. Reversible transporters can be employed to keep auxin levels constant or augment its uptake in cell types where importers are not present (Swarup et al. 2001; Jones et al. 2009; Yang and Murphy 2009). Intracellular partitioning can both influence the effective cytoplasmic concentration and expose auxin molecules to various enzymatic environments for conjugation or degradation.

## 3.1 Conditional ABCB Transport Responds to Auxin Concentration

ABCB4 was originally identified as the most similar *Arabidopsis* homologue to the *Coptis japonica* ATP-dependent berberine influx transporter CjMDR1 (71 % amino acid identity) (Shitan et al. 2003). Despite the fact that *Arabidopsis* does not produce any isoquinoline alkaloids, this homology is greater than the 60 % amino acid sequence identity ABCB4 shares with ABCB1 (Terasaka et al. 2005). The sequence of ABCB4 was also predicted to diverge substantially from other ABCB efflux proteins in the loop region adjoining the first conserved nucleotide-binding domain as well as in a unique coiled-coil interactive domain at its N-terminus (Terasaka et al. 2005). Compilation of characterization data revealed that ABCB4 is a root-specific transporter that functions in shootward epidermal transport of auxin from the root apex, primary and lateral root elongation, and regulation of auxin movement into root hair cells. However, early studies often led to incongruent results that were tissue specific and highly dependent on growth and treatment conditions (Santelia et al. 2005; Terasaka et al. 2005; Cho et al. 2007; Lewis et al. 2007). Despite the fact that ABCB4 belongs to a family of active transporters primarily known for their efflux action, ABCB4 auxin efflux is conditional, mediating import at very low IAA concentrations but reversing rapidly to stronger export activity with increased internal IAA levels (Yang and Murphy 2009; Kim et al. 2010; Kubeš et al. 2012). This homeostatic role is consistent with the need to balance auxin streams in root epidermal cells and would be expected as a plasma membrane complement to short PIN function at the ER (Mravec et al. 2009; Ding et al. 2012; see below and Chap. 4).

### 3.2 *In Silico Modeling Supports ABCB Homeostatic Function*

ABC transporters, regardless of the direction in which they transport substrates, have a basic conserved structure of two transmembrane domains (TMDs) and two nucleotide-binding domains (NBDs) (Dawson and Locher 2006). Using these NBDs, ATP is expended in a "power stroke" that drives the rearrangement of the TMDs between open and closed conformations (Hopfner et al. 2000). But confined to this general structure, from where does the capacity for import arise? Indeed, AtABCB4 has a closer alignment with efflux transporters MmABCB1 and Sav1866 than with other ABC importers (Bailly et al. 2012). The explanation lies in the differences between the charge potentials of the binding pocket. When AtABCB1 faces the cytosol, it presents an environment that is predominantly negative at the opening of the cavity and weakens in charge nearer the interior binding regions. AtABCB19 also displays a similar distribution of negative to neutral electrostatic potentials. Strikingly, AtABCB4 presents the opposite charge, having a neutral to positive electrostatic surface spanning the entire binding pocket (Bailly et al. 2012) (Fig. 5.3b). This strongly suggests that these proteins have significantly diverged and adopted specialized functions. A model could be proposed in which ABCB4, due to its altered binding potential, does not engage auxin when cytoplasmic concentrations are low but rather switches conformation to export either other cargo or remains independent of substrate entirely (Procko et al. 2009). This would result in empty auxin-binding sites being exposed to the apoplast, and a net uptake activity could result upon restoration to the open, cytosolic-facing conformation (Aller et al. 2009). This scenario would account for auxin efflux activity when challenged with greater concentration of cytosolic auxin, as well as the observed lack of saturable influx kinetics (Dawson and Locher 2006; Yang and Murphy 2009). In addition to electrostatic potential differences, ABCB4 displays three other notable structural traits that are not present in ABCB1 or ABCB19. A shift in the hydrophobic region of transmembrane helix 4 would change its positioning in respect to the plasma membrane and thus significantly alter the distances and interactions between the second intercellular loop and its NBD, resulting in altered binding and transport properties (Yang and Murphy 2009). Docking simulations also identified two additional coiled coil domains for ABCB4, one of which is shared with ABCB14 guard cell malate importer and CjMDR1 putative berberine importer (Shitan et al. 2003; Lee et al. 2008; Yang and Murphy 2009) (Figs. 5.3c). These N-terminal domains may be interaction sites in which other proteins could further shift the positioning of the transmembrane helices.

In addition to the activities of ABCB4, it was expected that another protein would share this conditional function. Most plant ABCB members exist with a paralog, and ABCB21 indeed shares 68 % nucleotide identity and 79 % amino acid identity with ABCB4. These proteins are grouped in clade II of the phylogenic tree of P-glycoproteins, which is distinct from clade I where ABCB1 and ABCB19 are

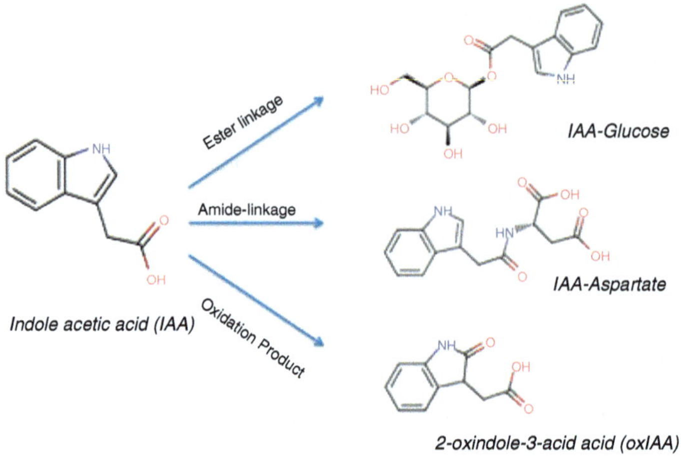


**Fig. 5.4** Inactivation of auxin. To attenuate the signal, auxin can undergo reversible or irreversible modification. Conjugates of auxin are found mainly in one of three forms: ester-linked to sugars, amide-linked to amino acids, or amide-linked to peptides or proteins. Many conjugation products can be hydrolyzed to reactivate auxin. Other modifications, such as oxidation, are terminal

classified (Knoller et al. 2010). Characterization of ABCB21 revealed an NPA-sensitive, plasma membrane-localized auxin transporter with strong expression in the abaxial side of cotyledons, junctions of aerial lateral organs, and root pericycle cells adjacent to the protoxylem poles. However, likeness of ABCB21 auxin transport to that of ABCB4 was not as strong as anticipated, as time course experiments with low concentration (300 nM) of IAA showed a different pattern of seedling response to exogenously added IAA. This suggests a different physiological role for ABCB21 from that of ABCB4. When cytoplasmic IAA concentration was increased by preloading IAA into yeast cells however, IAA uptake activity by ABCB21 was abolished. This same effect is seen in the activity of ABCB4 and suggests that ABCB21 also functions as a facultative auxin transporter in plant cells (Kamimoto et al. 2012).

### 3.3 "Short" PIN and PIN-Like Proteins Act in ER Compartmentalization of Auxin

In contrast to "long" PIN plasma membrane efflux facilitators, PIN5, 6, and 8 encode "short" proteins with a reduced or absent central hydrophilic loop (see Chap. 4). Unlike their longer, plasma membrane-localized family members, short PINs reside in endomembrane structures where they are hypothesized to function in the homeostatic compartmentalization of auxin (Mravec et al. 2009). This sequestration within the ER would both reduce the pool of auxin available for cell-to-cell transport and alter intracellular perception and nuclear signaling. Although the motive force for auxin efflux via short PINs is not known, recent studies have unveiled the important implications of their activity. Overexpression of PIN5, for example, leads to a dramatic shift in the profile of auxin metabolites. Upon induction of overexpressed PIN5, levels of free IAA and IAA-glucosyl ester nearly vanish, while there is increased accumulation of amino acid–auxin conjugates (Mravec et al. 2009). This implicates an unexpected role for PIN5 in controlling the metabolic fate of intracellular auxin.

On the other hand, PIN6 and PIN8 have been shown to act antagonistically to PIN5 in directional auxin efflux (Ding et al. 2012; Sawchuk et al. 2013). In contrast to PIN5 overexpression studies, when PIN8 is overexpressed, elevated levels of free IAA and ester-conjugated IAA are observed (Dal Bosco et al. 2012). It is yet unknown how these complementary activities are regulated, although varying hypotheses have been proposed. One suggestion is that PIN6 and PIN8 may serve to move auxin from the lumen of the ER to the nucleus for signaling and thus regulate auxin-dependent transcriptional activity (Dal Bosco et al. 2012). Alternatively, these PINs could have different affinities for alternate auxins or auxin conjugates and form a more complex regulatory network for control of intracellular auxin levels (Sawchuk et al. 2013). Characterization of these PINs points to the possibility of highly specialized and conditional functions. PIN8 has been shown to specifically accumulate in pollen and functions in the development of the pollen tube and auxin homeostasis of the male gametophyte (Ganguly et al. 2010; Ding et al. 2012; Dal Bosco et al. 2012). PIN6 has been shown to act in floral development in *Arabidopsis* and maintains the auxin homeostasis required for proper nectary function (Bender et al. 2013). Studies of vein patterning and defects have led to the conclusions that PIN6 can act redundantly with PIN8 (Sawchuk et al. 2013).

A complicating factor is that *in silico* analyses indicate that some members of the PIN-LIKES (PILS) family exhibit a topology that would include a central hydrophilic loop similar to that of the PIN family (see Chap. 4). Characterization of these seven family members revealed that PILS localize to the ER and stimulate intracellular auxin accumulation, potentially contributing to the regulation of auxin metabolism via compartmentalization. The decreased levels of free IAA in PILS2 and PILS5 overexpressors and increase of auxin conjugates in double mutant *pils2pils5* are reminiscent of similarly altered PIN5 activity (Barbez et al. 2012).

## 4 Rectification: The Oxidation of Auxin Irreversibly Terminates Auxin Transport and Signaling

Ultimately, auxin that has been transported from cell to cell must be redirected or catabolized to terminate response processes in destination cells. Auxin can be reversibly conjugated for temporary inactivation or can be eliminated from the system via irreversible catabolism (5.4). The metabolites oxIAA and oxIAA-hexose (oxIAA-Hex) are the major degradation products of IAA and are not transported in polar streams (Östin et al. 1998; Kai et al. 2007; Novák et al. 2012; Kubeš et al. 2012). These oxidation products are terminally inactivated and no longer induce the expression of auxin-responsive genes, as tested with auxin-inducible reporters DR5rev:GFP and 2XD0:GUS (Peer et al. 2013). However, addition of oxIAA does activate IAA transport mediated by ABCB1 and 4 (Geisler et al. 2005; Kubeš et al. 2012; Peer et al. 2013).

## 5 Conclusions

In conclusion, the phytohormone auxin is found to function in a tremendously complex system that requires a profusion of interacting proteins and specifically defined lipid environments to be synthesized, perceived, and (particularly) transported. The level of refinement required to transport auxin distinctly for diverse plans such as polar gradients, organogenic concentrations, long-distance sinks, and intracellular sequestration is staggering. While we have a conceptual blueprint of auxin cellular transport to guide our research, the complexity of the cellular transport systems makes direct measurements difficult. Compartmentalization of auxin has an unaccounted influence on effective concentrations that will greatly impact the results of intercellular transport models.

Compared to the well-defined ATP-driven export activity of ABCB proteins and the $H^+$ symport activity of AUX1/LAX proteins, PIN function remains ill defined at the molecular level. Currently, there exist no crystal structures for PIN proteins or their close homologues, and the best conceptual models available are inspired by distantly related microbial transporters (Galvan-Ampudia and Offringa 2007; Peer and Murphy 2007). Determining the true structure of PIN proteins may shed some light on putative substrates capable of maintaining a sufficient gradient for this purpose.

## References

Aller SG, Yu J, Ward A, Weng Y, Chittaboina S, Zhuo R, Harrell PM, Trinh YT, Zhang Q, Urbatsch IL, Chang G (2009) Structure of P-glycoprotein reveals a molecular basis for polyspecific drug binding. Science 323:1718–1722

Bailly A, Sovero V, Vincenzetti V, Santelia D, Bartnik D, Koenig BW, Mancuso S, Martinoia E, Geisler M (2008) Modulation of P-glycoproteins by auxin transport inhibitors is mediated by interaction with immunophilins. J Biol Chem 283:21817–21826

Bailly A, Yang H, Martinoia E, Geisler M, Murphy AS (2012) Plant lessons: exploring ABCB functionality through structural modeling. Front Plant Sci 2:108

Bainbridge K, Guyomarch S, Bayer E, Swarup R, Bennett M, Mandel T, Kuhlemeier C (2008) Auxin influx carriers stabilize phyllotactic patterning. Genes Dev 22:810–823

Bandyopadhyay A, Blakeslee JJ, Lee OR, Mravec J, Sauer M, Titapiwatanakun B, Makam SN, Bouchard R, Geisler M, Martinoia E, Friml J, Peer WA, Murphy AS (2007) Interactions of PIN and PGP auxin transport mechanisms. Biochem Soc Trans 35:137–141

Barbez E, Kubeš M, Rolčík J, Béziat C, Pěnčík A, Wang B, Rosquete MR, Zhu J, Dobrev PI, Lee Y, Zažímalová E, Petrášek J, Geisler M, Friml J, Kleine-Vehn J (2012) A novel putative auxin carrier family regulates intracellular auxin homeostasis in plants. Nature 485:119–122

Bayer EM, Smith RS, Mandel T, Nakayama N, Sauer M, Prusinkiewicz P, Kuhlemeier C (2009) Integration of transport-based models for phyllotaxis and midvein formation. Genes Dev 23:373–384

Bender RL, Fekete ML, Klinkenberg PM, Hampton M, Bauer B, Malecha M, Lindgren K, Maki JA, Perera MA, Nikolau BJ, Carter CJ (2013) PIN6 is required for nectary auxin response and short stamen development. Plant J 74:893–904

Benjamins R, Scheres B (2008) Auxin: the looping star in plant development. Annu Rev Plant Biol 59:443–465

Benjamins R, Quint A, Weijers D, Hooykaas P, Offringa R (2001) The PINOID protein kinase regulates organ development in Arabidopsis by enhancing polar auxin transport. Development 128:4057–4067

Benkova E, Michniewicz M, Sauer M, Teichmann T, Seifertova D, Jurgens G, Friml J (2003) Local, efflux-dependent auxin gradients as a common module for plant organ formation. Cell 115:591–602

Bhalerao RP, Eklof J, Ljung K, Marchant A, Bennett M, Sandberg G (2002) Shoot-derived auxin is essential for early lateral root emergence in Arabidopsis seedlings. Plant J 29:325–332

Blakeslee JJ, Peer WA, Murphy AS (2005) Auxin transport. Curr Opin Plant Biol 8:494–500

Blakeslee JJ, Bandyopadhyay A, Lee OR, Mravec J, Titapiwatanakun B, Sauer M, Makam SN, Cheng Y, Bouchard R, Adamec J, Geisler M, Nagashima A, Sakai T, Martinoia E, Friml J, Peer WA, Murphy AS (2007) Interactions among PIN-FORMED and P-glycoprotein auxin transporters in Arabidopsis. Plant Cell 19:131–147

Blilou I, Xu J, Wildwater M, Willemsen V, Paponov I, Friml J, Heidstra R, Aida M, Palme K, Scheres B (2005) The PIN auxin efflux facilitator network controls growth and patterning in Arabidopsis roots. Nature 433:39–44

Bouchard R, Bailly A, Blakeslee JJ, Oehring SC, Vincenzetti V, Lee OR, Paponov I, Palme K, Mancuso S, Murphy AS, Schulz B, Geisler M (2006) Immunophilin-like TWISTED DWARF1 modulates auxin efflux activities of Arabidopsis P-glycoproteins. J Biol Chem 281:30603–30612

Carland FM, Fujioka S, Takatsuto S, Yoshida S, Nelson T (2002) The identification of CVP1 reveals a role for sterols in vascular patterning. Plant Cell 14:2045–2058

Chao DY, Gable K, Chen M, Baxter I, Dietrich CR, Cahoon EB, Guerinot ML, Lahner B, Lu S, Markham JE, Morrissey J, Han G, Gupta SD, Harmon JM, Jaworski JG, Dunn TM, Salt DE (2011) Sphingolipids in the root play an important role in regulating the leaf ionome in Arabidopsis thaliana. Plant Cell 23:1061–1081

Chen R, Hilson P, Sedbrook J, Rosen E, Caspar T, Masson PH (1998) The Arabidopsis thaliana AGRAVITROPIC 1 gene encodes a component of the polar-auxin-transport efflux carrier. Proc Natl Acad Sci 95:15112–15117

Cheng Y, Dai X, Zhao Y (2006) Auxin biosynthesis by the YUCCA flavin monooxygenases controls the formation of floral organs and vascular tissues in Arabidopsis. Genes Dev 20:1790–1799

Cheng Y, Dai X, Zhao Y (2007) Auxin synthesized by the YUCCA flavin monooxygenases is essential for embryogenesis and leaf formation in Arabidopsis. Plant Cell 19:2430–2439

Cho M, Lee SH, Cho HT (2007) P-Glycoprotein4 displays auxin efflux transporter like action in Arabidopsis root hair cells and tobacco cells. Plant Cell 19:3930–3943

Cholodny N (1927) Wuchshormone und Tropismen bei den Pflanzen. Biol Zent Bl 47:604–626

Christie JM, Yang H, Richter GL, Sullivan S, Thomson CE, Lin J, Titapiwatanakun B, Ennis M, Kaiserli E, Lee OR, Adamec J, Peer WA, Murphy AS (2011) phot1 inhibition of ABCB19 primes lateral auxin fluxes in the shoot apex required for phototropism. PLoS Biol 9:e1001076

Dai X, Mashiguchi K, Chen Q, Kasahara H, Kamiya Y, Ojha S, DuBois J, Ballou D, Zhao Y (2013) The biochemical mechanism of auxin biosynthesis by an Arabidopsis YUCCA flavin-containing monooxygenase. J Biol Chem 288:1448–1457

Dal Bosco C, Dovzhenko A, Liu X, Woerner N, Rensch T, Eismann M, Eimer S, Hegermann J, Paponov IA, Ruperti B, Heberle-Bors E, Touraev A, Cohen JD, Palme K (2012) The endoplasmic reticulum localized PIN8 is a pollen-specific auxin carrier involved in intracellular auxin homeostasis. Plant J 71:860–870

Darwin C (1880) The power of movement in plants. John Murray, London, p 474

Dawson RJ, Locher KP (2006) Structure of a bacterial multidrug ABC transporter. Nature 443:180–185

Del Pozo JC, Boniotti MB, Gutierrez C (2002) Arabidopsis E2Fc functions in cell division and is degraded by the ubiquitin-SCF(AtSKP2) pathway in response to light. Plant Cell 14:3057–3071

Del Pozo JC, Diaz-Trivino S, Cisneros N, Gutierrez C (2006) The balance between cell division and endoreplication depends on E2FC- DPB, transcription factors regulated by the ubiquitin–SCFSKP2A pathway in Arabidopsis. Plant Cell 18:2224–2235

Dharmasiri N, Dharmasiri S, Estelle M (2005) The F-box protein TIR1 is an auxin receptor. Nature 435:441–445

Dharmasiri S, Swarup R, Mockaitis K, Dharmasiri N, Singh SK, Kowalchyk M, Marchant A, Mills S, Sandberg G, Bennett MJ, Estelle M (2006) AXR4 is required for localization of the auxin influx facilitator AUX1. Science 312:1218–1220

Dhonukshe P, Huang F, Galvan-Ampudia CS, Mahonen AP, Kleine-Vehn J, Xu J, Quint A, Prasad K, Friml J, Scheres B, Offringa R (2010) Plasma membrane-bound AGC3 kinases phosphorylate PIN auxin carriers at TPRXS(N/S) motifs to direct apical PIN recycling. Development 137:3245–3255

Ding Z, Wang B, Moreno I, Duplakova N, Simon S, Carraro N, Reemmer J, Pencik A, Chen X, Tejos R, Skupa P, Pollmann S, Mravec J, Petrasek J, Zazimalova E, Honys D, Rolcik J, Murphy A, Orellana A, Geisler M, Friml J (2012) ER-localized auxin transporter PIN8 regulates auxin homeostasis and male gametophyte development in Arabidopsis. Nat Commun 3:941

Dudler R, Hertig C (1992) Structure of an mdr-like gene from Arabidopsis thaliana: evolutionary implications. J Biol Chem 267:5882–5888

Esmon CA, Tinsley AG, Ljung K, Sandberg G, Hearne LB, Liscum E (2006) A gradient of auxin and auxin-dependent transcription precedes tropic growth responses. Proc Natl Acad Sci USA 103:236–241

Friml J, Wisniewska J, Benkova E, Mendgen K, Palme K (2002) Lateral relocation of auxin efflux regulator PIN3 mediates tropism in Arabidopsis. Nature 415:806–809

Friml J, Vieten A, Sauer M, Weijers D, Schwarz H, Hamann T, Offringa R, Jürgens G (2003) Efflux-dependent auxin gradients establish the apical-basal axis of Arabidopsis. Nature 426:147–153

Friml J, Yang X, Michniewicz M, Weijers D, Quint A, Tietz O, Benjamins R, Ouwerkerk PB, Ljung K, Sandberg G, Hooykaas PJ, Palme K, Offringa R (2004) A PINOID-dependent binary switch in apical-basal PIN polar targeting directs auxin efflux. Science 306:862–865

Fu Y, Gu Y, Zheng Z, Wasteneys G, Yang Z (2005) Arabidopsis interdigitating cell growth requires two antagonistic pathways with opposing action on cell morphogenesis. Cell 120:687–700

Fu Y, Xu T, Zhu L, Wen M, Yang Z (2009) A ROP GTPase signaling pathway controls cortical microtubule ordering and cell expansion in Arabidopsis. Curr Biol 19:1827–1832

Galvan-Ampudia CS, Offringa R (2007) Plant evolution: AGC kinases tell the auxin tale. Trends Plant Sci 12:541–547

Ganguly A, Lee SH, Cho M, Lee OR, Yoo H, Cho HT (2010) Differential auxin-transporting activities of PIN-FORMED proteins in Arabidopsis root hair cells. Plant Physiol 153:1046–1061

Geisler M, Kolukisaoglu HU, Bouchard R, Billion K, Berger J, Saal B, Frangne N, Koncz-Kalman Z, Koncz C, Dudler R, Blakeslee JJ, Murphy AS, Martinoia E, Schulz B (2003) TWISTED DWARF1, a unique plasma membrane-anchored immunophilin-like protein, interacts with Arabidopsis multidrug resistance-like transporters AtPGP1 and AtPGP19. Mol Biol Cell 14:4238–4249

Geisler M, Blakeslee JJ, Bouchard R, Lee OR, Vincenzetti V, Bandyopadhyay A, Titapiwatanakun B, Peer WA, Bailly A, Richards EL, Ejendal KF, Smith AP, Baroux C, Grossniklaus U, Muller A, Hrycyna CA, Dudler R, Murphy AS, Martinoia E (2005) Cellular efflux of auxin catalyzed by the Arabidopsis MDR/PGP transporter AtPGP1. Plant J 44:179–194

Goto N, Starke M, Kranz AR (1987) Effect of gibberellins on flower development of the pin-formed mutant of *Arabidopsis thaliana*. Arabidopsis Inf Serv 23:66–71

Grunewald W, Friml J (2010) The march of the PINs: developmental plasticity by dynamic polar targeting in plant cells. EMBO J 29:2700–2714

Guilfoyle TJ, Suzich JA, Lindberg M (1986) Synthesis of 5S rRNA and putative tRNAs in nuclei isolated from wheat embryos. Plant Mol Biol 7:95–104

Hager A (2003) Role of the plasma membrane H+-ATPase in auxin-induced elongation growth: historical and new aspects. J Plant Res 116:483–505

Harper RM, Stowe-Evans EL, Luesse DR, Muto H, Tatematsu K, Watahiki MK, Yamamoto K, Liscum E (2000) The NPH4 locus encodes the auxin response factor ARF7, a conditional regulator of differential growth in aerial Arabidopsis tissue. Plant Cell 12:757–770

Henrichs S, Wang B, Fukao Y, Zhu J, Charrier L, Bailly A, Oehring SC, Linnert M, Weiwad M, Endler A, Nanni P, Pollmann S, Mancuso S, Schulz A, Geisler M (2012) Regulation of ABCB1/PGP1-catalysed auxin transport by linker phosphorylation. EMBO J 31:2965–2980

Hopfner KP, Karcher A, Shin DS, Craig L, Arthur LM, Carney JP, Tainer JA (2000) Structural biology of Rad50 ATPase: ATP-driven conformational control in DNA double-strand break repair and the ABC-ATPase superfamily. Cell 101:789–800

Hosek P, Kubes M, Lankova M, Dobrev PI, Klima P, Kohoutova M, Petrasek J, Hoyerova K, Jirina M, Zazimalova E (2012) Auxin transport at cellular level: new insights supported by mathematical modelling. J Exp Bot 63:3815–3827

Huang F, Zago MK, Abas L, van Marion A, Galvan-Ampudia CS, Offringa R (2010) Phosphorylation of conserved PIN motifs directs Arabidopsis PIN1 polarity and auxin transport. Plant Cell 22:1129–1142

Jones AR, Kramer EM, Knox K, Swarup R, Bennett MJ, Lazarus CM, Leyser HM, Grierson CS (2009) Auxin transport through non-hair cells sustains root-hair development. Nat Cell Biol 11:78–84

Jonsson H, Heisler MG, Shapiro BE, Meyerowitz EM, Mjolsness E (2006) An auxin-driven polarized transport model for phyllotaxis. Proc Natl Acad Sci USA 103:1633–1638

Jurado S, Diaz-Trivino S, Abraham Z, Manzano C, Gutierrez C, del Pozo C (2008) SKP2A, an F-box protein that regulates cell division, is degraded via the ubiquitin pathway. Plant J 53:828–841

Jurado S, Abraham Z, Manzano C, Lopez-Torrejon G, Pacios LF, Del Pozo JC (2010) The Arabidopsis cell cycle F-box protein SKP2A binds to auxin. Plant Cell 22:3891–3904

Kai K, Horita J, Wakasa K, Miyagawa H (2007) Three oxidative metabolites of indole-3-acetic acid from Arabidopsis thaliana. Phytochemistry 68:1651–1663

Kamimoto Y, Terasaka K, Hamamoto M, Takanashi K, Fukuda S, Shitan N, Sugiyama A, Suzuki H, Shibata D, Wang B, Pollmann S, Geisler M, Yazaki K (2012) Arabidopsis ABCB21 is a facultative auxin importer/exporter regulated by cytoplasmic auxin concentration. Plant Cell Physiol 53:2090–2100

Kamphausen T, Fanghänel J, Neumann D, Schulz B, Rahfeld JU (2002) Characterization of Arabidopsis thaliana AtFKBP42 that is membrane-bound and interacts with Hsp90. Plant J 32:263–276

Katekar GK, Geissler AE (1977) Auxin transport inhibitors. III. Chemical requirements of a class of auxin transport inhibitors. Plant Physiol 60:826–829

Kepinski S, Leyser O (2005) The Arabidopsis F-box protein TIR1 is an auxin receptor. Nature 435:446–451

Kim JY, Henrichs S, Bailly A, Vincenzetti V, Sovero V, Mancuso S, Pollmann S, Kim D, Geisler M, Nam HG (2010) Identification of an ABCB/P-glycoprotein-specific inhibitor of auxin transport by chemical genomics. J Biol Chem 285:23309–23317

Kleine-Vehn J, Dhonukshe P, Swarup R, Bennett M, Friml J (2006) Subcellular trafficking of the Arabidopsis auxin influx carrier AUX1 uses a novel pathway distinct from PIN1. Plant Cell 18:3171–3181

Knöller AS, Blakeslee JJ, Richards EL, Peer WA, Murphy AS (2010) Brachytic2/ZmABCB1 functions in IAA export from intercalary meristems. J Exp Bot 61:3689–3696

Kögl F, Haagen-Smith AJ (1931) Über die Chemie des Wuchsstoffs K. Akad. Wetenschap. Amsterdam. Proc Sect Sci 34:1411–1416

Kramer EM (2004) PIN and AUX/LAX proteins: their role in auxin accumulation. Trends Plant Sci 9:578–582

Kubeš M, Yang H, Richter GL, Cheng Y, Mlodzinska E, Wang X, Blakeslee JJ, Carraro N, Petrasek J, Zazimalova E, Hoyerova K, Peer WA, Murphy AS (2012) The Arabidopsis concentration- dependent influx/efflux transporter ABCB4 regulates cellular auxin levels in the root epidermis. Plant J 69:640–654

Langowski L, Ruzicka K, Naramoto S, Kleine-Vehn J, Friml J (2010) Trafficking to the outer polar domain defines the root-soil interface. Curr Biol 20:904–908

Lee M, Choi Y, Burla B, Kim YY, Jeon B, Maeshima M, Yoo JY, Martinoia E, Lee Y (2008) The ABC transporter AtABCB14 is a malate importer and modulates stomatal response to CO2. Nat Cell Biol 10:1217–1223

Leng ER, Vineyard ML (1951) Dwarf and short plants. Maize Genet Coop Newsl 25:31–32

Lewis DR, Miller ND, Splitt BL, Wu G, Spalding EP (2007) Separating the roles of acropetal and basipetal auxin transport on gravitropism with mutations in two Arabidopsis multidrug resistance-like ABC transporter genes. Plant Cell 19:1838–1850

Lin R, Wang H (2005) Two homologous ATP-binding cassette transporter proteins, AtMDR1 and AtPGP1, regulate Arabidopsis photomorphogenesis and root development by mediating polar auxin transport. Plant Physiol 138:949–964

Lin D, Cao L, Zhou Z, Zhu L, Ehrhardt D, Yang Z, Fu Y (2013) Rho GTPase signaling activates microtubule severing to promote microtubule ordering in Arabidopsis. Curr Biol 23:290–297

Liscum E, Briggs WR (1995) Mutations in the NPH1 locus of Arabidopsis disrupt the perception of phototropic stimuli. Plant Cell 7:473–485

Ljung K, Östin A, Lioussanne L, Sandberg G (2001) Developmental regulation of indole-3-acetic acid turnover in Scots pine seedlings. Plant Physiol 125:464–475

Ljung K, Hull AK, Celenza J, Yamada M, Estelle M, Normanly J, Sandberg G (2005) Sites and regulation of auxin biosynthesis in Arabidopsis roots. Plant Cell 17:1090–1104

Löbler M, Klämbt D (1985) Auxin-binding protein from coleoptile membranes of corn (Zea mays L.). I. Purification by immunological methods and characterization. J Biol Chem 260:9848–9853

Luckie DB, Wilterding JH, Krha M, Krouse ME (2003) CFTR and MDR: ABC transporters with homologous structure but divergent function. Curr Genomics 4:225–235

Marchant A, Bhalerao R, Casimiro I, Eklöf J, Casero PJ, Bennett M, Sandberg G (2002) AUX1 promotes lateral root formation by facilitating indole-3-acetic acid distribution between sink and source tissues in the Arabidopsis seedling. Plant Cell 14:589–597

Markham JE, Molino D, Gissot L, Bellec Y, Hematy K, Marion J, Belcram K, Palauqui JC, Satiat-Jeunemaitre B, Faure JD (2011) Sphingolipids containing very-long-chain fatty acids define a secretory pathway for specific polar plasma membrane protein targeting in Arabidopsis. Plant Cell 23:2362–2378

Matsuda S, Kajizuka T, Kadota A, Nishimura T, Koshiba T (2011) NPH3- and PGP-like genes are exclusively expressed in the apical tip region essential for blue-light perception and lateral auxin transport in maize coleoptiles. J Exp Bot 62:3459–3466

Men S, Boutte Y, Ikeda Y, Li X, Palme K, Stierhof YD, Hartmann MA, Moritz T, Grebe M (2008) Sterol-dependent endocytosis mediates post-cytokinetic acquisition of PIN2 auxin efflux carrier polarity. Nat Cell Biol 10:237–244

Merks RMH, Van de Peer Y, Inzé D, Beemster GTS (2007) Canalization without flux sensors: a traveling-wave hypothesis. Trends Plant Sci 12:384–390

Michniewicz M, Zago MK, Abas L, Weijers D, Schweighofer A, Meskiene I, Heisler MG, Ohno C, Zhang J, Huang F, Schwab R, Weigel D, Meyerowitz EM, Luschnig C, Offringa R, Friml J (2007) Antagonistic regulation of PIN phosphorylation by PP2A and PINOID directs auxin flux. Cell 130:1044–1056

Mitchison GJ (1980) A model for vein formation in higher plants. Proc R Soc Lond B Boil Sci 207:79–109

Mravec J, Kubes M, Bielach A, Gaykova V, Petrasek J, Skupa P, Chand S, Benkova E, Zazımalova E, Friml J (2008) Interaction of PIN and PGP transport mechanisms in auxin distribution-dependent development. Development 135:3345–3354

Mravec J, Skůpa P, Bailly A, Hoyerová K, Křeček P, Bielach A, Petrášek J, Zhang J, Gaykova V, Stierhof YD, Dobrev PI, Schwarzerová K, Rolčík J, Seifertová D, Luschnig C, Benkova E, Zažímalová E, Geisler M, Friml J (2009) Subcellular homeostasis of phytohormone auxin is mediated by the ER-localized PIN5 transporter. Nature 459:1136–1140

Müller A, Guan C, Galweiler L, Tanzler P, Huijser P, Marchant A, Parry G, Bennett M, Wisman E, Palme K (1998) AtPIN2 defines a locus of Arabidopsis for root gravitropism control. EMBO J 17:6903–6911

Mullet JE, Klein RR, Klein PE (2002) Sorghum bicolor—an important species for comparative grass genomics and a source of beneficial genes for agriculture. Curr Opin Plant Biol 5:118–121

Multani DS, Briggs SP, Chamberlin MA, Blakeslee JJ, Murphy AS, Johal GS (2003) Loss of an MDR transporter in compact stalks of maize br2 and sorghum dw3 mutants. Science 302:81–84

Murphy AS, Hoogner KR, Peer WA, Taiz L (2002) Identification, purification, and molecular cloning of N-1-naphthylphthalmic acid-binding plasma membrane-associated aminopeptidases from Arabidopsis. Plant Physiol 128:935–950

Nagashima A, Suzuki G, Uehara Y, Saji K, Furukawa T, Koshiba T, Sekimoto M, Fujioka S, Kuroha T, Kojima M, Sakakibara H, Fujisawa N, Okada K, Sakai T (2008) Phytochromes and cryptochromes regulate the differential growth of Arabidopsis hypocotyls in both a PGP19-dependent and a PGP19-independent manner. Plant J 53:516–529

Noh B, Murphy AS, Spalding EP (2001) Multidrug resistance-like genes of Arabidopsis required for auxin transport and auxin-mediated development. Plant Cell 13:2441–2454

Noh B, Bandyopadhyay A, Peer WA, Spalding EP, Murphy AS (2003) Enhanced gravi- and phototropism in plant mdr mutants mislocalizing the auxin efflux protein PIN1. Nature 423:999–1002

Novák O, Hényková E, Sairanen I, Kowalczyk M, Pospíšil T, Ljung K (2012) Tissue-specific profiling of the Arabidopsis thaliana auxin metabolome. Plant J 72:523–536

Östin A, Kowalyczk M, Bhalerao RP, Sandberg G (1998) Metabolism of indole-3-acetic acid in Arabidopsis. Plant Physiol 118:285–296

Pan J, Fujioka S, Peng J, Chen J, Li G, Chen R (2009) The E3 ubiquitin ligase SCFTIR1/AFB and membrane sterols play key roles in auxin regulation of endocytosis, recycling, and plasma membrane accumulation of the auxin efflux transporter PIN2 in Arabidopsis thaliana. Plant Cell 21:568–580

Parry G, Delbarre A, Marchant A, Swarup R, Napier R, Perrot-Rechenmann C, Bennett MJ (2001) Novel auxin transport inhibitors phenocopy the auxin influx carrier mutation aux1. Plant J 25:399–406

Peer WA, Murphy AS (2007) Flavonoids and auxin transport: modulators or regulators? Trends Plant Sci 12:556–563
Peer WA, Blakeslee JJ, Yang H, Murphy AS (2011) Seven things we think we know about auxin transport. Mol Plant 4:487–504
Peer WA, Cheng Y, Murphy AS (2013) Evidence of oxidative attenuation of auxin signalling. J Exp Bot 64:2629–2639
Peret B, Swarup K, Ferguson A, Seth M, Yang Y, Dhondt S, James N, Casimiro I, Perry P, Syed A, Yang H, Reemmer J, Venison E, Howells C, Perez-Amador MA, Yun J, Alonso J, Beemster GT, Laplaze L, Murphy A, Bennett MJ, Nielsen E, Swarup R (2012) AUX/LAX genes encode a family of auxin influx transporters that perform distinct functions during Arabidopsis development. Plant Cell 24:2874–2885
Petersson SV, Johansson AI, Kowalczyk M, Makoveychuk A, Wang JY, Moritz T, Grebe M, Benfey PN, Sandberg G, Ljung K (2009) An auxin gradient and maximum in the Arabidopsis root apex shown by high-resolution cell-specific analysis of IAA distribution and synthesis. Plant Cell 21:1659–1668
Procko E, O'Mara ML, Bennett WFD, Tieleman DP, Gaudet R (2009) The mechanism of ABC transporters: general lessons from structural and functional studies of an antigenic peptide transporter. FASEB J 23:1287–1302
Radwanski ER, Last RL (1995) Tryptophan biosynthesis and metabolism: biochemical and molecular genetics. Plant Cell 7:921–934
Rahman A, Takahashi M, Shibasaki K, Wu S, Inaba T, Tsurumi S, Baskin TI (2010) Gravitropism of Arabidopsis thaliana roots requires the polarization of PIN2 toward the root tip in meristematic cortical cells. Plant Cell 22:1762–1776
Raven JA (1975) Transport of indoleacetic acid in plant cells in relation to pH and electrical potential gradients, and its significance for polar IAA transport. New Phytol 74:163–172
Reinhardt D, Pesce ER, Stieger P, Mandel T, Baltensperger K, Bennett M, Traas J, Friml J, Kuhlemeier C (2003) Regulation of phyllotaxis by polar auxin transport. Nature 426:255–260
Robert S, Kleine-Vehn J, Barbez E, Sauer M, Paciorek T, Baster P, Vanneste S, Zhang J, Simon S, Čovanová M, Hayashi K, Dhonukshe P, Yang Z, Bednarek SY, Jones AM, Luschnig C, Aniento F, Zažimalová E, Friml J (2010) ABP1 mediates auxin inhibition of clathrin-dependent endocytosis in Arabidopsis. Cell 143:111–121
Roudier F, Gissot L, Beaudoin F, Haslam R, Michaelson L, Marion J, Molino D, Lima A, Bach L, Morin H, Tellier F, Palauqui JC, Bellec Y, Renne C, Miquel M, Dacosta M, Vignard J, Rochat C, Markham JE, Moreau P, Napier J, Faure JD (2010) Very-long-chain fatty acids are involved in polar auxin transport and developmental patterning in Arabidopsis. Plant Cell 22:364–375
Rubery PH, Sheldrake AR (1974) Carrier-mediated auxin transport. Planta 118:101–121
Ruzicka K, Strader LC, Bailly A, Yang H, Blakeslee J, Langowski L, Nejedla E, Fujita H, Itoh H, Syono K, Hejatko J, Gray WM, Martinoia E, Geisler M, Bartel B, Murphy AS, Friml J (2010) Arabidopsis PIS1 encodes the ABCG37 transporter of the auxinic compounds including the auxin precursor indole-3-butyric acid. Proc Natl Acad Sci USA 107:10749–10753
Sachs T (1969) Polarity and the induction of organized vascular tissues. Ann Bot Lond 33:263
Sakai T, Kagawa T, Kasahara M, Swartz TE, Christie JM, Briggs WR, Wada M, Okada K (2001) Arabidopsis nph1 and npl1: Blue light receptors that mediate both phototropism and chloroplast relocation. Proc Natl Acad Sci USA 98:6969–6974
Santelia D, Vincenzetti V, Azzarello E, Bovet L, Fukao Y, Duchtig P, Mancuso S, Martinoia E, Geisler M (2005) MDR-like ABC transporter AtPGP4 is involved in auxin-mediated lateral root and root hair development. FEBS Lett 579:5399–5406
Santner AA, Watson JC (2006) The WAG1 and WAG2 protein kinases negatively regulate root waving in Arabidopsis. Plant J 45:752–764
Sauer M, Balla J, Luschnig C, Wisniewska J, Reinohl V, Friml J, Benkova E (2006) Canalization of auxin flow by Aux/IAA-ARF-dependent feedback regulation of PIN polarity. Genes Dev 20:2902–2911

Sawchuk MG, Edgar A, Scarpella E (2013) Patterning of leaf vein networks by convergent auxin transport pathways. PLoS Genet 9:e1003294

Scarpella E, Marcos D, Friml J, Berleth T (2006) Control of leaf vascular patterning by polar auxin transport. Genes Dev 20:1015–1027

Shitan N, Bazin I, Dan K, Obata K, Kigawa K, Ueda K, Sato F, Forestier C, Yazaki K (2003) Involvement of CjMDR1, a plant multidrug-resistance-type ATP-binding cassette protein, in alkaloid transport in Coptis japonica. Proc Natl Acad Sci USA 100:751–756

Sidler M, Hassa P, Hasan S, Ringli C, Dudler R (1998) Involvement of an ABC transporter in a developmental pathway regulating hypocotyl cell elongation in the light. Plant Cell 10:1623–1636

Smith RS, Guyomarch S, Mendel T, Reinhardt D, Kuhlemeier C (2006) A plausible model of phyllotaxis. Proc Natl Acad Sci 103:1301–1306

Stein OL (1955) Rates of leaf initiation in two mutants of Zea mays, dwarf1 and brachytic2. Am J Bot 42:885–892

Stepanova AN, Robertson-Hoyt J, Yun J, Benavente LM, Xie DY, Dolezal K, Schlereth A, Jurgens G, Alonso JM (2008) TAA1-mediated auxin biosynthesis is essential for hormone crosstalk and plant development. Cell 133:177–191

Stoma S, Lucas M, Chopard J, Schaedel M, Traas J, Godin C (2008) Flux-based transport enhancement as a plausible unifying mechanism for auxin transport in meristem development. PLoS Comput Biol 4:e1000207

Sugawara S, Hishiyama S, Jikumaru Y, Hanada A, Nishimura T, Koshiba T, Zhao Y, Kamiya Y, Kasahara H (2009) Biochemical analyses of indole-3-acetaldoxime-dependent auxin biosynthesis in Arabidopsis. Proc Natl Acad Sci USA 106:5430–5435

Swarup R, Friml J, Marchant A, Ljung K, Sandberg G, Palme K, Bennett M (2001) Localization of the auxin permease AUX1 suggests two functionally distinct hormone trans- port pathways operate in the Arabidopsis root apex. Genes Dev 15:2648–2653

Swarup K, Benkova E, Swarup R, Casimiro I, Peret B, Yang Y, Parry G, Nielsen E, De Smet I, Vanneste S, Levesque MP, Carrier D, James N, Calvo V, Ljung K, Kramer E, Roberts R, Graham N, Marillonnet S, Patel K, Jones JD, Taylor CG, Schachtman DP, May S, Sandberg G, Benfey P, Friml J, Kerr I, Beeckman T, Laplaze L, Bennett MJ (2008) The auxin influx carrier LAX3 promotes lateral root emergence. Nat Cell Biol 10:946–954

Tanaka K, Fujimura-Kamada K, Yamamoto T (2011) Functions of phospholipid flippases. J Biochem 149:131–143

Tao Y, Ferrer JL, Ljung K, Pojer F, Hong F, Long JA, Li L, Moreno JE, Bowman ME, Ivans LJ, Cheng Y, Lim J, Zhao Y, Ballare CL, Sandberg G, Noel JP, Chory J (2008) Rapid synthesis of auxin via a new tryptophan-dependent pathway is required for shade avoidance in plants. Cell 133:164–176

Tarling EJ, Edwards PA (2011) ATP binding cassette transporter G1 (ABCG1) is an intracellular sterol transporter. Proc Natl Acad Sci USA 108:19719–19724

Teale WD, Paponov IA, Palme K (2006) Auxin in action: signalling, transport and the control of plant growth and development. Nat Rev Mol Cell Biol 7:847–859

Terasaka K, Blakeslee JJ, Titapiwatanakun B, Peer WA, Bandyopadhyay A, Makam SN, Lee OR, Richards EL, Murphy AS, Sato F, Yazaki K (2005) PGP4, an ATP binding cassette P-glycoprotein, catalyzes auxin transport in Arabidopsis thaliana roots. Plant Cell 17:2922–2939

Theologis A, Huynh TV, Davis RW (1985) Rapid induction of specific mRNAs by auxin in pea epicotyl tissue. J Mol Biol 183:53–68

Titapiwatanakun B, Blakeslee JJ, Bandyopadhyay A, Yang H, Mravec J, Sauer M, Cheng Y, Adamec J, Nagashima A, Geisler M, Sakai T, Friml J, Peer WA, Murphy AS (2009) ABCB19/PGP19 stabilises PIN1 in membrane microdomains in Arabidopsis. Plant J 57:27–44

Vernoux T, Besnard F, Traas J (2010) Auxin at the shoot apical meristem. Cold Spring Harb Perspect Biol 2:a001487

Verrier PJ, Bird D, Burla B, Dassa E, Forestier C, Geisler M, Klein M, Kolukisaoglu U, Lee Y, Martinoia E, Murphy A, Rea PA, Samuels L, Schulz B, Spalding EJ, Yazaki K, Theodoulou FL (2008) Plant ABC proteins-unified nomenclature and updated inventory. Trends Plant Sci 13:151–159

Vieten A, Vanneste S, Wisniewska J, Benková E, Benjamins R, Beeckman T, Luschnig C, Friml J (2005) Functional redundancy of PIN proteins is accompanied by auxin-dependent cross-regulation of PIN expression. Development 132:4521–4531

Wang B, Bailly A, Zwiewka M, Henrichs S, Azzarello E, Mancuso S, Maeshima M, Friml J, Schulz A, Geisler M (2013) Arabidopsis TWISTED DWARF1 functionally interacts with auxin exporter ABCB1 on the root plasma membrane. Plant Cell 25:202–214

Went FW, Thimann KV (1937) Phytohormones. MacMillan, New York

Willemsen V, Friml J, Grebe M, van den Toorn A, Palme K, Scheres B (2003) Cell polarity and PIN protein positioning in Arabidopsis require STEROL METHYLTRANSFERASE1 function. Plant Cell 15:612–625

Willige BC, Ahlers S, Zourelidou M, Barbosa ICR, Demarsy E, Trevisan M, Davis PA, Roelfsema MRG, Hangarter R, Fankhauser C, Schwechheimer C (2013) D6PK AGCVIII kinases are required for auxin transport and phototropic hypocotyl bending in Arabidopsis. Plant Cell 25:1674–1688

Wu G, Lewis DR, Spalding EP (2007) Mutations in Arabidopsis multidrug resistance-like ABC transporters separate the roles of acropetal and basipetal auxin transport in lateral root development. Plant Cell 19:1826–1837

Wu G, Otegui MS, Spalding EP (2010) The ER-localized TWD1 immunophilin is necessary for localization of multidrug resistance-like proteins required for polar auxin transport in Arabidopsis roots. Plant Cell 22:3295–3304

Yamada M, Greenham K, Prigge MJ, Jensen PJ, Estelle M (2009) The TRANSPORT INHIBITOR RESPONSE2 gene is required for auxin synthesis and diverse aspects of plant development. Plant Physiol 151:168–179

Yang H, Murphy AS (2009) Functional expression and characterization of Arabidopsis ABCB, AUX 1 and PIN auxin transporters in Schizosaccharomyces pombe. Plant J 59:179–191

Yang Y, Hammes UZ, Taylor CG, Schachtman DP, Nielsen E (2006) High-affinity auxin transport by the AUX1 influx carrier protein. Curr Biol 16:1123–1127

Yang H, Richter GL, Wang X, Młodzińska E, Carraro N, Ma G, Jenness M, Chao D, Peer WA, Murphy AS (2013) Sterols and sphingolipids differentially function in trafficking of the Arabidopsis ABCB19 auxin transporter. Plant J 74:37–47

Zažímalová E, Murphy AS, Yang H, Hoyerová K, Hošek P (2010) Auxin transporters-why so many? Cold Spring Harb Perspect Biol 2:a001552

Zourelidou M, Müller I, Willige BC, Nill C, Jikumaru Y, Li H, Schwechheimer C (2009) The polarly localised D6 PROTEIN KINASE is required for efficient auxin transport in Arabidopsis thaliana. Development 136:627–636

# Chapter 6
# Auxin Receptors and Perception

Richard M. Napier

**Abstract** Auxin perception initiates auxin action and is the key step for controlled plant growth and development. The term perception implies two linked processes: auxin binding and the biological reaction to that binding event. In human terms we might describe these steps as recognition and interpretation. Auxin signalling has become a collective description for all the various mechanisms driving interpretation and often may be only indirectly and distantly connected to perception. Other chapters will describe how plants interpret auxin perception and the multitude of auxin-driven responses we recognise during development. In this chapter, we will examine what has been learnt about auxin-binding sites and link these observations to early signalling events, the immediate consequences of binding. Signal amplification progresses through, for example, genetic cascades will not be considered here because they are covered in other chapters, particularly under Auxin and Plant Development (see Chaps. 7–14). Signals linked to perception by biochemical events are considered briefly because these open up the discussion about the probability of additional classes of receptor, including ABP1.

## 1 Introduction

It is tempting to consider auxin to be only indole-3-acetic acid (IAA). Certainly, all known auxin responses can be initiated by IAA and it is ubiquitous in green plants. Auxin is also produced by algae and some bacteria, but responses to IAA in these organisms are either absent or ill-defined and this chapter will not deal further with how auxin might be perceived in these systems. Some other naturally occurring

R.M. Napier (✉)
School of Life Sciences, University of Warwick, Coventry CV4 7AL, UK
e-mail: Richard.Napier@warwick.ac.uk

E. Zažímalová et al. (eds.), *Auxin and Its Role in Plant Development*,
DOI 10.1007/978-3-7091-1526-8_6, © Springer-Verlag Wien 2014 

molecules, such as indole-3-butyric acid (IBA), may be functional auxins in some higher plant genera. Nevertheless, despite the huge diversity of known responses to auxin it appears that IAA is 'The Endogenous Auxin'. Future analytical science may prove this title to be unsustainable, but for the purposes of this review, the starting point will be that IAA is the auxin ligand. It is astonishing how one small molecule can initiate such a rich diversity of responses.

The Endogenous Auxin has been supplemented by many synthetic auxin agonists and a number of them are highly valued as agrochemicals. The pursuit of these compounds has provided a useful chemical biology arsenal and much structure–activity data to inform us about the site or sites into which IAA binds. This information will be considered in the context of a number of binding site models.

Auxin-binding proteins were sought for many years and associated with varying levels of speculation about their functionality before protein crystallography produced convincing evidence of auxin binding. Many putative auxin-binding proteins have been reported and then passed over (Napier et al. 2002). Few remain to claim the title of auxin receptor. Principal amongst these are Transport Inhibitor Resistant 1 (TIR1) and its paralogues, the Auxin F Box proteins (AFBs). Convincing proof of TIR1's activity as a receptor was presented in 2005 (Dharmasiri et al. 2005a, b; Kepinski and Leyser 2005) and 2 years later the crystal structure of TIR1 was solved with IAA in the binding pocket (Tan et al. 2007). Prior to this, the structure of Auxin-Binding Protein 1 (ABP1) had been solved with the auxin 1–naphthylacetic acid (1-NAA) bound in its binding pocket (Woo et al. 2002). More recently, a further F-Box protein (SKP2), distinct from TIR1, has been associated with auxin binding (Jurado et al. 2010), although no structural evidence is available yet.

With a diversity of potential auxin receptors there is clearly a platform from which the diversity of auxin responses may be explained. The different binding sites also offer one explanation for the wide dynamic range of different auxin responses. For example, typical responses in aerial tissues require micromolar concentrations of auxin, yet inhibition of primary root growth in seedlings (for example) peaks in the nanomolar range. Nevertheless, as noted above, much response diversity arises from the multitude of interactors for TIR1 and the AFBs. Is the AFB family of receptors sufficient to account for all responses to auxin? This chapter will focus on auxin binding and the resulting immediate early events and consider how the various auxin-binding proteins may contribute to responsiveness towards auxin.

## 2 TIR1 Is an Auxin Receptor: Discovery

With the development of Arabidopsis as a tool for molecular genetics came the description of a number of auxin-resistant mutants. As these were mapped and cloned it became clear that some of the genes were linked to ubiquitin-mediated protein breakdown (e.g. *AXR1*, an ubiquitin E2 conjugating enzyme; Leyser et al. 1993) and others were transcriptional regulators which were known to have rapid turnover times (the Aux/IAA proteins: Abel et al. 1994). It also became clear that many auxin responses were mediated by derepression of transcription and that an SCF-type ubiquitin E3 ligase complex was intimately involved in regulating these auxin responses (Ruegger et al. 1998). Auxin was shown to promote a reversible interaction between the E3 protein TIR1 and the degron domain of Aux/IAA proteins, but only in 2005 was it demonstrated that TIR1 was also the site to which auxin binds (Dharmasiri et al. 2005a; Kepinski and Leyser 2005). In these key papers affinity-tagged versions of either TIR1 or an Aux/IAA protein were shown to precipitate each other in the presence of IAA. More convincing still, purified TIR1 bound IAA reversibly in a radiolabel binding assay, and for the radiolabel assay to work, both TIR1 and Aux/IAA protein were needed. The competition dissociation curves suggested affinities for IAA of between 20–80 nM (Kepinski and Leyser 2005) and 84 nM (Dharmasiri et al. 2005a). These values were consistent with expectations from physiological assays. Differential affinities for a series of familiar auxins were also demonstrated. Coupled with the accumulating knowledge about components of the $SCF^{TIR1}$ complex and its regulation, a picture of the mechanism of action for TIR1 has been developed (Fig. 6.1).

## *2.1 The Structure of TIR1 and Auxin Binding*

Auxin binding is a precondition of Aux/IAA binding. By implication this is a two-step process, but binding activity has only been recorded in the presence of all three partners. Several studies now consider TIR1 and Aux/IAA protein to be auxin co-receptors given this mutual dependence (e.g. Calderón Villalobos et al. 2012). Therefore, although we might consider Aux/IAA binding as the start of signalling, its association with TIR1 and auxin needs to be considered part of perception. Turnover of the Aux/IAA pool then becomes the first step of signalling in this pathway.

The structure of TIR1 was solved soon after its description as a receptor (Tan et al. 2007). In order to obtain active protein, TIR1 was co-expressed with ASK1 in a eukaryotic system (baculovirus-induced expression in insect cells in tissue culture). Also present in the crystal were IAA, and Aux/IAA as a degron peptide. The F-box domain of TIR1 associates with ASK1 and the leucine-rich repeat (LRR) domain of TIR1 forms a locked horseshoe-like structure made up of LRR units (Fig. 6.2), also described as like the head of a mushroom with the F-box domain as the stem (Tan et al. 2007). The loops extending from LRRs 2, 12 and 14 are integral to the auxin and degron-binding sites along with residues lining the concave surface inside the solenoid. The binding site for auxin is a tight, mostly hydrophobic pocket within and on the inner face of the horseshoe (Fig. 6.2c).

In addition to the binding site for auxin, the structure revealed inositol-6-phosphate (IP6) bound centrally (Fig. 6.2). It interacts with key residues involved in forming the auxin pocket, but not with auxin itself. Mutagenesis of the IP6-coordinating residues disrupts auxin-dependent co-receptor assembly

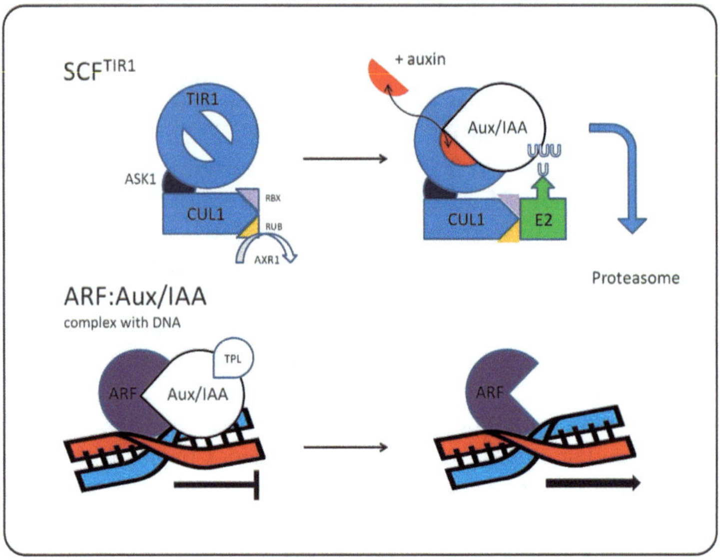


**Fig. 6.1** Overview of the mechanism of activation of transcription in the presence of auxin. In the absence of auxin (*left side*) Aux/IAA proteins saturate dimer associations with Auxin Response Factor (ARF) proteins, repressing transcription of auxin-regulated genes. The protein Topless (TPL) may also associate with the repressor complex. On addition of auxin (*right side*), TIR1 binds both auxin and Aux/IAA proteins, leading to poly-ubiquitination of the Aux/IAAs and their subsequent degradation in the 26S proteasome. The reduced pool of Aux/IAAs allows transcription to proceed.
The $SCF^{TIR1}$ complex includes ASK1 (Arabidopsis SKP1) to which the F-box domain of TIR1 binds, and CUL1 (Cullin 1) which is known as the scaffold of the complex and links ASK1 to the E2 ubiquitin transferase. CUL1 is regulated by reversible addition of RUB (Related to Ubiquitin) and this is mediated by a RUBylation complex (of which AXR1 is the E1 activating enzyme and RCE the likely E3; Dharmasiri et al. 2003). The catalytic activity of the $SCF^{TIR1}$ E2 requires the Ring finger domain protein RBX. Once activated, the E2 transfers ubiquitins to adjacent substrate Aux/IAAs. Once polyubiquitinated, Aux/IAAs are targeted for the proteasome (dos Santos et al. 2009)

(Calderón Villalobos et al. 2012), but no role for IP6 as a mediator of auxin action has been identified. It is likely that IP6 acts structurally, anchoring the LRR horseshoe and the base of the auxin-binding site.

Classification of auxins is discussed below, but two key features of active auxins are a hydrophobic ring system and a carboxylic acid group. The most visually attractive and influential model of the auxin-binding site before the availability of crystal structures was the hydrophobic platform model (Katekar 1979). Katekar's hydrophobic platform accommodated the indolic rings of IAA and placed the carboxylate oxygens perpendicular to the plane of the platform and spaced from it by a distance approximately equal to the space occupied by the methyl carbon of the acetate side chain. In considering the structure determined for TIR1, it was seen that the hydrophobic platform is contributed by a series of hydrophobic residues

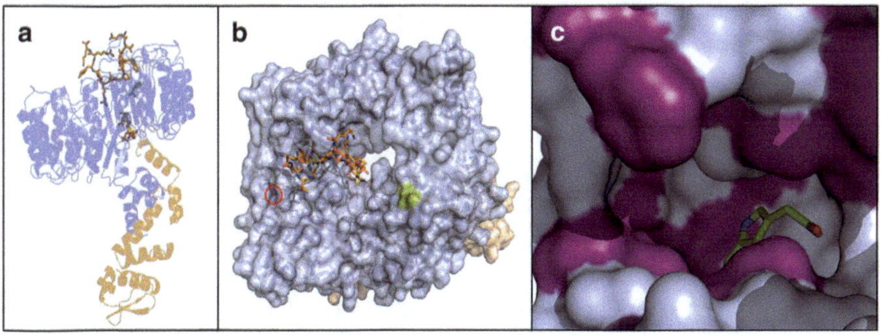


**Fig. 6.2** TIR1 topology and its interaction with ASK1 and IAA. Ribbon images of TIR1 (*blue*) and ASK1 (*brown*) in profile (**a**), and TIR1 from above as a space-filling model (**b**). The hydrophobic binding pocket for IAA (**c**). The ligand IAA (*green*), the substrate degron peptide (*orange*) and IP6 (*yellow*) molecules are shown as skeleton structures in each view. It is seen that the pocket into which IAA binds appears quite deep (**b** and **c**) against the inner side of the LRR solenoid. The Aux/IAA peptide covers IAA effectively, as well as a considerable area around the auxin-binding site (**b**). Many of the residues lining the biding pocket (**c**) are hydrophobic (*purple*). In (**b**), the residues D170 (*pale green surface*) and M473 (*red circle* because the residue itself is buried) have been found to affect Aux/IAA protein binding (Yu et al. 2013). Rendered in PyMOL from PDB 2P1Q

lining the pocket. The hydrophobic space defines a volume and some steric barriers, but does not fully restrict how ring systems align, allowing some promiscuity for auxin structures. Conversely, binding for the carboxylate is specified precisely by two polar residues: an arginine (R403) and a serine (S438) from adjacent LRR strands lining the inner face of the solenoid (Fig. 6.3b). Importantly, R403 is positioned by cofactor IP6. The acetate side chain of IAA is twisted at the α-methyl, turning the bound carboxylate away from the plane of the indole ring and so the oxygens do not sit as if astride the plane of the ring (Katekar 1979), but to the side (Figs. 6.2c and 6.3b, c). This conformation corresponds well with an earlier prediction of ligand orientation based on structure–activity analysis (Kaethner 1977).

The binding of auxin to TIR1 was shown to create a second binding site, a site for the degron motif carried by Aux/IAA proteins. At its core, the degron motif is hydrophobic (I/V, VGWPPV) and the bound auxin contributes its own large hydrophobic platform lining the base of a pocket (Fig. 6.2c). The tryptophan at the heart of the degron (W86 of Aux/IAA7) aligns end to face with the bound IAA and with the adjacent proline (P88; Fig. 6.3a) completely covers and traps the ligand. To allow this tight twist the degron adopts a *cis*-bond in the peptide backbone, rather than the more common *trans* peptide bond conformation, giving -$W_{cis}P_{trans}$P- (Fig. 6.3a). Much effort has been put into seeking a prolyl isomerase activity associated with auxin and $SCF^{TIR1}$ action (e.g. Kepinski and Leyser 2004), but no convincing evidence of any such activity has been reported so far. It is possible that TIR1 itself catalyses the isomerization on degron binding, a hypothesis again reminiscent of the conformational change binding model of Kaethner

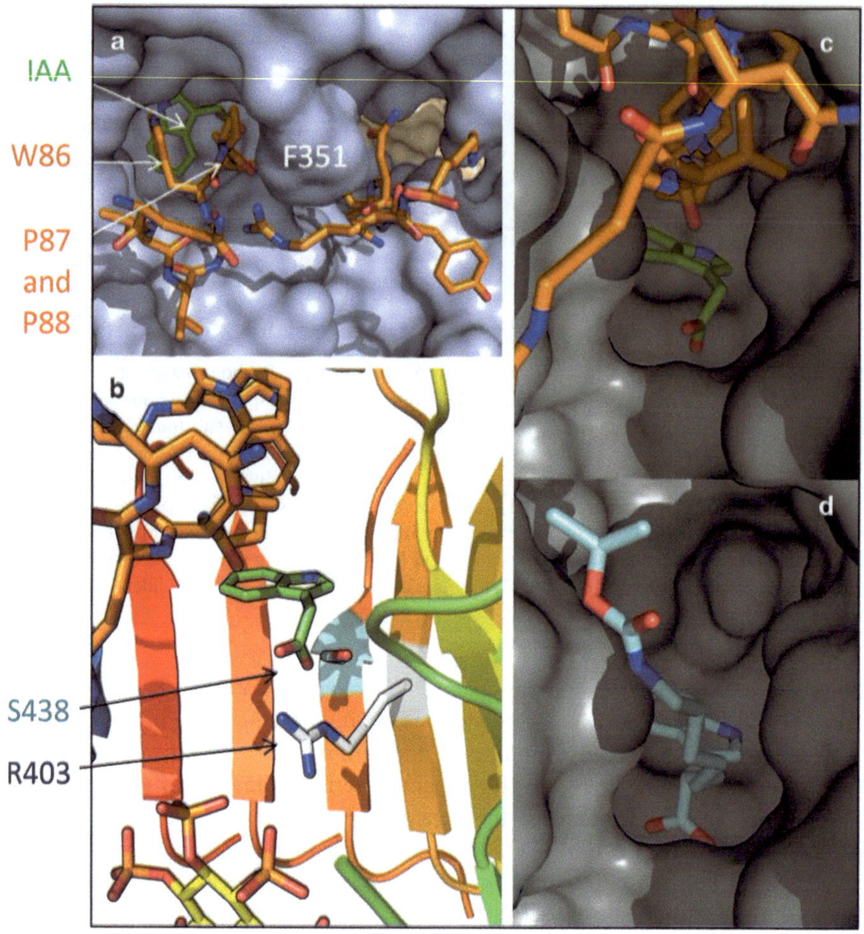


**Fig. 6.3** Details of the IAA and degron-binding pocket. (**a**) The degron is folded intimately against the hydrophobic surfaces of its binding site on TIR1, trapping IAA (*green*). The -$W_{cis}$P-twist is seen between W86 and P87 (labelled). Phenylalanine F351 is also indicated (Hao and Yang 2010). (**b**) The two residues which are hydrogen bonded to the carboxylate oxygens of IAA, S438 and R403, are provided by adjacent β-sheets from the LRRs at the side of the solenoid. IP6 is not directly associated with IAA binding. (**c**) Bound IAA seen from a side view of the pocket with the degron peptide above. (**d**) The same view as in (**c**), but with the anti-auxin α-tert-butoxycarbonylaminohexyl-indole-3-acetic acid in the binding site. The bulky side chain extends from the methyl carbon into the volume that would otherwise be occupied by the degron peptide (Hayashi et al. 2008). Rendered in PyMOL from PDB 2P1Q and 3C6N

(1977). Molecular dynamic simulations of the TIR1 binding pocket residues have also suggested that two phenylalanines (F82 and F351) act as 'conformational inducers', with their phenyl rings pivoting in response to IAA binding, both trapping IAA and contributing hydrophobic surfaces to complete the binding site for Aux/IAA degrons (Fig. 6.3a). All these observations indicate the prescience of Kaethner's hypotheses.

## 2.2 Auxin Antagonists

The role auxin plays in creating a new surface for degron binding has been described as molecular glue (Tan et al. 2007). Clearly both TIR1 and Aux/IAA contribute to trapping auxin, justifying the description of them as co-receptors. Nevertheless, the key ligand-selective residues lie inside the TIR1 binding site. Given that records of auxin binding in the absence of Aux/IAA have proved elusive, it is likely that the affinity of the binding pocket in TIR1 is much lower than the nanomolar figures reported for the complex. This has implications for the development of anti-auxins. Historically, a few compounds have been found to behave antagonistically to IAA. Principal amongst these has been *p*-chlorophenoxyisobutyric acid (PCIB). The structure of the TIR1 complex gave the possibility of rational design of anti-auxins. It was recognised that a compound binding into the TIR1 pocket, but then preventing association of the degron, would act as a competitive antagonist. A series of indoles and naphthalenes with bulky side groups substituted onto the α methyl was synthesised (Fig. 6.3) and physiological tests confirmed their activity as anti-auxins (Hayashi et al. 2008, 2012).

## 2.3 Aux/IAAs as Co-receptors

The section above suggests that auxin specificity resides with TIR1. However, some selectivity in the auxin response is contributed by the Aux/IAA co-receptor (Lokerse and Weijers 2009; Calderón Villalobos et al. 2012). The Aux/IAA family numbers 29 members in *Arabidopsis* and most undergo rapid turnover in the presence of auxin (Abel et al. 1994). The core degron motif is highly conserved, and variants display considerable changes in half-life (Ramos et al. 2001; Dreher et al. 2006). Indeed, many of the early auxin-resistant mutant genes were found to encode altered degron sequences and this helped reveal the vital role ubiquitin-mediated proteolysis plays in auxin action (reviewed in Chapman and Estelle 2009). Particularly instructive were the *axr3* lines (the protein is also known as IAA17), with the core -VGWPPV- mutated to e.g. -VGWPLV- in axr3.1 and with its half-life extended sevenfold from 80 min to 550 min (Ouellet et al. 2001). The longer half-life is consistent with the phenotype of this gain-of-function mutant which shows auxin-like phenotypes in the absence of applied auxin, a hypermorphic response (Rouse et al. 1998). It is worth noting that such phenotypes may also be expected from plants treated with anti-auxins because anti-auxins should extend Aux/IAA half-lives by preventing ubiquitination and proteolysis (Oono et al. 2003). This indicates a potential application for them as herbicides, although no chemical biology screen has yet knowingly yielded a commercially useful anti-auxin.

Viewing Fig. 6.3 it is clear that there is more to Aux/IAA degron binding than the contribution of trapping IAA by -WPP-. As noted above, adjacent degron residues are generally hydrophobic and interact with hydrophobic residues around the neck of the pocket, but the contributions of further residues should not be overlooked. Arginine R93 and tyrosine Y92 appear to cover a large surface area on the solenoid face (Fig. 6.2b), whilst arginine R90 penetrates deep into its centre. It seems likely that these residues contribute considerably to the free energy of binding. However, the structure and most of the binding experimentation have been done with degron peptides and although biochemical results agree with genetic and physiological findings, a note of caution should be added. The full-length Aux/IAA protein may not present the same loop to TIR1 and so the roles of most degron residues must remain hypothetical until we know the structure of full-length Aux/IAA proteins and understand the kinetics of these interactions.

Two mutations found from screening TIR1 with full-length Aux/IAA protein in a yeast 2-hybrid experiment have been reported (Fig. 6.2b; Yu et al. 2013). These mutations enhance the Aux/IAA–TIR1 interaction leading to enhanced Aux/IAA degradation. The two mutation sites sit at opposite sides on the outer face of the TIR1 solenoid, distant from the auxin-binding pocket. Their distance from auxin- and degron-binding sites reinforces the point that the core degron may not mediate Aux/IAA binding alone. Indeed, measured binding affinities for Aux/IAA7 and its domain II degron peptide were found to vary by 20-fold (Calderón Villalobos et al. 2012).

Given that many residues across the degron as well as interaction sites from full-length Aux/IAAs contribute to co-receptor interaction, it is remarkable that a small molecule (IAA) binding at the base of the small, deep binding pocket changes the character of TIR1 so much that it then drives forward so much of plant biology.

## *2.4 Dimerization of Aux/IAAs and ARFs; The Start of Signalling*

The Aux/IAA proteins are modular and in addition to the degron, domains III and IV are conserved within the family and shared with similarly conserved C-terminal domains in the Auxin Response Factor proteins (ARFs; Tiwari et al. 2003). Dimerization is promiscuous within and between these two families, particularly between activator ARFs and Aux/IAAs (Fig. 6.1). It is clear that the multiplicity of possible interactors (29 Aux/IAAs with 23 ARFs—of which five are activators, the rest repressors of transcription) could contribute to the diversity of auxin responses, and specific pairings might deliver certain selective responses. However, generally there do not appear to be more- or less-favoured Aux/IAA-ARF partnerships when options are presented ex vivo, such as in yeast two-hybrid experiments.

Much more convincing are the models based on selectivity delivered by zones of restricted expression (Rademacher et al. 2011; Vernoux et al. 2011) and for expression to be determined by position and time as well as by auxin concentration (Vernoux et al. 2011).

In terms of auxin perception, it appears that in the absence of auxin, Aux/IAAs are present in abundance. The appearance of auxin leads to co-receptor assembly and the rapid degradation of Aux/IAAs. As the free concentration of Aux/IAAs falls, ARF–Aux/IAA dimers dissociate, freeing transcription (Fig. 6.1). Some of the earliest changes in expression induced by auxin are new *Aux/IAA* transcripts, giving an active and sensitive feedback loop. The signalling outcomes of concerted expression (and repression), partner interactions and localised auxin gradients in the shoot apical meristem are described beautifully in Vernoux's paper (Vernoux et al. 2011; see also Chap. 10).

## 2.5 The TIR1 Family

The TIR1 family in *Arabidopsis thaliana* includes Auxin F-Box proteins (AFBs) 1–5. All bind IAA to promote co-receptor assembly with Aux/IAA degrons (Calderón Villalobos et al. 2012). The two most distinct clades are represented by TIR1 and AFB5 and yet these two bind a range of degron sequences with similar preferences (Calderón Villalobos et al. 2012). However, TIR1 and AFBs are not all equivalent. In *Arabidopsis* roots TIR1 and AFB2 are dominant and neither AFB1 nor AFB2 can replace TIR1 in *tir1* plants (Parry et al. 2009). Further, a striking feature of AFB5 is its enhanced affinity over TIR1 for synthetic picolinate auxins such as Picloram (Walsh et al. 2006; Lee et al. 2014) and a structural model for the AFB5 auxin-binding pocket has been developed (Calderón Villalobos et al. 2012). A second dramatic difference between TIR1 and AFB5 is the co-receptor dissociation rates, with AFB5 complexes falling apart far more rapidly than TIR1 complexes (Lee et al. 2014). Whether this makes AFB5 more efficient at turning over Aux/IAAs or more likely to yield mono-ubiquitinated products (which will not be recognised by the proteasome) is not known. Expression in yeast reporter systems has suggested that TIR1 and AFB2 do support more rapid degradation of Aux/IAAs than AFB1 and AFB3 (Havens et al. 2012), but unfortunately the biophysics and genetic reporter assays have not yet crossed over sufficiently to cover all TIR1/AFBs.

Two interesting observations arise from the distinctiveness of AFB5 from TIR1. Whilst IAA acts as a good ligand in both, the very different binding site specificity may suggest that additional endogenous ligands await discovery. Secondly, 'spray and pray' chemical biology screens have discovered the picolinates, but knowledge-led design can now start to contribute molecule leads for novel site-selective auxins and site-selective anti-auxins.

# 3 The Cell Cycle Protein SKP2

An F-box protein known as S-phase Kinase-associated Protein 2A (SKP2A) has been reported to bind auxin (Jurado et al. 2010). The protein was shown to bind radiolabelled IAA in a saturable and specific manner (estimated $K_{dIAA} = 0.2\ \mu M$), although expression was from *E. coli* which raises some concerns over whether it was folded authentically. Other plant F-box proteins require expression in eukaryotic cells to acquire functionality (such as TIR1; Dharmasiri et al. 2005a; Kepinski and Leyser 2005; Tan et al. 2007). There is little homology with TIR1 except in the common domains of F-box and LRRs, but the LRR motif allowed modelling of this part of SKP2A onto the TIR1 structure and a putative binding site was identified. Auxin was shown to induce ubiquitin-mediated degradation of SKP2A and a model was presented for how this linked auxin signalling to the cell cycle and division. SKP2 seems to be involved in auxin signalling, but its role in auxin perception is far less certain.

# 4 ABP1 as an Auxin Receptor

Auxin-binding protein1 (ABP1) has a long history as a candidate receptor. Like TIR1, it needs to be expressed and purified from eukaryotic cells (baculovirus system) to hold activity and the structure has been solved with auxin bound (Woo et al. 2002). What has proved more elusive for ABP1 has been its mechanism of action and with no known mechanism the claim that it is a receptor has been infirm. With TIR1/AFBs apparently in control of auxin-mediated gene expression, were additional receptors necessary? However, others argued that because auxin action via TIR1 requires degradation of Aux/IAAs, transcription, and translation (and estimates for this series are in excess of 10 min), then more rapid responses and responses which persist in *tir1/afb* multiple mutants may indicate additional, non-TIR1 pathways (Badescu and Napier 2006). Recent reports suggest a functional requirement for ABP1 and its signalling pathway.

## *4.1 The Structure of ABP1 and Auxin Binding*

The auxin-binding site in ABP1 shares some general features with that in TIR1, but some specific details are quite different. The protein is in the cupin superfamily which is defined by the way the polypeptide folds into a β-jellyroll barrel (Fig. 6.4; Woo et al. 2002). At the heart of the barrel is a zinc ion. This zinc coordinates the carboxylate group of bound auxins, which may be compared with the arginine and serine hydrogen bonds of TIR1 (Fig. 6.3). In common with TIR1, the hydrophobic auxin ring system (1-NAA in the ABP1 crystal) sits in a hydrophobic pocket and

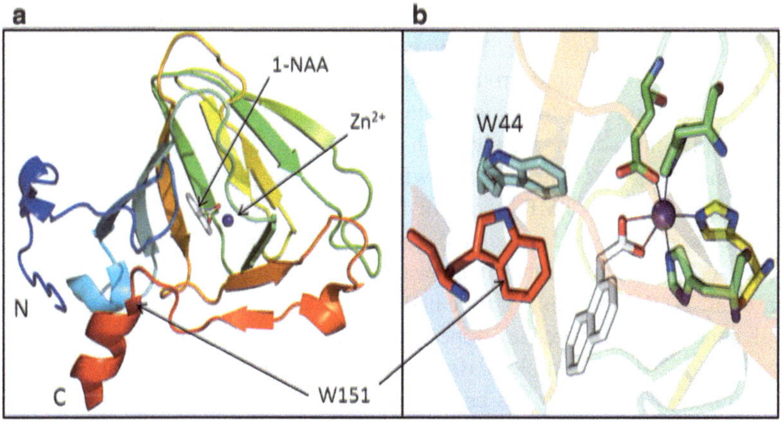


**Fig. 6.4** The auxin-binding site in ABP1. (**a**) Auxin binds inside the cupin barrel. (**b**) The carboxylic acid group of 1-NAA coordinates a zinc ion which is held in place by a group of histidines and a glutamic acid residue. There is an end-to-face interaction with W151 and this is one of several hydrophobic interactions around the auxinic ring system. The *rainbow colouring system* is used to code the structure from N- (*Blue*) to C-terminus (*red*). W151 sits at the start of the C-terminal helix and auxin binding may induce repositioning of this residue which would reorient the helix (Bertosa et al. 2008). Rendered in PyMOL from 1LRH

packs end to face with a tryptophan, W151, in the maize sequence, contributed by the protein's C-terminal α-helix. This tryptophan–auxin electronic interaction may be mechanistically significant because molecular dynamic simulations have suggested that the C-terminal helix moves as a consequence of auxin binding (Bertosa et al. 2008). The crystallography data show little or no molecular movement between ligand-free and ligand-bound ABP1 (Woo et al. 2002), but the calculations allow a piston-like movement of the whole terminal helix driven primarily by repositioning W151 against the bound auxin. Such a movement is consistent with earlier immunological evidence using purified ABP1 which suggested that the C-terminus undergoes a ligand-induced conformational change (Napier and Venis 1990).

## 4.2 *Signalling from ABP1*

Early work linked ABP1 with rapid plasma membrane ion fluxes and cell elongation (reviewed in Napier et al. 2002); more recent work has linked it with a range of developmental responses (Shi and Yang 2011; Scherer 2011; Xu et al. 2014). It is understood that all the known physiological responses linked to ABP1 are initiated on the outside of the plasma membrane, yet many of these responses require that the signal is passed across the membrane into the cell. A membrane-spanning docking protein has been proposed (Barbier-Brygoo et al. 1989; Tromas et al. 2010) and changes in exposure of the C-terminal residues in ABP1 could clearly modulate the activity of a signalling partner. Early efforts to identify the docking protein identified a membrane-anchored protein known as C-terminal peptide-binding protein 1 (CBP1; Shimomura 2006). Unfortunately CBP1 also lacked the capacity to signal across the plasma membrane, although it may help. The first integral membrane proteins shown to associate directly with ABP1 are plasma membrane-localised transmembrane kinase (TMK) receptor-like kinases (Xu et al. 2014).

Unlike the gene family associated with TIR1, *ABP1* is alone in the *A. thaliana* genome. A T-DNA insertion mutant was isolated and shown to be embryo lethal (Chen et al. 2001) and this complicated genetic analysis for some time. Latterly, several genetic tools were developed to circumvent the experimental challenge (Tromas et al. 2010; Xu et al. 2010; Robert et al. 2010; Effendi et al. 2011). As a result, a range of responses has been linked to ABP1 action, but featuring consistently has been the regulation of PIN protein internalisation via clathrin-mediated endocytosis at the plasma membrane to change polar auxin transport. Importantly, PINs become internalised within 5 min and the response is sustained in *tir1/afb* mutants showing that these responses are not downstream of TIR1/AFB signalling. Further, the abundance of ABP1 was found to correlate with the loss or gain of response giving confidence to the proposal that ABP1 promotes clathrin-dependent endocytosis of PIN1. Auxin binding to ABP1 inhibits this activity, at least in roots (Robert et al. 2010).

In leaf tissue auxin promotes interdigitation of pavement cells and this is dependent on activation of ROP/Rac GTPases (Xu et al. 2010). In the same weak *abp1-5* mutant line used by others (Robert et al. 2010), rapid ROP activation was absent. This and other observations led to the conclusion that ABP1 activates ROP2 and this leads to inhibition of PIN1 internalisation, somewhat different to the situation in roots. Importantly, the leaf pavement cell interdigitation response led to the link with plasma membrane TMKs, because loss of these trans-membrane kinases led to a similar lack of interdigitation (and other responses) as loss of ABP1 (Xu et al. 2014). A direct and auxin-dependent association between TMKs and ABP1 was demonstrated by co-immunoprecipitation, and the association was shown to be with the extracellular domain of the TMK (Xu et al. 2014). Exactly how ABP1 interacts to modulate transmembrane kinase activity remains to be determined, but a signalling pathway now links extracellular ABP1 with auxin-dependent activation of ROP GTPase signalling.

Additional rapid responses to auxin have been reported, some associated with ABP1. These include calcium fluxes, phospholipase A2 activation and kinase activation (Scherer 2011). A metaphor for a two-receptor system has been given, painting TIR1/AFBs as cell-selective inducers of auxin activity, ABP1 as a more generic throttle (Scherer 2011). The outstanding challenge is to explain how ABP1 works. There are suggestions of binding site specific inducers or inhibitors based on structural understanding of the distinct sites discussed above (Hayashi et al. 2012). Such pharmacological tools are anticipated eagerly.

## 5 Models of Auxin Binding and Structure–Activity Relationships

Pharmacological models of auxins and of auxin-binding sites have been developed with the wealth of structure–activity data collected over many years. These need some reinterpretation now that we know the bioassay responses measured are mediated by a family, if not a collection of receptors on top of the effects contributed by auxin transport, metabolism and compartmentation. Indeed, some models have recognised the multifactorial nature of the system inputs (Tomić et al 1998; Ferro et al 2006, 2010) and have developed auxin classifications accordingly. Nevertheless, the advent of receptor-specific assays (Hayashi et al. 2012; Lee et al. 2014) will permit far more accurate classification systems and far better design-led pharmacological chemistry. It remains necessary to be alert to the demands of, for example, Lipinski's rule of five, systemic delivery and uptake, etc. (Lamberth et al. 2013), but there is every chance that in the future useful auxinic agrochemicals will be designed, not found.

## 6 Auxin Perception, Signalling Diversity and Signalling Sensitivity

It has been noted that there is a large physiologically active dose–response range for auxin, with responses in primary roots sensitive to low nanomolar auxin concentrations, whereas other responses (typically in aerial tissues) have half maximal doses in the mid micromolar range. Classical ligand binding theory describes 80 % of the binding within a concentration range one log unit above and below the affinity of the site, a far more restricted dose dependence than that seen for auxins. Two distinct receptors offer an explanation for a wider dynamic range, yet it is clear that TIR1/AFBs are the initiators of many auxin-induced responses in both root and aerial tissues. The observation that different pairings of TIR1 and Aux/IAA had different affinities (Calderón Villalobos et al. 2012) and that these varied over several log concentration units has helped explain how a diversity of responses can be driven through one class of receptor and support a wide dynamic sensitivity range.

A range of TIR1 co-receptor kinetics does not exclude the possibility that ABP1 might also contribute to modulate the signal outputs (Scherer 2011). It is becoming apparent that ABP1's role may be linked to rapid changes in the capacity for polar auxin transport, hence moderating the input signal to the TIR1/AFB system. It is clear that active auxin transport mediates plant development by controlling temporal, environmental and spatial auxin gradients (see Chaps. 4, 5, 8, 10, 14, 15). Associated changes in membrane area and protein complement might account for several of the rapid membrane-linked responses on record. Plant morphogenesis appears to be driven by a two-receptor system.

# 7 Concluding Remarks

The mechanisms of auxin perception still hold some secrets. The role bound auxin plays in the folding of co-receptor degrons is a key issue and a much more detailed understanding of binding site selectivities for the TIR1/AFBs should allow us to define what it takes to be an auxin. Some of the same understanding may allow us to design novel agonists and antagonists, as well as tools to help distinguish between signalling from different receptor systems.

**Acknowledegments** I am grateful to BBSRC for funding (grant reference BB/F014651/1) and to Alexander J. Fullwood for creating all the images of protein structures.

# References

Abel S, Oeller PW, Theologis A (1994) Early auxin-induced genes encode short-lived nuclear proteins. Proc Natl Acad Sci USA 91:326–330

Badescu G, Napier RM (2006) Receptors for auxin: will it all end in TIRs? Trends Plant Sci 11:217–223

Barbier-Brygoo H, Ephritikhine G, Klambt D, Ghislain M, Guern J (1989) Functional evidence for an auxin receptor at the plasmalemma of tobacco mesophyll protoplasts. Proc Natl Acad Sci USA 86:891–895

Bertoša B, Kojić-Prodić B, Wade RC, Tomić S (2008) Mechanism of auxin interaction with auxin binding protein (ABP1): a molecular dynamics simulation study. Biophys J 94:27–37. doi:10.1529/biophysj.107.109025, PMCID: PMC2134879

Calderón Villalobos LI, Lee S, De Oliveira C, Ivetac A, Brandt W, Armitage L, Sheard LB, Tan X, Parry G, Mao H, Zheng N, Napier R, Kepinski S, Estelle M (2012) A combinatorial TIR1/AFB-Aux/IAA co-receptor system for differential sensing of auxin. Nat Chem Biol 8:477–485. doi:10.1038/nchembio.926

Chapman EJ, Estelle M (2009) Mechanism of auxin-regulated gene expression in plants. Annu Rev Genet 43:265–285. doi:10.1146/annurev-genet-102108-134148

Chen JG, Ullah H, Young JC, Sussman MR, Jones AM (2001) ABP1 is required for organized cell elongation and division in Arabidopsis embryogenesis. Genes Dev 15:902–911

Dharmasiri S, Dharmasiri N, Hellmann H, Estelle M (2003) The RUB/Nedd8 conjugation pathway is required for early development in Arabidopsis. EMBO J 22:1762–1770

Dharmasiri N, Dharmasiri S, Estelle M (2005a) The F-box protein TIR1 is an auxin receptor. Nature 435:441–445

Dharmasiri N, Dharmasiri S, Weijers D, Lechner E, Yamada M, Hobbie L, Ehrismann JS, Jurgens G, Estelle M (2005b) Plant development is regulated by a family of auxin receptor F box proteins. Dev Cell 9:109–119

dos Santos MF, Memelink J, Offringa R (2009) Auxin-induced, SCFTIR1-mediated polyubiquitination marks AUX/IAA proteins for degradation. Plant J 59:100–109

Dreher KA, Brown J, Saw RE, Callis J (2006) The Arabidopsis Aux/IAA protein family has diversified in degradation and auxin responsiveness. Plant Cell 18:699–714

Effendi Y, Rietz S, Fischer U, Scherer GF (2011) The heterozygous abp1/ABP1 insertional mutant has defects in functions requiring polar auxin transport and in regulation of early auxin-induced genes. Plant J 65:282–294

Ferro N, Gallegos A, Bultinck P, Jacobsen HJ, Carbó-Dorca R, Reinard T (2006) Coulomb and overlap self-similarities: a comparative selectivity analysis of structure-function relationships for auxin-like molecules. J Chem Inf Model 46:1751–1762

Ferro N, Bredow T, Jacobsen H-J, Reinard T (2010) Route to novel auxin: auxin chemical space toward biological correlation carriers. Chem Rev 110:4690–4708

Hao G-F, Yang G-F (2010) The role of Phe82 and Phe351 in auxin-induced substrate perception by TIR1 ubiquitin ligase: a novel insight from molecular dynamics simulations. PLoS One 5 (5):e10742

Havens KA, Guseman JM, Jang SS, Pierre-Jerome E, Bolten N, Klavins E, Nemhauser JL (2012) A synthetic approach reveals extensive tunability of auxin signaling. Plant Physiol 160:135–142

Hayashi K-I, Tan X, Zheng N, Hatate T, Kimura Y, Kepinski S, Nozaki H (2008) Small-molecule agonists and antagonists of F-box protein-substrate interactions in auxin perception and signaling. Proc Natl Acad Sci USA 105:5632–5637

Hayashi K-I, Neve J, Hirose M, Kuboki A, Shimada Y, Kepinski S, Nozaki H (2012) Rational design of an auxin antagonist of the SCF TIR1 auxin receptor complex. ACS Chem Biol 7:590–598

Jurado S, Abraham Z, Manzano C, López-Torrejón G, Pacios LF, Del Pozo JC (2010) The Arabidopsis cell cycle F-box protein SKP2A binds to auxin. Plant Cell 22:3891–3904

Kaethner TM (1977) Conformational change theory for auxin structure-activity relationships. Nature 267:19–23

Katekar GF (1979) Auxins: on the nature of the receptor site and molecular requirements for auxin activity. Phytochemistry 18:223–233

Kepinski S, Leyser O (2004) Auxin-induced $SCF^{TIR1}$-Aux/IAA interaction involves stable modification of the $SCF^{TIR1}$ complex. Proc Natl Acad Sci USA 101:12381–12386

Kepinski S, Leyser O (2005) The Arabidopsis F-box protein TIR1 is an auxin receptor. Nature 435:446–451

Lamberth C, Jeanmart S, Luksch T, Plant A (2013) Current challenges and trends in the discovery of agrochemicals. Science 341:742–746

Lee S, Sundaram S, Armitage L, Evans JP, Hawkes T, Kepinski S, Ferro N, Napier RM (2014) Defining binding efficiency and specificity of auxins for SCFTIR1/AFB-Aux/IAA co-receptor complex formation. ACS Chem Biol 9(3):673–682

Leyser HM, Lincoln CA, Timpte C, Lammer D, Turner J, Estelle M (1993) Arabidopsis auxin-resistance gene AXR1 encodes a protein related to ubiquitin-activating enzyme E1. Nature 364:161–164

Lokerse AS, Weijers D (2009) Auxin enters the matrix—assembly of response machineries for specific outputs. Curr Opin Plant Biol 12:520–526

Napier RM, Venis MA (1990) Monoclonal antibodies detect an auxin-induced conformational change in the auxin-binding protein. Planta 182:313–318

Napier RM, David KM, Perrot-Rechenmann C (2002) A short history of auxin-binding proteins. Plant Mol Biol 49:339–348

Oono Y, Ooura C, Rahman A, Aspuria ET, Hayashi K, Tanaka A, Uchimiya H (2003) *p*-Chlorophenoxyisobutyric acid impairs auxin response in Arabidopsis root. Plant Physiol 113:1135–1147

Ouellet F, Overvoorde PJ, Theologis A (2001) IAA17/AXR3: biochemical insight into an auxin mutant phenotype. Plant Cell 13:829–841

Parry G, Calderón Villalobos LI, Prigge M, Peret B, Dharmasiri S, Itoh H, Lechner E, Gray WM, Bennett M, Estelle M (2009) Complex regulation of the TIR1/AFB family of auxin receptors. Proc Natl Acad Sci USA 106:22540–22545

Rademacher EH, Möller B, Lokerse AS, Llavata-Peris CI, van den Berg W, Weijers D (2011) A cellular expression map of the Arabidopsis AUXIN RESPONSE FACTOR gene family. Plant J 68:597–606

Ramos J, Zenser N, Leyser O, Callis J (2001) Rapid degradation of auxin/indoleacetic acid proteins requires conserved amino acids of domain II and is proteasome dependent. Plant Cell 13:2349–2360
Robert S, Kleine_Vehn J et al (2010) ABP1 mediates auxin inhibition of clathrin-dependent endocytosis in *Arabidopsis*. Cell 143:111–121
Rouse D, Mackay P, Stirnberg P, Estelle M, Leyser O (1998) Changes in auxin response from mutations in an AUX/IAA gene. Science 279:1371–1373. doi:10.1126/science.279.5355.1371
Ruegger M, Dewey E, Gray WM, Hobbie L, Turner J, Estelle M (1998) The TIR1 protein of Arabidopsis functions in auxin response and is related to human SKP2 and yeastGrr1p. Genes Dev 12:198–207, PMCID: PMC316440
Scherer GFE (2011) AUXIN-BINDING-PROTEIN1, the second auxin receptor: what is the significance of a two-receptor concept in plant signal transduction? J Exp Bot 62:3339–3357. doi:10.1093/jxb/err033
Shi J-H, Yang Z-B (2011) Is ABP1 a receptor yet? Mol Plant 4:635–640
Shimomura S (2006) Identification of a glycosylphosphatidylinositol-anchored plasma membrane protein interacting with the C-terminus of auxin-binding protein1. Plant Mol Biol 60:663–677
Tan X, Calderon-Villalobos LIA, Sharon M, Zheng C, Robinson CV, Estelle M, Zheng N (2007) Mechanism of auxin perception by the TIR1 ubiquitin ligase. Nature 446:640–645
Tiwari SB, Hagen G, Guilfoyle T (2003) The roles of auxin response factor domains in auxin-responsive transcription. Plant Cell 15:533–543
Tomić S, Gabdoulline RR, Kojić-Prodić B, Wade RC (1998) Classification of auxin plant hormones by interaction property similarity indices. J Comput Aided Mol Des 12:63–79
Tromas A, Popanov I, Perrot-Rechenmann C (2010) AUXIN-BINDING PROTEIN1 is required for differential responses mediating root growth. PLoS One 4:436–446
Vernoux T, Brunoud G et al (2011) The auxin signalling network translates dynamic input into robust patterning at the shoot apex. Mol Syst Biol 7:508. doi:10.1038/msb.2011.39
Walsh TA, Neal R, Merlo AO, Honma M, Hicks GR, Wolff K, Matsumura W, Davies JP (2006) Mutations in an auxin receptor homolog AFB5 and in SGT1b confer resistance to synthetic picolinate auxins and not to 2,4-dichlorophenoxyacetic acid or indole-3-acetic acid in Arabidopsis. Plant Physiol 142:542–552
Woo EJ, Marshall J, Bauly J, Chen JG, Venis M, Napier RM, Pickersgill RW (2002) Crystal structure of auxin-binding protein 1 in complex with auxin. EMBO J 21:2877–2885
Xu T, Wen M, Nagawa S, Fu Y, Chen JG, Wu MJ, Perrot-Rechenmann C, Friml J, Jones AM, Yang Z (2010) Cell surface- and rho GTPase-based auxin signaling controls cellular interdigitation in Arabidopsis. Cell 143:99–110. doi:10.1016/j.cell.2010.09.003
Xu T, Dai N, Chen J, Nagawa S, Cao M, Li H, Zhou Z, Chen X, De Rycke R, Rakusová H, Wang W, Jones AM, Friml J, Patterson SE, Bleecker AB, Yang Z (2014) Cell surface ABP1-TMK auxin-sensing complex activates ROP GTPase signalling. Science 343:1025–1028
Yu H, Moss BL, Jang SS, Prigge M, Klavins E, Nemhauser JL, Estelle M (2013) Mutations in the TIR1 auxin receptor that increase affinity for auxin/indole-3-acetic acid proteins result in auxin hypersensitivity. Plant Physiol 162:295–303. doi:10.1104/pp. 113.215582

# Part II
# Auxin and Plant Development

# Chapter 7
# The Interplay Between Auxin and the Cell Cycle During Plant Development

**Marlies J.F. Demeulenaere and Tom Beeckman**

**Abstract** The essential role of auxin for cell proliferation in plants is well known. Both auxin signaling and cell cycle regulation have been studied elaborately, but less is known about the connection between these processes. Recent studies report on the first molecular pathways that have been found to directly link auxin levels to the regulation of cell cycle activity. Here, we discuss the general effect of auxin on cell cycle progression and then zoom in on the interplay between auxin and the cell cycle during root development in *Arabidopsis thaliana*. At the root tip, an auxin gradient maintains the correct organization of the ground tissue layers and controls the size of the root apical meristem. During auxin-induced lateral root initiation *LATERAL ORGAN BOUNDARIES-DOMAIN* transcription factors are upregulated and control reactivation of the cell cycle and cell specification, both of which are needed for proper lateral root initiation. Auxin-induced lateral root initiation-like pathways are also involved in cell cycle reactivation during the formation of nematode feeding sites, nitrogen-fixing nodules and callus tissue, pointing to the existence of one common auxin–cell cycle module to initiate new organs in plants.

## 1 Introduction

Auxin has been known for a long time as an important regulator of cell proliferation and raised auxin levels are generally considered as a prerequisite for cell division competence. Although recently much knowledge has been gained about both auxin

---

M.J.F. Demeulenaere • T. Beeckman (✉)
Department of Plant Systems Biology, VIB, Technologiepark 927, 9052 Gent, Belgium

Department of Plant Biotechnology and Bioinformatics, Ghent University, Technologiepark 927, 9052 Gent, Belgium
e-mail: tom.beeckman@psb.vib-ugent.be

E. Zažímalová et al. (eds.), *Auxin and Its Role in Plant Development*,
DOI 10.1007/978-3-7091-1526-8_7, © Springer-Verlag Wien 2014

signaling and the regulation of the plant cell cycle, many questions remain unanswered concerning the connection between both processes. During the last years, several studies, mainly on root development, have elucidated molecular pathways that link cellular auxin levels to regulation of the cell cycle machinery. Here, we first give an overview of the general effects of auxin on the plant cell cycle and then discuss the interaction between auxin and cell cycle activity during plant developmental processes. We focus on post-embryonic root development in *Arabidopsis thaliana* (*Arabidopsis*) and particularly on the reactivation of the cell cycle during auxin-induced lateral root initiation.

## 2 The Cell Cycle Machinery

A regular mitotic cell cycle consists of four different phases. From the G1 gap phase, cells progress toward the S phase, during which DNA is replicated. After a second gap phase, G2, cells divide into two daughter cells at the M phase. The three phases from G1 to G2 are collectively called the interphase. During the gap phases the cells get ready for the upcoming DNA synthesis or mitosis and repair mechanisms are activated when needed. This makes the G1/S and G2/M boundaries important cell cycle checkpoints, which is also shown by the arrest of cells in either G1 or G2 during stressful conditions (Tardieu and Granier 2000; Van't Hof 1985). The switches between the different phases are controlled by the presence of different CDK–CYCLIN complexes. Despite the conservation of the core cell cycle mechanisms in all eukaryotic organisms, there are major differences in the total number of cell cycle regulators. In plants, many more regulators are involved compared to animal and yeast, with over 80 cell cycle proteins known to date in *Arabidopsis* (Menges et al. 2005; Vandepoele et al. 2002; Van Leene et al. 2010). Before we zoom in on how auxin interferes with cell cycle regulation, we first give a summary of the molecular players of the plant cell cycle (Fig. 7.1). For a more profound overview, however, we refer the readers to the review by Inzé and De Veylder (2006).

CYCLIN-DEPENDENT KINASES (CDKs) are the main players in the regulation of the cell cycle. As stated by their name, these serine/threonine kinases are functionally dependent on CYCLINS with which they form protein complexes. The transcription and degradation of CYCLINS is the main mechanism driving cells from one phase of the cell cycle to the next. Most plant D-type CYCLINS (CYCD) are constitutively expressed during the cell cycle, but mainly function during the transition from G1 to S. A-type CYCLINS (CYCA) are most abundant from S to M phase and the majority of B-type CYCLINS (CYCB) are specific to G2 and M phase (Ito 2000; Menges et al. 2005).

Further regulation of the core cell cycle machinery is established by binding of CDK inhibitory proteins to the CDK–CYCLIN complexes. In plants, these inhibitors are known as ICKs (INTERACTOR/INHIBITOR OF CDK) and until now two classes have been identified, being the KIP-RELATED PROTEINS (KRPs) and the

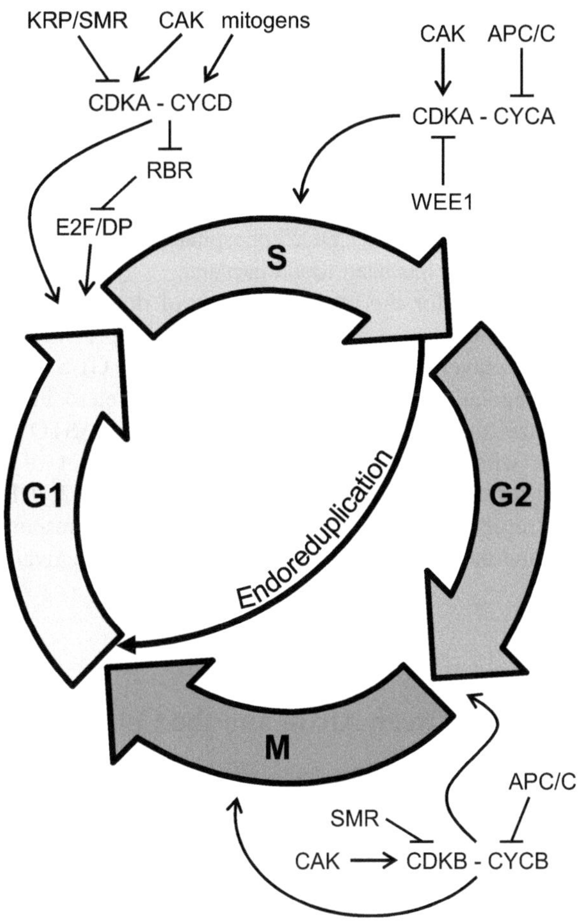


**Fig. 7.1** Progression of the plant cell cycle is regulated by the activity of CDK–CYC complexes. The transition from the G1 to S phase is mainly controlled by CDKA–CYCD complexes, which become more abundant in the presence of mitogens due to upregulation of *CYCD* expression. CDKA–CYCD activity inhibits RBR through phosphorylation. Subsequently, E2F/DP activity is released from the inhibition by RBR, and S-phase specific genes are induced. During the S phase, CDKA–CYCA complexes are most important. At the G2/M transition, CDKAs are replaced by CDKBs, which mainly function in collaboration with B-type CYCLINS. For CDKs to be active, they need to be phosphorylated by CAKs, while binding by KRPs or SMRs inactivates them. During S and G2 the activity of CDKAs can also be inhibited by the WEE1 kinase. The APC/C complex is involved in the phase-specific degradation of CYCA and CYCB

SIAMESE (RELATED) (SIM and SMR) proteins (Churchman et al. 2006; Peres et al. 2007). More levels of control arise from protein turnover, which is mainly executed by the SKP1/CULLIN/F-BOX (SCF)-related complex and the ANAPHASE-PROMOTING COMPLEX/CYCLOSOME (APC/C), with the latter specifically functioning during the progression through and exit from mitosis

(Heyman and De Veylder 2012; Vodermaier 2004). CDK activity can also be switched on and off by changes in the phosphorylation status of the CDK itself. CDK-ACTIVATING KINASES (CAKs) mediate the phosphorylation of a conserved threonine residue in the T-loop of the CDKs, which is crucial for CDK activity (Umeda et al. 2005). The WEE1 kinase is responsible for inactivating phosphorylation of the CDKs during S and G2 (Cools et al. 2011; De Schutter et al. 2007; Sorrell et al. 2002; Sun et al. 1999). In most eukaryotes, the activity of WEE1 can be counteracted by the CDC25 phosphatase (Landrieu et al. 2004), but no functional *CDC25* gene has been found in plants.

CDKs are responsible for the phosphorylation of downstream targets, which leads to the breakdown of phase-specific proteins from the previous cell cycle phase and the activation of players in the upcoming phase. At the G1/S transition much of this downstream regulation happens through E2F/DP heterodimer transcription factor activity (Inzé and De Veylder 2006). RETINOBLASTOMA-RELATED (RBR) associates with the E2F/DPs and is a direct target of CDKs. In its unphosphorylated form RBR represses the transcription of E2F/DP target genes. During the G1/S transition, CYCLIN D-dependent CDKs are responsible for RBR phosphorylation, and thus inactivation, thereby enabling the activation of E2F/DP targets.

# 3 The Interplay Between Auxin and the Cell Cycle

## *3.1 Auxin Is Needed for Progression Through the Cell Cycle*

From studies with tissue cultures, it has been known for many decades that both auxin and cytokinin are important to maintain cell divisions. Furthermore, the ratio between these two plant hormones is a decisive factor to discriminate between root and shoot development from callus tissue and to decide between cell growth and differentiation on the one hand and cell division on the other hand (Murashige and Skoog 1962; Tao and Verbelen 1996). Much of the early work on the effect of auxin and cytokinin on cell cycle progression has been carried out on a diverse set of samples from several plant species such as tobacco, pea and *Arabidopsis*. As a consequence, the apparent discrepancies between the obtained results might have been caused by different characteristics of the used tissues (Bayliss 1985). Also later studies with hormonal treatments *in planta* did not always generate a uniform picture, most likely because of differences in the endogenous hormone levels and transcriptional networks that are active in the samples that were studied (John 2007). As a conclusion from the early experiments on auxin and cytokinin as plant growth regulators, it can be said that both of them are needed to maintain cells in a proliferation competent state. Pinpointing specific cell cycle stages during which these hormones are acting is difficult, since they seem to have a rather general effect on the progression of the cell cycle (see Chap. 12).

### 3.2 Auxin and Regulation of the Core Cell Cycle Machinery

Many reports demonstrate the influence of auxin treatments on the expression of core cell cycle genes such as *CDKA;1* (Hemerly et al. 1993; John et al. 1993), *CYCB1;1* (Ferreira et al. 1994b), *CYCA2;1* (Burssens et al. 2000), *CYCD3* (Murray et al. 1998; Soni et al. 1995), and *E2Fb* (Magyar et al. 2005). Although auxin treatment increases the expression levels of CDKs (Zhang et al. 1996), it is not able to induce CDK activity. Cotreatment with cytokinin is needed to dephosphorylate and thereby activate the mitotic CDKs (John et al. 1993; Orchard et al. 2005; Zhang et al. 1996, 2005). The effect of auxin on gene regulation is often established through the activity of AUXIN RESPONSE FACTORS (ARFs). These transcription factors bind their target sequences through recognition of auxin response elements (AuxREs, Ulmasov et al. 1995, 1997). For most of the cell cycle genes that have been shown to be transcriptionally regulated by auxin, the involvement of direct binding by ARFs has not been investigated. Many cell cycle genes do however contain AuxREs in their upstream sequence (Table 7.1). The functionality of these elements would need to be tested before a direct link between auxin signaling and the regulation of these genes can be made.

Next to the transcriptional regulation of their gene expression, auxin also targets protein levels of core cell cycle genes. In the case of E2Fb, auxin both increases the expression levels and stabilizes the protein (Magyar et al. 2005). Auxin also binds directly to SKP2A (Jurado et al. 2010), which is an F-box protein that functions in an SCF complex to regulate the ubiquitin-dependent proteolysis of E2Fc and DPb (del Pozo et al. 2002, 2006). The interaction between SKP2A and auxin stimulates both the degradation of SKP2A itself and of its targets (Jurado et al. 2010). The latter is very reminiscent of the TRANSPORT INHIBITOR RESPONSE 1 (TIR1)-AUXIN/INDOLE-3-ACETIC ACID INDUCIBLE (Aux/IAA) auxin signaling pathway (del Pozo and Manzano 2013), in which binding of auxin to the F-box protein TIR1 leads to the degradation of the Aux/IAA proteins and as such activates downstream auxin signaling via ARFs. For more details on the auxin signaling mechanism, we refer readers to the dedicated chapter in this book (see Chap. 6).

*SKP2B*, the closest homolog of *SKP2A*, is also regulated by auxin. Auxin has been shown to promote histone H3 acetylation in the promoter region of *SKP2B*, which increases its transcription levels (Manzano et al. 2012). KRP1 was reported to be targeted for degradation by SKP2B (Ren et al. 2008), but probably SKP2B also targets positive cell cycle regulators, because it negatively regulates cell proliferation in both apical and lateral root meristems (Manzano et al. 2012).

Modulation of calcium levels is another strategy used by auxin to influence the progression through the cell cycle. Auxin induces a rapid increase of cytosolic $Ca^{2+}$ concentration (Monshausen 2012; Monshausen et al. 2011; Shishova and Lindberg 2010). Through the activity of calcium-dependent protein kinases and phosphatases, these increased $Ca^{2+}$ levels can impact on cell cycle progression. A KRP protein from *Medicago* was shown to have higher activity after calcium-dependent phosphorylation (Pettkó-Szandtner et al. 2006) and regulation of RBR activity by PP2A phosphatases might also depend on $Ca^{2+}$ binding to the PP2As (Dudits et al. 2011).

**Table 7.1** Potential binding sites for AUXIN RESPONSE FACTORS in cell cycle genes

| Gene name | AGI ID | AuxRE TGTCTC/ GAGACA | AuxRE core TGTC/GACA |
|---|---|---|---|
| *CDKA;1* | AT3g48750 | – | 497, 340, 324, 303, 56 |
| *CDKB1;1* | AT3g54180 | – | 95 |
| *CDKB1;2* | AT2g38620 | – | – |
| *CDKB2;1* | AT1g76540 | – | 84, 76 |
| *CDKB2;2* | AT1g20930 | – | 334, 239, 194, 159, 142, 34 |
| *CDKC;1* | AT5g10270 | – | 935, 696, 518, 378, 225, 198 |
| *CDKC;2* | AT5g64960 | – | 120 |
| *CDKD;1/CAK3* | AT1g73690 | – | 781, 740, 417 |
| *CDKD;2/CAK4* | AT1g66750 | – | 873, 353 |
| *CDKD;3/CAK2* | AT1g18040 | – | – |
| *CDKE;1* | AT5g63610 | – | 35 |
| *CDKF;1/CAK1* | AT4g28980 | – | 441, 112 |
| *CDKG;1* | AT5g63370 | – | – |
| *CDKG;2* | AT1g67580 | – | 832, 646, 345, 222, 209, 193, 5 |
| *CKS1* | AT2g27960 | 161 | – |
| *CKS2* | AT2g27970 | – | 75, 19 |
| *CYCA1;1* | AT1g44110 | 648, 140 | 851, 771 |
| *CYCA1;2* | AT1g77390 | – | 481, 332, 263, 63, 23, 8 |
| *CYCA2;1* | AT5g25380 | – | 782, 767, 728, 550, 377, 340, 292 |
| *CYCA2;2* | AT5g11300 | – | 774, 744, 668, 642, 539, 351, 91 |
| *CYCA2;3* | AT1g15570 | – | 686, 640, 340, 213, 97, 36 |
| *CYCA2;4* | AT1g80370 | 35 | 794, 560, 516, 362 |
| *CYCA3;1* | AT5g43080 | 96 | 426, 334, 305 |
| *CYCA3;2* | AT1g47210 | – | 988, 881, 731, 717, 332, 325, 288, 276, 268, 261 |
| *CYCA3;3* | AT1g47220 | 127 | 743, 681, 573, 471 |
| *CYCA3;4* | AT1g47230 | – | 985, 823, 810, 804, 781, 733, 718, 648, 510, 488, 478, 444, 372, 106 |
| *CYCB1;1* | AT4g37490 | 859 | 526, 289, 136 |
| *CYCB1;2* | AT5g06150 | – | 355, 237 |
| *CYCB1;3* | AT3g11520 | – | 921, 877, 832, 321, 250 |
| *CYCB1;4* | AT2g26760 | 981, 759 | 928, 237 |
| *CYCB1;5* | AT1g34460 | 365, 65 | 819, 733, 441, 347, 315, 305, 179 |
| *CYCB2;1* | AT2g17620 | 440 | 905, 817, 787, 686, 553, 444, 280, 181 |
| *CYCB2;2* | AT4g35620 | – | 854, 739, 295, 185 |
| *CYCB2;3* | AT1g20610 | 666 | 856, 763, 471, 228, 78 |
| *CYCB2;4* | AT1g76310 | – | 832, 644, 388 |
| *CYCB2;5* | AT1g20590 | – | 797, 713, 458, 411, 326, 305, 296 |
| *CYCB3;1* | AT1g16330 | – | 182 |
| *CYCC1;1* | AT5g48640 | – | 970, 325, 256, 51 |
| *CYCC1;2* | AT5g48630 | – | – |
| *CYCD1;1* | AT1g70210 | – | – |
| *CYCD2;1* | AT2g22490 | – | 872, 857, 831, 717, 420, 229, 124 |
| *CYCD3;1* | AT4g34160 | – | 251, 219, 54 |
| *CYCD3;2* | AT5g67260 | 609, 156 | 362, 263, 166 |

(continued)

**Table 7.1** (continued)

| Gene name | AGI ID | AuxRE TGTCTC/ GAGACA | AuxRE core TGTC/GACA |
|---|---|---|---|
| *CYCD3;3* | AT3g50070 | – | 997, 958, 870, 509, 366, 208, 154 |
| *CYCD4;1* | AT5g65420 | – | 641, 434, 264, 71 |
| *CYCD4;2* | AT5g10440 | – | 939, 319, 151 |
| *CYCD5;1* | AT4g37630 | – | 659, 610, 514, 260 |
| *CYCD6;1* | AT4g03270 | 920 | 926, 899, 19 |
| *CYCD7;1* | AT5g02110 | – | 691, 538, 173, 92 |
| *CYCH;1* | AT5g27620 | – | 771, 755, 743, 654, 467, 308, 221, 89, 57 |
| *CYCL1* | AT2g26430 | 835, 727, 496 | 717, 417, 14 |
| *DEL1* | AT3g48160 | – | 518, 428, 341 |
| *DEL2* | AT5g14960 | 425, 50, 7 | 563, 488, 334, 169 |
| *DEL3* | AT3g01330 | – | 491 |
| *DPa* | AT5g02470 | 768 | 921, 840, 749, 597 |
| *DPb* | AT5g03415 | – | – |
| *E2Fa* | AT2g36010 | – | 692, 604, 584 |
| *E2Fb* | AT5g22220 | 478 | 574, 559, 544, 400 |
| *E2Fc* | AT1g47870 | – | 869, 690, 296, 35 |
| *KRP1* | AT2g23430 | 330 | 842, 836, 694, 515, 466, 458, 445, 355, 281, 183 |
| *KRP2* | AT3g50630 | – | 970, 909, 876, 666, 620, 589, 539, 184, 67 |
| *KRP3* | AT5g48820 | – | 889, 749, 340 |
| *KRP4* | AT2g32710 | – | 503, 466, 323, 229 |
| *KRP5* | AT3g24810 | – | 988, 925, 847, 519, 99, 76, 61, 16 |
| *KRP6* | AT3g19150 | – | 507, 457, 267, 73, 63 |
| *KRP7* | AT1g49620 | 407 | 625, 548, 362, 310, 50 |
| *RBR* | AT3g12280 | – | 284, 141, 22 |
| *SIM* | AT5g04470 | – | 704, 605, 563 |
| *SMR1* | AT3g10525 | – | 393, 354 |
| *SMR2* | AT1g08180 | 632 | 828, 788, 623, 600, 523, 446, 264, 198, 26 |
| *SMR3* | AT5g02420 | 10 | 854, 390, 208, 141, 107 |
| *SMR4* | AT5g02220 | – | 687, 375, 318, 12 |
| *SMR5* | AT1g07500 | – | 343 |
| *SMR6* | AT5g40460 | – | 872, 705, 469 |
| *SMR8* | AT1g10690 | – | 846, 787, 691, 365, 242, 167 |
| *SMR11* | AT2g28330 | – | 839, 817, 752, 508, 32 |
| *WEE1* | AT1g02970 | – | – |

This table gives an overview of the presence of auxin response elements (AuxREs) in the upstream region of core cell cycle regulators. The gene list is based on the work of Menges et al. (2005), Vandepoele et al. (2002), and Van Leene et al. (2010). Since variations of the canonical AuxRE sequence have also been shown to be targeted by ARFs, we present the position of both AuxREs (TGTCTC) and their core sequence (TGTC, Ulmasov et al. 1999; Walcher and Nemhauser 2012; Donner et al. 2009). This overview is limited to 1 kb upstream regions and does not exclude the presence of functional AuxREs in other parts of the promoters. One kb sequences upstream of the transcriptional start site were obtained from PLAZA 2.5 (Van Bel et al. 2012). Shorter sequences were extracted if the upstream gene was situated within less than 1 kb. The position of the elements is indicated as the number of nucleotides upstream of the transcriptional start site. *CKS* CDK-SUBUNIT, *DEL* DP-E2F-LIKE

### 3.3 Auxin and Endoreduplication

When plant cells start to differentiate, they often switch from the mitotic cell cycle to the endocycle. During the endocycle, or endoreduplication, the cell proceeds from the S phase directly to G1, without passing through G2 and M (Fig. 7.1). This results in a doubling of the DNA content of the nucleus. This process can occur several times, yielding cells with 4C, 8C, 16C, and even higher ploidy levels instead of the usual 2C in *Arabidopsis*. The biological relevance of endoreduplication has long remained elusive, but recent progress indicates that the cell cycle switch to endoreduplication is essential for proper plant development and for responses to the changing environment (De Veylder et al. 2011).

Different effects of auxin on the endocycle have been reported. In tobacco cell cultures depleted of cytokinin, auxin was shown to induce cell elongation that was accompanied by endoreduplication (Valente et al. 1998). High auxin levels have also been correlated with the induction of endoreduplication in differentiating tissues, for example during endosperm development (Sabelli et al. 2007), in apricot and tomato fruits (Bradley and Crane 1955; Chevalier 2007) and in cultured *Petunia* tissue (Liscum and Hangarter 1991). The work of Ishida et al. (2010) on *Arabidopsis* root tips showed however an opposite effect of auxin on cell ploidy levels and on the transition from meristematic cells to differentiating cells in the root apical meristem. Their data indicate that high auxin levels drive cells through the mitotic cell cycle, while the switch to endocycle is only made when less auxin is present. This switch was correlated with a decreased transcription of M phase-specific genes such as *CYCA2;3* and *CYCB1;1* (Ishida et al. 2010). Earlier work from Magyar et al. (2005) showed that auxin stabilizes E2Fb and induces its expression, which in turn stimulates both the G1/S and G2/M transition. As such auxin promotes cell proliferation instead of cell cycle exit and endoreduplication.

This dual effect of high auxin concentrations on endoreduplication clearly shows the importance of the plant developmental context. Depending on the presence of mitosis promoting factors such as cytokinin, auxin stimulates either cell proliferation or endoreduplication. Either way, auxin accumulation enhances the G1/S transition. When concentrations of mitosis-specific factors are high enough, cells will divide. If this is not the case, they will switch to endoreduplication. Since auxin is involved in the regulation of both S and M phase-specific genes, the auxin level itself might be involved in making the difference. This is in accordance with a proposed mechanism for cell cycle progression that depends on an increasing CDK activity from G1/S to G2/M (De Veylder et al. 2011). As such, cells in which auxin is not able to induce high enough CDK activity to cross the G2/M border would start the endocycle. Nevertheless, it is clear that not only the absolute levels of auxin and cytokinin matter for a balance between mitosis and the endocycle, but the ratio between these two plant hormones is as important.

# 4 An Auxin Gradient at the Root Tip Maintains Pattern and Size of the Apical Meristem

The transcription factors SHORT ROOT (SHR) and SCARECROW (SCR) function as a heterodimer at the root apical stem cell niche to establish the cortex and endodermis ground tissue layers. The continuous production of these layers depends on two successive asymmetric divisions in the cortex-endodermis initial cells and their daughter cells. These divisions are controlled by SHR and SCR and their interaction with RBR. RBR binds to SCR, thereby restricting the functionality of the SHR–SCR heterodimer to induce the formative divisions (Cruz-Ramírez et al. 2012). One of the direct targets of SHR and SCR, amongst other cell cycle genes, is *CYCD6;1* (Sozzani et al. 2010). The CDKB;1–CYCD6;1 complex was shown to phosphorylate RBR, which leads to its inactivation (Cruz-Ramírez et al. 2012). In this way a feedforward loop is established that maintains a bistable circuit in which SHR–SCR activity is either high or low, with high activity inducing asymmetric cell division (Cruz-Ramírez et al. 2012). Auxin can impact on this circuit by its ability to induce *CYCD6;1* expression, although in an SHR–SCR dependent way (Cruz-Ramírez et al. 2012). As such the radial information provided by the cell type-specific expression of SHR and SCR and the longitudinal gradient of auxin at the root tip (Grieneisen et al. 2007, see Chap. 5) together restrict formative divisions to the cortex-endodermis initial cells and their daughter cells. Thereby the correct patterning of the ground tissue is guaranteed. Moreover, Weimer et al. (2012) showed that RBR not only inhibits the transcription of cell cycle genes, but also of genes that are needed for asymmetric divisions and cell fate acquisition. Depending on the activity level of CDKA;1, the RBR-controlled inhibition of only the cell cycle genes or of both cell cycle and specification genes is abrogated. In this way, the asymmetric divisions of the cortex-endodermis daughter cells only occur when CDKA;1 activity is high enough (Weimer et al. 2012).

As described earlier, the auxin gradient at the root tip also controls the size of the root apical meristem. High auxin levels inhibit the transition from the mitotic cell cycle to the endocycle and the coupled transition from meristematic to differentiating cells by affecting the transcription of cell cycle genes (Ishida et al. 2010). The PLETHORA (PLT) transcription factors are involved in maintaining the root meristem size (Aida et al. 2004; Galinha et al. 2007), and their activity can be linked to the auxin gradient. PLT1 and PLT2 are also responsible for correct patterning of the root stem cell niche, in parallel with SCR and SHR. Both *PLT* genes are inducible by auxin via an ARF-dependent pathway (Aida et al. 2004). Their expression follows the auxin gradient at the root tip and determines the size of the meristem: high PLT activity close to the quiescent center maintains stem cell identity, medium levels are linked to mitotic activity of the stem cell daughters, and low levels are correlated with cellular differentiation (Aida et al. 2004; Galinha et al. 2007).

# 5 Auxin Regulates Cell Cycle Reactivation During Lateral Root Initiation

Two important features of lateral root initiation (LRI) have made it a favorable model system to study the interaction between auxin and the cell cycle. First, the inductive effect of auxin treatment on the formation of lateral roots has been known for a long time (Torrey 1950). An increase of endogenous auxin levels results in an overproliferation of lateral roots as well (Boerjan et al. 1995; Celenza et al. 1995; Delarue et al. 1998; King et al. 1995). Second, lateral roots originate from differentiated pericycle cells. The mitotic cell cycle thus needs to be reactivated in these cells to allow the development of new meristematic regions for the production of lateral root primordia. During the last decades many studies have been performed to unravel the tight link between auxin signaling and cell proliferation during the initiation of lateral roots in *Arabidopsis*. After a general introduction to lateral root development, we will give an overview of the interplay between auxin and the cell cycle during LRI.

## *5.1 Lateral Root Development in* Arabidopsis

In *Arabidopsis*, lateral roots originate from those pericycle cells that are located opposite the two xylem strands of the vascular tissue. These cells undergo a series of divisions which lead to the formation of a dome-shaped primordium that emerges through the overlaying tissues (Malamy and Benfey 1997). The first division is asymmetric and is thought to follow respecification of the cells. The primordium becomes an independent lateral root when a new meristem has been established at its tip (Laskowski et al. 1995). When treated with auxin, all xylem pole pericycle cells can be stimulated into LRI, but under normal conditions only a subset of these cells will actually start to proliferate (Himanen et al. 2002). This specification process is preceded by the occurrence of auxin response oscillations in the basal meristem of the main root tip (De Rybel et al. 2010; De Smet et al. 2007; Moreno-Risueno et al. 2010). The basal meristem is the region immediately shootward of the root apical meristem where both cell division and cell elongation occur. The auxin response oscillations take place in the protoxylem strands and induce priming of the neighboring pericycle cells. Hereafter, auxin response maxima are maintained as static spots of expression of the *pDR5::luciferase* auxin response marker (Moreno-Risueno et al. 2010). These spots are referred to as prebranch sites and predict the location where later a lateral root might, but not necessarily will be initiated (Moreno-Risueno et al. 2010; Van Norman et al. 2013). In the prebranch sites, some primed pericycle cells will get specified as lateral root founder cells higher up in the root. Only these specified founder cells will eventually give rise to new lateral root primordia. The moment when the founder cells start to divide is what we refer to as lateral root initiation. It is preceded by increased auxin levels and the occurrence of an auxin signaling maximum in the lateral root founder cells (Benková et al. 2003; Dubrovsky et al. 2008).

From the above, it is clear that auxin is involved in almost every single step of lateral root development, during the formation of prebranch sites, founder cell specification, and primordium initiation, but also later during development and shaping of the primordium. In what follows we only discuss LRI, which coincides with the reactivation of the cell cycle in the founder cells.

## 5.2 *Pericycle Cells Must Maintain the Capacity to Divide*

As mentioned before, in *Arabidopsis* lateral roots originate specifically from pericycle cells that are located next to the two xylem poles that are present in the diarch vasculature. These cells thus have to remain capable of proceeding through mitotic cell division cycles, in contrast to most other root cells. The active transcription of *CDKA;1* in all pericycle cells indicates that they still have the competence to divide, also in mature tissues (Hemerly et al. 1993; Martinez et al. 1992).

One of the main morphological differences that were noted in different species is the shorter length of the root cell type that produces lateral roots compared to other root cells. Lloret et al. (1989) measured the length of both xylem and phloem pole-associated pericycle cells in onion, pea, and carrot and found that the cell files from which lateral roots originate always contained the shortest cells. Similar results were obtained in radish and *Arabidopsis* (Beeckman et al. 2001; Dubrovsky et al. 2000; Laskowski et al. 1995). The data from Lloret et al. (1989) indicate that also non-lateral root founder cells still divide after having left the meristematic region of the root tip. As such LRI would rather be regulated by a switch from symmetric to asymmetric divisions than by the reactivation of the mitotic cell cycle (Lloret et al. 1989). This hypothesis is in accordance with the data from Dubrovsky et al. (2000) who have shown the occurrence of symmetric, proliferative divisions in cells in between initiating primordia in *Arabidopsis*. However, when LRI is induced, for example by auxin treatment, in mature parts of the root that are located further away from the root meristem, a reactivation of the cell cycle is needed (Laskowski et al. 1995).

## 5.3 *A Lateral Root Induction System Synchronizes Auxin-Induced Cell Cycle Reactivation*

A great part of the studies on cell cycle regulation during LRI has been done in lateral root inducing conditions. Since lateral root formation occurs in an acropetal manner, different developmental stages of primordia can be found along the length of the primary root. As such it is hard to pinpoint the cells wherein a specific developmental stage of LRI is occurring. The small number of cells involved in early lateral root development makes these studies even harder. By inducing the initiation of lateral roots with auxin treatment, it is possible to synchronize the development of lateral root primordia over the entire length of the root. This greatly facilitates the study of specific steps during lateral root formation.

A commonly used method to obtain such synchronization of lateral root development and the associated cell cycle reactivation is the lateral root induction system (LRIS) developed by Himanen et al. (2002). Plants are germinated on the auxin transport inhibitor *N*-1-naphthylphthalamic acid (NPA) to block the formation of lateral roots. Subsequent treatment with the synthetic auxin 1-naphthalene acetic acid (NAA) triggers the simultaneous development of lateral roots at both strands of xylem pole pericycle cells along the entire main root. This system has been used frequently to study the cell cycle reactivation during auxin-induced LRI (for example by de Almeida Engler et al. 2009; Himanen et al. 2002, 2004; Vanneste et al. 2005) and the main findings will be summarized below.

As discussed before, the initiation of lateral roots in the LRIS differs from non-induced initiation, because the latter does not occur in fully matured pericycle cells. A second difference can be found in the cell cycle phase in which the cells are residing before being recruited for LRI. Blakely and Evans (1979) have shown that in radish seedling roots all pericycle cells arrest in G2 and can be stimulated by auxin to reactivate the cell cycle. An arrest in G2 can also clarify the very fast response to auxin treatment, with divisions already occurring 2 h after the start of the treatment (Blakely and Evans 1979). The experiments of Beeckman et al. (2001) also showed evidence for the occurrence of LRI in cells that are residing in the G2 phase. In the LRIS in *Arabidopsis*, however, LRI starts from G1 cells (Himanen et al. 2002, 2004), while recently Jansen et al. (2013a) demonstrated a G2 start using an adapted LRIS in maize.

## 5.4 Auxin-Induced Cell Cycle Activity During Lateral Root Initiation

During naturally occurring and auxin-induced LRI, many cell cycle genes can be found that show changes in their transcriptional regulation. Most of these changes, such as upregulation of A, B, and D-type *CYCLINS*, *E2Fa*, *DPa*, and *CDKB* as well as downregulation of *KRP1* and *KRP2* (Beeckman et al. 2001; de Almeida Engler et al. 2009; De Veylder et al. 1999; Doerner et al. 1996; Ferreira et al. 1994a; Himanen et al. 2002, 2004; Vanneste et al. 2005), are linked to progression of the cell cycle.

Auxin not only represses the transcription of *KRP2* but also lowers its protein levels (Himanen et al. 2002; Richard et al. 2002; Sanz et al. 2011). KRP2 prevents formative divisions in the pericycle and overexpression reduces the number of lateral roots (Himanen et al. 2002). Correspondingly, its expression can be seen at sites where no lateral roots are to be formed, such as phloem pole pericycle cells and xylem associated pericycle cells opposite developing primordia (Himanen et al. 2002), suggesting a role in the strict bilateral origin of lateral roots in *Arabidopsis*. Sanz et al. (2011) reported that KRP2 binds CYCD2;1, thereby probably keeping the CDKA;1–CYCD2;1 complexes inactive. KRP1 is likely to act in a similar way to KRP2 during LRI. It also inhibits auxin-induced LRI, is downregulated by auxin, and interacts with CDKA;1–CYCD2;1 complexes (Himanen et al. 2002; Ren et al. 2008). Repression of KRP activity and induction of positive cell cycle regulators by auxin can thus account for a reactivation of the cell cycle.

It has however become clear that an induction of cell proliferation is not enough to stimulate lateral root development. Overexpression of the G1/S regulators *CYCD3;1* or *E2Fa/DPa* or of the mitotic *CYCB1;1* induces divisions in the pericycle, but these are symmetric and do not lead to lateral root primordium establishment (De Smet et al. 2010; Doerner et al. 1996; Vanneste et al. 2005). Auxin signaling is thus responsible for cell specification next to reactivation of the cell cycle. Both processes were shown to act through the SOLITARY ROOT (SLR)/IAA14-ARF7-ARF19 auxin signaling module (Fukaki et al. 2002, 2005, 2006; Okushima et al. 2005; Vanneste et al. 2005; Wilmoth et al. 2005).

## 5.5 Auxin-Induced Cell Specification During Lateral Root Initiation

Specification of cells is often correlated with asymmetric cell divisions (De Smet and Beeckman 2011). This is also the case during LRI, in which the first division of the founder cells is preceded by a nuclear migration that leads to the asymmetric nature of the division. Evidence has been found that auxin-dependent expression of the *LATERAL ORGAN BOUNDARIES-DOMAIN/ASYMMETRIC LEAVES2-LIKE* (*LBD/ASL*) gene family is involved in establishing the polarity of these divisions.

Five members of the *LBD/ASL* family have been shown to be transcriptionally regulated by ARF7/19 and to be expressed during lateral root development, being *LBD16/ASL18*, *LBD17/ASL15*, *LBD18/ASL20*, *LBD29/ASL16*, and *LBD33/ASL24* (hereafter referred to as *LBD16*, *LBD17*, *LBD18*, *LBD29*, and *LBD33*, Goh et al. 2012; Okushima et al. 2005). Okushima et al. (2007) showed that ARF7 and ARF19 induce the expression of *LBD16* and *LBD29* through direct binding to AuxREs in their promoter regions. LBD16 activity in the founder cells was later found to be required for nuclear migration and to establish the asymmetric nature of the first divisions during LRI (Goh et al. 2012). Overexpression of *LBD16*, *LBD18*, or *LBD29*, but not *LBD33*, can induce the formation of lateral roots in the *arf7arf19* lateral rootless mutant background (Goh et al. 2012; Lee et al. 2009; Okushima et al. 2007). Moreover, LBD18 and LBD33 were shown to function as a heterodimer to induce the expression of *E2Fa* in lateral root founder cells by

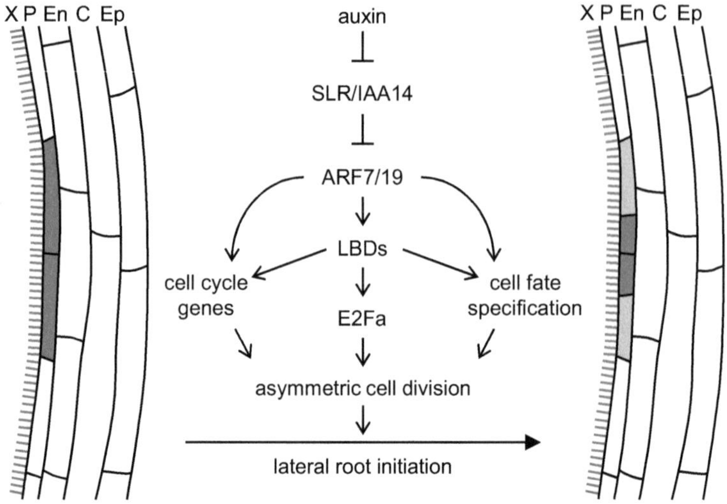


**Fig. 7.2** LBD transcription factors regulate cell cycle activity and cell specification during lateral root initiation. During auxin-induced lateral root initiation the cell cycle must be reactivated and the founder cells need to be respecified. Both processes depend on auxin signaling via SLR/IAA14 and ARF7/19. LBD transcription factors are directly targeted by ARF7/19 and subsequently also directly induce *E2Fa* expression. Both ARFs and LBDs probably also target other genes that are involved in cell cycle activity and cell fate specification. As such an asymmetric cell division is established that ensures proper lateral root initiation. *Gray*-colored cells have elevated auxin response levels. *X* xylem, *P* pericycle, *En* endodermis, *C* cortex, *Ep* epidermis

binding directly to its promoter (Berckmans et al. 2011). LBD29 levels have also been linked to changes in the expression of several cell cycle genes (Feng et al. 2012). Altogether, these data indicate that the *LBD* genes play an important role during LRI by linking auxin signaling downstream of ARF7 and ARF19 to cell cycle activity and cell specification (Fig. 7.2).

## 6 Cell Cycle Reactivation During the Formation of Feeding Sites, Nodules and Callus Has Similarities with Lateral Root Initiation

There are two processes known in plants that depend on the reactivation of the cell cycle in a similar way as is the case for LRI, namely the production of feeding cells by plant-parasitic nematodes and nitrogen-fixing nodule formation in leguminous plants. During both processes the cell cycle is activated in differentiated root cells and in most cases this occurs in the proximity of the xylem poles (Gheysen and Mitchum 2011; Goverse et al. 2000; Grunewald et al. 2009b). Moreover, the development of these organs is linked to an increased auxin response during the early events (Grunewald et al. 2009a, b; Takanashi et al. 2011; van Noorden et al. 2007), similar to the accumulation of auxin in founder cells before LRI (Benková et al. 2003; Dubrovsky et al. 2008). Concordantly, a large number of proteins were shown to be coregulated during auxin treatment and nodule-inducing *Rhizobium* infection (van Noorden et al. 2007).

Both nematodes and *Rhizobium* bacteria affect auxin distribution, as such generating high auxin concentrations at the sites of infection (Gheysen and Mitchum 2011; Grunewald et al. 2009a, b; Hewezi and Baum 2013; see Chap. 18). This leads, amongst others, to the upregulation of genes involved in cell proliferation. *CDKA;1*, *CDKB1;1*, *CYCA2;1*, and *CYCB1;1* are upregulated during nematode infection, at the site where a feeding cell will be established (de Almeida Engler et al. 1999; Niebel et al. 1996). In the legume *Medicago*, *CYCA2;2* was shown to be upregulated in the proliferating cells involved in nodule formation (Roudier et al. 2003). *KRP1* and *KRP2*, inhibitors of LRI (see higher, Himanen et al. 2002), also have an inhibitory effect on the formation of feeding sites and nodules (Ren et al. 2008; Vieira et al. 2013). All this points to an important role for auxin during the formation of nodules and nematode feeding sites in plant roots. Since the same cell types are involved as for LRI, it can be assumed that similar pathways are activated to translate the auxin signal into stimulation of the cell cycle machinery. This idea is enforced by the fact that nematode infection is correlated with the production of lateral roots at feeding sites (Goverse et al. 2000).

Another process with high similarity to LRI is the in vitro regeneration of plants through callus formation, which also requires elevated auxin levels (Gordon et al. 2007). Callus induction from both root and shoot explants was shown to follow a lateral-root-development pathway (Sugimoto et al. 2010) and callus from root and hypocotyl explants is initiated from xylem-pole-associated pericycle cells (Atta et al. 2009), which are also responsible for LRI. Moreover, xylem-pole-pericycle-like cells were found to be involved in callus formation from cotyledons and petals (Sugimoto et al. 2010).

Four of the LBD transcription factors functioning during LRI are also required for callus induction (Fan et al. 2012; Xu et al. 2012). In line with the data on LRI, upregulation of *LBD16*, *17*, *18*, and *29* during callus formation was shown to be dependent on auxin signaling via ARF7 and ARF19 (Fan et al. 2012). ABERRANT LATERAL ROOT FORMATION 4 (ALF4) is another protein that has been shown to be essential during LRI as well as callus induction (Celenza et al. 1995; DiDonato et al. 2004; Sugimoto et al. 2010). ALF4 functions independently of auxin signaling and seems to have a general effect on cell proliferation competence in different tissues and plant organs (Celenza et al. 1995; Chupeau et al. 2013; DiDonato et al. 2004; Vanneste et al. 2005).

Clearly, a common auxin-dependent pathway is used for the initiation of lateral roots, the induction of feeding sites and nodules and the formation of callus. The distinction between these processes most probably depends on the combination of elevated auxin levels with other growth regulators. An example of this is the switch from callus proliferation to root regeneration when the auxin-to-cytokinin ratio in the growth medium is increased.

## 7 Perspectives

From all the studies that have been done to clarify the interplay between auxin and the cell cycle, it is clear that a tight link exists between cellular auxin levels and the ability of cells to maintain a proliferative status. For some of the developmental processes that we have described here, factors bridging these auxin levels with cell cycle regulatory genes have been elucidated. There are of course many more processes in which both auxin signaling and cell cycle activity play prominent roles, for example during embryogenesis, the initiation of shoot-derived organs, leaf development, vasculature and cambium development, and lateral root formation beyond the initiation stage. In the future, we expect more links between auxin and cell cycle regulation to be revealed. It will be interesting to find out if identical pathways are employed in distinct plant organs. This seems possible, since, similar to their function in the root tip, SHR, SCR, RBR, and CYCD6;1 activity have been correlated and linked to progression of the cell cycle during leaf development in *Arabidopsis* (Dhondt et al. 2010). Due to the extensiveness of some of the gene families involved in linking auxin to the cell cycle during root development (for example the *Aux/IAA*, *ARF*, *CDK*, *CYCLIN* and *LBD* families), it is also plausible that other members of these families will be found to perform similar functions in other tissues. It is however very well possible that other genes will also emerge as signal transducers between auxin and cell proliferation.

Most of the results described here were generated from research on *Arabidopsis*. Comparisons with other species will tell us more about the evolutionary conservation of these pathways. Studies with a recently developed lateral root induction system for maize (Jansen et al. 2013b) will certainly shed more light on this aspect. The first results point toward conserved mechanisms between mono- and dicotyledonous plants concerning auxin-induced lateral root initiation (Jansen et al. 2013a).

**Acknowledgments** M.J.F.D. is indebted to the Agency for Innovation by Science and Technology in Flanders (IWT) for a predoctoral grant. Work in the lab of T.B. is supported by the Interuniversity Attraction Poles Programme IAP7/29 from the Belgian Federal Science Policy Office.

## References

Aida M, Beis D, Heidstra R, Willemsen V, Blilou I, Galinha C, Nussaume L, Noh Y-S, Amasino R, Scheres B (2004) The *PLETHORA* genes mediate patterning of the *Arabidopsis* root stem cell niche. Cell 119(1):109–120. doi:10.1016/j.cell.2004.09.018

Atta R, Laurens L, Boucheron-Dubuisson E, Guivarc'h A, Carnero E, Giraudat-Pautot V, Rech P, Chriqui D (2009) Pluripotency of Arabidopsis xylem pericycle underlies shoot regeneration from root and hypocotyl explants grown *in vitro*. Plant J 57(4):626–644. doi:10.1111/j.1365-313X.2008.03715.x

Bayliss MW (1985) Regulation of the cell division cycle in cultured plant cells. In: Bryant JA, Francis D (eds) The cell division cycle in plants. Cambridge University Press, Cambridge, pp 157–177

Beeckman T, Burssens S, Inzé D (2001) The peri-*cell*-cycle in *Arabidopsis*. J Exp Bot 52(Spec Issue):403–411

Benková E, Michniewicz M, Sauer M, Teichmann T, Seifertová D, Jürgens G, Friml J (2003) Local, efflux-dependent auxin gradients as a common module for plant organ formation. Cell 115(5):591–602

Berckmans B, Vassileva V, Schmid SPC, Maes S, Parizot B, Naramoto S, Magyar Z, Alvim Kamei CL, Koncz C, Bögre L, Persiau G, De Jaeger G, Friml J, Simon R, Beeckman T, De Veylder L (2011) Auxin-dependent cell cycle reactivation through transcriptional regulation of *Arabidopsis E2Fa* by lateral organ boundary proteins. Plant Cell 23(10):3671–3683. doi:10.1105/tpc.111.088377

Blakely LM, Evans TA (1979) Cell dynamics studies on the pericycle of radish seedling roots. Plant Sci Lett 14(1):79–83

Boerjan W, Cervera M-T, Delarue M, Beeckman T, Dewitte W, Bellini C, Caboche M, Van Onckelen H, Van Montagu M, Inzé D (1995) *superroot*, a recessive mutation in Arabidopsis, confers auxin overproduction. Plant Cell 7(9):1405–1419

Bradley MV, Crane JC (1955) The effect of 2,4,5-trichlorophenoxyacetic acid on cell and nuclear size and endopolyploidy in parenchyma of apricot fruits. Am J Bot 42(3):273–281

Burssens S, de Almeida Engler J, Beeckman T, Richard C, Shaul O, Ferreira P, Van Montagu M, Inzé D (2000) Developmental expression of the *Arabidopsis thaliana CycA2;1* gene. Planta 211(5):623–631

Celenza JL Jr, Grisafi PL, Fink GR (1995) A pathway for lateral root formation in *Arabidopsis thaliana*. Genes Dev 9(17):2131–2142

Chevalier C (2007) Cell cycle control and fruit development. In: Inzé D (ed) Cell cycle control and plant development, vol 32, Annual plant reviews. Blackwell, Oxford, UK

Chupeau M-C, Granier F, Pichon O, Renou J-P, Gaudin V, Chupeau Y (2013) Characterization of the early events leading to totipotency in an *Arabidopsis* protoplast liquid culture by temporal transcript profiling. Plant Cell. doi:10.1105/tpc.113.109538

Churchman ML, Brown ML, Kato N, Kirik V, Hülskamp M, Inzé D, De Veylder L, Walker JD, Zheng Z, Oppenheimer DG, Gwin T, Churchman J, Larkin JC (2006) SIAMESE, a plant-specific cell cycle regulator, controls endoreplication onset in *Arabidopsis thaliana*. Plant Cell 18(11):3145–3157. doi:10.1105/tpc.106.044834

Cools T, Iantcheva A, Weimer AK, Boens S, Takahashi N, Maes S, Van den Daele H, Van Isterdael G, Schnittger A, De Veylder L (2011) The *Arabidopsis thaliana* checkpoint kinase WEE1 protects against premature vascular differentiation during replication stress. Plant Cell 23(4):1435–1448. doi:10.1105/tpc.110.082768

Cruz-Ramírez A, Díaz-Triviño S, Blilou I, Grieneisen VA, Sozzani R, Zamioudis C, Miskolczi P, Nieuwland J, Benjamins R, Dhonukshe P, Caballero-Pérez J, Horvath B, Long Y, Mähönen AP, Zhang H, Xu J, Murray JAH, Benfey PN, Bako L, Marée AFM, Scheres B (2012) A bistable circuit involving SCARECROW-RETINOBLASTOMA integrates cues to inform asymmetric stem cell division. Cell 150(5):1002–1015. doi:10.1016/j.cell.2012.07.017

de Almeida Engler J, De Vleesschauwer V, Burssens S, Celenza JL Jr, Inzé D, Van Montagu M, Engler G, Gheysen G (1999) Molecular markers and cell cycle inhibitors show the importance of cell cycle progression in nematode-induced galls and syncytia. Plant Cell 11(5):793–808

de Almeida Engler J, De Veylder L, De Groodt R, Rombauts S, Boudolf V, De Meyer B, Hemerly A, Ferreira P, Beeckman T, Karimi M, Hilson P, Inzé D, Engler G (2009) Systematic analysis of cell-cycle gene expression during Arabidopsis development. Plant J 59(4):645–660. doi:10.1111/j.1365-313X.2009.03893.x

De Rybel B, Vassileva V, Parizot B, Demeulenaere M, Grunewald W, Audenaert D, Van Campenhout J, Overvoorde P, Jansen L, Vanneste S, Möller B, Wilson M, Holman T, Van Isterdael G, Brunoud G, Vuylsteke M, Vernoux T, De Veylder L, Inzé D, Weijers D, Bennett

MJ, Beeckman T (2010) A novel Aux/IAA28 signaling cascade activates GATA23-dependent specification of lateral root founder cell identity. Curr Biol 20(19):1697–1706. doi:10.1016/j.cub.2010.09.007, S0960-9822(10)01088-2 [pii]

De Schutter K, Joubès J, Cools T, Verkest A, Corellou F, Babiychuk E, Van Der Schueren E, Beeckman T, Kushnir S, Inzé D, De Veylder L (2007) *Arabidopsis* WEE1 kinase controls cell cycle arrest in response to activation of the DNA integrity checkpoint. Plant Cell 19(1):211–225. doi:10.1105/tpc.106.045047

De Smet I, Beeckman T (2011) Asymmetric cell division in land plants and algae: the driving force for differentiation. Nat Rev Mol Cell Biol 12(3):177–188. doi:10.1038/nrm3064

De Smet I, Tetsumura T, De Rybel B, Frei dit Frey N, Laplaze L, Casimiro I, Swarup R, Naudts M, Vanneste S, Audenaert D, Inzé D, Bennett MJ, Beeckman T (2007) Auxin-dependent regulation of lateral root positioning in the basal meristem of *Arabidopsis*. Development 134(4):681–690

De Smet I, Lau S, Voß U, Vanneste S, Benjamins R, Rademacher EH, Schlereth A, De Rybel B, Vassileva V, Grunewald W, Naudts M, Levesque MP, Ehrismann JS, Inzé D, Luschnig C, Benfey PN, Weijers D, Van Montagu MCE, Bennett MJ, Jürgens G, Beeckman T (2010) Bimodular auxin response controls organogenesis in *Arabidopsis*. Proc Natl Acad Sci U S A 107(6):2705–2710. doi:10.1073/pnas.0915001107

De Veylder L, de Almeida Engler J, Burssens S, Manevski A, Lescure B, Van Montagu M, Engler G, Inzé D (1999) A new D-type cyclin of *Arabidopsis thaliana* expressed during lateral root primordia formation. Planta 208(4):453–462

De Veylder L, Larkin JC, Schnittger A (2011) Molecular control and function of endoreplication in development and physiology. Trends Plant Sci 16(11):624–634. doi:10.1016/j.tplants.2011.07.001

del Pozo JC, Manzano C (2013) Auxin and the ubiquitin pathway. Two players-one target: the cell cycle in action. J Exp Bot. doi:10.1093/jxb/ert363

del Pozo JC, Boniotti MB, Gutierrez C (2002) Arabidopsis E2Fc functions in cell division and is degraded by the ubiquitin-$SCF^{AtSKP2}$ pathway in response to light. Plant Cell 14(12):3057–3071

del Pozo JC, Diaz-Trivino S, Cisneros N, Gutierrez C (2006) The balance between cell division and endoreplication depends on E2FC-DPB, transcription factors regulated by the ubiquitin-$SCF^{SKP2A}$ pathway in *Arabidopsis*. Plant Cell 18(9):2224–2235. doi:10.1105/tpc.105.039651

Delarue M, Prinsen E, Van Onckelen H, Caboche M, Bellini C (1998) *Sur2* mutations of *Arabidopsis thaliana* define a new locus involved in the control of auxin homeostasis. Plant J 14(5):603–611

Dhondt S, Coppens F, De Winter F, Swarup K, Merks RMH, Inzé D, Bennett MJ, Beemster GTS (2010) SHORT-ROOT and SCARECROW regulate leaf growth in Arabidopsis by stimulating S-phase progression of the cell cycle. Plant Physiol 154(3):1183–1195. doi:10.1104/pp.110.158857

DiDonato RJ, Arbuckle E, Buker S, Sheets J, Tobar J, Totong R, Grisafi P, Fink GR, Celenza JL (2004) *Arabidopsis ALF4* encodes a nuclear-localized protein required for lateral root formation. Plant J 37(3):340–353. doi:10.1046/j.1365-313X.2003.01964.x

Doerner P, Jørgensen J-E, You R, Steppuhn J, Lamb C (1996) Control of root growth and development by cyclin expression. Nature 380(6574):520–523. doi:10.1038/380520a0

Donner TJ, Sherr I, Scarpella E (2009) Regulation of preprocambial cell state acquisition by auxin signaling in *Arabidopsis* leaves. Development 136(19):3235–3246. doi:10.1242/dev.037028

Dubrovsky JG, Doerner PW, Colón-Carmona A, Rost TL (2000) Pericycle cell proliferation and lateral root initiation in Arabidopsis. Plant Physiol 124(4):1648–1657

Dubrovsky JG, Sauer M, Napsucialy-Mendivil S, Ivanchenko MG, Friml J, Shishkova S, Celenza J, Benková E (2008) Auxin acts as a local morphogenetic trigger to specify lateral root founder cells. Proc Natl Acad Sci U S A 105(25):8790–8794. doi:10.1073/pnas.0712307105

Dudits D, Ábrahám E, Miskolczi P, Ayaydin F, Bilgin M, Horváth GV (2011) Cell-cycle control as a target for calcium, hormonal and developmental signals: the role of phosphorylation in the retinoblastoma-centred pathway. Ann Bot 107(7):1193–1202. doi:10.1093/aob/mcr038

Fan M, Xu C, Xu K, Hu Y (2012) LATERAL ORGAN BOUNDARIES DOMAIN transcription factors direct callus formation in *Arabidopsis* regeneration. Cell Res 22(7):1169–1180. doi:10.1038/cr.2012.63

Feng Z, Sun X, Wang G, Liu H, Zhu J (2012) LBD29 regulates the cell cycle progression in response to auxin during lateral root formation in *Arabidopsis thaliana*. Ann Bot 110(1):1–10. doi:10.1093/aob/mcs019

Ferreira P, Hemerly A, de Almeida Engler J, Bergounioux C, Burssens S, Van Montagu M, Engler G, Inzé D (1994a) Three discrete classes of *Arabidopsis* cyclins are expressed during different intervals of the cell cycle. Proc Natl Acad Sci U S A 91(24):11313–11317

Ferreira PCG, Hemerly AS, de Almeida Engler J, Van Montagu M, Engler G, Inzé D (1994b) Developmental expression of the Arabidopsis cyclin gene *cyc1At*. Plant Cell 6(12):1763–1774

Fukaki H, Tameda S, Masuda H, Tasaka M (2002) Lateral root formation is blocked by a gain-of-function mutation in the *SOLITARY-ROOT/IAA14* gene of *Arabidopsis*. Plant J 29(2):153–168

Fukaki H, Nakao Y, Okushima Y, Theologis A, Tasaka M (2005) Tissue-specific expression of stabilized SOLITARY-ROOT/IAA14 alters lateral root development in Arabidopsis. Plant J 44 (3):382–395. doi:10.1111/j.1365-313X.2005.02537.x

Fukaki H, Taniguchi N, Tasaka M (2006) PICKLE is required for SOLITARY-ROOT/IAA14-mediated repression of ARF7 and ARF19 activity during Arabidopsis lateral root initiation. Plant J 48(3):380–389. doi:10.1111/j.1365-313X.2006.02882.x

Galinha C, Hofhuis H, Luijten M, Willemsen V, Blilou I, Heidstra R, Scheres B (2007) PLETHORA proteins as dose-dependent master regulators of *Arabidopsis* root development. Nature 449(7165):1053–1057. doi:10.1038/nature06206

Gheysen G, Mitchum MG (2011) How nematodes manipulate plant development pathways for infection. Curr Opin Plant Biol 14(4):415–421. doi:10.1016/j.pbi.2011.03.012

Goh T, Joi S, Mimura T, Fukaki H (2012) The establishment of asymmetry in *Arabidopsis* lateral root founder cells is regulated by LBD16/ASL18 and related LBD/ASL proteins. Development 139(5):883–893. doi:10.1242/dev.071928

Gordon SP, Heisler MG, Reddy GV, Ohno C, Das P, Meyerowitz EM (2007) Pattern formation during de novo assembly of the *Arabidopsis* shoot meristem. Development 134(19):3539–3548. doi:10.1242/dev.010298

Goverse A, de Almeida Engler J, Verhees J, van der Krol S, Helder J, Gheysen G (2000) Cell cycle activation by plant parasitic nematodes. Plant Mol Biol 43(5–6):747–761

Grieneisen VA, Xu J, Marée AFM, Hogeweg P, Scheres B (2007) Auxin transport is sufficient to generate a maximum and gradient guiding root growth. Nature 449(7165):1008–1013. doi:10.1038/nature06215

Grunewald W, Cannoot B, Friml J, Gheysen G (2009a) Parasitic nematodes modulate PIN-mediated auxin transport to facilitate infection. PLoS Pathog 5(1):e1000266. doi:10.1371/journal.ppat.1000266

Grunewald W, van Noorden G, Van Isterdael G, Beeckman T, Gheysen G, Mathesius U (2009b) Manipulation of auxin transport in plant roots during *Rhizobium* symbiosis and nematode parasitism. Plant Cell 21(9):2553–2562. doi:10.1105/tpc.109.069617

Hemerly AS, Ferreira P, de Almeida Engler J, Van Montagu M, Engler G, Inzé D (1993) *cdc2a* expression in Arabidopsis is linked with competence for cell division. Plant Cell 5(12):1711–1723. doi:10.1105/tpc.5.12.1711

Hewezi T, Baum TJ (2013) Manipulation of plant cells by cyst and root-knot nematode effectors. Mol Plant Microbe Interact 26(1):9–16. doi:10.1094/MPMI-05-12-0106-FI

Heyman J, De Veylder L (2012) The anaphase-promoting complex/cyclosome in control of plant development. Mol Plant 5(6):1182–1194. doi:10.1093/mp/sss094

Himanen K, Boucheron E, Vanneste S, de Almeida Engler J, Inzé D, Beeckman T (2002) Auxin-mediated cell cycle activation during early lateral root initiation. Plant Cell 14(10):2339–2351. doi:10.1105/tpc.004960

Himanen K, Vuylsteke M, Vanneste S, Vercruysse S, Boucheron E, Alard P, Chriqui D, Van Montagu M, Inzé D, Beeckman T (2004) Transcript profiling of early lateral root initiation. Proc Natl Acad Sci U S A 101(14):5146–5151. doi:10.1073/pnas.0308702101

Inzé D, De Veylder L (2006) Cell cycle regulation in plant development. Annu Rev Genet 40:77–105. doi:10.1146/annurev.genet.40.110405.090431

Ishida T, Adachi S, Yoshimura M, Shimizu K, Umeda M, Sugimoto K (2010) Auxin modulates the transition from the mitotic cycle to the endocycle in *Arabidopsis*. Development 137(1):63–71. doi:10.1242/dev.035840

Ito M (2000) Factors controlling cyclin B expression. Plant Mol Biol 43(5–6):677–690

Jansen L, Hollunder J, Roberts I, Forestan C, Fonteyne P, Van Quickenborne C, Zhen R-G, McKersie B, Parizot B, Beeckman T (2013a) Comparative transcriptomics as a tool for the identification of root branching genes in maize. Plant Biotechnol J. doi:10.1111/pbi.12104

Jansen L, Parizot B, Beeckman T (2013b) Inducible system for lateral roots in *Arabidopsis thaliana* and maize. Methods Mol Biol 959:149–158. doi:10.1007/978-1-62703-221-6_9

John PCL (2007) Hormonal regulation of cell cycle progression and its role in development. In: Inzé D (ed) Cell cycle control and plant development. Blackwell, Oxford, pp 311–334

John PCL, Zhang K, Dong C, Diederich L, Wightman F (1993) $p34^{cdc2}$ related proteins in control of cell cycle progression, the switch between division and differentiation in tissue development, and stimulation of division by auxin and cytokinin. Aust J Plant Physiol 20(5):503–526

Jurado S, Abraham Z, Manzano C, López-Torrejón G, Pacios LF, Del Pozo JC (2010) The *Arabidopsis* cell cycle F-box protein SKP2A binds to auxin. Plant Cell 22(12):3891–3904. doi:10.1105/tpc.110.078972

King JJ, Stimart DP, Fisher RH, Bleecker AB (1995) A mutation altering auxin homeostasis and plant morphology in Arabidopsis. Plant Cell 7(12):2023–2037. doi:10.1105/tpc.7.12.2023

Landrieu I, da Costa M, De Veylder L, Dewitte F, Vandepoele K, Hassan S, Wieruszeski J-M, Corellou F, Faure J-D, Van Montagu M, Inzé D, Lippens G (2004) A small CDC25 dual-specificity tyrosine-phosphatase isoform in *Arabidopsis thaliana*. Proc Natl Acad Sci U S A 101(36):13380–13385. doi:10.1073/pnas.0405248101

Laskowski MJ, Williams ME, Nusbaum HC, Sussex IM (1995) Formation of lateral root meristems is a two-stage process. Development 121(10):3303–3310

Lee HW, Kim NY, Lee DJ, Kim J (2009) *LBD18/ASL20* regulates lateral root formation in combination with *LBD16/ASL18* downstream of *ARF7* and *ARF19* in Arabidopsis. Plant Physiol 151(3):1377–1389. doi:10.1104/pp.109.143685

Liscum E, Hangarter RP (1991) Manipulation of ploidy level in cultured haploid *Petunia* tissue by phytohormone treatments. J Plant Physiol 138(1):33–38

Lloret PG, Casero PJ, Pulgarín A, Navascués J (1989) The behaviour of two cell populations in the pericycle of *Allium cepa*, *Pisum sativum*, and *Daucus carota* during early lateral root development. Ann Bot 63:465–475

Magyar Z, De Veylder L, Atanassova A, Bakó L, Inzé D, Bögre L (2005) The role of the *Arabidopsis* E2FB transcription factor in regulating auxin-dependent cell division. Plant Cell 17(9):2527–2541. doi:10.1105/tpc.105.033761

Malamy JE, Benfey PN (1997) Organization and cell differentiation in lateral roots of *Arabidopsis thaliana*. Development 124(1):33–44

Manzano C, Ramirez-Parra E, Casimiro I, Otero S, Desvoyes B, De Rybel B, Beeckman T, Casero P, Gutierrez C, del Pozo JC (2012) Auxin and epigenetic regulation of *SKP2B*, an F-box that represses lateral root formation. Plant Physiol 160(2):749–762. doi:10.1104/pp.112.198341

Martinez MC, Jørgensen J-E, Lawton MA, Lamb CJ, Doerner PW (1992) Spatial pattern of *cdc2* expression in relation to meristem activity and cell proliferation during plant development. Proc Natl Acad Sci U S A 89(16):7360–7364

Menges M, de Jager SM, Gruissem W, Murray JAH (2005) Global analysis of the core cell cycle regulators of Arabidopsis identifies novel genes, reveals multiple and highly specific profiles of expression and provides a coherent model for plant cell cycle control. Plant J 41(4):546–566. doi:10.1111/j.1365-313X.2004.02319.x

Monshausen GB (2012) Visualizing $Ca^{2+}$ signatures in plants. Curr Opin Plant Biol 15(6):677–682. doi:10.1016/j.pbi.2012.09.014

Monshausen GB, Miller ND, Murphy AS, Gilroy S (2011) Dynamics of auxin-dependent $Ca^{2+}$ and pH signaling in root growth revealed by integrating high-resolution imaging with automated computer vision-based analysis. Plant J 65(2):309–318. doi:10.1111/j.1365-313X.2010.04423.x

Moreno-Risueno MA, Van Norman JM, Moreno A, Zhang J, Ahnert SE, Benfey PN (2010) Oscillating gene expression determines competence for periodic *Arabidopsis* root branching. Science 329(5997):1306–1311. doi:10.1126/science.1191937

Murashige T, Skoog F (1962) A revised medium for rapid growth and bio assays with Tobacco tissue cultures. Physiol Plant 15:473–497

Murray JAH, Freeman D, Greenwood J, Huntley R, Makkerh J, Riou-Khamlichi C, Sorrell DA, Cockcroft C, Carmichael JP, Soni R, Shah ZH (1998) Plant D cyclins and retinoblastoma protein homologues. In: Francis D, Dudits D, Inzé D (eds) Plant cell division. Portland Press, London, pp 99–127

Niebel A, de Almeida Engler J, Hemerly A, Ferreira P, Inzé D, Van Montagu M, Gheysen G (1996) Induction of *cdc2a* and *cyc1At* expression in *Arabidopsis thaliana* during early phases of nematode-induced feeding cell formation. Plant J 10(6):1037–1043

Okushima Y, Overvoorde PJ, Arima K, Alonso JM, Chan A, Chang C, Ecker JR, Hughes B, Lui A, Nguyen D, Onodera C, Quach H, Smith A, Yu G, Theologis A (2005) Functional genomic analysis of the *AUXIN RESPONSE FACTOR* gene family members in *Arabidopsis thaliana*: unique and overlapping functions of *ARF7* and *ARF19*. Plant Cell 17(2):444–463. doi:10.1105/tpc.104.028316

Okushima Y, Fukaki H, Onoda M, Theologis A, Tasaka M (2007) ARF7 and ARF19 regulate lateral root formation via direct activation of *LBD*/*ASL* genes in *Arabidopsis*. Plant Cell 19 (1):118–130. doi:10.1105/tpc.106.047761

Orchard CB, Siciliano I, Sorrell DA, Marchbank A, Rogers HJ, Francis D, Herbert RJ, Suchomelova P, Lipavska H, Azmi A, Van Onckelen H (2005) Tobacco BY-2 cells expressing fission yeast *cdc25* bypass a G2/M block on the cell cycle. Plant J 44(2):290–299. doi:10.1111/j.1365-313X.2005.02524.x

Peres A, Churchman ML, Hariharan S, Himanen K, Verkest A, Vandepoele K, Magyar Z, Hatzfeld Y, Van Der Schueren E, Beemster GTS, Frankard V, Larkin JC, Inzé D, De Veylder L (2007) Novel plant-specific cyclin-dependent kinase inhibitors induced by biotic and abiotic stresses. J Biol Chem 282(35):25588–25596. doi:10.1074/jbc.M703326200

Pettkó-Szandtner A, Mészáros T, Horváth GV, Bakó L, Csordás-Tóth É, Blastyák A, Zhiponova M, Miskolczi P, Dudits D (2006) Activation of an alfalfa cyclin-dependent kinase inhibitor by calmodulin-like domain protein kinase. Plant J 46(1):111–123. doi:10.1111/j.1365-313X.2006.02677.x

Ren H, Santner A, del Pozo JC, Murray JAH, Estelle M (2008) Degradation of the cyclin-dependent kinase inhibitor KRP1 is regulated by two different ubiquitin E3 ligases. Plant J 53(5):705–716. doi:10.1111/j.1365-313X.2007.03370.x

Richard C, Lescot M, Inzé D, De Veylder L (2002) Effect of auxin, cytokinin, and sucrose on cell cycle gene expression in *Arabidopsis thaliana* cell suspension cultures. Plant Cell Tissue Organ Cult 69(2):167–176. doi:10.1023/A:1015241709145

Roudier F, Fedorova E, Lebris M, Lecomte P, Györgyey J, Vaubert D, Horvath G, Abad P, Kondorosi A, Kondorosi E (2003) The *Medicago* species A2-type cyclin is auxin regulated and involved in meristem formation but dispensable for endoreduplication-associated developmental programs. Plant Physiol 131(3):1091–1103. doi:10.1104/pp.102.011122

Sabelli PA, Nguyen H, Larkins BA (2007) Cell cycle and endosperm development. In: Inzé D (ed) Cell cycle control and plant development, vol 32, Annual plant reviews. Blackwell, Oxford, UK

Sanz L, Dewitte W, Forzani C, Patell F, Nieuwland J, Wen B, Quelhas P, De Jager S, Titmus C, Campilho A, Ren H, Estelle M, Wang H, Murray JAH (2011) The *Arabidopsis* D-type cyclin CYCD2;1 and the inhibitor ICK2/KRP2 modulate auxin-induced lateral root formation. Plant Cell 23(2):641–660. doi:10.1105/tpc.110.080002

Shishova M, Lindberg S (2010) A new perspective on auxin perception. J Plant Physiol 167 (6):417–422. doi:10.1016/j.jplph.2009.12.014

Soni R, Carmichael JP, Shah ZH, Murray JAH (1995) A family of cyclin D homologs from plants differentially controlled by growth regulators and containing the conserved retinoblastoma protein interaction motif. Plant Cell 7(1):85–103

Sorrell DA, Marchbank A, McMahon K, Dickinson JR, Rogers HJ, Francis D (2002) A *WEE1* homologue from *Arabidopsis thaliana*. Planta 215(3):518–522. doi:10.1007/s00425-002-0815-4

Sozzani R, Cui H, Moreno-Risueno MA, Busch W, Van Norman JM, Vernoux T, Brady SM, Dewitte W, Murray JAH, Benfey PN (2010) Spatiotemporal regulation of cell-cycle genes by SHORTROOT links patterning and growth. Nature 466(7302):128–132. doi:10.1038/nature09143

Sugimoto K, Jiao Y, Meyerowitz EM (2010) *Arabidopsis* regeneration from multiple tissues occurs via a root development pathway. Dev Cell 18(3):463–471. doi:10.1016/j.devcel.2010.02.004

Sun Y, Dilkes BP, Zhang C, Dante RA, Carneiro NP, Lowe KS, Jung R, Gordon-Kamm WJ, Larkins BA (1999) Characterization of maize (*Zea mays* L.) Wee1 and its activity in developing endosperm. Proc Natl Acad Sci U S A 96(7):4180–4185

Takanashi K, Sugiyama A, Yazaki K (2011) Involvement of auxin distribution in root nodule development of *Lotus japonicus*. Planta 234(1):73–81. doi:10.1007/s00425-011-1385-0

Tao W, Verbelen J-P (1996) Switching on and off cell division and cell expansion in cultured mesophyll protoplasts of tobacco. Plant Sci 116:107–115

Tardieu F, Granier C (2000) Quantitative analysis of cell division in leaves: methods, developmental patterns and effects of environmental conditions. Plant Mol Biol 43(5–6):555–567

Torrey JG (1950) The induction of lateral roots by indoleacetic acid and root decapitation. Am J Bot 37:257–264

Ulmasov T, Liu Z-B, Hagen G, Guilfoyle TJ (1995) Composite structure of auxin response elements. Plant Cell 7(10):1611–1623. doi:10.1105/tpc.7.10.1611

Ulmasov T, Murfett J, Hagen G, Guilfoyle TJ (1997) Aux/IAA proteins repress expression of reporter genes containing natural and highly active synthetic auxin response elements. Plant Cell 9(11):1963–1971. doi:10.1105/tpc.9.11.1963

Ulmasov T, Hagen G, Guilfoyle TJ (1999) Dimerization and DNA binding of auxin response factors. Plant J 19(3):309–319

Umeda M, Shimotohno A, Yamaguchi M (2005) Control of cell division and transcription by cyclin-dependent kinase-activating kinases in plants. Plant Cell Physiol 46(9):1437–1442. doi:10.1093/pcp/pci170

Valente P, Tao W, Verbelen J-P (1998) Auxins and cytokinins control DNA endoreduplication and deduplication in single cells of tobacco. Plant Sci 134:207–215

Van Bel M, Proost S, Wischnitzki E, Movahedi S, Scheerlinck C, Van de Peer Y, Vandepoele K (2012) Dissecting plant genomes with the PLAZA comparative genomics platform. Plant Physiol 158(2):590–600. doi:10.1104/pp.111.189514

Van Leene J, Hollunder J, Eeckhout D, Persiau G, Van De Slijke E, Stals H, Van Isterdael G, Verkest A, Neirynck S, Buffel Y, De Bodt S, Maere S, Laukens K, Pharazyn A, Ferreira PCG, Eloy N, Renne C, Meyer C, Faure J-D, Steinbrenner J, Beynon J, Larkin JC, Van de Peer Y, Hilson P, Kuiper M, De Veylder L, Van Onckelen H, Inzé D, Witters E, De Jaeger G (2010)

Targeted interactomics reveals a complex core cell cycle machinery in *Arabidopsis thaliana*. Mol Syst Biol 6:397. doi:10.1038/msb.2010.53

van Noorden GE, Kerim T, Goffard N, Wiblin R, Pellerone FI, Rolfe BG, Mathesius U (2007) Overlap of proteome changes in *Medicago truncatula* in response to auxin and *Sinorhizobium meliloti*. Plant Physiol 144(2):1115–1131. doi:10.1104/pp.107.099978

Van Norman JM, Xuan W, Beeckman T, Benfey PN (2013) To branch or not to branch: the role of pre-patterning in lateral root formation. Development 140(21):4301–4310. doi:10.1242/dev.090548

Van't Hof J (1985) Control points within the cell cycle. In: Bryant JA, Francis D (eds) The cell division cycle in plants. Cambridge University Press, Cambridge, pp 1–13

Vandepoele K, Raes J, De Veylder L, Rouzé P, Rombauts S, Inzé D (2002) Genome-wide analysis of core cell cycle genes in Arabidopsis. Plant Cell 14(4):903–916

Vanneste S, De Rybel B, Beemster GTS, Ljung K, De Smet I, Van Isterdael G, Naudts M, Iida R, Gruissem W, Tasaka M, Inzé D, Fukaki H, Beeckman T (2005) Cell cycle progression in the pericycle is not sufficient for SOLITARY ROOT/IAA14-mediated lateral root initiation in *Arabidopsis thaliana*. Plant Cell 17(11):3035–3050. doi:10.1105/tpc.105.035493

Vieira P, Escudero C, Rodiuc N, Boruc J, Russinova E, Glab N, Mota M, De Veylder L, Abad P, Engler G, de Almeida Engler J (2013) Ectopic expression of Kip-related proteins restrains root-knot nematode-feeding site expansion. New Phytol 199(2):505–519. doi:10.1111/nph.12255

Vodermaier HC (2004) APC/C and SCF: controlling each other and the cell cycle. Curr Biol 14 (18):R787–R796. doi:10.1016/j.cub.2004.09.020

Walcher CL, Nemhauser JL (2012) Bipartite promoter element required for auxin response. Plant Physiol 158(1):273–282. doi:10.1104/pp.111.187559

Weimer AK, Nowack MK, Bouyer D, Zhao X, Harashima H, Naseer S, De Winter F, Dissmeyer N, Geldner N, Schnittger A (2012) Retinoblastoma related1 regulates asymmetric cell divisions in Arabidopsis. Plant Cell 24(10):4083–4095. doi:10.1105/tpc.112.104620

Wilmoth JC, Wang S, Tiwari SB, Joshi AD, Hagen G, Guilfoyle TJ, Alonso JM, Ecker JR, Reed JW (2005) NPH4/ARF7 and ARF19 promote leaf expansion and auxin-induced lateral root formation. Plant J 43(1):118–130. doi:10.1111/j.1365-313X.2005.02432.x

Xu K, Liu J, Fan M, Xin W, Hu Y, Xu C (2012) A genome-wide transcriptome profiling reveals the early molecular events during callus initiation in *Arabidopsis* multiple organs. Genomics 100 (2):116–124. doi:10.1016/j.ygeno.2012.05.013

Zhang K, Letham DS, John PCL (1996) Cytokinin controls the cell cycle at mitosis by stimulating the tyrosine dephosphorylation and activation of $p34^{cdc2}$-like H1 histone kinase. Planta 200 (1):2–12

Zhang K, Diederich L, John PCL (2005) The cytokinin requirement for cell division in cultured *Nicotiana plumbaginifolia* cells can be satisfied by yeast Cdc25 protein tyrosine phosphatase. Implications for mechanisms of cytokinin response and plant development. Plant Physiol 137 (1):308–316. doi:10.1104/pp.104.051938

# Chapter 8
# Auxin on the Road Navigated by Cellular PIN Polarity

**Pawel Baster and Jiří Friml**

**Abstract** The generation of asymmetry, at both cellular and tissue level, is one of the most essential capabilities of all eukaryotic organisms. It mediates basically all multicellular development ranging from embryogenesis and de novo organ formation till responses to various environmental stimuli. In plants, the awe-inspiring number of such processes is regulated by phytohormone auxin and its directional, cell-to-cell transport. The mediators of this transport, PIN auxin transporters, are asymmetrically localized at the plasma membrane, and this polar localization determines the directionality of intercellular auxin flow. Thus, auxin transport contributes crucially to the generation of local auxin gradients or maxima, which instruct given cell to change its developmental program. Here, we introduce and discuss the molecular components and cellular mechanisms regulating the generation and maintenance of cellular PIN polarity, as the general hallmarks of cell polarity in plants.

## 1 Introduction

The emergence of multicellularity during the evolution of species had its inevitable repercussions. Efficient intercellular communication was one of such. In other words, to achieve a mutual goal, neighboring cells (single elements of the system) needed to perceive and transduce externally or internally generated signals. As a result, the multicellular organism, as a whole, should be able to translate these

P. Baster • J. Friml (✉)
Department of Plant Systems Biology, VIB, 9052 Gent, Belgium

Department of Plant Biotechnology and Genetics, Ghent University, 9052 Gent, Belgium

Institute of Science and Technology Austria (IST Austria), 3400 Klosterneuburg, Austria
e-mail: Jiri.FRIML@ist.ac.at

E. Zažímalová et al. (eds.), *Auxin and Its Role in Plant Development*,
DOI 10.1007/978-3-7091-1526-8_8, 

signals into a developmental response. This basic necessity for efficient internal communication underlies the origin of the small signaling molecules, termed hormones, present in both plants and animals (Alberts et al. 2007). While the essential role of hormones in cell-to-cell communication is evident in both cases, the response of the organism into which the hormonal signal is translated, diverge between two kingdoms. In animal species, solutions based on a modulation of behavior were primarily promoted (Davies 2004). In contrast, plants, due to the sessile nature of their lifestyle, developed a remarkable repertoire of mechanisms which allow them, through fine-tuning of metabolism or body shaping, to adjust and survive in ever changing and often adverse environments (Tanaka et al. 2006). These mechanisms are facilitated not only by intercellular communication but also by tightly regulated cell division, morphogenesis, and differentiation. Importantly, most, if not all of them, are based on cell polarization and repolarization which guide tissue and organ patterning and thus underpin basic shape and functionality of an organism. The phenomenon of cell polarity can be reflected in various aspects like disproportional growth or asymmetrical distribution of the cellular components. The term itself, however, is much broader and in largest sense applies to the generation of any asymmetry in relation to an axis (Sauer and Friml 2004; Geldner 2009).

Despite its complexity, plant development, with its remarkable polarization-based flexibility, is coordinated most prominently by a single phytohormone—auxin—which serves itself as a polarizing cue (Chap. 14; Berleth and Sachs 2001; Sauer et al. 2006; Leyser 2011). Moreover, generation of local auxin accumulations (gradients) plays an essential role in plethora of crucial events during plant development, like embryogenesis (Chap. 9; Friml et al. 2003; Schlereth et al. 2010), organogenesis (Chaps. 10 and 11; Benková et al. 2003), phyllotaxis (Reinhardt et al. 2003), root meristem organization (Sabatini et al. 1999; Friml et al. 2002a), root stem cell differentiation (Ding and Friml 2010), or vascular tissue patterning (Scarpella et al. 2006). Although essentially all plant tissues possess the capacity for auxin biosynthesis (Mano and Nemoto 2012; Ljung 2013) and metabolism (Ludwig-Müller 2011; Ruiz Rosquete et al. 2012), it is the directional auxin transport that most significantly contributes for establishing of auxin gradients across plant tissues (Tanaka et al. 2006). The existence of such a cell-to-cell transport of auxin (see Chap. 5), which is the unique feature among other phytohormones, was predicted in the mid-1970s by so-called chemiosmotic model (Rubery and Sheldrake 1974; Raven 1975). This model postulated an auxin efflux from the cell, facilitated by polarly localized exporters, as a critical step during intercellular auxin translocation. These predictions were spectacularly confirmed by characterization of the broad spectrum of developmental phenotypes caused by mutations in the *PIN-FORMED* (*PIN*) genes (Gälweiler et al. 1998; Luschnig et al. 1998; Okada et al. 1991). Basically all different phenotypes found in various *pin* mutants could be mimicked by treatments with auxin transport inhibitors (Vieten et al. 2007). Transport assays from both plant and heterologous systems, provided later, shown that PIN proteins indeed mediate auxin export from the cells (Petrášek et al. 2006). The intercellular auxin transport, beside of PINs, rely on the coordinated activity of two other transporter families. These are AUXIN RESISTANT1/LIKE AUX1 (AUX1/LAX) and MULTIDRUG RESISTANCE/PHOSPHOGLYCOPROTEIN/ATP-BINDING CASSETTE OF B-TYPE (MDR/PGP/ABCB) proteins, facilitating influx and efflux of auxin from the cell, respectively (Bennett et al. 1996; Noh et al. 2001; Kramer 2004; Yang et al. 2006; Mravec et al. 2008; Swarup et al. 2008; Verrier et al. 2008). Despite the fact that multiple components are involved, a critical control of the directionality of auxin flux is attributed to the efflux activity of the PIN transporters at their highly defined, polar, subcellular domains (Wiśniewska et al. 2006; Blakeslee et al. 2007; Titapiwatanakun et al. 2009). PIN family consist of eight members, most of which (PIN1, 2, 3, 4, 6, and 7) dependent on the tissue or developmental context exhibit plasma membrane (PM) localization restricted mainly to the apical (shootward; shoot-apex-facing) or basal (rootward; root-apex-facing) side of the cell (Zažímalová et al. 2007). Notably, also AUX/LAXs and ABCBs, which serve as an additional source of auxin for PIN-mediated transport (Geisler et al. 2005; Mravec et al. 2008; Christie et al. 2011; Kubeš et al. 2012), in some cases display asymmetric distribution (Swarup et al. 2001; Panikashvili et al. 2007; McFarlane et al. 2010).

Although some similarities can be found (Geldner 2009), in general, the mechanisms underlying cell polarity in plants differ from those characterized in animals (Tepass et al. 2001; Humbert et al. 2006; Wells et al. 2006; Chen et al. 2010). It seems that in both animal and plant systems the delivery of protein to the place of action by subcellular trafficking is equally important (Dudu et al. 2004; Altschuler et al. 2008; Geldner 2009; Shivas et al. 2010). On the other hand, the most prominent trafficking-based determinants of polarity found in animals, like CRUMBS, SCRIBBLE, and PAR complexes, are missing in plant genomes (Geldner 2009). Additionally, counterparts of so-called tight junctions which in animals serve as diffusion barriers, dividing PM of epithelial cells into apical and baso-lateral, polar domains (Giepmans and van Ijzendoorn 2009), are missing in majority of plant cell types. A similar structure is present in plants in form of ‘Casparian Strip’, belts of specialized cell wall material generating an extracellular diffusion barrier, found exclusively in endodermis (Roppolo et al. 2011). Polar trafficking pathways described in plants cells appear to be more complex than those found in animals. Besides apical and basal PM domains, characteristic for animal epidermal cells, also outer- and inner-lateral domains, with corresponding polar cargos, can be found in similar cell types in plants (Miwa et al. 2007; Langowski et al. 2010; Takano et al. 2010). What is more, the differences are reflected not only on the cellular but also on the tissue level. Plants in contrast to animals cannot use the mechanism of invasive, cell-migration-based tissue patterning due to the presence of the rigid extracellular matrix, cell wall, encapsulating plant cells and making them immobile (Dettmer and Friml 2011).

Considering the aforementioned differences it is not surprising that alternative, to animal, solutions were promoted by evolution in plant kingdom. In this chapter, these plant-specific mechanisms for cellular polarization are discussed. The polarity of cellular components is mainly considered in context of PIN proteins, due to their essential role in auxin-mediated plant development. Moreover, in case of PIN family enough molecular components and polarity generating/maintaining signals are described for drawing a comprehensive and interesting overview. First, subcellular trafficking machinery relevant for PIN polarity is characterized. Next, cell structure and cargo-related determinants for PIN targeting and maintenance at the polar domain are presented. Finally, the feedback mechanisms for PIN polarization are discussed. It should be taken into account that such a categorization, due to frequently redundant nature of the biological processes, is largely subjective and serves mainly for presentation purposes.

## 2 Long Journey with Unsure Destination: Trafficking for PIN Polarization

Polar localization of PIN auxin transporters, restricted to the specific side of the cell relies, among other mechanisms, on the function of complicated network, encompassing multiple bypassing and interconnected pathways collectively referred to as 'intracellular trafficking' (Paul and Frigerio 2007; Bassham et al. 2008; Robinson et al. 2008; Irani and Russinova 2009; Žárský et al. 2009). Auxin transporters, subjected to intracellular trafficking, originate from de novo synthesis and are delivered to the PM through so-called anterograde route. This mode of intracellular transport generally involves sequential steps including protein folding in endoplasmic reticulum (ER), translocation through the cis- and trans-cisternae of the Golgi Apparatus (GA) with final arrival to the cell surface (Vitale and Denecke 1999; Matheson et al. 2006).

### *2.1 Constitutive PIN Cycling for Rapid Repolarizations*

Contrary to the usually presented rather static snapshot pictures, the polar localization of PIN proteins is in reality very dynamic. Once delivered to the cell surface, PINs undergo continuous shuttling between PM and intracellular compartments by rounds of internalization (endocytosis) and polar recycling (exocytosis). These processes are jointly referred to as 'constitutive endocytic cycling' (Geldner et al. 2001; Dhonukshe et al. 2007; Kleine-Vehn et al. 2011). It is not entirely clear to which extent the initial secretion of de novo synthesized PINs to the PM occurs in polar fashion (Langowski and Friml, unpublished) or whether the endocytic cycling is generating the polar distribution following apolar secretion (Dhonukshe et al. 2008a).

Nevertheless, the first station reached by internalized PINs on their endocytic trafficking route from the PM is early endosome (EE). This subcellular compartment is critical as it is the intersection between secretory and endocytic routes and, in plants, originates from and associates with trans-Golgi network (TGN) (Dettmer et al. 2006; Lam et al. 2007; Viotti et al. 2010). Thus, sorting of vesicular cargo for various subcellular destinations occurs at EE/TGN (Liu et al. 2002; Dettmer et al. 2006). Once destined for recycling, PINs are translocated from EE/TGN to the hypothetical compartment called recycling endosome (RE), where they fall under the control of ADP-Ribosylation Factor GTPase (ARF-GTPase) machinery. ARF-GTPases, by recruitment of vesicle coat proteins and organizing cytoskeleton at membrane surfaces, control vesicle trafficking. Their spatiotemporal activity is determined by the antagonistic activity of ARF guanine nucleotide exchange factors (ARF-GEFs) and ARF-GTPase activating proteins (ARF-GAPs), which are activating and deactivating ARF-GTPase complex, respectively (D'Souza-Schorey and Chavrier 2006). GNOM, a protein belonging to the golgi-associated, brefeldin-A resistant guanine nucleotide exchange factor (GBF) class of ARF-GEFs, most prominently controls the polar recycling of PINs to the PM (Fig. 8.1; Geldner et al. 2003). This protein, and more specifically its Sec7 domain, is a target of fungal toxin brefeldin A (BFA), which by inhibiting GNOM-mediated exocytosis, causes reversible intracellular accumulation of constitutively endocytosed proteins and aggregation of TGN into so-called BFA compartments or BFA bodies. This effect serves as a tool to visualize the constitutive cycling of PM proteins (Geldner et al. 2001). Interestingly, GNOM controls preferentially PIN recycling to the basal side of the cell, whereas pathway by which PINs are targeted to the apical domain is, most likely, additionally controlled by uncharacterized, BFA-insensitive ARF-GEF (Kleine-Vehn et al. 2008a, b).

Besides ARF-GTPases, also RabA1B, a member of small Rab-GTPase family and INTERACTOR OF CONSTITUTIVE ACTIVE ROP1 (ICR1), an effector of RHO OF PLANTS1 (ROP1) RAC-GTPase, were recently associated with defective PIN recycling (Fig. 8.1; Hazak et al. 2010; Feraru et al. 2012). Importantly, genetic interference with ICR1 results with severely disturbed polarity of PIN1 and PIN2, as well as various developmental defects (Hazak et al. 2010). Moreover, ICR1 was shown to interact with Sec3A (Lavy et al. 2007), one of the exocyst complex (EC) components. EC is known to participate in the extensive fusion of exocytic vesicles at specific sites of PM during so-called polarized exocytosis. Polarized exocytosis is controlled, among others, by Rho GTPases (Žárský et al. 2009). Interestingly, another subunit of the EC, Exo70, influences polar auxin transport through the regulation PIN1 and PIN2 recycling (Drdová et al. 2013).

The fundamental role of the cellular scaffolding and cytoskeleton for intracellular PIN trafficking should be also underlined here. PIN constitutive cycling seems to depend mainly on actin filaments since pharmacological interference with the integrity of this component abolishes internalization and recycling of PINs. On the other hand, microtubules are essential for both PIN trafficking in dividing cells as well as for maintenance of the general polarity of the cell (Geldner et al. 2001; Friml et al. 2002b; Petrášek and Schwarzerová 2009; Dhonukshe et al. 2008b; Boutté et al. 2006; Kleine-Vehn et al. 2006, 2008b, c; Kleine-Vehn and Friml 2008). The components of cytoskeleton not only serve as an orientation cues,

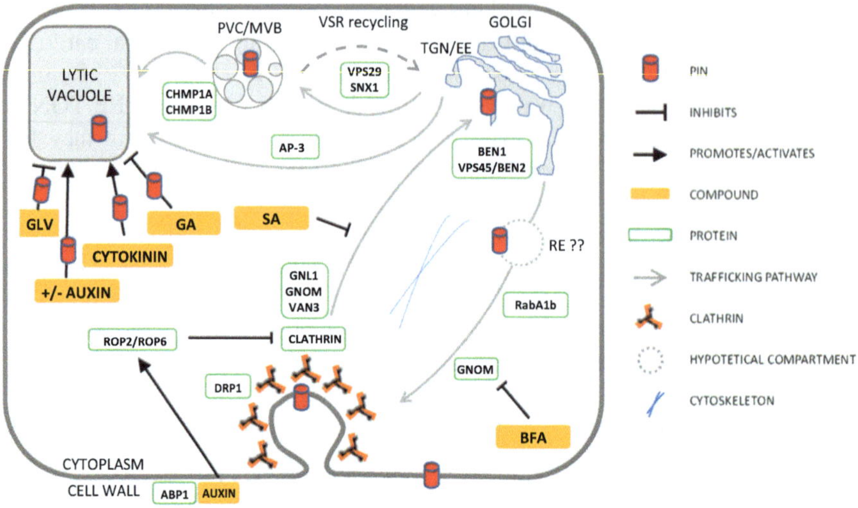


**Fig. 8.1** Intracellular trafficking for polarization of PINs. Auxin, when extracellularly bound to ABP1, rapidly inhibits clathrin-mediated PIN endocytosis through mutually exclusive ROP2/ROP6 signaling. Formation of the PIN-containing, clathrin-coated vesicles, in case of cell plate formation, requires the function of DRP1. Internalization of cargo vesicles from the plasma membrane is mediated by Rab5 GTPase ARA7, ARF-GEFs GNOM, GNL1, and ARF-GAP VAN3, inhibited by SA and proceeds along cytoskeletal cell scaffold. Early endocytic trafficking of PINs requires function of VPS45 BEN2 and ARF-GEF BEN1. PIN recycling depends on RabA1b GTPase and GNOM, a target of BFA-mediated inhibition of exocytosis. PIN vacuolar targeting route passing through PVC/MVB includes the regulation by retromer complex components VPS29 and SNX1, as well as ESCRT components CHMP1A and CHMP1B. PVC/MVB-bypassing route for regulating vacuolar function is mediated by AP-3 complex. Cytokinin and long-term, above- or below-optimal auxin levels reduce the membrane abundance of PINs by promoting their turnover. In contrast, vacuolar targeting of PINs is inhibited by GA and GLV peptides

according to which PINs polarize (Heisler et al. 2010), but also provide guidance for vesicle trafficking (Voigt et al. 2005) (Fig. 8.1).

The functional significance of an energy-demanding process such as constitutive cycling still remains unclear. Evidently, this process provides the means for fast and de novo synthesis-independent repositioning of PIN auxin transporters, in response to internal and external cues. It, thus, allows developmental flexibility conditioned by rapid redirection of auxin fluxes within tissues (Michniewicz et al. 2007a; Vanneste and Friml 2009). Another, attractive scenario highlights a possible analogy between auxin efflux and neurotransmitter release. In this scenario, PIN transporters localized at the surface of constitutively cycling intracellular vesicles would mediate uptake of auxin from the cytosol into these vesicles and after their arrival and fusion with the PM, auxin would be released from the cell, similarly to the synaptic release of neurotransmitters (Friml and Palme 2002; Baluška et al. 2003). Such a hypothesis was to some extent supported by experiments based on auxin immunolocalization and manipulation of secretory pathway using phospholipase

Dζ2 mutant (Schlicht et al. 2006; Mancuso et al. 2007). The validation of the 'neurotransmitter' hypothesis would provide a connection between the effect of established inhibitors of auxin transport on both vesicle trafficking and on PIN-dependent auxin transport (Geldner et al. 2001; Dhonukshe et al. 2008b). Finally, auxin transporters were proposed to have an additional function, similar to known, dual function receptor/transporters (Hertel 1983; Foti et al. 2004; Holler and Dikic 2004). In this scenario, endocytic cycling, analogously to the situation in animals, would serve as a way to transduce the signal and to regenerate receptors during ligand-dependent endocytosis. It is important to note that all the interpretations for functionality of constitutive cycling in the process of auxin transport are not mutually exclusive, but the latter two ('neurotransmitter' and transporter/receptor) remain rather speculative.

On the other hand, changes in PIN polarity have been observed in courses of many developmental processed including embryogenesis (Friml et al. 2003), organogenesis (Benková et al. 2003; Reinhardt et al. 2003), fruit development (Sorefan et al. 2009), vascular tissue formation and regeneration (Scarpella et al. 2006; Balla et al. 2011), as well as in response to light (Ding et al. 2011) or gravity (Kleine-Vehn et al. 2010; Rakusova et al. 2011). Such a dynamic translocation of polar cargos from one cell side to another via recycling endosomes is called transcytosis (Tuma and Hubbard 2003; Kleine-Vehn et al. 2008a, b) and it appears that plants evolved to utilize this mechanism to adjust their development in response to different cues via redirecting of PIN-dependent auxin fluxes.

## 2.2 *Early Endocytic Processes for PIN Polarization*

The first step of endocytic recycling is the internalization from the PM. The most prominent route of PIN internalization occurs through the creation of the membrane curvature from which upon scission the intracellular vesicle is formed, during clathrin-mediated endocytosis (CME) (Fig. 8.1). Clathrin is a self-assembling protein, recruited to the membranes where it contributes to membrane deformation and serves as a vesicular coat constituent (Chen et al. 2011). The fundamental role of CME in PIN trafficking and thus its essential contribution to generation of intrinsic polar localization of PINs is well characterized. It was demonstrated by pharmacological and genetic interference with the clathrin function (Dhonukshe et al. 2007; Kitakura et al. 2011). The notion that auxin efflux carriers from the PIN family are cargos of CME was further supported by identification of DYNAMIN-RELATED PROTEIN 1 (DRP1) as associated with PIN1 and important for its endocytosis (Fig. 8.1; Mravec et al. 2011). In plants, the precise function of the dynamin superfamily, represented among others by DRP1, is not well characterized. It is speculated that these proteins might regulate membrane dynamics by modulation of its scission and tubulation (Praefcke and McMahon 2004; Heymann and Hinshaw 2009). Importantly, proteins belonging to this family were previously implicated in plant CME (Konopka et al. 2008; Fujimoto et al. 2010) and interference with DRP1 function results in altered distribution of PIN1 in dividing cells which eventually leads to a range of developmental phenotypes related to defective auxin transport (Mravec et al. 2011).

Interestingly, ARF-GEF GNOM, function of which is typically associated with recycling to the PM, was also implicated in the regulation of endocytosis, based on its partial localization at the PM and PIN endocytosis defects observed in *gnom* knock-down mutants (Fig. 8.1; Naramoto et al. 2010). In addition, another GBF subfamily member GNOM LIKE1 (GNL1) and ARF-GAP—VASCULAR NETWORK DEFECTIVE 3 (VAN3)—have been functionally associated with PM and endocytic processes, corroborating the notion that ARF GTPase machinery is involved in endocytosis (Fig. 8.1; Teh and Moore 2007; Naramoto et al. 2010).

Following internalization from the PM, the early endocytic trafficking of PINs has been shown to rely on the BFA-VISUALIZED ENDOCYTIC TRAFFICKING DEFECTIVE1/HOPM INTERACTOR7 (BEN1/MIN7/BIG5) ARF-GEF, belonging to BFA-inhibited guanine nucleotide exchange protein (BIG) subfamily (Fig. 8.1; Tanaka et al. 2009). PIN internalization is regulated also by Rab5/ARA7, a member of Rab GTPase family (Fig. 8.1; Dhonukshe et al. 2008a) and an universal component of membrane fusion in eukaryotes—VACUOLAR PROTEIN SORTING45 (VPS45/BEN2) (Fig. 8.1; Tanaka et al. 2013). Genetic interference with all aforementioned components regulating endocytosis or early endocytic trafficking leads, besides perturbations in PIN trafficking and polarity, to significant developmental aberrations like embryogenesis and organogenesis defects, reduced growth and apical dominance, leaf venation pattern discontinuity, and root meristem disorganization. These observation links early endocytic processes to PIN polarity and auxin-mediated development.

## 2.3 *Late Endocytic Trafficking for PIN Abundance at the Cell Surface*

Certain proportion of internalized PINs, based on signals which are not fully understood, is targeted for the late endocytic pathway. This mode of transport originates generally at EE/TGN, where proteins are sorted, proceeds through late endosomes (LE), prevacuolar compartments/multivesicular bodies (PVC/MVB) and terminates at the final destination of membrane proteins—the lytic vacuole. This subcellular route eventually results with protein degradation. Vacuolar targeting defines additional mechanism, by which polar localization of PINs and specifically the aspect of their membrane abundance can be controlled (Müller et al. 2007; Scheuring et al. 2011). As mentioned above, plant trafficking machinery encompasses multiple bypassing, often unidirectional transport routes. Accordingly, PINs can be retrieved from late endocytic pathway and thus avoid degradation. This is accomplished by the Retromer Complex (RC) capacity for retrieval of certain trafficking components like VACUOLAR SORTING RECEPTORS (VSR) from PVC to EE/TGN (Arighi et al. 2004; Seaman 2005). Both the core component of this complex—VACUOLAR PROTEIN SORTING29 (VPS29) and its prominent interactor—SORTIN NEXIN1 (SNX1) were shown to control the rate of PINs progression toward the vacuole (Fig. 8.1; Kleine-Vehn et al. 2008c; Nodzyński et al. 2013). Such a mechanism is in agreement with widely accepted and evolutionary conserved function of RC (Arighi et al. 2004; Seaman 2005; Shimada et al. 2006). Notably, a more unorthodox function, related more directly to PIN polarity, has been also proposed for RC at the level of early endocytic recycling (Jaillais et al. 2006, 2007).

The significance of the final steps during late endocytic trafficking of PIN proteins for their polar localization and abundance is not to be underestimated. The endosomal-sorting complexes required for transport (ESCRT) machinery, controlling the formation of internal vesicles within PVC/MVB, which upon fusion are released into vacuolar lumen, appears to play important role for PIN polarization (Winter and Hauser 2006; Wollert et al. 2009; Scheuring et al. 2011). Interfering with the function of CHARGED MULTIVESICULAR BODY PROTEIN/ CHROMATIN MODIFYING PROTEIN 1A and 1B (CHMP1A/CHMP1B), components of ESCRT machinery, leads to severe developmental defects including seedling lethality. These phenotypes were associated with inaccurately generated auxin distribution correlated with the ectopic (PVC/MVB and vacuolar membranes) PIN localization (Fig. 8.1; Spitzer et al. 2009). Also an alternative, PVC-bypassing, late endocytic pathway, dependent on the ADAPTOR PROTEIN (AP) Complex 3 subunits β and δ, generally regulates vacuolar function and thus PIN degradation rate, although it does not affect PIN polarity or abundance at the PM (Fig. 8.1; Feraru et al. 2010; Zwiewka et al. 2011).

# 3 Where to Go and Why to Stay? Cues and Cellular Requirements for Pin Polarity

## *3.1 Cargo-Based Determinants for Polar PIN Localization*

One of the most important initial findings concerning the determination of PIN polarity was derived from the ectopic PIN expression in particular cell types. The PIN2 promoter-driven expression of PIN1 targeted this protein predominantly to the basal side of root epidermal cells contrasting to the native apical PIN2 localization in the same cells. Consecutive introduction of the fluorescent tag into certain place within central hydrophilic loop of PIN1 was sufficient to cause basal-to-apical switch in PIN1 localization. This was a clear indication that some determinants of polar PIN localization are encoded intrinsically in PIN amino acid sequence (Wiśniewska et al. 2006). This sequence-based instruction turned out to be the phosphorylation status of specific serine residues located within PIN central hydrophilic loop (Huang et al. 2010; Zhang et al. 2010). Current model postulates that

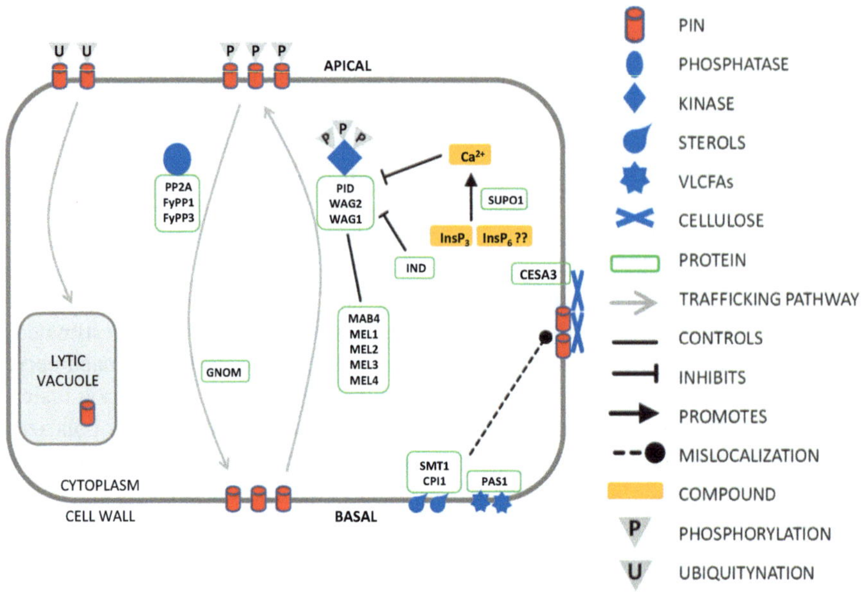


**Fig. 8.2** Posttranslational modifications and cellular requirements for polarization of PINs. PINs dephosphorylated by PP2A, FyPP1, or FyPP3 phosphatases are preferentially recruited to the basal, GNOM-dependent, BFA-sensitive pathway, whereas phosphorylation by PID, WAG1, or WAG2 kinases targets PINs into apical cell site independently of GNOM function. The transcription of PID kinase is controlled by IND transcription factor. The activity of the kinases is influenced also by MAB4, its homologs MELs and by $InsP_3$ (or alternatively $InsP_6$) that regulate cellular $Ca^{2+}$ levels with the assistance of SUPO1. The polar localization of PINs depends on the sterol and VLCFA composition of the membrane controlled by SMT1, CPI1, and PAS1, respectively. Membrane abundance of PINs, regulated by their vacuolar targeting, depends on the ubiquitination status of the protein. Cellulose content of the cell wall regulated by CESA3 contributes to the maintenance of PIN polarity

dephosphorylated PINs are preferentially recruited to the basal, GNOM-dependent and BFA-sensitive pathway, whereas phosphorylation targets PINs into the apical cell side, independently of GNOM function (Fig. 8.2; Friml et al. 2004; Michniewicz et al. 2007b; Kleine-Vehn et al. 2009).

Readjustments of PIN phosphorylation status rely on the antagonistic activity of protein kinases and phosphatases. AGC3 protein kinases PINOID (PID) and its homologs WAVY ROOT GROWTH1/2 (WAG1/WAG2) (Benjamins et al. 2001; Friml et al. 2004; Santner and Watson 2006; Dhonukshe et al. 2010) phosphorylate, while PROTEIN PHOSPHATASE2A (PP2A) (Muday and DeLong 2001; Michniewicz et al. 2007b; Ballesteros et al. 2013) dephosphorylate PINs (Fig. 8.2). Along with PID and WAG proteins, also other kinases such as D6 protein kinase (Zourelidou et al. 2009) or CDPK-RELATED KINASE5 (Rigó et al. 2013) can phosphorylate PIN proteins and regulate their function, but their exact role is less clear. On the other hand, phosphatase subunit PP2AA interacts with another Ser/Thr protein phosphatase, FyPP1, to form functional holoenzyme. FyPP1 and its close homolog FyPP3 were reported to interact with and directly dephosphorylate PINs (Fig. 8.2; Dai et al. 2012). Importantly, fluctuations of PIN phosphorylation status both above and below certain native threshold lead eventually to severe developmental aberrations like defective embryogenesis and patterning of shoot apical meristem and root (Christensen et al. 2000; Benjamins et al. 2001; Friml et al. 2004; Michniewicz et al. 2007b; Dhonukshe et al. 2010; Li et al. 2011; Dai et al. 2012).

Given the crucial role of PID-mediated PIN phosphorylation in PIN polarity, it is important to highlight that the regulation of PID activity, on various levels, has also impact on polar PIN localization and thus on auxin fluxes. Calcium ($Ca^{2+}$), which is one of the most ubiquitous secondary messengers in eukaryotes, appears to be the prominent part of such a system for the regulation of PID kinase activity (Fig. 8.2). Early experiments in animal and yeast systems have associated the appearance of cytosolic $Ca^{2+}$ with the phospholipase C (PLC)-generated inositol trisphosphate ($InsP_3$). PLC signaling is known to be important for various biological processes including cell division and differentiation (Michell 2008). The plant field has followed this paradigm upon an observation that $InsP_3$ is able to trigger the release of $Ca^{2+}$ from the cellular storage compartments (Blatt et al. 1990; Gilroy et al. 1990; Krinke et al. 2007; Tang et al. 2007). Controversially, up to date no unambiguous $InsP_3$-activated $Ca^{2+}$ channel could be identified in plants (Testerink and Munnik 2011). Interestingly, there are indications that $InsP_6$ can function as a signaling molecule, triggering $Ca^{2+}$ release with a much higher potency than $InsP_3$ (Lemtiri-Chlieh et al. 2003). Moreover, $InsP_3$ when microinjected into plant can be rapidly converted into $InsP_6$ (Munnik and Testerink 2009), explaining the earlier observations of $InsP_3$ being able to release $Ca^{2+}$. PID kinase was shown to be regulated by both phospholipid and $Ca^{2+}$ signaling. Some PID interactors bind $Ca^{2+}$ (Benjamins et al. 2003; Zegzouti et al. 2006). Moreover, *suppressor of PIN1 overexpression*-1 (*supo-1*) mutant encoding ALTERED EXPRESSION OF APX2 8/FIERY1/HIGH EXPRESSION OF OSMOTICALLY RESPONSIVE GENES2/ROTUNDA1 (ALX8/FRY1/HOS2/RON1/SAL1) inositol polyphosphate 1-phosphatase also with 3′(2′),5′-bisphosphate nucleotidase activity exhibits aberrant PIN polarization presumably as a result of changed PID activity. It was proposed that this effect is a result of altered content of cytosolic $Ca^{2+}$ in a mutant due to disturbed $InsP_3$ metabolism (Fig. 8.2; Zhang et al. 2011a). Although in light of recent findings, an alternative explanation, in which $InsP_6$ would be the signal activating $Ca^{2+}$ cascade, seems more plausible (Munnik and Nielsen 2011).

Another mode of PID activity regulation applies to its transcription. INDEHISCENT (IND), a basic helix–loop–helix transcription factor through negative regulation of PID and WAG2 expression influences polar localization of PIN1 and PIN3 proteins (Fig. 8.2). Interestingly, in this case local auxin minimum, which is required for valve margin formation in *Arabidopsis* fruit, is not properly established. Consequently, abnormal fruits which fail to open and thus do not disperse seeds are observed in *ind* mutant (Sorefan et al. 2009). Another regulator of PID activity is ENHANCER OF PINOID/MACCHI-BOU4 (ENP/MAB4) (Treml et al. 2005; Furutani et al. 2007) and its close homologs MAB4/ENP/

NPY1-LIKEs (MEL1, MEL2, MEL3, and MEL4) that are known to influence polar PIN localization (Fig. 8.2; Furutani et al. 2011). MAB4 encodes NON-PHOTOTROPIC HYPOCOTYL3-like (NPH3)-like protein, which associates with light activated kinase PHOTOTROPIN1 (PHOT1), a blue-light receptor (Motchoulski and Liscum 1999). NPH3 has been shown to modulate PIN2 trafficking in root phototropism (Wan et al. 2012).

Regulation of PID activity provides also an entry point for various external signals, such as light or gravity, to which plants are able to dynamically respond by modulation of their growth. For example, during hypocotyl gravitropic response, PIN3 and PIN7 polarize to the bottom side of gravity-sensing endodermal cells and mediate the differential auxin accumulation at the lower side of hypocotyl for asymmetric bending and growth (Rakusova et al. 2011). Accordingly, blue-light-dependent signaling cascade causes polarization of PIN3 away from the light during phototropic response in hypocotyl endodermal cells. This coincides with an establishment of auxin maximum at the shaded side of an organ (Ding et al. 2011). Importantly, in both cases PIN repolarization is initiated by differential recruitment into GNOM-mediated trafficking pathway and depends on the PID-mediated phosphorylation status of the PIN protein (Ding et al. 2011; Rakusova et al. 2011).

Beside PIN phosphorylation, another posttranslational protein modification appears to be instructive for PIN localization. The destabilization of PINs from the PM and their sorting for vacuolar delivery was associated with the linking of the polyubiquitin chains to specific lysine residues within PIN2 central hydrophilic loop (Fig. 8.2; Leitner et al. 2012). Moreover, PIN2 degradation was shown to be dependent on the 26S proteasome, the universal proteolysis complex of eukaryotic organisms, which targets ubiquitinated, typically soluble proteins (Sieberer et al. 2000; Abas et al. 2006). Although recent data clearly show that ubiquitination is a crucial part of PIN abundance control (Abas et al. 2006; Leitner et al. 2012), it remains unclear how proteasome activity can contribute to the degradation of PM proteins such as PINs, which were shown to be targeted to the lytic vacuoles (Kleine-Vehn et al. 2008c; Laxmi et al. 2008; Shirakawa et al. 2009; Marhavý et al. 2011; Baster et al. 2013).

## 3.2 Cell Structural Determinants of PIN Polarity Maintenance

The cues and mechanisms described in the previous paragraph dealt mainly with targeting of PINs to their respective polar domains but not with their maintenance there. As mentioned before, crucial polarity trafficking components as well as 'tight junctions', which in animals limit migration of the surface proteins between polar domains of epithelial cells, are absent in the plant kingdom. On the other hand, plant cells are surrounded by cell wall, a structure absent in animal cells. It appears that in order to preserve abundance and asymmetry of proteins within fluid PM, in comparison to animals, plants developed both overlapping and alternative mechanisms. The migration of transmembrane proteins within the lipid bilayer can serve as an example of polarity regulation mechanism which is common for plants and animals. Interestingly, certain fraction of membrane localized PINs, as reported previously, appears to be surprisingly stagnant (Dhonukshe et al. 2008a; Men et al. 2008; Kleine-Vehn et al. 2011). This phenomenon was proposed to be linked with processes that actually immobilize this fraction of PINs within specific structures at the PM, called clusters, which have been detected by super-resolution microscopy approaches (Kleine-Vehn et al. 2011). The nature of these highly immobile structures is not entirely clear; however, their appearance might be related to specific sterol and lipid composition of the PM (Kleine-Vehn et al. 2011; Men et al. 2008; Roudier et al. 2010; Carland et al. 2010; Martinière et al. 2012). In fact, the *sterol methyl transferase 1* (*smt1*) mutant, function of which is required for appropriate synthesis and composition of major membrane sterols (Diener et al. 2000), is characterized by defective polar auxin transport correlating with mislocalization of PIN1 and PIN3 proteins (Fig. 8.2; Willemsen et al. 2003). Similarly, the improper reestablishment of PIN2 polarity following cytokinesis as a consequence of defective PIN2 endocytosis was reported for sterol biosynthesis, *cyclopropylsterol isomerase1-1 (cpi1-1)* mutant (Fig. 8.2; Men et al. 2008). The involvement of sterols in polar distribution of PIN proteins is additionally supported by the fact that the internalized PIN2 co-localizes with the sterol marker filipin and a prolonged disruption of membrane sterols by filipin treatments reduces the heterogeneity and polar localization of PIN2 in the PM (Grebe et al. 2003; Kleine-Vehn et al. 2006, 2011).

Apart from sterols, some other molecular components that physically scaffold structure of the membrane appear very important for maintaining PIN polarity. Sphingolipids, membrane constituents and signaling molecules (Dickson et al. 2006), were shown to influence cell polarity (Hoekstra et al. 2003; Nyasae et al. 2003). The immunophilin-like protein PASSTICINO1 (PAS1) (Bach et al. 2008; Roudier et al. 2010) is involved in the biosynthesis and metabolism of very-long-chain fatty acids (VLCFAs), one of the composites of sphingolipids. In case of *pas1* mutant, patterning defects at the cellular level were attributed to altered auxin distribution during key events in plant life. Disturbed formation of auxin gradients was associated with abnormal PM distribution of PIN1 protein due to defective VLCFA synthesis (Fig. 8.2; Roudier et al. 2010). Additionally, PIN1 abundance in its polar domain was shown to be stabilized by interaction with PGP1/PGP19 ABCB transporters (Titapiwatanakun et al. 2009). Notably, the efflux activity of these ABCB transporters at the PM is positively regulated by PID, most likely through direct phosphorylation. In this case, another immunophilin—TWISTED DWARF (TWD) by interaction with PID appears to decrease ABCB activity at the cell surface (Bouchard et al. 2006; Henrichs et al. 2012; Wang et al. 2012, 2013).

Finally, recent reports suggest that not only structure of the PM but also the integrity of the cell wall is required for maintenance of PIN polarity. Such a notion was suggested upon characterization of *regulator of PIN polarity3 (repp3)* mutant (Feraru et al. 2011), exhibiting defects in localization of ectopically expressed PIN1. The mutation responsible for *repp* mutant phenotype was found in the gene coding for CELLULOSE SYNTHASE CATALYTIC SUBUNIT3/CONSTITUTIVE EXPRESSION OF VSP1/ISOXABEN RESISTANT1/ECTOPIC LIGNIN1 (CESA3/CEV1/IXR1/ELI1). CESA3 is a part of the enzymatic complex required for synthesis of 1,4 glucans, molecules which are able to associate to form cellulose microfibrils and thus scaffold cell wall (Fig. 8.2; Richmond and Somerville 2000; Ellis and Turner 2001; Scheible et al. 2001; Caño-Delgado et al. 2003; Desprez et al. 2007). Additionally, pharmacologically induced cell wall degradation or inhibition of the cell wall biosynthesis resulted in similar phenotypes as in case of *repp3* mutant. Interestingly, plasmolysis-based experiments forcing detachment of the polar domain from the cell wall suggested that the mechanisms immobilizing PIN-containing PM clusters might relate to cellulose-based connections between the polar domain and the cell wall (Feraru et al. 2011; Martinière et al. 2012).

## 4 Transport in Loops: Hormonal Feedback Regulations of PIN Polarity

### *4.1 Auxin Feedback on PIN-Dependent Auxin Transport*

Polar distribution of PIN auxin transporters can be regulated through plethora of controlling mechanisms with various molecular players involved. Intriguingly, another layer of complexity for auxin-driven plant development emerges from the self-organizing abilities of auxin transport. Already early experiments implied that auxin-induced changes could instruct capacity and directionality of auxin flow and thus auxin would have the ability to shape its own transport (Sachs 1981, 1991). Indeed, the effect of auxin on PIN amounts and PIN localization and thus existence of multiple feedback mechanisms at various levels has been validated experimentally. One of such mechanisms is a well-characterized nucleus-based $SCF^{TIR1/AFB}$-dependent auxin signaling (Dharmasiri and Estelle 2004; Dharmasiri et al. 2005a, b; Kepinski and Leyser 2005; Chapman and Estelle 2009), involved in the regulation of PIN transcription (Peer et al. 2004; Vieten et al. 2005; Heisler et al. 2005; Scarpella et al. 2006). $SCF^{TIR1/AFB}$-dependent auxin signaling appears to have a double role in transport feedback, controlling both PIN transcription and abundance at the PM by promoting, upon prolonged auxin exposure, PIN vacuolar targeting for degradation (Baster et al. 2013). The gravitropic response of the roots serve as an example of the process facilitated by such a dual mechanism in which fluctuations of auxin above or below certain physiological threshold through ($SCF^{TIR1/AFB}$)-dependent signaling mediate PIN degradation (Fig. 8.1; Abas et al. 2006; Kleine-Vehn et al. 2008c; Baster et al. 2013). In addition, this transcriptional signaling is also required to feedback on PIN polarity and thus directionality of auxin transport in both root- and shoot-based model systems (Sauer et al. 2006; Balla et al. 2011).

The auxin feedback loop which has recently drawn considerable amount of attention, due to its proposed contribution to the polarization of auxin transporters, is the non-transcriptional auxin effect on PIN endocytosis. Indeed auxin, rapidly upon application, inhibits PIN internalization and promotes the retention of PINs at the PM correlating with increased auxin efflux capacity (Paciorek et al. 2005). Although the mechanism is still largely elusive, it was proposed that auxin, when extracellulary bound to AUXIN-BINDING PROTEIN1 (ABP1), executes this inhibitory function through dynamic activation of mutually exclusive RHO OF PLANTS2 and 6 (ROP2/ROP6) pathways downstream of ABP1 (Fig. 8.1). This mechanism is functionally important in developmental processes like patterning of the leaf epidermis or root gravitropism (Robert et al. 2010; Xu et al. 2010; Nagawa et al. 2012; Chen et al. 2012; Lin et al. 2012).

Additional complexity of the system, which depends on auxin-mediated feedback and regulates plant development, emerges from the fact that some of its elements are interconnected. For example CME, activity of which was shown to be nonuniform throughout the root meristem, directly influences auxin-responsive gene expression. Such a mechanism is based on the positive autoregulatory feedback mediated by BREVIS RADIX (BRX). This plant-specific transcription factor, through auxin-regulated PM-to-nucleus transfer and subsequent transcriptional activation of certain auxin response factor targets, controls the cell elongation and proliferation in the root tip. Therefore, the differential pattern of endocytosis splits the transcriptional auxin signaling within the root meristem and might thus provide additional positional information to interpret auxin gradients (Mouchel et al. 2004; Santuari et al. 2011).

Notably, the experimental approaches aiming to explain the principles of auxin feedback-mediated plant development are more and more prominently supported by computational models. For example, recently proposed 'extracellular receptor-based polarization' (ERP) model integrates transcription-based intracellular feedback mechanisms with a competitive utilization of auxin receptors in the cell exterior for the spatial regulation of PIN internalization (Wabnik et al. 2010, 2011). Given the fact that during various developmental processes, PINs can behave differentially, polarizing either toward or away from the auxin source (Grieneisen et al. 2007; Blilou et al. 2005; Benjamins and Scheres 2008; Kleine-Vehn et al. 2008b), it is worth mentioning that ERP model proposes mechanistic principles explaining these contrasting self-organizing properties of auxin transport (Wabnik et al. 2010, 2011). However, it remains to be seen whether this largely theoretical model corresponds to the biological reality. This question will be solved only after the molecular mechanism underlying the polarization of PINs and auxin transport will be elucidated.

### 4.2 Other Hormonal Regulations of PIN-Dependent Auxin Transport

It seems that not only auxin can shape the capacity and directionality of its transport. Other hormones, by influencing the PM stability of PIN auxin transporters, can be also integrated into the PIN-dependent auxin distribution network. Most of the plant hormones have been shown to regulate transcription of numerous genes downstream of their corresponding signaling pathways, thus many of them directly or indirectly influence also the transcription of *PIN* genes. Such an effect is well characterized in case of cytokinin and ethylene, both of which have also multiple developmental functions (Swarup et al. 2007; Růžička et al. 2007, 2009; Dello Ioio et al. 2008; Zhang et al. 2011b; Bishopp et al. 2011; Liu et al. 2013).

Other hormones prominently modulate PIN activity by posttranscriptional regulation. For example, a stabilization of PINs at the PM by interference with their endocytosis was observed when plants were subjected to pharmacologically or genetically induced accumulation of the plant hormone salicylic acid (SA) (Fig. 8.1; Du et al. 2013). In contrast, gibberellic acid (GA) deficiency, observed in GA biosynthesis mutants, promotes degradation of PIN proteins, whereas treatment with GA increases PIN protein stability by inhibiting PIN vacuolar trafficking. This mechanism appears to be important for correct gravitropic response of the root (Fig. 8.1; Willige et al. 2011; Löfke et al. 2013). The stabilization of PIN2 at the membrane resulting in the perturbations in root gravitropism could be also observed in both *Arabidopsis* plants overexpressing *GOLVEN* (*GLV*) genes encoding for small secretory peptides of ROOT GROWTH FACTOR (RGF) family as well as upon exogenous applications of such compounds (Fig. 8.1; Matsuzaki et al. 2010; Whitford et al. 2012). In contrast to these stabilizing effects, cytokinin promotes vacuolar trafficking of PINs thus destabilizing them from the PM. Functionality of this mechanism was demonstrated during lateral root organogenesis (Fig. 8.1; Marhavý et al. 2011). These frequent observations on the effects of various signaling pathways converging at the regulation of PIN-dependent auxin distribution network are in line with the model that PIN-mediated asymmetric auxin distribution functions as a versatile mechanism integrating multiple internal and external signals (Vanneste and Friml 2009).

## 5 Conclusions and Perspectives

In recent years, there has been a significant advance achieved in our understanding of the basic rules governing generation, maintenance, and refining of the cellular polarity in plants. It becomes clear how polar protein localization at the subcellular scale can underpin basic functionality of an organism and instruct its development. In particular, the studying of asymmetric distribution displayed by PIN auxin transporters allowed us to deepen the knowledge about polarity. At the same time, it provided excellent means to comprehend auxin transport machinery and its contribution, by establishment of asymmetric auxin distribution, to the regulation of various developmental processes. An important open question remains: to which extent cellular mechanisms and molecular components of the PIN polar targeting machinery can be translated into proteins localized at different polar domains in plants. Another important aspect is the evolution of auxin transport machinery and thus mechanisms governing polarity generation in more ancient plant species. For this, the examination of evolutionary older than Angiosperms plant species like moss or algae, which is rapidly gaining popularity in the plant field, will hopefully significantly contribute to our understanding of the polarity phenomenon in the close future.

## References


Abas L, Benjamins R, Malenica N, Paciorek T, Wiśniewska J, Moulinier-Anzola JC, Sieberer T, Friml J, Luschnig C (2006) Intracellular trafficking and proteolysis of the *Arabidopsis* auxin-efflux facilitator PIN2 are involved in root gravitropism. Nat Cell Biol 8:249–256

Alberts B, Johnson A, Lewis J, Raff M, Roberts K, Walter P (2007) Molecular biology of the cell, 5th edn. Garland, New York

Altschuler SJ, Angenent SB, Wang Y, Wu LF (2008) On the spontaneous emergence of cell polarity. Nature 454:886–889

Arighi CN, Hartnell LM, Aguilar RC, Haft CR, Bonifacino JS (2004) Role of the mammalian retromer in sorting of the cation-independent mannose 6-phosphate receptor. J Cell Biol 165:123–133

Bach L, Michaelson LV, Haslam R, Bellec Y, Gissot L, Marion J, Da Costa M, Boutin JP, Miquel M, Tellier F, Domergue F, Markham JE, Beaudoin F, Napier JA, Faure JD (2008) The plant very long chain hydroxy fatty Acyl-CoA dehydratase PASTICCINO2 is essential and limiting for plant development. Proc Natl Acad Sci U S A 105:14727–14731

Balla J, Kalousek P, Reinöhl V, Friml J, Procházka S (2011) Competitive canalization of PIN-dependent auxin flow from axillary buds controls pea bud outgrowth. Plant J 65:571–577

Ballesteros I, Domínguez T, Sauer M, Paredes P, Duprat A, Rojo E, Sanmartín M, Sánchez-Serrano JJ (2013) Specialized functions of the PP2A subfamily II catalytic subunits PP2A-C3 and PP2A-C4 in the distribution of auxin fluxes and development in Arabidopsis. Plant J 73:862–872

Baluška F, Samaj J, Menzel D (2003) Polar transport of auxin: carrier-mediated flux across the plasma membrane or neurotransmitter-like secretion? Trends Cell Biol 13:282–285

Bassham DC, Brandizzi F, Otegui MS, Sanderfoot AA (2008) The secretory system of *Arabidopsis*. Arabidopsis Book 6:e0116

Baster P, Robert S, Kleine-Vehn J, Vanneste S, Kania U, Grunewald W, De Rybel B, Beeckman T, Friml J (2013) SCF(TIR1/AFB)-auxin signaling regulates PIN vacuolar trafficking and auxin fluxes during root gravitropism. EMBO J 32:260–274

Benjamins R, Scheres B (2008) Auxin: the looping star in plant development. Annu Rev Plant Biol 59:443–465

Benjamins R, Quint A, Weijers D, Hooykaas P, Offringa R (2001) The PINOID protein kinase regulates organ development in Arabidopsis by enhancing polar auxin transport. Development 128:4057–4067

Benjamins R, Ampudia CSG, Hooykaas PJJ, Offringa R (2003) PINOID-mediated signaling involves calcium-binding proteins. Plant Physiol 132:1623–1630

Benková E, Michniewicz M, Sauer M, Teichmann T, Seifertová D, Jürgens G, Friml J (2003) Local, efflux-dependent auxin gradients as a common module for plant organ formation. Cell 115:591–602

Bennett MJ, Marchant A, Green HG, May ST, Ward SP, Millner PA, Walker AR, Schulz B, Feldmann KA (1996) Arabidopsis AUX1 gene: a permease-like regulator of root gravitropism. Science 273:948–950

Berleth T, Sachs T (2001) Plant morphogenesis: long-distance coordination and local patterning. Curr Opin Plant Biol 4:57–62

Bishopp A, Help H, El-Showk S, Weijers D, Scheres B, Friml J, Benková E, Mähönen AP, Helariutta Y (2011) A mutually inhibitory interaction between auxin and cytokinin specifies vascular pattern in roots. Curr Biol 21:917–926

Blakeslee JJ, Bandyopadhyay A, Lee OR, Mravec J, Titapiwatanakun B, Sauer M, Makam SN, Cheng Y, Bouchard R, Adamec J, Geisler M, Nagashima A, Sakai T, Martinoia E, Friml J, Peer WA, Murphy AS (2007) Interactions among PIN-FORMED and P-glycoprotein auxin transporters in Arabidopsis. Plant Cell 19:131–147

Blatt MR, Thiel G, Trentham DR (1990) Reversible inactivation of K+ channels of Vicia stomatal guard cells following the photolysis of caged inositol 1,4,5-trisphosphate. Nature 346:766–769

Blilou I, Xu J, Wildwater M, Willemsen V, Paponov I, Friml J, Heidstra R, Aida M, Palme K, Scheres B (2005) The PIN auxin efflux facilitator network controls growth and patterning in *Arabidopsis* roots. Nature 433:39–44

Bouchard R, Bailly A, Blakeslee JJ, Oehring SC, Vincenzetti V, Lee OR, Paponov I, Palme K, Mancuso S, Murphy AS, Schulz B, Geisler M (2006) Immunophilin-like TWISTED DWARF1 modulates auxin efflux activities of *Arabidopsis* P-glycoproteins. J Biol Chem 281:30603–30612

Boutté Y, Crosnier MT, Carraro N, Traas J, Satiat-Jeunemaitre B (2006) The plasma membrane recycling pathway and cell polarity in plants: studies on PIN proteins. J Cell Sci 119:1255–1265

Caño-Delgado A, Penfield S, Smith C, Catley M, Bevan M (2003) Reduced cellulose synthesis invokes lignification and defense responses in *Arabidopsis thaliana*. Plant J 34:351–362

Carland F, Fujioka S, Nelson T (2010) The sterol methyltransferases SMT1, SMT2, and SMT3 influence *Arabidopsis* development through nonbrassinosteroid products. Plant Physiol 153:741–756

Chapman EJ, Estelle M (2009) Mechanism of auxin-regulated gene expression in plants. Annu Rev Genet 43:265–285

Chen CL, Gajewski KM, Hamaratoglu F, Bossuyt W, Sansores-Garcia L, Tao C, Halder G (2010) The apical-basal cell polarity determinant Crumbs regulates Hippo signaling in *Drosophila*. Proc Natl Acad Sci U S A 107:15810–15815

Chen X, Irani NG, Friml J (2011) Clathrin-mediated endocytosis: the gateway into plant cells. Curr Opin Plant Biol 14:674–682

Chen X, Naramoto S, Robert S, Tejos R, Löfke C, Lin D, Yang Z, Friml J (2012) ABP1 and ROP6 GTPase signaling regulate clathrin-mediated endocytosis in *Arabidopsis* roots. Curr Biol 22:1326–1332

Christensen SK, Dagenais N, Chory J, Weigel D (2000) Regulation of auxin response by the protein kinase PINOID. Cell 100:469–478

Christie JM, Yang H, Richter GL, Sullivan S, Thomson CE, Lin J, Titapiwatanakun B, Ennis M, Kaiserli E, Lee OR, Adamec J, Peer WA, Murphy AS (2011) phot1 inhibition of ABCB19 primes lateral auxin fluxes in the shoot apex required for phototropism. PLoS Biol 9(6): e1001076

D'Souza-Schorey C, Chavrier P (2006) ARF proteins: roles in membrane traffic and beyond. Nat Rev Mol Cell Biol 7:347–358

Dai M, Zhang C, Kania U, Chen F, Xue Q, McCray T, Li G, Qin G, Wakeley M, Terzaghi W, Wan J, Zhao Y, Xu J, Friml J, Deng XW, Wang H (2012) A PP6-type phosphatase holoenzyme directly regulates PIN phosphorylation and auxin efflux in Arabidopsis. Plant Cell 24:2497–2514

Davies PJ (2004) Plant hormones: biosynthesis, signal transduction, action! Kluwer Academic, London

Dello Ioio R, Nakamura K, Moubayidin L, Perilli S, Taniguchi M, Morita MT, Aoyama T, Costantino P, Sabatini S (2008) A genetic framework for the control of cell division and differentiation in the root meristem. Science 322:1380–1384

Desprez T, Juraniec M, Crowell EF, Jouy H, Pochylova Z, Parcy F, Höfte H, Gonneau M, Vernhettes S (2007) Organization of cellulose synthase complexes involved in primary cell wall synthesis in *Arabidopsis thaliana*. Proc Natl Acad Sci U S A 104:15572–15577

Dettmer J, Friml J (2011) Cell polarity in plants: when two do the same, it is not the same. Curr Opin Cell Biol 23:686–696

Dettmer J, Hong-Hermesdorf A, Stierhof Y-D, Schumacher K (2006) Vacuolar H+-ATPase activity is required for endocytic and secretory trafficking in *Arabidopsis*. Plant Cell 18:715–730

Dharmasiri N, Estelle M (2004) Auxin signaling and regulated protein degradation. Trends Plant Sci 9:302–308

Dharmasiri N, Dharmasiri S, Estelle M (2005a) The F-box protein TIR1 is an auxin receptor. Nature 435:441–445

Dharmasiri N, Dharmasiri S, Weijers D, Lechner E, Yamada M, Hobbie L, Ehrismann JS, Jurgens G, Estelle M (2005b) Plant development is regulated by a family of auxin receptor F box proteins. Dev Cell 9:109–119

Dhonukshe P, Aniento F, Hwang I, Robinson DG, Mravec J, Stierhof YD, Friml J (2007) Clathrin-mediated constitutive endocytosis of PIN auxin efflux carriers in *Arabidopsis*. Curr Biol 17:520–527

Dhonukshe P, Tanaka H, Goh T, Ebine K, Mähönen AP, Prasad K, Blilou I, Geldner N, Xu J, Uemura T, Chory J, Ueda T, Nakano A, Scheres B, Friml J (2008a) Generation of cell polarity in plants links endocytosis, auxin distribution and cell fate decisions. Nature 456:962–966

Dhonukshe P, Grigoriev I, Fischer R, Tominaga M, Robinson DG, Hasek J, Paciorek T, Petrásek J, Seifertová D, Tejos R, Meisel LA, Zazímalová E, Gadella TW Jr, Stierhof YD, Ueda T, Oiwa K, Akhmanova A, Brock R, Spang A, Friml J (2008b) Auxin transport inhibitors impair vesicle motility and actin cytoskeleton dynamics in diverse eukaryotes. Proc Natl Acad Sci U S A 105:4489–4494

Dhonukshe P, Huang F, Galvan-Ampudia CS, Mähönen AP, Kleine-Vehn J, Xu J, Quint A, Prasad K, Friml J, Scheres B, Offringa R (2010) Plasma membrane-bound AGC3 kinases phosphorylate PIN auxin carriers at TPRXS(N/S) motifs to direct apical PIN recycling. Development 137:3245–3255

Dickson RC, Sumanasekera C, Lester RL (2006) Functions and metabolism of sphingolipids in *Saccharomyces cerevisiae*. Prog Lipid Res 45:447–465

Diener AC, Li H, Zhou W, Whoriskey WJ, Nes WD, Fink GR (2000) Sterol methyltransferase 1 controls the level of cholesterol in plants. Plant Cell 12:853–870

Ding Z, Friml J (2010) Auxin regulates distal stem cell differentiation in Arabidopsis roots. Proc Natl Acad Sci U S A 107:12046–12051

Ding Z, Galván-Ampudia CS, Demarsy E, Łangowski Ł, Kleine-Vehn J, Fan Y, Morita MT, Tasaka M, Fankhauser C, Offringa R, Friml J (2011) Light-mediated polarization of the PIN3 auxin transporter for the phototropic response in Arabidopsis. Nat Cell Biol 13:447–452

Drdová EJ, Synek L, Pečenková T, Hála M, Kulich I, Fowler JE, Murphy AS, Žárský V (2013) The exocyst complex contributes to PIN auxin efflux carrier recycling and polar auxin transport in *Arabidopsis*. Plant J 73:709–719

Du Y, Tejos R, Beck M, Himschoot E, Li H, Robatzek S, Vanneste S, Friml J (2013) Salicylic acid interferes with clathrin-mediated endocytic protein trafficking. Proc Natl Acad Sci U S A 110 (19):7946–7951

Dudu V, Pantazis P, Gonzalez-Gaitan M (2004) Membrane traffic during embryonic development: epithelial formation, cell fate decisions and differentiation. Curr Opin Cell Biol 16:407–414

Ellis C, Turner JG (2001) The *Arabidopsis* mutant cev1 has constitutively active jasmonate and ethylene signal pathways and enhanced resistance to pathogens. Plant Cell 13:1025–1033

Feraru E, Paciorek T, Feraru MI, Zwiewka M, De Groodt R, De Rycke R, Kleine-Vehn J, Friml J (2010) The AP-3 B adaptin mediates the biogenesis and function of lytic vacuoles in *Arabidopsis*. Plant Cell 22:2812–2824

Feraru E, Feraru MI, Kleine-Vehn J, Martinière A, Mouille G, Vanneste S, Vernhettes S, Runions J, Friml J (2011) PIN polarity maintenance by the cell wall in *Arabidopsis*. Curr Biol 21:338–343

Feraru E, Feraru MI, Asaoka R, Paciorek T, De Rycke R, Tanaka H, Nakano A, Friml J (2012) BEX5/RabA1b regulates trans-Golgi network-to-plasma membrane protein trafficking in Arabidopsis. Plant Cell 24:3074–3086

Foti M, Moukil MA, Dudognon P, Carpentier JL (2004) Insulin and IGF-1 receptor trafficking and signaling. Novartis Found Symp 262:125–141

Friml J, Palme K (2002) Polar auxin transport – old questions and new concepts? Plant Mol Biol 49:273–284

Friml J, Benková E, Blilou I, Wisniewska J, Hamann T, Ljung K, Woody S, Sandberg G, Scheres B, Jürgens G, Palme K (2002a) AtPIN4 mediates sink-driven auxin gradients and root patterning in Arabidopsis. Cell 108:661–673

Friml J, Wiśniewska J, Benková E, Mendgen K, Palme K (2002b) Lateral relocation of auxin efflux regulator PIN3 mediates tropism in Arabidopsis. Nature 415:806–809

Friml J, Vieten A, Sauer M, Weijers D, Schwarz H, Hamann T, Offringa R, Jürgens G (2003) Efflux-dependent auxin gradients establish the apical-basal axis of *Arabidopsis*. Nature 426:147–153

Friml J, Yang X, Michniewicz M, Weijers D, Quint A, Tietz O, Benjamins R, Ouwerkerk PB, Ljung K, Sandberg G, Hooykaas PJ, Palme K, Offringa R (2004) APINOID-dependent binary switch in apical-basal PIN polar targeting directs auxin efflux. Science 306:862–865

Fujimoto M, Arimura S, Ueda T, Takanashi H, Hayashi Y, Nakano A, Tsutsumi N (2010) *Arabidopsis* dynamin-related proteins DRP2B and DRP1A participate together in clathrin-coated vesicle formation during endocytosis. Proc Natl Acad Sci U S A 107:6094–6099

Furutani M, Kajiwara T, Kato T, Treml BS, Stockum C, Torres-Ruiz RA, Tasaka M (2007) The gene MACCHI-BOU 4/ENHANCER OF PINOID encodes a NPH3-like protein and reveals similarities between organogenesis and phototropism at the molecular level. Development 134:3849–3859

Furutani M, Sakamoto N, Yoshida S, Kajiwara T, Robert HS, Friml J, Tasaka M (2011) Polarlocalized NPH3-like proteins regulate polarity and endocytosis of PIN-FORMED auxin efflux carriers. Development 138:2069–2078

Gälweiler L, Guan C, Müller A, Wisman E, Mendgen K, Yephremov A, Palme K (1998) Regulation of polar auxin transport by AtPIN1 in *Arabidopsis* vascular tissue. Science 282:2226–2230

Geisler M, Blakeslee JJ, Bouchard R, Lee OR, Vincenzetti V, Bandyopadhyay A, Titapiwatanakun B, Peer WA, Bailly A, Richards EL, Ejendal KF, Smith AP, Baroux C, Grossniklaus U, Müller A, Hrycyna CA, Dudler R, Murphy AS, Martinoia E (2005) Cellular efflux of auxin catalyzed by the Arabidopsis MDR/PGP transporter AtPGP1. Plant J 44:179–194

Geldner N (2009) Cell polarity in plants: a PARspective on PINs. Curr Opin Plant Biol 12:42–48

Geldner N, Friml J, Stierhof YD, Jürgens G, Palme K (2001) Auxin transport inhibitors block PIN1 cycling and vesicle trafficking. Nature 413:425–428

Geldner N, Anders N, Wolters H, Keicher J, Kornberger W, Muller P, Delbarre A, Ueda T, Nakano A, Jürgens G (2003) The *Arabidopsis* GNOM ARF-GEF mediates endosomal recycling, auxin transport, and auxin-dependent plant growth. Cell 112:219–230

Giepmans BN, van Ijzendoorn SC (2009) Epithelial cell-cell junctions and plasma membrane domains. Biochim Biophys Acta 1788:820–831

Gilroy S, Read ND, Trewavas AJ (1990) Elevation of cytoplasmic calcium by caged calcium or caged inositol triphosphate initiates stomatal closure. Nature 346:769–771

Grebe M, Xu J, Mobius W, Ueda T, Nakano A, Geuze HJ, Rook MB, Scheres B (2003) *Arabidopsis* sterol endocytosis involves actin-mediated trafficking via ARA6-positive early endosomes. Curr Biol 13:1378–1387

Grieneisen VA, Xu J, Marée AF, Hogeweg P, Scheres B (2007) Auxin transport is sufficient to generate a maximum and gradient guiding root growth. Nature 449:1008–1013

Hazak O, Bloch D, Poraty L, Sternberg H, Zhang J, Friml J, Yalovsky S (2010) A rho scaffold integrates the secretory system with feedback mechanisms in regulation of auxin distribution. PLoS Biol 8(1):e1000282

Heisler MG, Ohno C, Das P, Sieber P, Reddy GV, Long JA, Meyerowitz EM (2005) Patterns of auxin transport and gene expression during primordium development revealed by live imaging of the *Arabidopsis* inflorescence meristem. Curr Biol 15:1899–1911

Heisler MG, Hamant O, Krupinski P, Uyttewaal M, Ohno C, Jönsson H, Traas J, Meyerowitz EM (2010) Alignment between PIN1 polarity and microtubule orientation in the shoot apical meristem reveals a tight coupling between morphogenesis and auxin transport. PLoS Biol 8 (10):e1000516

Henrichs S, Wang B, Fukao Y, Zhu J, Charrier L, Bailly A, Oehring SC, Linnert M, Weiwad M, Endler A, Nanni P, Pollmann S, Mancuso S, Schulz A, Geisler M (2012) Regulation of ABCB1/PGP1-catalysed auxin transport by linker phosphorylation. EMBO J 31:2965–2980

Hertel R (1983) The mechanism of auxin transport as a model for auxin action. Z Pflanzenphysiol 112:53–67

Heymann JAW, Hinshaw JE (2009) Dynamins at a glance. J Cell Sci 122:3427–3431

Hoekstra D, Maier O, van der Wouden JM, Slimane TA, van IJzendoorn SC (2003) Membrane dynamics and cell polarity: the role of sphingolipids. J Lipid Res 44:869–877

Holler D, Dikic I (2004) Receptor endocytosis via ubiquitin-dependent and -independent pathways. Biochem Pharmacol 67:1013–1017

Huang F, Zago MK, Abas L, van Marion A, Galván-Ampudia CS, Offringa R (2010) Phosphorylation of conserved PIN motifs directs *Arabidopsis* PIN1 polarity and auxin transport. Plant Cell 22:1129–1142

Humbert PO, Dow LE, Russell SM (2006) The Scribble and Par complexes in polarity and migration: friends or foes? Trends Cell Biol 16:622–630

Irani NG, Russinova E (2009) Receptor endocytosis and signaling in plants. Curr Opin Plant Biol 12:653–659

Jaillais Y, Fobis-Loisy I, Miège C, Rollin C, Gaude T (2006) AtSNX1 defines an endosome for auxin-carrier trafficking in *Arabidopsis*. Nature 443:106–109

Jaillais Y, Santambrogio M, Rozier F, Fobis-Loisy I, Miège C, Gaude T (2007) The retromer protein VPS29 links cell polarity and organ initiation in plants. Cell 130:1057–1070

Kepinski S, Leyser O (2005) The *Arabidopsis* F-box protein TIR1 is an auxin receptor. Nature 435:446–451

Kitakura S, Vanneste S, Robert S, Löfke C, Teichmann T, Tanaka H, Friml J (2011) Clathrin mediates endocytosis and polar distribution of PIN auxin transporters in Arabidopsis. Plant Cell 23:1920–1931

Kleine-Vehn J, Friml J (2008) Polar targeting and endocytic recycling in auxin-dependent plant development. Annu Rev Cell Dev Biol 24:447–473

Kleine-Vehn J, Dhonukshe P, Swarup R, Bennett M, Friml J (2006) Subcellular trafficking of the *Arabidopsis* auxin influx carrier AUX1 uses a novel pathway distinct from PIN1. Plant Cell 18:3171–3181

Kleine-Vehn J, Dhonukshe P, Sauer M, Brewer PB, Wiśniewska J, Paciorek T, Benková E, Friml J (2008a) ARF GEF-dependent transcytosis and polar delivery of PIN auxin carriers in Arabidopsis. Curr Biol 18:526–531

Kleine-Vehn J, Łangowski Ł, Wisniewska J, Dhonukshe P, Brewer PB, Friml J (2008b) Cellular and molecular requirements for polar PIN targeting and transcytosis in plants. Mol Plant 1:1056–1066
Kleine-Vehn J, Leitner J, Zwiewka M, Sauer M, Abas L, Luschnig C, Friml J (2008c) Differential degradation of PIN2 auxin efflux carrier by retromer-dependent vacuolar targeting. Proc Natl Acad Sci U S A 105:17812–17817
Kleine-Vehn J, Huang F, Naramoto S, Zhang J, Michniewicz M, Offringa R, Friml J (2009) PIN auxin efflux carrier polarity is regulated by PINOID kinase-mediated recruitment into GNOM-independent trafficking in *Arabidopsis*. Plant Cell 21:3839–3849
Kleine-Vehn J, Ding Z, Jones AR, Tasaka M, Morita MT, Friml J (2010) Gravity-induced PIN transcytosis for polarization of auxin fluxes in gravity-sensing root cells. Proc Natl Acad Sci U S A 107:22344–22349
Kleine-Vehn J, Wabnik K, Martinière A, Łangowski Ł, Willig K, Naramoto S, Leitner J, Tanaka H, Jakobs S, Robert S, Luschnig C, Govaerts W, Hell SW, Runions J, Friml J (2011) Recycling, clustering, and endocytosis jointly maintain PIN auxin carrier polarity at the plasma membrane. Mol Syst Biol 25:540
Konopka CA, Backues SK, Bednarek SY (2008) Dynamics of *Arabidopsis* dynamin-related protein 1C and a clathrin light chain at the plasma membrane. Plant Cell 20:1363–1380
Kramer EM (2004) PIN and AUX/LAX proteins: their role in auxin accumulation. Trends Plant Sci 9:578–582
Krinke O, Novotná Z, Valentová O, Martinec J (2007) Inositol trisphosphate receptor in higher plants: is it real? J Exp Bot 58:361–376
Kubeš M, Yang H, Richter GL, Cheng Y, Młodzińska E, Wang X, Blakeslee JJ, Carraro N, Petrášek J, Zažímalová E, Hoyerová K, Peer WA, Murphy AS (2012) The Arabidopsis concentration-dependent influx/efflux transporter ABCB4 regulates cellular auxin levels in the root epidermis. Plant J 69:640–654
Lam SK, Siu CL, Hillmer S, Jang S, An G, Robinson DG, Jiang L (2007) Rice SCAMP1 defines clathrin-coated, trans-golgi-located tubular-vesicular structures as an early endosome in tobacco BY-2 cells. Plant Cell 19:296–319
Langowski L, Růzicka K, Naramoto S, Kleine-Vehn J, Friml J (2010) Trafficking to the outer polar domain defines the root-soil interface. Curr Biol 20:904–908
Lavy M, Bloch D, Hazak O, Gutman I, Poraty L, Sorek N, Sternberg H, Yalovsky S (2007) A Novel ROP/RAC effector links cell polarity, root-meristem maintenance, and vesicle trafficking. Curr Biol 17:947–952
Laxmi A, Pan J, Morsy M, Chen R (2008) Light plays an essential role in intracellular distribution of auxin efflux carrier PIN2 in Arabidopsis thaliana. PLoS One 3(1):e1510
Leitner J, Petrášek J, Tomanov K, Retzer K, Pařezová M, Korbei B, Bachmair A, Zažímalová E, Luschnig C (2012) Lysine63-linked ubiquitylation of PIN2 auxin carrier protein governs hormonally controlled adaptation of Arabidopsis root growth. Proc Natl Acad Sci U S A 109:8322–8327
Lemtiri-Chlieh F, MacRobbie EA, Webb AA, Manison NF, Brownlee C, Skepper JN, Chen J, Prestwich GD, Brearley CA (2003) Inositol hexakisphosphate mobilizes an endomembrane store of calcium in guard cells. Proc Natl Acad Sci U S A 100:10091–10095
Leyser O (2011) Auxin, self-organisation, and the colonial nature of plants. Curr Biol 21:R331–R337
Li H, Lin D, Dhonukshe P, Nagawa S, Chen D, Friml J, Scheres B, Guo H, Yang Z (2011) Phosphorylation switch modulates the interdigitated pattern of PIN1 localization and cell expansion in *Arabidopsis* leaf epidermis. Cell Res 21:970–978
Lin D, Shingo N, Chen J, Chen X, Cao L, Xu T, Li H, Dhonukshe P, Friml J, Scheres B, Fu Y, Yang Z (2012) A ROP GTPase-dependent auxin signaling pathway inhibits PIN2 endocytosis in *Arabidopsis* roots. Curr Biol 22:1319–1325

Liu T, Mirschberger C, Chooback L, Arana Q, Dal Sacco Z, MacWilliams H, Clarke M (2002) Altered expression of the 100 kDa subunit of the *Dictyostelium* vacuolar proton pump impairs enzyme assembly, endocytic function and cytosolic pH regulation. J Cell Sci 115:1907–1918

Liu J, Mehdi S, Topping J, Friml J, Lindsey K (2013) Interaction of PLS and PIN and hormonal crosstalk in Arabidopsis root development. Front Plant Sci 4:75

Ljung K (2013) Auxin metabolism and homeostasis during plant development. Development 140:943–950

Löfke C, Zwiewka M, Heilmann I, Van Montagu MC, Teichmann T, Friml J (2013) Asymmetric gibberellin signaling regulates vacuolar trafficking of PIN auxin transporters during root gravitropism. Proc Natl Acad Sci U S A 110:3627–3632

Ludwig-Müller J (2011) Auxin conjugates: their role for plant development and in the evolution of land plants. J Exp Bot 62:1757–1773

Luschnig C, Gaxiola RA, Grisafi P, Fink GR (1998) EIR1, a root-specific protein involved in auxin transport, is required for gravitropism in *Arabidopsis thaliana*. Genes Dev 12:2175–2218

Mancuso S, Marras AM, Mugnai S, Schlicht M, Zársky V, Li G, Song L, Xue HW, Baluska F (2007) Phospholipase Dζ2 drives vesicular secretion of auxin for its polar cell-cell transport in the transition zone of the root apex. Plant Signal Behav 2:240–244

Mano Y, Nemoto K (2012) The pathway of auxin biosynthesis in plants. J Exp Bot 63:2853–2872

Marhavý P, Bielach A, Abas L, Abuzeineh A, Duclercq J, Tanaka H, Parezová M, Petrásek J, 917 Friml J, Kleine-Vehn J, Benková E (2011) Cytokinin modulates endocytic trafficking of PIN1 918 auxin efflux carrier to control plant organogenesis. Dev Cell 21:796–804

Martinière A, Lavagi I, Nageswaran G, Rolfe DJ, Maneta-Peyret L, Luu DT, Botchway SW, Webb SEW, Mongrand S, Maurel C, Martin-Fernandez ML, Kleine-Vehn J, Friml J, Moreau P, Runions J (2012) The cell wall constrains lateral diffusion of plant plasma-membrane proteins. Proc Natl Acad Sci U S A 109:12805–12810

Matheson LA, Hanton SL, Brandizzi F (2006) Traffic between the plant endoplasmic reticulum and Golgi apparatus: to the Golgi and beyond. Curr Opin Plant Biol 9:601–609

Matsuzaki Y, Ogawa-Ohnishi M, Mori A, Matsubayashi Y (2010) Secreted peptide signals required for maintenance of root stem cell niche in *Arabidopsis*. Science 329:1065–1067

McFarlane HE, Shin JJ, Bird DA, Samuels AL (2010) *Arabidopsis* ABCG transporters, which are required for export of diverse cuticular lipids, dimerize in different combinations. Plant Cell 22:3066–3075

Men S, Boutté Y, Ikeda Y, Li X, Palme K, Stierhof Y-D, Hartmann M-A, Moritz T, Grebe M (2008) Sterol-dependent endocytosis mediates post-cytokinetic acquisition of PIN2 auxin efflux carrier polarity. Nat Cell Biol 10:237–244

Michell RH (2008) Inositol derivatives: evolution and functions. Nat Rev Mol Cell Biol 9:151–161

Michniewicz M, Brewer PB, Friml J (2007a) Polar auxin transport and asymmetric auxin distribution. Arabidopsis Book 5:e0108

Michniewicz M, Zago MK, Abas L, Weijers D, Schweighofer A, Meskiene I, Heisler MG, Ohno C, Zhang J, Huang F, Schwab R, Weigel D, Meyerowitz EM, Luschnig C, Offringa R, Friml J (2007b) Antagonistic regulation of PIN phosphorylation by PP2A and PINOID directs auxin flux. Cell 130:1044–1056

Miwa K, Takano J, Omori H, Seki M, Shinozaki K, Fujiwara T (2007) Plants tolerant of high boron levels. Science 318:1417

Motchoulski A, Liscum E (1999) *Arabidopsis* NPH3: a NPH1 photoreceptor-interacting protein essential for phototropism. Science 286:961–964

Mouchel CF, Briggs GC, Hardtke CS (2004) Natural genetic variation in *Arabidopsis* identifies BREVIS RADIX, a novel regulator of cell proliferation and elongation in the root. Genes Dev 18:700–714

Mravec J, Kubes M, Bielach A, Gaykova V, Petrásek J, Skůpa P, Chand S, Benková E, Zazímalová E, Friml J (2008) Interaction of PIN and PGP transport mechanisms in auxin distribution-dependent development. Development 135:3345–3354

Mravec J, Petrášek J, Li N, Boeren S, Karlova R, Kitakura S, Pařezová M, Naramoto S, Nodzyński T, Dhonukshe P, Bednarek SY, Zažímalová E, de Vries S, Friml J (2011) Cell plate restricted association of DRP1A and PIN proteins is required for cell polarity establishment in *Arabidopsis*. Curr Biol 21:1055–1060
Muday GK, DeLong A (2001) Polar auxin transport: controlling where and how much. Trends Plant Sci 6:535–542
Müller J, Mettbach U, Menzel D, Šamaj J (2007) Molecular dissection of endosomal compartments in plants. Plant Physiol 145:293–304
Munnik T, Nielsen E (2011) Green light for polyphosphoinositide signals in plants. Curr Opin Plant Biol 14:489–497
Munnik T, Testerink C (2009) Plant phospholipid signaling: 'in a nutshell'. J Lipid Res 50:260–265
Nagawa S, Xu T, Lin D, Dhonukshe P, Zhang X, Friml J, Scheres B, Fu Y, Yang Z (2012) ROP GTPase-dependent actin microfilaments promote PIN1 polarization by localized inhibition of clathrin-dependent endocytosis. PLoS Biol 10(4):e1001299
Naramoto S, Kleine-Vehn J, Robert S, Fujimoto M, Dainobu T, Paciorek T, Ueda T, Nakano A, Van Montagu MCE, Fukuda H, Friml J (2010) ADP-ribosylation factor machinery mediates endocytosis in plant cells. Proc Natl Acad Sci U S A 107:21890–21895
Nodzyński T, Feraru MI, Hirscha S, De Rycke R, Niculaesa C, Boerjana W, Van Leene J, De Jaeger G, Vanneste S, Friml J (2013) Retromer subunits VPS35A and VPS29 mediate prevacuolar compartment (PVC) function in Arabidopsis. Mol Plant 6:1849–1862
Noh B, Murphy AS, Spalding EP (2001) Multidrug resistance-like genes of *Arabidopsis* required for auxin transport and auxin-mediated development. Plant Cell 13:2441–2454
Nyasae LK, Hubbard AL, Tuma PL (2003) Transcytotic efflux from early endosomes is dependent on cholesterol and glycosphingolipids in polarized hepatic cells. Mol Biol Cell 14:2689–2705
Okada K, Ueda J, Komaki MK, Bell CJ, Shimura Y (1991) Requirement of the auxin polar transport system in early stages of *Arabidopsis* floral bud formation. Plant Cell 3:677–684
Paciorek T, Zažímalová E, Ruthardt N, Petrášek J, Stierhof YD, Kleine-Vehn J, Morris DA, Emans N, Jürgens G, Geldner N, Friml J (2005) Auxin inhibits endocytosis and promotes its own efflux from cells. Nature 435:1251–1256
Panikashvili D, Savaldi-Goldstein S, Mandel T, Yifhar T, Franke RB, Höfer R, Schreiber L, Chory J, Aharoni A (2007) The *Arabidopsis* DESPERADO/AtWBC11 transporter is required for cutin and wax secretion. Plant Physiol 145:1345–1360
Paul MJ, Frigerio L (2007) Coated vesicles in plant cells. Semin Cell Dev Biol 18:471–478
Peer WA, Bandyopadhyay A, Blakeslee JJ, Makam SN, Chen RJ, Masson PH, Murphy AS (2004) Variation in expression and protein localization of the PIN family of auxin efflux facilitator proteins in flavonoid mutants with altered auxin transport in Arabidopsis thaliana. Plant Cell 16:1898–1911
Petrášek J, Schwarzerová K (2009) Actin and microtubule cytoskeleton interactions. Curr Opin Plant Biol 12:728–734
Petrášek J, Mravec J, Bouchard R, Blakeslee JJ, Abas M, Seifertová D, Wiśniewska J, Tadele Z, Kubeš M, Čovanová M, Dhonukshe P, Skůpa P, Benková E, Perry L, Křeček P, Lee OR, Fink GR, Geisler M, Murphy AS, Luschnig C, Zažímalová E, Friml J (2006) PIN proteins perform a rate-limiting function in cellular auxin efflux. Science 312:914–918
Praefcke GJK, McMahon HT (2004) The dynamin superfamily: universal membrane tubulation and fission molecules? Nat Rev Mol Cell Biol 5:133–147
Rakusova H, Gallego-Bartolomé J, Vanstraelen M, Robert HS, Alabadí D, Blázquez MA, Benková E, Friml J (2011) Polarization of PIN3-dependent auxin transport for hypocotyl gravitropic response in *Arabidopsis thaliana*. Plant J 67:817–826
Raven JA (1975) Transport of indoleacetic acid in plant cells in relation to pH and electrical potential gradients, and its significance for polar IAA transport. New Phytol 74:163–172
Reinhardt D, Pesce ER, Stieger P, Mandel T, Baltensperger K, Bennett M, Traas J, Friml J, Kuhlemeier C (2003) Regulation of phyllotaxis by polar auxin transport. Nature 426:255–260

Richmond TA, Somerville CR (2000) The cellulose synthase superfamily. Plant Physiol 124:495–498
Rigó G, Ayaydin F, Tietz O, Zsigmond L, Kovács H, Páy A, Salchert K, Darula Z, Medzihradszky KF, Szabados L, Palme K, Koncz C, Cséplo A (2013) Inactivation of plasma membrane-localized CDPK-RELATED KINASE5 decelerates PIN2 exocytosis and root gravitropic response in Arabidopsis. Plant Cell 25(5):1592–1608
Robert S, Kleine-Vehn J, Barbez E, Sauer M, Paciorek T, Baster P, Vanneste S, Zhang J, Simon S, Čovanová M, Hayashi K, Dhonukshe P, Yang Z, Bednarek SY, Jones AM, Luschnig C, Aniento F, Zažímolová E, Friml J (2010) ABP1 mediates auxin inhibition of clathrin-dependent endocytosis in *Arabidopsis*. Cell 143:111–121
Robinson DG, Jiang L, Schumacher K (2008) The endosomal system of plants: charting new and familiar territories. Plant Physiol 147:1482–1492
Roppolo D, De Rybel B, Tendon VD, Pfister A, Alassimone J, Vermeer JE, Yamazaki M, Stierhof YD, Beeckman T, Geldner N (2011) A novel protein family mediates Casparian strip formation in the endodermis. Nature 473:380–383
Roudier F, Gissot L, Beaudoin F, Haslam R, Michaelson L, Marion J, Molino D, Lima A, Bach L, Morin H, Tellier F, Palauqui JC, Bellec Y, Renne C, Miquel M, Dacosta M, Vignard J, Rochat C, Markham JE, Moreau P, Napier J, Faure JD (2010) Very-long-chain fatty acids are involved in polar auxin transport and developmental patterning in Arabidopsis. Plant Cell 22:364–375
Rubery PH, Sheldrake AR (1974) Carrier-mediated auxin transport. Planta 118:101–121
Ruiz Rosquete M, Barbez E, Kleine-Vehn J (2012) Cellular auxin homeostasis: gatekeeping is housekeeping. Mol Plant 5:772–786
Růžička K, Ljung K, Vanneste S, Podhorská R, Beeckman T, Friml J, Benková E (2007) Ethylene regulates root growth through effects on auxin biosynthesis and transport-dependent auxin distribution. Plant Cell 19:2197–2212
Růžička K, Šimášková M, Duclercq J, Petrášek J, Zažímalová E, Simon S, Friml J, Van Montagu MC, Benková E (2009) Cytokinin regulates root meristem activity via modulation of the polar auxin transport. Proc Natl Acad Sci U S A 106:4284–4289
Sabatini S, Beis D, Wolkenfelt H, Murfett J, Guilfoyle T, Malamy J, Benfey P, Leyser O, Bechtold N, Weisbeek P, Scheres B (1999) An auxin-dependent distal organizer of pattern and polarity in the Arabidopsis root. Cell 99:463–472
Sachs T (1981) The control of the patterned differentiation of vascular tissues. Adv Bot Res 9:151–262
Sachs T (1991) Cell polarity and tissue patterning in plants. Development 113:83–93
Santner AA, Watson JC (2006) The WAG1 and WAG2 protein kinases negatively regulate root waving in *Arabidopsis*. Plant J 45:752–764
Santuari L, Scacchi E, Rodriguez-Villalon A, Salinas P, Dohmann EM, Brunoud G, Vernoux T, Smith RS, Hardtke CS (2011) Positional information by differential endocytosis splits auxin response to drive Arabidopsis root meristem growth. Curr Biol 21:1918–1923
Sauer M, Friml J (2004) The matryoshka dolls of plant polarity. Development 131:5774–5775
Sauer M, Balla J, Luschnig C, Wisniewska J, Reinöhl V, Friml J, Benková E (2006) Canalization of auxin flow by Aux/IAA-ARF-dependent feedback regulation of PIN polarity. Genes Dev 20:2902–2911
Scarpella E, Marcos D, Friml J, Berleth T (2006) Control of leaf vascular patterning by polar auxin transport. Genes Dev 20:1015–1027
Scheible WR, Eshed R, Richmond T, Delmer D, Somerville C (2001) Modifications of cellulose synthase confer resistance to isoxaben and thiazolidinone herbicides in *Arabidopsis* Ixr1 mutants. Proc Natl Acad Sci U S A 98:10079–10084
Scheuring D, Viotti C, Krüger F, Künzl F, Sturm S, Bubeck J, Hillmer S, Frigerio L, Robinson DG, Pimpl P, Schumacher K (2011) Multivesicular bodies mature from the trans-Golgi network/ early endosome in *Arabidopsis*. Plant Cell 23:3463–3481

Schlereth A, Möller B, Liu W, Kientz M, Flipse J, Rademacher EH, Schmid M, Jürgens G, Weijers D (2010) MONOPTEROS controls embryonic root initiation by regulating a mobile transcription factor. Nature 464:913–916

Schlicht M, Strnad M, Scanlon MJ, Mancuso S, Hochholdinger F, Palme K, Volkmann D, Menzel D, Baluska F (2006) Auxin immunolocalization implicates vesicular neurotransmitter-like mode of polar auxin transport in root apices. Plant Signal Behav 1:122–133

Seaman MNJ (2005) Recycle your receptors with retromer. Trends Cell Biol 15:68–75

Shimada T, Koumoto Y, Li L, Yamazaki M, Kondo M, Nishimura M, Hara-Nishimura I (2006) AtVPS29, a putative component of a retromer complex, is required for the efficient sorting of seed storage proteins. Plant Cell Physiol 47:1187–1194

Shirakawa M, Ueda H, Shimada T, Nishiyama C, Hara-Nishimura I (2009) Vacuolar SNAREs function in the formation of the leaf vascular network by regulating auxin distribution. Plant Cell Physiol 50:1319–1328

Shivas JM, Morrison HA, Bilder D, Skop AR (2010) Polarity and endocytosis: reciprocal regulation. Trends Cell Biol 20:445–452

Sieberer T, Seifert GJ, Hauser MT, Grisafi P, Fink GR, Luschnig C (2000) Post-transcriptional control of the *Arabidopsis* auxin efflux carrier EIR1 requires AXR1. Curr Biol 10:1595–1598

Sorefan K, Girin T, Liljegren SJ, Ljung K, Robles P, Galván-Ampudia CS, Offringa R, Friml J, Yanofsky MF, Østergaard L (2009) A regulated auxin minimum is required for seed dispersal in *Arabidopsis*. Nature 459:583–586

Spitzer C, Reyes FC, Buono R, Sliwinski MK, Haas TJ, Otegui MS (2009) The ESCRT-related CHMP1A and B proteins mediate multivesicular body sorting of auxin carriers in *Arabidopsis* and are required for plant development. Plant Cell 21:749–766

Swarup R, Friml J, Marchant A, Ljung K, Sandberg G, Palme K, Bennett M (2001) Localization of the auxin permease AUX1 suggests two functionally distinct hormone transport pathways operate in the *Arabidopsis* root apex. Genes Dev 15:2648–2653

Swarup R, Perry P, Hagenbeek D, Van Der Straeten D, Beemster GT, Sandberg G, Bhalerao R, Ljung K, Bennett MJ (2007) Ethylene upregulates auxin biosynthesis in *Arabidopsis* seedlings to enhance inhibition of root cell elongation. Plant Cell 19:2186–2196

Swarup K, Benkova E, Swarup R, Casimiro I, Peret B, Yang Y, Parry G, Nielsen E, De Smet I, Vanneste S, Levesque MP, Carrier D, James N, Calvo V, Ljung K, Kramer E, Roberts R, Graham N, Marillonnet S, Patel K (2008) The auxin influx carrier LAX3 promotes lateral root emergence. Nat Cell Biol 10:946–954

Takano J, Tanaka M, Toyoda A, Miwa K, Kasai K, Fuji K, Onouchi H, Naito S, Fujiwara T (2010) Polar localization and degradation of Arabidopsis boron transporters through distinct trafficking pathways. Proc Natl Acad Sci U S A 107:5220–5225

Tanaka H, Dhonukshe P, Brewer PB, Friml J (2006) Spatiotemporal asymmetric auxin distribution: a means to coordinate plant development. Cell Mol Life Sci 63:2738–2754

Tanaka H, Kitakura S, De Rycke R, De Groodt R, Friml J (2009) Fluorescence imaging-based screen identifies ARF GEF component of early endosomal trafficking. Curr Biol 19:391–397

Tanaka H, Kitakura S, Rakusová R, Uemura T, Feraru MI, De Rycke R, Robert S, Kakimoto T, Friml J (2013) Cell polarity and patterning by PIN trafficking through early endosomal compartments in *Arabidopsis*. PLoS Genet 9(5):e1003540

Tang RH, Han S, Zheng H, Cook CW, Choi CS, Woerner TE, Jackson RB, Pei ZM (2007) Coupling diurnal cytosolic Ca2+ oscillations to the CAS-IP3 pathway in *Arabidopsis*. Science 315:1423–1426

Teh OK, Moore I (2007) An ARF-GEF acting at the golgi and in selective endocytosis in polarized plant cells. Nature 448:493–496

Tepass U, Tanentzapf G, Ward R, Fehon R (2001) Epithelial cell polarity and cell junctions in *Drosophila*. Annu Rev Genet 35:747–784

Testerink C, Munnik T (2011) Molecular, cellular and physiological responses to phosphatidic acid formation in plants. J Exp Bot 62:2349–2361

Titapiwatanakun B, Blakeslee JJ, Bandyopadhyay A, Yang H, Mravec J, Sauer M, Cheng Y, Adamec J, Nagashima A, Geisler M, Sakai T, Friml J, Peer WA, Murphy AS (2009) ABCB19/PGP19 stabilizes PIN1 in membrane microdomains in *Arabidopsis*. Plant J 57:27–44

Treml BS, Winderl S, Radykewicz R, Herz M, Schweizer G, Hutzler P, Glawischnig E, Ruiz RA (2005) The gene ENHANCER OF PINOID controls cotyledon development in the *Arabidopsis* embryo. Development 132:4063–4074

Tuma PL, Hubbard AL (2003) Transcytosis: crossing cellular barriers. Physiol Rev 83:871–932

Vanneste S, Friml J (2009) Auxin: a trigger for change in plant development. Cell 136:1005–1016

Verrier PJ, Bird D, Burla B, Dassa E, Forestier C, Geisler M, Klein M, Kolukisaoglu Ü, Lee Y, Martinoia E, Murphy A, Rea PA, Samuels L, Schulz B, Spalding EP, Yazaki K, Theodoulou FL (2008) Plant ABC proteins—a unified nomenclature and updated inventory. Trends Plant Sci 13:151–159

Vieten A, Vanneste S, Wiśniewska J, Benková E, Benjamins R, Beeckman T, Luschnig C, Friml J (2005) Functional redundancy of PIN proteins is accompanied by auxin-dependent cross-regulation of PIN expression. Development 132:4521–4531

Vieten A, Sauer M, Brewer PB, Friml J (2007) Molecular and cellular aspects of auxin transport-mediated development. Trends Plant Sci 12:160–168

Viotti C, Bubeck J, Stierhof YD, Krebs M, Langhans M, van den Berg W, van Dongen W, Richter S, Geldner N, Takano J, Jürgens G, de Vries SC, Robinson DG, Schumacher K (2010) Endocytic and secretory traffic in *Arabidopsis* merge in the trans-Golgi network/early endosome, an independent and highly dynamic organelle. Plant Cell 22:1344–1357

Vitale A, Denecke J (1999) The endoplasmic reticulum – gateway of the secretory pathway. Plant Cell 11:615–628

Voigt B, Timmers AC, Šamaj J, Müller J, Baluška F, Menzel D (2005) GFP-FABD2 fusion construct allows in vivo visualization of the dynamic actin cytoskeleton in all cells of *Arabidopsis* seedlings. Eur J Cell Biol 84:595–608

Wabnik K, Kleine-Vehn J, Balla J, Sauer M, Naramoto S, Reinöhl V, Merks RMH, Govaerts W, Friml J (2010) Emergence of tissue polarization from synergy of intracellular and extracellular auxin signaling. Mol Syst Biol 6:447

Wabnik K, Govaerts W, Friml J, Kleine-Vehn J (2011) Feedback models for polarized auxin transport: an emerging trend. Mol Biosyst 7:2352

Wan Y, Jasik J, Wang L, Hao H, Volkmann D, Menzel D, Mancuso S, Baluška F, Lin J (2012) The signal transducer NPH3 integrates the phototropin1 photosensor with PIN2-based polar auxin transport in *Arabidopsis* root phototropism. Plant Cell 24:551–565

Wang B, Henrichs S, Geisler M (2012) The AGC kinase, PINOID, blocks interactive ABCB/PIN auxin transport. Plant Signal Behav 7(12):1515–1517 (Epub ahead of print)

Wang B, Bailly A, Zwiewka M, Henrichs S, Azzarello E, Mancuso S, Maeshima M, Friml J, Schulz A, Geisler M (2013) *Arabidopsis* TWISTED DWARF1 functionally interacts with auxin exporter ABCB1 on the root plasma membrane. Plant Cell 25:202–214

Wells CD, Fawcett JP, Traweger A, Yamanaka Y, Goudreault M, Elder K, Kulkarni S, Gish G, Virag C, Lim C, Colwill K, Starostine A, Metalnikov P, Pawson T (2006) A Rich1/Amot complex regulates the Cdc42 GTPase and apical-polarity proteins in epithelial cells. Cell 125:535–548

Whitford R, Fernandez A, Tejos R, Pérez AC, Kleine-Vehn J, Vanneste S, Drozdzecki A, Leitner J, Abas L, Aerts M, Hoogewijs K, Baster P, De Groodt R, Lin Y-C, Storme V, Van de Peer Y, Beeckman T, Madder A, Devreese B, Luschnig C, Friml J, Hilson P (2012) GOLVEN secretory peptides regulate auxin carrier turnover during plant gravitropic responses. Dev Cell 22:678–685

Willemsen V, Friml J, Grebe M, van den Toorn A, Palme K, Scheres B (2003) Cell polarity and PIN protein positioning in *Arabidopsis* require STEROL METHYLTRANSFERASE1 function. Plant Cell 15:612–625

Willige BC, Isono E, Richter R, Zourelidou M, Schwechheimer C (2011) Gibberellin regulates PIN-FORMED abundance and is required for auxin transport-dependent growth and development in *Arabidopsis thaliana*. Plant Cell 23:2184–2195

Winter V, Hauser MT (2006) Exploring the ESCRTing machinery in eukaryotes. Trends Plant Sci 11:115–123

Wiśniewska J, Xu J, Seifertová D, Brewer PB, Růžička K, Blilou I, Rouquié D, Benková E, Scheres B, Friml J (2006) Polar PIN localization directs auxin flow in plants. Science 312:883

Wollert T, Wunder C, Lippincott-Schwartz J, Hurley JH (2009) Membrane scission by the ESCRT-III complex. Nature 458:172–177

Xu T, Wen M, Nagawa S, Fu Y, Chen JG, Wu MJ, Perrot-Rechenmann C, Friml J, Jones AM, Yang Z (2010) Cell surface- and rho GTPase-based auxin signaling controls cellular interdigitationin *Arabidopsis*. Cell 143:99–110

Yang Y, Hammes UZ, Taylor CG, Schachtman DP, Nielsen E (2006) High-affinity auxin transport by the AUX1 influx carrier protein. Curr Biol 16:1123–1127

Žárský V, Cvrčková F, Potocký M, Hála M (2009) Exocytosis and cell polarity in plants – exocyst and recycling domains. New Phytol 183:255–272

Zažímalová E, Krecek P, Skůpa P, Hoyerová K, Petrásek J (2007) Polar transport of the plant hormone auxin—the role of PIN-FORMED (PIN) proteins. Cell Mol Life Sci 64:1621–1637

Zegzouti H, Anthony RG, Jahchan N, Bögre L, Christensen SK (2006) Phosphorylation and activation of PINOID by the phospholipid signaling kinase 3-phosphoinositide-dependent protein kinase 1 (PDK1) in *Arabidopsis*. Proc Natl Acad Sci U S A 103:6404–6409

Zhang J, Nodzynski T, Pencík A, Rolcík J, Friml J (2010) PIN phosphorylation is sufficient to mediate PIN polarity and direct auxin transport. Proc Natl Acad Sci U S A 107:918–922

Zhang J, Vanneste S, Brewer PB, Michniewicz M, Grones P, Kleine-Vehn J, Löfke C, Teichmann T, Bielach A, Cannoot B, Hoyerová K, Chen X, Xue HW, Benková E, Zažímalová E, Friml J (2011a) Inositol trisphosphate-induced Ca2+ signaling modulates auxin transport and PIN polarity. Dev Cell 20:855–866

Zhang W, To JP, Cheng CY, Eric Schaller G, Kieber JJ (2011b) Type-A response regulators are required for proper root apical meristem function through post-transcriptional regulation of PIN auxin efflux carriers. Plant J 68:1–10

Zourelidou M, Müller I, Willige BC, Nill C, Jikumaru Y, Li H, Schwechheimer C (2009) The polarly localized D6 PROTEIN KINASE is required for efficient auxin transport in *Arabidopsis thaliana*. Development 136:627–636

Zwiewka M, Feraru E, Müller B, Hwang I, Feraru MI, Kleine-Vehn J, Weijers D, Friml J (2011) The AP-3 adaptor complex is required for vacuolar function in *Arabidopsis*. Cell Res 21:1711–1722

# Chapter 9
# Auxin Regulation of Embryo Development

**Alejandra Freire Rios, Saiko Yoshida, and Dolf Weijers**

**Abstract** Important steps in plant development are made shortly after fertilization. In a brief succession of cell divisions, the zygote is transformed into an embryo, a multicellular structure carrying all fundamental tissue types and the meristems. Hence, embryogenesis offers excellent opportunities to dissect the molecular control and cellular mechanisms underlying plant development. In the past decades, forward and reverse genetics studies have revealed that the plant hormone auxin plays a central role in the establishment of pattern and polarity in the *Arabidopsis* embryo. Here, we review the roles that localized auxin biosynthesis, directional transport and cell type-specific response play in embryo development. We focus on the molecular mechanisms, as well as the feedbacks that connect these disparate levels of regulation. Finally, we discuss the potential for hormonal cross-talk in auxin-dependent control of the key events during the earliest, formative phase of plant life.

## 1 Introduction: Early Plant Embryogenesis

Multicellular organisms begin life as a single zygote cell. While the animal embryo is a miniature form of the adult body and thus has a relatively complex structure, the plant mature embryo has a rather simple structure: an embryonic root, hypocotyl and one or two embryonic leaves. This miniature encompasses meristems in the shoot and root tips. The meristems will create all the other parts of the mature plant body after germination (Weigel and Jurgens 2002). In *Arabidopsis* embryos, these meristem primordia consist only of a few stem cells. These stem cells will divide

---

Alejandra Freire Rios and Saiko Yoshida contributed equally.

A. Freire Rios • S. Yoshida • D. Weijers (✉)
Laboratory of Biochemistry, Wageningen University, Dreijenlaan 3, 6703HA Wageningen, The Netherlands
e-mail: dolf.weijers@wur.nl

E. Zažímalová et al. (eds.), *Auxin and Its Role in Plant Development*,
DOI 10.1007/978-3-7091-1526-8_9, © Springer-Verlag Wien 2014

and spatially coordinate the acquisition of different cell identities during the post-embryonic generation of a functional body. Stem cell niches are an excellent example for the importance of spatial coordination in cell specification where, in order to keep a functional niche, stem and organizer cells need to be in direct vicinity (Scheres 2007). Additionally, as early embryos consist of few cells, yet different layers, tissues and organs are established, there is a general need for tightly coordinated development. In this chapter, we discuss how the plant hormone mediates the coordinated acquisition of cell types during this formative phase of *Arabidopsis* life.

Unlike in animal embryos, plant cells do not migrate during embryogenesis because they are surrounded by a rigid cell wall. Therefore, oriented cell division and directional expansion plays an important role in morphogenesis. Division patterns during embryogenesis have been studied and described based on the observations of sections (Scheres et al. 1994; Jürgens and Mayer 1994). After fertilization, the apical–basal polarity is forecast when the zygote divides asymmetrically to create an apical embryonic cell and a basal extra embryonic cell (Fig. 9.1). The embryonic cell further divides to generate eight embryonic cell organizer in two tiers. While the cells in the upper tier will generate the shoot (shoot apical meristem and cotyledons), the cells in the lower tier will make the hypocotyl, root and root apical meristem. Subsequently, all embryonic cells divide periclinally and generate inner and outer cells corresponding to the first establishment of a radial axis. Next, lower tier inner cells divide periclinally to generate the initials for ground tissue and vascular cells at early globular stage. On the other hand, the extra-embryonic cell divides to create the suspensor. Its uppermost cell is specified into hypophysis and then divides asymmetrically to create the upper lens-shaped cell and lower cell which will respectively become the initials for quiescent centre (QC) and columella cells in the root. From transition to heart stage, primordia of the two cotyledons are formed and the structure of embryo obtains bilateral symmetry. Thus, during this morphogenesis phase the basic body pattern is established and the meristem of shoot/root and embryonic organs are generated. Arguably, this morphogenetic phase is of great importance for the establishment of a new plant from a single fertilized egg cell. Hence, understanding of the molecular and cellular mechanisms underlying these events is a key goal for plant developmental biology. In the past decades, much progress has been made, and interestingly, the plant hormone auxin has surfaced repeatedly as a central regulator. Here, we will discuss the various aspects of auxin regulation that contribute to regulating its activity in embryo development.

## 2 Auxin Biosynthesis Pathways in Embryogenesis

The major natural auxin, indole-3-acetic acid (IAA), is biosynthesized from tryptophan via a two-step pathway (Fig. 9.2). Several key enzymes are known to be involved in this pathway (see Chap. 2). TRYPTOPHAN AMINOTRANSFERASE

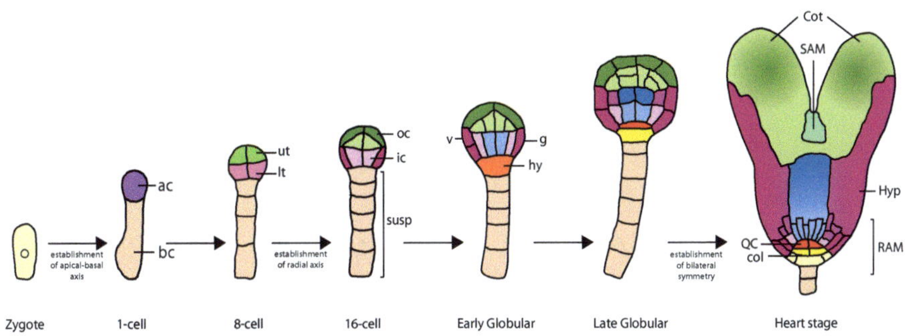


**Fig. 9.1** Embryo development in *Arabidopsis*. Sequential cell division events as observed in 2D sections of embryos starting in the zygote until the heart stage. Points in the process where symmetries are established are indicated. Cells in different colours will differentiate in different tissues/organs. *ac* apical cell, *bc* basal cell, *ut* upper tier, *lt* lower tier, *oc* outer cell, *ic* inner cell, *susp* suspensor, *v* vascular initial, *g* ground tissue initial, *hy* hypophysis, *Cot* cotyledon, *SAM* shoot apical meristem, *Hyp* hypocotyl, *RAM* root apical meristem, *QC* quiescent centre, *col* columella cells

OF ARABIDOPSIS 1 (TAA1) and its closest homologues TRYPTOPHAN AMINOTRANSFERASE RELATED 1,2 (TAR1, 2) are transaminases that convert tryptophan to indole-3-pyruvate (IPA) (Tao et al. 2008). YUCCA is a flavin monooxygenase that catalyses oxidative decarboxylation of IPA to produce IAA (Mashiguchi et al. 2011). Eleven YUC homologues are known in the *Arabidopsis* genome (Cheng et al. 2007). *TAA1* and some of the *YUC* genes are expressed during embryogenesis. *YUC3*, *YUC4* and *YUC9* are expressed in the suspensor from eight-cell stage on. At the globular stage, *YUC1* and *YUC4* are expressed in the cells around the future shoot apical meristem, while *YUC8* is expressed around the hypophysis. From late globular to heart stage, *YUC4* is not only expressed in apical cells around the shoot meristem but is also expressed in the basal hypophysis. *TAA1* is expressed in the apical epidermal cells from 16-cell stage (Robert et al. 2013). From transition to heart stage on, it is expressed in the L1 layer of shoot apical meristem (Cheng et al. 2007). Consistent with the TAA/TAR and YUC proteins acting in a linear biosynthetic pathway, higher-order mutants in each family lead to nearly indistinguishable phenotypes. Embryos of *yuc1 yuc4 yuc10 yuc11* quadruple mutant and the *taa1 tar1 tar2* triple mutant display abnormal cell division in embryonic cells. Seedlings of these mutants lack root/hypocotyl and often have aberrant number of cotyledons, which strongly suggests that auxin activity is required for the normal establishment of these embryonic structures (Stepanova et al. 2008; Cheng et al. 2007). Interestingly, these severe phenotypes cannot be rescued by ectopic production of auxin, which indicates the importance of spatial and temporal regulation of auxin local biosynthesis (Robert et al. 2013).

If local, rather than ubiquitous auxin biosynthesis is important for auxin-dependent embryo development, a key question is what activates the expression of the biosynthesis genes. Transcriptional regulators of the SHORT INTERNODES/STYLISH (SHI/STY) family have been identified as activators of *YUC4*

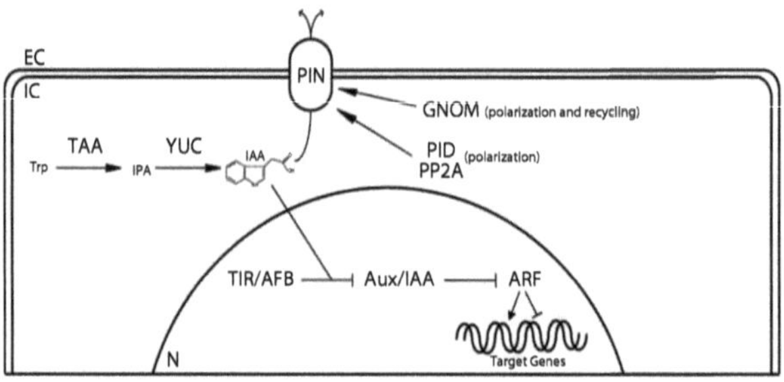


**Fig. 9.2** Core auxin pathways. After being synthesized by the TAA/YUC pathway, intracellular (IC) auxin can either be transported to the extracellular space (EC) or initiate the signalling cascade inside the nucleus (N). Auxin in the nucleus will allow the recognition of Aux/IAAs by the TIR/AFB complex. This will lead to Aux/IAA degradation, and hence de-repression of ARF activity. Transport of auxin is carried out by PIN proteins. The polarized position of PINs within the cells is controlled by proteins like PID, PP2A and GNOM

and *YUC8* (Sohlberg et al. 2006; Eklund et al. 2010a, b; Staldal et al. 2008). The expression pattern of *STY1* and *YUC* genes overlap (Kuusk et al. 2002) and *STY1* is expressed in the future cotyledon primordia of early globular embryos (Kuusk et al. 2002). Overexpression of *STY1* increases auxin biosynthesis, while in the *sty1 sty2* mutant the level of free IAA is low. This mutant displays severe defects in the development of leaf and floral organs, especially in the style. The style phenotype of *sty1 sty2* can be restored by exogenous application of auxin (Staldal et al. 2008). A dominant negative transgenic line of *STY1* does not make a shoot apical meristem, indicating that STY1-dependent auxin biosynthesis has a role in the formation of the shoot apical meristem during embryogenesis or its post-embryonic maintenance. STY1 is in turn activated by DORNRÖSCHEN-LIKE (DRNL), an AP2/ERF family transcription factor (Eklund et al. 2011). DRNL and its homologue DORNRÖSCHEN (DRN) interact with class III HD-ZIP family transcription factors PHAVOLUTA, PHABULOSA, REVOLUTA, CORONA and ATHB8 to regulate patterning of apical embryo (Chandler et al. 2007). As a quintuple mutant among these HD-ZIP transcription factor genes does not develop an embryonic shoot meristem (Emery et al. 2003), this suggests the outlines of a genetic network that directs morphogenesis of shoot apical meristem during embryogenesis and involves several interacting transcription factors to control local auxin biosynthesis. Interestingly, *DRN* as well as the HD-ZIP gene *ATHB8* are direct transcriptional target genes of the auxin response factor MONOPTEROS (see below; Cole et al. 2009; Donner et al. 2009), which suggests that this control network is not linear, but likely involves feedback regulation by auxin.

# 3 Auxin Redistribution Through Directional Transport

The expression pattern of auxin biosynthesis genes does not always match the locations that are marked by auxin response reporter genes such as DR5-GFP (Friml et al. 2003). This suggests that locally produced auxin should be transported to where auxin is needed. Indeed, expression analysis and protein localization, as well as genetic analysis supports an important role for the PIN auxin efflux facilitators in embryo development. The PIN proteins are well known to be involved in the efficient transport system of auxin (Vanneste and Friml 2009; see Chap. 5). Of the eight *Arabidopsis PIN* genes, four are expressed during embryogenesis (Friml et al. 2003). From the one-cell stage onward, the *PIN7* gene is expressed and its protein becomes polarly localized in the apical side of extra-embryonic cells. PIN1 is also expressed from one-cell stage and subsequent stages, although it does not show clear subcellular polarity until mid-globular stage. Auxin response activity (DR5-GFP reporter) is also detected in the embryonic cells, but not the extra-embryonic cells at these stages. In *pin7 and pin1 pin2 pin4 pin7* quadruple mutants, ectopic DR5-GFP expression is observed in the suspensor (Friml et al. 2003). Considering that few *YUC* genes are expressed in the suspensor, it suggests that PIN7 transports the biosynthesized auxin from suspensor to embryonic cells (Robert and Friml 2009). At globular stage, the polarity of PIN7 is changed from apical to basal in the suspensor cells, while PIN1 becomes basally polarized in the vascular initials. This likely promotes accumulation of auxin in the hypophysis. Consistently, strong DR5-GFP signal is observed in the hypophysis (Friml et al. 2003). At the transition stage, PIN1 polarity localizes towards the flanks of the apical embryonic cells. This promotes accumulation of auxin in the cotyledons primordial (Benkova et al. 2003). On the other hand, PIN4 is expressed in the hypophysis at the globular stage embryo. After the division of the hypophysis, it is expressed in the upper lens-shaped cell. PIN3 is expressed in the columella precursors in the heart stage embryo. The *pin7* mutant displays abnormal cell division as well as the quadruple mutants of *pin1 pin3 pin4 pin7* and *pin2 pin3 pin4 pin7* indicating that coordinated polar localization of PINs regulates embryo patterning (Blilou et al. 2005; Friml et al. 2003).

Recent work has suggested the existence of a connection between local auxin biosynthesis and polarization of PIN proteins (see Chap. 8). In auxin biosynthesis mutants, localization of PIN1 is apolar and its expression level is reduced in the later stage (Robert et al. 2013). The mechanisms underlying this link are unclear, but will likely involve control of the cellular mechanisms that target the polar localization of PIN proteins. The polar membrane localization of PIN proteins is the result of a continuous exocytosis/endocytosis cycle, where regulation can act on either step (Vanneste and Friml 2009). Several proteins are known as regulators of PIN trafficking and polarity (Fig. 9.2). A serine–threonine kinase, PINOID (PID) and PROTEIN PHOSPHATASE 2A (PP2A) antagonistically regulate polarity of PIN proteins by regulating their phosphorylation status (Bennett et al. 1995; Benjamins et al. 2001; Friml et al. 2004; Michniewicz et al. 2007). While PIN proteins are targeted to the apical plasma membrane by phosphorylation, PINs are targeted to the basal plasma membrane by dephosphorylation. PID, three PID homologues (PID2, WAG1 and WAG2) and PP2A are expressed during embryogenesis (Cheng et al. 2008; Michniewicz et al. 2007). As expected, mutations in any of these polarity regulators cause defects similar to loss of PIN proteins or auxin biosynthesis (Michniewicz et al. 2007; Benjamins et al. 2001).

The recycling of PIN proteins from endocytic vesicles (endosomes) to the plasma membrane requires GNOM, an ADP ribosylation factor-guanine exchange factor (Geldner et al. 2003; Steinmann et al. 1999). GNOM regulates the recycling of PIN1 protein by controlling its polar localization on the membrane (Geldner et al. 2003). GNOM is ubiquitously expressed (Geldner et al. 2004) and the mutant seedling is rootless and makes fused cotyledons (Mayer et al. 1993), which phenocopies the *pin1* mutant and the *pin1,3,4,7* quadruple mutant (Friml et al. 2003). The *gnom* mutant embryo defects appear from the zygote onward and the mutant embryo fails to establish an apical–basal axis (Mayer et al. 1993). At the globular stage, orientation of cell division plate and cell division pattern of the entire embryo become abnormal (Wolters et al. 2011). Establishment of the bilateral symmetric structure at heart stage is also disturbed and results in a ball-shaped embryo (Mayer et al. 1993). In *gnom* embryos, PIN1 is no longer polarized and auxin transport is reduced causing accumulation in the apical region of globular embryos (Wolters et al. 2011). When GNOM expression is driven by the promoter of a provascular gene, polar localization of PIN1 is restored, auxin accumulation in the apical embryo is reduced and formation of primary root is rescued. When GNOM is expressed in the hypophysis, primary root formation is also restored. This suggests that GNOM acts non-autonomously to regulate root apical meristem formation, which is consistent with its function in intercellular auxin transport. Furthermore, GNOM expression in apical epidermis restores the formation of two cotyledons. Thus GNOM-dependent polar auxin transport is important for the establishment of meristems (Wolters et al. 2011).

Studies on the HANABA TARANU (HAN) GATA transcription factor have revealed a critical role for PIN gene regulation in embryo patterning (Nawy et al. 2010). *HAN* is first expressed in the zygote, and expression is retained in all embryonic cells until the 16-cell stage. From globular to heart stage, *HAN* expression becomes restricted to the provasculature cells (Nawy et al. 2010; Zhao et al. 2004). HAN regulates the transcription of the genes regulating the development of the basal embryonic cells. Therefore, in the *han* mutant embryo, establishment of the apical–basal axis fails. The basal cells of *han* embryos from 16-cell stage resemble suspensor cells with large vacuoles and lower cell divisions. In these cells, marker genes of suspensor and hypophysis (*SUC3* and *WOX5*) are expressed, whereas *SHR*, which is normally expressed in provasculature of basal cells, disappears. The expression patterns of apical marker genes (*WUS* and *ML1*) are not affected in the *han* mutant. Distribution of auxin is also changed in the *han* mutant. While DR5-GFP is expressed in hypophysis and neighbouring suspensor cells in the wild-type embryo, the expression domain is expanded to basal embryonic cells in *han* mutant. The expression domain of PIN7 is also expanded to the basal embryonic cells, while the expression of PIN1 is restricted to the apical embryonic cells. Thus, the lack of root meristem in the *han* mutant is strongly correlated with the disruption of the establishment of an auxin maxima in hypophysis likely due to PIN gene misexpression (Nawy et al. 2010). Whether the regulation is direct remains to be determined, but this finding opens new avenues for understanding the regulation of PIN gene activity during embryo development.

In addition to the PIN proteins, other regulators of auxin transport have been identified. Notably, the PGP/ABCB transporters facilitate non-polar auxin transport (Geisler et al. 2005; see Chap. 5), but their activity has not yet been proven essential for normal embryo development (Mravec et al. 2008). Recently, a new family of auxin transporters, the PIN-LIKES (PILS) proteins, was shown to mediate intracellular auxin transport between cytosol and the endoplasmic reticulum (Barbez et al. 2012). PILS are auxin efflux carriers with a similar topology to PINs. The PILS family consists of seven proteins containing an auxin carrier domain and six (PILS 1, 2, 3, 5, 6 and 7) are localized to the endoplasmic reticulum (Barbez et al. 2012, see Chap. 4). Among the seven PILS, PILS2 and PILS5 are the most abundantly expressed in seedlings. Unlike PIN proteins, the family is conserved throughout the plant kingdom and even exists in algae, suggesting that PILS can be evolutionally older than PINs. PILS are uniformly expressed in various tissues and some of them are auxin inducible. Overexpression and loss of function of PILS2 and/or PILS5 affect hypocotyl and root growth, lateral root organogenesis and root hair length. It will be interesting to see if this novel mode of auxin partitioning is also important for embryo development.

## 4 Transcriptional Response to Auxin

After auxin biosynthesis and transport, hormone accumulation triggers transcriptional changes to affect cell division and identity. The auxin-mediated transcriptional responses are mainly controlled by the interaction of two families of plant transcriptional regulators: the auxin response factors (ARFs) and the Aux/IAA proteins. In a general mechanism of action, the Aux/IAA proteins, together with transcriptional co-repressors like TOPLESS (TPL), form a complex with the ARFs in auxin low levels. When auxin levels rise in the cell, the Aux/IAA proteins are targeted to the 26S-proteosome by an SCF E3 ubiquitin ligase complex. Upon degradation of Aux/IAAs, ARFs are released and can act activating or repressing their target genes (Fig. 9.2; reviewed in Chapman and Estelle 2009; see Chap. 6).

The *Arabidopsis* genome encodes 29 Aux/IAA proteins that share conserved domains. Domain I is necessary for transcriptional repression and it has been shown to recruit the TPL co-repressor in most of the Aux/IAAs (Causier et al. 2012; Szemenyei et al. 2008). Domain II contains the degron motif, a 13 amino acids sequence responsible for the Aux/IAAs' instability by mediating their interaction with the TIR1/AFB receptor (Dharmasiri et al. 2005; Kepinski and Leyser 2005). There might be other sequences outside Domain II contributing to this interaction;

for example, the affinity between the Aux/IAAs and the TIR1/AFB can drop dramatically when a KR conserved motif between Domain I and Domain II is mutated (Calderon Villalobos et al. 2012). Finally, domains III/IV in the C-terminal region, considered as the interaction domain, is thought to mediate homo- and heterodimerization between the Aux/IAAs and with the ARFs (Tiwari et al. 2003). There are also Aux/IAAs that lack one of the domains. They are considered non-canonical Aux/IAAs and their function remain unknown. Overexpression of a subclade of Aux/IAAs, which lack domain II (IAA20, IAA30 and IAA31), results in auxin-related phenotypes suggesting that they may interfere with endogenous ARF-Aux/IAA interactions (Sato and Yamamoto 2008).

The degradation of Aux/IAAs is mediated by the family of E3 ligases called **SCF$^{\mathbf{TIR1/AFB1–5}}$** which targets them for ubiquitination. The F-box protein TRANSPORT INHIBITOR RESPONSE 1 (TIR1) and related proteins AUXIN SIGNALLING F-BOX PROTEIN 1,2,3,4,5 (AFB 1–5) are the substrate receptor of the SCF (Dharmasiri et al. 2005; Kepinski and Leyser 2005). All the six members of the TIR1/AFB family are auxin receptors although they show individual distinct biochemical properties and biological functions (Parry et al. 2009). For the SCF to recognize the Aux/IAAs, the F-box protein needs to be directly bound to auxin. The structure of TIR1 has been determined in the presence of auxin and the degron peptide of IAA7, and it shows that its C-terminal 18 leucine-rich repeats (LRRs) is essential for Aux/IAA and auxin binding (Tan et al. 2007). There are no conformational changes of TIR1 upon auxin binding, which sits in a binding pocket underneath the Aux/IAA binding site (Calderon-Villalobos et al. 2010; Tan et al. 2007). Recent experiments using TIR1-ASK1 and IAA7 showed that both proteins act as co-receptors and are necessary and sufficient for auxin binding. With 29 Aux/IAAs and 6 TIR/AFBs, many qualitatively different co-receptor pairs may exist. Recent data suggests that this is the case. In a qualitative yeast two-hybrid assay, different receptor pairs showed different auxin dose–response and, furthermore, an inverse correlation between the Aux/IAAs' stability and the strength of the interactions. Also, quantitative biochemical assays (saturation and/or homologue competitive IAA-binding assays) showed that affinity of different co-receptor pairs for auxin ranges from 10 nM to >1 μM (Calderon Villalobos et al. 2012). Given that the complex TIR1/AFB–auxin–AUX/IAA is the first step towards Aux/IAA degradation, hence de-repression of ARFs; this complex formation may be a part of a control mechanism of differential auxin responses. While mutations in the degron of many Aux/IAA proteins have been reported to cause distinct auxin-related defects (Reed 2001), only few were shown to affect embryo development. The *iaa12/bdl* and *iaa13* mutations cause defects in root initiation (Hamann et al. 2002; Weijers et al. 2005), while *iaa18* mutants have defects in cotyledon formation (Ploense et al. 2009). In contrast, *iaa10* and *iaa11* mutations affect suspensor and hypophysis development (Rademacher et al. 2012). Rather than distinct protein capacities, these unique phenotypes likely reflect the highly specific gene expression patterns of Aux/IAA genes. Misexpression of the unrelated Aux/IAA protein iaa3/shy2 in the IAA12/BDL expression domain causes bdl-like root defects (Weijers et al. 2005).

As Aux/IAAs are degraded, ARFs are free to elicit gene expression response to auxin. There are 23 ARFs in *Arabidopsis* (Okushima et al. 2005), which adds to the possible combinatorial logic of auxin signalling considering that the members of this family also have different biochemical properties and biological functions. All auxin-dependent processes in the embryo seem to be mediated by ARF activity, as mutations in individual or multiple ARFs, or misexpression of the Aux/IAA proteins disrupt all these processes (Hardtke and Berleth 1998; Rademacher et al. 2012). ARFs have three protein domains. At the N-terminus there is the conserved DNA-binding domain, which binds to the auxin response elements in the promoter regions of the direct targets. The second domain, named middle region, is the non-conserved part and is proposed to determine the activity of the ARFs. ARFs have been classified as gene activators or repressors based on the amino acid composition of their middle region. Experiments in protoplasts showed that some ARFs with a relatively glutamine-rich MR (ARF5–8 and 19) can activate synthetic auxin promoters, and that ARFs with less glutamines (ARF1–4 and 9) can repress the same promoters. The rest of the ARFs have been arbitrarily classified based on these results (Tiwari et al. 2003). Since there is experimental data for some ARFs suggesting that they can act as both activators and repressors of different genes (Schlereth et al. 2010; Zhao et al. 2010), this classification might not be entirely accurate and the regulation of transcription by ARFs might be more complex. Indeed, triple mutants between ARF1, ARF2 and ARF6 show a phenotype that none of the single or double mutants shows (Rademacher et al. 2012). This suggests that redundancy among ARFs is not limited to close related members with similar domains III/IV.

Interestingly, recognition motifs for co-repressor proteins have been found in several ARFs (Ikeda and Ohme-Takagi 2009), and yeast two-hybrid assays have shown strong interactions between the transcriptional co-repressor TPL and ARF2, 9, 17 and 18; and putative weak interactions with ARF1, 3, 4 and 19 (Causier et al. 2012). Finally at the C-terminus, we find the Domains III/IV which mediates the ARFs interaction with the Aux/IAAs and with other ARFs (Tiwari et al. 2003).

## 5 Generating Cell Type-Specific Responses

Various developmental processes during embryogenesis are dependent on proper auxin response (Rademacher et al. 2012), and each of these is marked by activity of the DR5-GFP auxin response reporter (Friml et al. 2003). Therefore, whether or not there is an auxin response does not seem to define what developmental output is triggered. A key question is what defines the nature of the auxin output. A plausible explanation lies in the biochemically distinct properties of the transcriptional regulators. Importantly, these proteins are differentially expressed in embryos: different sets of Aux/IAAs and ARFs are expressed in different cell types. Tissue types are established in the embryo soon after fertilization and the differential expression of ARFs can already be observed at this level in *Arabidopsis*. An expression map of all ARFs at different embryonic developmental stages was recently described (Rademacher et al. 2011). All the embryonic cells express at least one *ARF* gene and most cell types express a unique combination. These expression patterns are dynamic, changing between developmental stages, suggesting that as the embryo gets more complex, and as the different cell types gain identities, different subsets of ARFs need to be active. In the octant stage two subsets of ARFs can be observed; ARFs expressed in all cells of pro-embryo and suspensor (ARF1, 6 and 18) and ARFs expressed only in the suspensor (ARF2, 9 and 13). During the globular stage, seven ARFs are expressed in the embryo in partially overlapping patterns; ARF1 and 18 are expressed in every cell, ARF6 is expressed in the basal tier and suspensor, ARF5/MP is strongly expressed in the lower tier of the pro-embryo, ARF13 is expressed in the suspensor and surrounding endosperm and finally ARF2 and 9 are expressed in the suspensor and in the protoderm of the lower tier of the pro-embryo. In heart stage, when cotyledons and meristems are established, more ARFs are expressed. Nine ARFs (1, 2, 4, 5, 6, 7, 10, 11 and 18) are expressed in the vascular cylinder. ARF3 is restricted to the abaxial side of the cotyledons. ARF9 and 10 are expressed in the protoderm. ARF1, 2, 6 and 18 are expressed in the quiescent centre (QC) and columella cells. ARF5 and 7 are expressed in the QC and ARF9 is expressed in the columella cells. Co-expression of ARFs in specific tissues may imply that they act in the same biological processes, but this still needs to be determined (Rademacher et al. 2011) (Fig. 9.3).

In any event, the cell type-specific ARF complements have the potential to generate unique cellular auxin output if the ARF proteins are not biochemically equivalent. In promoter-swap and misexpression experiments, it was shown that ARF16 cannot fully replace the embryonic function of ARF5/MP (Weijers et al. 2005). Furthermore, driving ARF9 from the ARF5/MP promoter enhances weak *mp* phenotypes (Rademacher et al. 2012), and while an *arf6* mutation enhances the *mp* phenotype, *arf1* mutation suppressed it (Rademacher et al. 2012). Finally, expression of MP in the suspensor from the ARF13 promoter interfered with suspensor development, while an extra ARF13 dose did not. These data together suggest that some ARFs are interchangeable (ARF6 and ARF5/MP), while others act differently (ARF16 and ARF5/MP) or even antagonistically (ARF1 or ARF9 and ARF5/MP). This supports a role for an ARF prepattern in establishing cell-specific auxin output.

## 5.1 Downstream Effectors of Auxin Signalling

The key to understanding the cellular mechanisms for auxin-dependent development lies in the identification of the genes that are controlled in each cell type. The only ARF targets identified in the context of the embryo are regulated by ARF5/MP. Three direct target genes of MP expressed in cells relevant for root initiation have been identified through a micro-array approach and were named *TARGET OF*

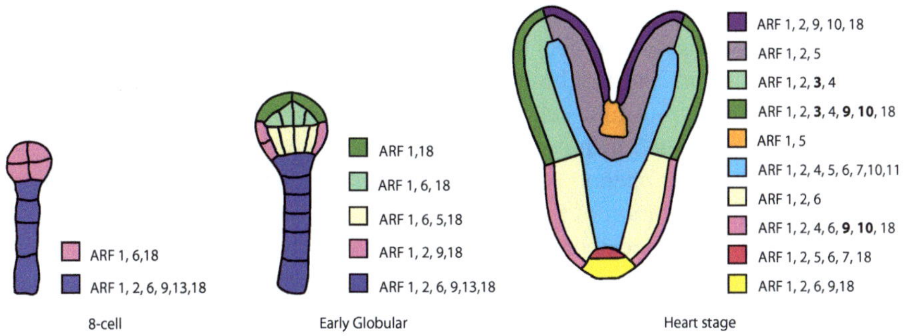


**Fig. 9.3** Expression of auxin response transcription factors during embryogenesis. Schematic depiction of gene expression patterns of *ARF* genes as determined by promoter-GFP reporters. Expression can be observed already early at embryogenesis. The number of *ARFs* expressed increases in time. Unique combinations of *ARFs* correlate with the identities obtained by each group of cells which are depicted in different *colours*

*MONOPTEROS* (*TMO*) (Schlereth et al. 2010). MP acts in inner, lower tier embryonic cells to promote root initiation, and does this in part non-cell-autonomously in the case of hypophysis specification (Weijers et al. 2006). All three TMO genes (*TMO3,5,7*) are expressed in the cells adjacent to the hypophysis where MP acts. *TMO3* encodes an AP2 transcription factor that becomes broadly expressed in later stages. *TMO5* and *TMO7* encode basic helix–loop–helix (bHLH) transcription factors. At later stages, *TMO5* is expressed specifically in vascular tissues and *TMO7* becomes restricted to the future root stem cells. These three TMOs are all individually able to partially rescue root initiation in a weak *mp* mutant when misexpressed, but the largest effect was seen with *TMO5* and *TMO7*. Since these TMOs are transcription factors, we can assume that root initiation is a process of successive transcriptional steps. It is to be noted that the TMO7 protein moves directionally to the hypophysis where it is presumed to act as a cofactor for other bHLH transcription factors and control hypophysis specification or division. The mobilization of this may be taken as the signal of intercellular communication used by MP to regulate hypophysis divisions (Schlereth et al. 2010).

In contrast to the mobile TMO7, TMO5 acts cell-autonomously in vascular cells. MP is required for oriented divisions in these cells that allow the development of a vascular bundle with more than 30 cell files from 4 precursor cells. TMO5 was recently shown to mediate these local, oriented divisions (De Rybel et al. 2013). TMO5 acts in a complex with its bHLH partner LONESOME HIGHWAY. The activity of the TMO5/LHW complex in promoting oriented division is normally restricted to a small domain by transcriptional control through MP (on TMO5) and other pathways that restrict LHW expression. When ectopically expressed, the TMO5/LHW dimer is able to trigger the same oriented division in a variety of cell types in the root. Hence, the diverse functions of MP in activating root formation bifurcate at the level of its *TMO* target genes.

Another identified direct target of MONOPTEROS is *DRN*. *DRN* encodes an AP2 transcription factor that acts redundantly with its paralogue *DRNL* upstream auxin polar transport and synthesis (see above). The expression of *DRN* can be observed from the two-to-four-cell stage in the embryo proper, then it focuses in the emerging cotyledons and then it gets restricted to the SAM at torpedo stage. Loss-of-function *drn* mutant phenotype affects the apical and the basal embryo domains. In *drn* mutants, localization of the PIN1 protein is altered, being randomly distributed instead of being located basally as in the wild type. This can already be observed at 32-cell stage, where it is found to be localized laterally. In this mutant, the expression of the auxin distribution and response reporter DR5-GFP at different embryonic stages is also abnormal. In the *drn* mutant, abnormal cell division can be observed from the globular stage onwards. This abnormal division affects specially cotyledon organogenesis. Also, phenotypes resembling the one of *mp* and *bdl* can be observed (Chandler et al. 2007; Cole et al. 2009), which suggests that MP acts in part by controlling PIN1 protein localization through the DRN/DRNL genes.

Auxin plays a fundamental role not only in determining the location of the distal stem niche but also in the specification of the QC and entire embryonic root. Downstream MP and other ARFs, other genes are activated. The AP2 putative transcription factors *PLETHORA* (*PLT*) are transcribed in response to auxin. In the embryo, ARF5-MP and ARF7-NHP4 are necessary for the transcription of *PLETHORA1* (*PLT1*) and *PLETHORA2* (*PLT2*), genes that act in the specification and maintenance of the QC and the stem cell niche (Aida et al. 2004). The expression of both PLT genes can be detected already at octant stage and it is restricted to the basal half of the embryo (Aida et al. 2004; Galinha et al. 2007). At globular stage, it is expressed in the provascular cells and the QC progenitor and later on it is expressed only in the QC and surrounding stem cells. It has been observed that misexpression of PLT1 and PLT2 leads to the development of ectopic roots, and that PLT activity is critical for embryonic root formation (Galinha et al. 2007). The regulation by MP and ARF7 may not be direct given the slow activation after auxin treatment (Aida et al. 2004). Recently, an intriguing aspect of PLT activity was revealed. The expression of the HD-ZIP III genes (see above), master regulators of the embryonic apical fate, is expanded to the root domain in *plt* mutants (Smith and Long 2010). PLT misexpression suppresses HD-Zip gene expression, and conversely, HD-Zip misexpression suppressed PLT expression. In these cases, misexpression induced the formation of a second shoot in the root domain (HD-Zip misexpression), or a second root in the shoot position (PLT misexpression). This suggests that part of the network downstream of auxin acts through PLT genes to suppress shoot development in the root pole (Smith and Long 2010).

## 5.2 Interactions of Auxin with Other Hormonal Pathways

Recent studies have shown that auxin interacts with other hormones to regulate developmental patterning and growth in various tissues (Marhavý et al. 2011; Dello Ioio et al. 2008; Mouchel et al. 2006; Shani et al. 2006; Zhao et al. 2010, see Chap. 12). Therefore, understanding how auxin controls development will have to include a description of its cross-talk with other hormones. We briefly review recent insights in cross-talks relevant to auxin-dependent embryo development.

While auxin promotes cytokinin signalling in the shoot apical meristem (Zhao et al. 2010), cytokinin acts antagonistically to auxin signalling in the root. It was shown that auxin promotes meristematic activity, whereas cytokinin promotes differentiation of stem cells in root. Furthermore, cytokinin regulates redistribution of auxin in the root apical meristem. A primary cytokinin-response transcription factor, ARR1, activates the gene SHY2/IAA3 (SHY2), a repressor of auxin signalling that negatively regulates the PIN genes (Dello Ioio et al. 2008). First evidence for auxin control of cytokinin signalling in embryo development has come from the analysis of the cytokinin signalling reporter pTCS-GFP (Muller and Sheen 2008). Its activity is first detected in the hypophysis and suspensor at the 16-cell stage embryo (Muller and Sheen 2008). After the division of the hypophysis, TCS expression is only maintained in the apical cell that is specified to become the quiescent centre, whereas it is repressed in the basal cell that becomes the distal root cap. Consistently, the negative regulators of cytokinin signalling, type-A *ARR7* and *15*, are upregulated in the basal hypophysis cell. Importantly, expression of these ARR genes is promoted by auxin. Furthermore, expression of DR5-GFP is maintained in the basal hypophysis cell while it is suppressed in apical hypophysis cell. These data suggest that auxin antagonizes cytokinin signalling via activating the type A ARRs. This appears to be biologically meaningful as altering cytokinin signalling in embryos through manipulating ARR gene expression causes defects in embryonic root formation similar to those found in several auxin-related mutants.

In the root vasculature, cytokinin signalling markers TCS-GFP and *ARR5* are expressed in procambial cells adjacent to the xylem, whereas auxin signalling markers such as DR5-GFP and *IAA2* promoters are expressed in the xylem. This indicates that cytokinin activity is correlated with procambium division while auxin promotes xylem differentiation (Bishopp et al. 2011). Indeed, cytokinin treatment inhibits protoxylem formation, and in the cytokinin receptor mutant *wol*, all the vascular cells differentiate into protoxylem (Mähönen et al. 2006). Mutation in the ARABIDOPSIS HISTIDINE PHOSPHOTRANSFER PROTEIN 6 (AHP6) gene was able to restore phloem and procambium in *wol* mutant. AHP6 belongs to a family of histidine phosphotransfer proteins that transduce cytokinin signal. However, AHP6 has a mutation in a conserved histidine residue, which is a target of phosphorylation of AHPs. Therefore, AHP6 cannot participate in phosphotransfer and is considered as pseudo-AHP. In the *ahp6* mutant, differentiation of protoxylem is disrupted. However, the phenotype is restored by expressing cytokinin oxidase CKX2 from the *AHP6* promoter, which is expressed in protoxylem and pericycle cells (Mähönen et al. 2006). The expression domain of auxin signalling markers overlaps with that of *AHP6* expression and auxin promotes transcription of AHP6, likely mediated by MP (Bishopp et al. 2011). Thus, *AHP6* is an auxin inducible, negative regulator of cytokinin signalling that promotes protoxylem differentiation (Mähönen et al. 2006).

In procambium cells adjacent to the xylem, PIN1 was localized on both basal and lateral side of the plasma membrane. This lateral localization of PIN1 promotes auxin transport from procambium to protoxylem. Accumulated auxin in the protoxylem likely creates a bisymmetric AHP6 expression domain. During embryogenesis, *AHP6* is expressed in the cotyledon tips of the heart stage embryo and the expression domain migrates from cotyledon to vasculature cells. At the same time, expression pattern of the auxin signalling marker, IAA2 changes from symmetric to bisymmetric in vasculature cells. This suggests that AHP6 is required to the establishment of bisymmetric pattern of protoxylem during embryogenesis (Bishopp et al. 2011). Likely, these examples are the first of many more that show intimate connections between auxin and cytokinin, as well as perhaps other hormones, that dynamically control cell fate and division during embryo development.

## 6 Concluding Remarks

The morphogenetic potential of the plant hormone auxin has been discovered many decades ago (Skoog and Miller), but the mechanisms by which it controls embryo development have only been revealed in the last decade. In this chapter, we discussed the developmental progression of early embryogenesis and reviewed which steps are under auxin control. We show that a network involving the regulation of auxin biosynthesis, transport and cell type-specific response allows this generic hormone to control a variety of processes during embryogenesis. While the outlines of this network have been drafted, important questions remain. These include how local biosynthesis is activated, how PIN protein polarity regulation leads to precise auxin accumulation patterns and, finally, how these accumulation patterns in turn trigger the activation of specific sets of developmental effector genes. With the current pace of progress, we anticipate that the next years will have much in store for our understanding of how this hormone directs multicellular plant development.

**Acknowledgements** We thank all members of our group for inspiring discussions. Work by the authors is funded by grants from the Netherlands Organization for Scientific research (NWO; ECHO grant 711.011.002 and ALW grant 820.02.019).

# References


Aida M, Beis D, Heidstra R, Willemsen V, Blilou I, Galinha C, Nussaume L, Noh YS, Amasino R, Scheres B (2004) The PLETHORA genes mediate patterning of the Arabidopsis root stem cell niche. Cell 119(1):109–120. doi:10.1016/j.cell.2004.09.018

Barbez E, Kubes M, Rolcik J, Beziat C, Pencik A, Wang B, Rosquete MR, Zhu J, Dobrev PI, Lee Y, Zazimalova E, Petrasek J, Geisler M, Friml J, Kleine-Vehn J (2012) A novel putative auxin carrier family regulates intracellular auxin homeostasis in plants. Nature 485(7396):119–122. doi:10.1038/nature11001

Benjamins R, Quint A, Weijers D, Hooykaas P, Offringa R (2001) The PINOID protein kinase regulates organ development in Arabidopsis by enhancing polar auxin transport. Development 128(20):4057–4067

Benkova E, Michniewicz M, Sauer M, Teichmann T, Seifertova D, Jurgens G, Friml J (2003) Local, efflux-dependent auxin gradients as a common module for plant organ formation. Cell 115(5):591–602

Bennett SRM, Alvarez J, Bossinger G, Smyth DR (1995) Morphogenesis in pinoid mutants of Arabidopsis thaliana. Plant J 8(4):505–520. doi:10.1046/J.1365-313x.1995.8040505.X

Bishopp A, Help H, El-Showk S, Weijers D, Scheres B, Friml J, Benkova E, Mahonen AP, Helariutta Y (2011) A mutually inhibitory interaction between auxin and cytokinin specifies vascular pattern in roots. Curr Biol 21(11):917–926. doi:10.1016/j.cub.2011.04.017

Blilou I, Xu J, Wildwater M, Willemsen V, Paponov I, Friml J, Heidstra R, Aida M, Palme K, Scheres B (2005) The PIN auxin efflux facilitator network controls growth and patterning in Arabidopsis roots. Nature 433(7021):39–44. doi:10.1038/nature03184

Calderon-Villalobos LI, Tan X, Zheng N, Estelle M (2010) Auxin perception – structural insights. Cold Spring Harb Perspect Biol 2(7):a005546. doi:10.1101/cshperspect.a005546

Calderon Villalobos LI, Lee S, De Oliveira C, Ivetac A, Brandt W, Armitage L, Sheard LB, Tan X, Parry G, Mao H, Zheng N, Napier R, Kepinski S, Estelle M (2012) A combinatorial TIR1/AFB-Aux/IAA co-receptor system for differential sensing of auxin. Nat Chem Biol 8(5):477–485. doi:10.1038/nchembio.926

Causier B, Ashworth M, Guo W, Davies B (2012) The TOPLESS interactome: a framework for gene repression in Arabidopsis. Plant Physiol 158(1):423–438. doi:10.1104/pp. 111.186999

Chandler JW, Cole M, Flier A, Grewe B, Werr W (2007) The AP2 transcription factors DORNROSCHEN and DORNROSCHEN-LIKE redundantly control Arabidopsis embryo patterning via interaction with PHAVOLUTA. Development 134(9):1653–1662. doi:10.1242/dev.001016

Chapman EJ, Estelle M (2009) Mechanism of auxin-regulated gene expression in plants. Annu Rev Genet 43:265–285. doi:10.1146/annurev-genet-102108-134148

Cheng Y, Dai X, Zhao Y (2007) Auxin synthesized by the YUCCA flavin monooxygenases is essential for embryogenesis and leaf formation in Arabidopsis. Plant Cell 19(8):2430–2439. doi:10.1105/tpc.107.053009

Cheng Y, Qin G, Dai X, Zhao Y (2008) NPY genes and AGC kinases define two key steps in auxin-mediated organogenesis in Arabidopsis. Proc Natl Acad Sci U S A 105(52):21017–21022. doi:10.1073/pnas.0809761106

Cole M, Chandler J, Weijers D, Jacobs B, Comelli P, Werr W (2009) DORNROSCHEN is a direct target of the auxin response factor MONOPTEROS in the Arabidopsis embryo. Development 136(10):1643–1651. doi:10.1242/dev.032177

De Rybel B, Moller B, Yoshida S, Grabowicz I, Barbier de Reuille P, Boeren S, Smith RS, Borst JW, Weijers D (2013) A bHLH complex controls embryonic vascular tissue establishment and indeterminate growth in Arabidopsis. Dev Cell 24(4):426–437. doi:10.1016/j.devcel.2012.12.013

Dello Ioio R, Nakamura K, Moubayidin L, Perilli S, Taniguchi M, Morita MT, Aoyama T, Costantino P, Sabatini S (2008) A genetic framework for the control of cell division and differentiation in the root meristem. Science 322:1380–1384

Dharmasiri N, Dharmasiri S, Estelle M (2005) The F-box protein TIR1 is an auxin receptor. Nature 435(7041):441–445. doi:10.1038/nature03543
Donner TJ, Sherr I, Scarpella E (2009) Regulation of preprocambial cell state acquisition by auxin signaling in Arabidopsis leaves. Development 136(19):3235–3246. doi:10.1242/dev.037028
Eklund DM, Staldal V, Valsecchi I, Cierlik I, Eriksson C, Hiratsu K, Ohme-Takagi M, Sundstrom JF, Thelander M, Ezcurra I, Sundberg E (2010a) The Arabidopsis thaliana STYLISH1 protein acts as a transcriptional activator regulating auxin biosynthesis. Plant Cell 22(2):349–363. doi:10.1105/tpc.108.064816
Eklund DM, Thelander M, Landberg K, Staldal V, Nilsson A, Johansson M, Valsecchi I, Pederson ER, Kowalczyk M, Ljung K, Ronne H, Sundberg E (2010b) Homologues of the Arabidopsis thaliana SHI/STY/LRP1 genes control auxin biosynthesis and affect growth and development in the moss Physcomitrella patens. Development 137(8):1275–1284. doi:10.1242/dev.039594
Eklund DM, Cierlik I, Staldal V, Claes AR, Vestman D, Chandler J, Sundberg E (2011) Expression of Arabidopsis SHORT INTERNODES/STYLISH family genes in auxin biosynthesis zones of aerial organs is dependent on a GCC box-like regulatory element. Plant Physiol 157(4):2069–2080. doi:10.1104/pp. 111.182253
Emery JF, Floyd SK, Alvarez J, Eshed Y, Hawker NP, Izhaki A, Baum SF, Bowman JL (2003) Radial patterning of Arabidopsis shoots by class III HD-ZIP and KANADI genes. Curr Biol 13 (20):1768–1774
Friml J, Vieten A, Sauer M, Weijers D, Schwarz H, Hamann T, Offringa R, Jurgens G (2003) Efflux-dependent auxin gradients establish the apical-basal axis of Arabidopsis. Nature 426 (6963):147–153. doi:10.1038/nature02085
Friml J, Yang X, Michniewicz M, Weijers D, Quint A, Tietz O, Benjamins R, Ouwerkerk PBF, Ljung K, Sandberg G, Hooykaas PJJ, Palme K, Offringa R (2004) A PINOID-dependent binary switch in apical-basal PIN polar targeting directs auxin efflux. Science 306(5697):862–865. doi:10.1126/Science.1100618
Galinha C, Hofhuis H, Luijten M, Willemsen V, Blilou I, Heidstra R, Scheres B (2007) PLETHORA proteins as dose-dependent master regulators of Arabidopsis root development. Nature 449(7165):1053–1057. doi:10.1038/nature06206
Geisler M, Blakeslee JJ, Bouchard R, Lee OR, Vincenzetti V, Bandyopadhyay A, Titapiwatanakun B, Peer WA, Bailly A, Richards EL, Ejendal KF, Smith AP, Baroux C, Grossniklaus U, Muller A, Hrycyna CA, Dudler R, Murphy AS, Martinoia E (2005) Cellular efflux of auxin catalyzed by the Arabidopsis MDR/PGP transporter AtPGP1. Plant J 44 (2):179–194. doi:10.1111/j.1365-313X.2005.02519.x
Geldner N, Anders N, Wolters H, Keicher J, Kornberger W, Muller P, Delbarre A, Ueda T, Nakano A, Jurgens G (2003) The Arabidopsis GNOM ARF-GEF mediates endosomal recycling, auxin transport, and auxin-dependent plant growth. Cell 112(2):219–230. doi:10.1016/S0092-8674(03)00003-5
Geldner N, Richter S, Vieten A, Marquardt S, Torres-Ruiz RA, Mayer U, Jurgens G (2004) Partial loss-of-function alleles reveal a role for GNOM in auxin transport-related, post-embryonic development of Arabidopsis. Development 131(2):389–400. doi:10.1242/dev.00926
Hamann T, Benkova E, Baurle I, Kientz M, Jurgens G (2002) The Arabidopsis BODENLOS gene encodes an auxin response protein inhibiting MONOPTEROS-mediated embryo patterning. Genes Dev 16(13):1610–1615. doi:10.1101/gad.229402
Hardtke CS, Berleth T (1998) The Arabidopsis gene MONOPTEROS encodes a transcription factor mediating embryo axis formation and vascular development. EMBO J 17(5):1405–1411. doi:10.1093/emboj/17.5.1405
Ikeda M, Ohme-Takagi M (2009) A novel group of transcriptional repressors in Arabidopsis. Plant Cell Physiol 50(5):970–975. doi:10.1093/pcp/pcp048
Jürgens G, Mayer U (1994) Arabidopsis. In: Bard J (ed) A colour atlas of developing embryos. Wolfe, London, pp 7–21

Kepinski S, Leyser O (2005) The Arabidopsis F-box protein TIR1 is an auxin receptor. Nature 435 (7041):446–451. doi:10.1038/nature03542

Kuusk S, Sohlberg JJ, Long JA, Fridborg I, Sundberg E (2002) STY1 and STY2 promote the formation of apical tissues during Arabidopsis gynoecium development. Development 129 (20):4707–4717

Mashiguchi K, Tanaka K, Sakai T, Sugawara S, Kawaide H, Natsume M, Hanada A, Yaeno T, Shirasu K, Yao H, McSteen P, Zhao Y, Hayashi K, Kamiya Y, Kasahara H (2011) The main auxin biosynthesis pathway in Arabidopsis. Proc Natl Acad Sci U S A 108(45):18512–18517. doi:10.1073/pnas.1108434108

Marhavy P, Bielach A, Abas L, Abuzeineh A, Duclercq J, Tanaka H, Pařezová M, Petrásek J, Friml J, Kleine-Vehn J, Benková E (2011) Cytokinin modulates endocytic trafficking of PIN1 auxin efflux carrier to control plant organogenesis. Dev Cell 21:796–804

Mähönen AP, Bishopp A, Higuchi M, Nieminen KM, Kinoshita K, Törmäkangas K, Ikeda Y, Oka A, Kakimoto T, Helariutta Y (2006) Cytokinin signaling and its inhibitor AHP6 regulate cell fate during vascular development. Science 311:94–98

Mayer U, Buttner G, Jurgens G (1993) Apical-basal pattern-formation in the Arabidopsis embryo – studies on the role of the gnom gene. Development 117(1):149–162

Michniewicz M, Zago MK, Abas L, Weijers D, Schweighofer A, Meskiene I, Heisler MG, Ohno C, Zhang J, Huang F, Schwab R, Weigel D, Meyerowitz EM, Luschnig C, Offringa R, Friml J (2007) Antagonistic regulation of PIN phosphorylation by PP2A and PINOID directs auxin flux. Cell 130(6):1044–1056. doi:10.1016/j.cell.2007.07.033

Mouchel CF, Osmont KS, Hardtke CS (2006) BRX mediates feedback between brassinosteroid levels and auxin signalling in root growth. Nature 443:458–461

Mravec J, Kubes M, Bielach A, Gaykova V, Petrasek J, Skupa P, Chand S, Benkova E, Zazimalova E, Friml J (2008) Interaction of PIN and PGP transport mechanisms in auxin distribution-dependent development. Development 135(20):3345–3354. doi:10.1242/dev.021071

Muller B, Sheen J (2008) Cytokinin and auxin interaction in root stem-cell specification during early embryogenesis. Nature 453(7198):1094–1097. doi:10.1038/nature06943

Nawy T, Bayer M, Mravec J, Friml J, Birnbaum KD, Lukowitz W (2010) The GATA factor HANABA TARANU is required to position the proembryo boundary in the early Arabidopsis embryo. Dev Cell 19(1):103–113. doi:10.1016/J.Devcel.2010.06.004

Okushima Y, Overvoorde PJ, Arima K, Alonso JM, Chan A, Chang C, Ecker JR, Hughes B, Lui A, Nguyen D, Onodera C, Quach H, Smith A, Yu G, Theologis A (2005) Functional genomic analysis of the AUXIN RESPONSE FACTOR gene family members in Arabidopsis thaliana: unique and overlapping functions of ARF7 and ARF19. Plant Cell 17(2):444–463. doi:10.1105/tpc.104.028316

Parry G, Calderon-Villalobos LI, Prigge M, Peret B, Dharmasiri S, Itoh H, Lechner E, Gray WM, Bennett M, Estelle M (2009) Complex regulation of the TIR1/AFB family of auxin receptors. Proc Natl Acad Sci U S A 106(52):22540–22545. doi:10.1073/pnas.0911967106

Ploense SE, Wu MF, Nagpal P, Reed JW (2009) A gain-of-function mutation in IAA18 alters Arabidopsis embryonic apical patterning. Development 136(9):1509–1517. doi:10.1242/dev.025932

Rademacher EH, Moller B, Lokerse AS, Llavata-Peris CI, van den Berg W, Weijers D (2011) A cellular expression map of the Arabidopsis AUXIN RESPONSE FACTOR gene family. Plant J 68(4):597–606. doi:10.1111/j.1365-313X.2011.04710.x

Rademacher EH, Lokerse AS, Schlereth A, Llavata-Peris CI, Bayer M, Kientz M, Freire Rios A, Borst JW, Lukowitz W, Jurgens G, Weijers D (2012) Different auxin response machineries control distinct cell fates in the early plant embryo. Dev Cell 22(1):211–222. doi:10.1016/j.devcel.2011.10.026

Reed JW (2001) Roles and activities of Aux/IAA proteins in Arabidopsis. Trends Plant Sci 6 (9):420–425

Robert HS, Friml J (2009) Auxin and other signals on the move in plants. Nat Chem Biol 5(5):325–332. doi:10.1038/nchembio.170

Robert HS, Grones P, Stepanova AN, Robles LM, Lokerse AS, Alonso JM, Weijers D, Friml J (2013) Local auxin sources orient the apical-basal axis in *Arabidopsis* embryos. Curr Biol 23 (24):2506–2512

Sato A, Yamamoto KT (2008) Overexpression of the non-canonical Aux/IAA genes causes auxin-related aberrant phenotypes in Arabidopsis. Physiol Plant 133(2):397–405. doi:10.1111/j.1399-3054.2008.01055.x

Scheres B (2007) Stem-cell niches: nursery rhymes across kingdoms. Nat Rev Mol Cell Biol 8 (5):345–354. doi:10.1038/nrm2164

Scheres B, Wolkenfelt H, Willemsen V, Terlouw M, Lawson E, Dean C, Weisbeek P (1994) Embryonic origin of the *Arabidopsis* primary root and root meristem initials. Development 120:2475–2487

Schlereth A, Moller B, Liu W, Kientz M, Flipse J, Rademacher EH, Schmid M, Jurgens G, Weijers D (2010) MONOPTEROS controls embryonic root initiation by regulating a mobile transcription factor. Nature 464(7290):913–916. doi:10.1038/nature08836

Shani E, Yanai O, Ori N (2006) The role of hormones in shoot apical meristem function. Curr Opin Plant Biol 9:484–489

Smith ZR, Long JA (2010) Control of Arabidopsis apical-basal embryo polarity by antagonistic transcription factors. Nature 464(7287):423–426. doi:10.1038/nature08843

Sohlberg JJ, Myrenas M, Kuusk S, Lagercrantz U, Kowalczyk M, Sandberg G, Sundberg E (2006) STY1 regulates auxin homeostasis and affects apical-basal patterning of the Arabidopsis gynoecium. Plant J 47(1):112–123. doi:10.1111/j.1365-313X.2006.02775.x

Staldal V, Sohlberg JJ, Eklund DM, Ljung K, Sundberg E (2008) Auxin can act independently of CRC, LUG, SEU, SPT and STY1 in style development but not apical-basal patterning of the Arabidopsis gynoecium. New Phytol 180(4):798–808. doi:10.1111/j.1469-8137.2008.02625.x

Steinmann T, Geldner N, Grebe M, Mangold S, Jackson CL, Paris S, Galweiler L, Palme K, Jurgens G (1999) Coordinated polar localization of auxin efflux carrier PIN1 by GNOM ARF GEF. Science 286(5438):316–318

Stepanova AN, Robertson-Hoyt J, Yun J, Benavente LM, Xie DY, Dolezal K, Schlereth A, Jurgens G, Alonso JM (2008) TAA1-mediated auxin biosynthesis is essential for hormone crosstalk and plant development. Cell 133(1):177–191. doi:10.1016/j.cell.2008.01.047

Szemenyei H, Hannon M, Long JA (2008) TOPLESS mediates auxin-dependent transcriptional repression during Arabidopsis embryogenesis. Science 319(5868):1384–1386. doi:10.1126/science.1151461

Tan X, Calderon-Villalobos LIA, Sharon M, Zheng C, Robinson CV, Estelle M, Zheng N (2007) Mechanism of auxin perception by the TIR1 ubiquitin ligase. Nature 446(7136):6

Tao Y, Ferrer JL, Ljung K, Pojer F, Hong F, Long JA, Li L, Moreno JE, Bowman ME, Ivans LJ, Cheng Y, Lim J, Zhao Y, Ballare CL, Sandberg G, Noel JP, Chory J (2008) Rapid synthesis of auxin via a new tryptophan-dependent pathway is required for shade avoidance in plants. Cell 133(1):164–176. doi:10.1016/j.cell.2008.01.049

Tiwari SB, Hagen G, Guilfoyle T (2003) The roles of auxin response factor domains in auxin-responsive transcription. Plant Cell 15(2):533–543

Vanneste S, Friml J (2009) Auxin: a trigger for change in plant development. Cell 136(6):1005–1016. doi:10.1016/j.cell.2009.03.001

Weigel D, Jurgens G (2002) Stem cells that make stems. Nature 415(6873):751–754. doi:10.1038/415751a

Weijers D, Benkova E, Jager KE, Schlereth A, Hamann T, Kientz M, Wilmoth JC, Reed JW, Jurgens G (2005) Developmental specificity of auxin response by pairs of ARF and Aux/IAA transcriptional regulators. EMBO J 24(10):1874–1885. doi:10.1038/sj.emboj.7600659

Weijers D, Schlereth A, Ehrismann JS, Schwank G, Kientz M, Jurgens G (2006) Auxin triggers transient local signaling for cell specification in Arabidopsis embryogenesis. Dev Cell 10 (2):265–270. doi:10.1016/j.devcel.2005.12.001

Wolters H, Anders N, Geldner N, Gavidia R, Jurgens G (2011) Coordination of apical and basal embryo development revealed by tissue-specific GNOM functions. Development 138(1):117–126. doi:10.1242/Dev.059147

Zhao YX, Medrano L, Ohashi K, Fletcher JC, Yu H, Sakai H, Meyerowitz EM (2004) HANABA TARANU is a GATA transcription factor that regulates shoot apical meristem and flower development in Arabidopsis. Plant Cell 16(10):2586–2600. doi:10.1105/Tpc.104.024869

Zhao Z, Andersen SU, Ljung K, Dolezal K, Miotk A, Schultheiss SJ, Lohmann JU (2010) Hormonal control of the shoot stem-cell niche. Nature 465(7301):1089–1092. doi:10.1038/nature09126

# Chapter 10
# Auxin, Chief Architect of the Shoot Apex

**Benoît Landrein and Teva Vernoux**

**Abstract** Plants have a unique capacity for continuous postembryonic development linked to the existence of permanent stem cell niches, located in specialized tissues called meristems. The activity of the shoot apical meristem (SAM), which is located at the tip of stems and branches, allows for the continuous production of all aerial organs that will develop as lateral shoots, leaves, or flowers. As it defines the number, type, and position of lateral primordia, the SAM is at the basis of plant architecture and its activity can be modulated by both internal and environmental cues. Successive initiations of new organ primordia occur in the meristem following very precise spatiotemporal patterns, called phyllotaxis. The maintenance of the meristem over time is thus expected to require precise spatiotemporal control of cell fate to allow for the continuous emergence of new primordia at precise positions and the maintenance of the stem cell niche. Signaling initiated by the plant hormone auxin plays a central role in the control of cell identities during organogenesis and in the dynamics of phyllotaxis. We first describe the structure and function of the SAM, focusing on the model species *Arabidopsis thaliana*. We then discuss the central role played by auxin in the coordination of cellular behaviors and cell identities in the SAM, and thus in providing the primary instructions for phyllotaxis and for elaborating the shoot architecture. Finally, we illustrate how the gene network downstream of auxin and mechanical properties of tissues participate in controlling morphogenesis and phyllotaxis dynamics.

---

B. Landrein
Laboratoire de Reproduction et Développement des Plantes, CNRS, INRA, ENS Lyon, UCBL, Université de Lyon, 46 Allée d'Italie, 69364 Lyon Cedex 07, France

Laboratoire Joliot-Curie, CNRS, ENS Lyon, Université de Lyon, 46 Allée d'Italie, 69364 Lyon Cedex 07, France

T. Vernoux (✉)
Laboratoire de Reproduction et Développement des Plantes, CNRS, INRA, ENS Lyon, UCBL, Université de Lyon, 46 Allée d'Italie, 69364 Lyon Cedex 07, France
e-mail: teva.vernoux@ens-lyon.fr

E. Zažímalová et al. (eds.), *Auxin and Its Role in Plant Development*,
DOI 10.1007/978-3-7091-1526-8_10, 

# 1 From the Stem Cell Niche to Organogenesis at the Shoot Apical Meristem

The SAM presents a very organized tissular structure; it can be subdivided into different zones and layers which are maintained throughout the life of the plant (Fig. 10.1a, b; Bowman and Eshed 2000). First, the stem cells can be found at the top of the SAM in the central zone and are primarily involved in meristem maintenance. In the surrounding peripheral zone, new lateral organs (leaves, lateral shoots, or flowers) are generated following specific spatiotemporal patterns called phyllotaxis. Higher growth and division of cells that are recruited in the organs result in the formation of a bump that will develop as the new lateral organ. The area between the growing bump and the meristem will rapidly form a creased-shape boundary that allows for the physical separation between the organ and the meristem. The meristem can also be divided into layers whose integrity largely depends on the orientation of division planes within the structure (Fig. 10.1b): the epidermal layer or L1, the L2 that together with L1 forms the tunica, and the L3 that forms the corpus of the meristem. The organization of the SAM into zones and layers is maintained during the life of the plant, suggesting a strict control of the integrity of the tissue alongside the continuous generation of new organs.

During the last two decades, a large amount of master genes involved in the control of meristem activity have been isolated and have allowed the fine dissection of the mechanisms underlying both the maintenance of the meristem structure and the emergence of the organs. Most of this work was carried in the model *Arabidopsis thaliana* and we only present briefly some of the most important master regulators of the SAM functions in this species (Fig. 10.1c and for more extensive reviews, see Barton 2010; Chandler 2012). Both the formation of the SAM during embryogenesis and its maintenance during the post-embryonic life of the plant rely on the activity of the homeodomain transcription factor *SHOOTMERISTEMLESS* (*STM*). *STM* belongs to the *KNOX* gene family and is expressed throughout the meristem except in the early developing primordia where it is downregulated (Long et al. 1996). *WUSCHEL* (*WUS*), another homeodomain transcription factor, is more specifically involved in controlling the maintenance of the stem cell niche in the meristem. *WUS* is expressed in a small area called organizing center and located below the central zone, and prevent the differentiation of the stem cells above non-cell autonomously (Mayer et al. 1998; Yadav et al. 2011). WUS acts in a different pathway than STM (Endrizzi et al. 1996; Lenhard et al. 2002), but overexpression of the two genes together is sufficient to induce SAM identity, even in root tissues (Brand et al. 2002; Gallois et al. 2002). The control of the size of the central zone involves a feedback loop between *WUSCHEL* and *CLAVATA* (*CLV*) genes (Brand et al. 2000; Schoof et al. 2000). *CLV3* encodes a precursor for a small peptide and its expression is induced in the upper part of the central zone by *WUSCHEL* (Kondo et al. 2006). This peptide diffuse broadly in the meristem and act non-autonomously to restrict *WUSCHEL* expression to the organizing

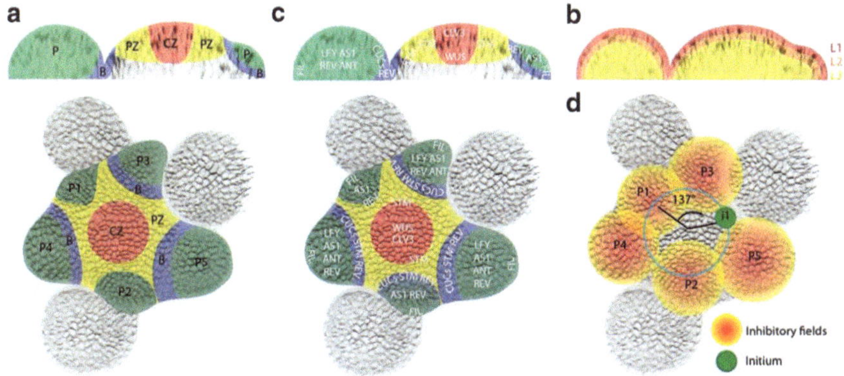


**Fig. 10.1** Structural and functional organization of the SAM of *Arabidopsis*. On the *top*, orthogonal views, at the *bottom*, top view of the SAM. (**a**) Structural zonation of the SAM. (CZ) Central zone; (PZ) peripheral zone; (P) primordia; and (B) boundaries. (**b**) Organization into layers of the SAM. (**c**) Spatiotemporal pattern of meristematic genes. Note that the timing of *AS1* activation in the developing primordia is speculative. (**d**) Illustration of a model of organogenesis based on the inhibitory field theory. Each primordium is locally inhibiting the development of new organs at its vicinity. A new initium (*green dot*) is appearing on the *blue circle* at the position where the effect on inhibitory fields is the lowest

center through binding to the *CLAVATA1-2* receptor complex (Nimchuk et al. 2011; Fletcher et al. 1999; Ogawa et al. 2008 and for a recent review, see Aichinger et al. 2012).

A large set of genes is also known to control the initiation, the development and the identity (leaf, lateral meristem, or flower) of the lateral primordia. As pointed out above, cells incorporated in the primordium lose their meristematic identity through the repression of *STM* expression (Long et al. 1996). This allows induction of the expression of the *MYB* transcription factor *ASYMMETRIC LEAVES 1* (*AS1*) that is normally repressed by *STM*. *AS1* acts in a feedback loop to inhibit the expression of several other *KNOX* genes, a process that is thought to be important for the correct distinction between organ and meristem identity (Byrne et al. 2000; Phelps-Durr et al. 2005). As the primordium starts to emerge, a complex network is set to control the size, the identity, and the polarity of the growing organs. Among these players, the transcription factor *LEAFY* (*LFY*) is necessary for inducing floral identity (Parcy et al. 1998; Weigel et al. 1992). Another transcription factor, *AINTEGUMENTA* (*ANT*) controls proliferation in the organs and therefore determines their final size (Mizukami and Fischer 2000). During flower development *ANT* acts redundantly with another member of the *AINTEGUMENTA-LIKE/PLETHORA* family, *AIL6/PLT3*, the double mutant showing severe defects in floral development (Krizek 2009; Yamaguchi et al. 2013). Genetic evidence further indicates that *ANT* and *AIL6/PLT3* are also required for flower initiation together with *LFY* in inflorescence meristems (Yamaguchi et al. 2013). Indeed, the *ant lfy ail6* triple mutant shows a drastic reduction in the production of organs, the organs produced being only filamentous structures. The establishment of abaxial–adaxial polarity requires notably the antagonist activity of adaxial *class III HD-ZIP* transcription factors [*PHABULOSA*, *PHAVOLUTA*, and *REVOLUTA* (REV)] and of the different members of the abaxial *YABBY* family of transcription factor including *FILAMENTOUS FLOWERS* (*FIL*) (for reviews on organ polarity, see Bowman and Floyd 2008). Finally, a specific set of genes is expressed in the boundary to allow for the correct separation between the developing organ and the meristem. Amongst these, the three members of the *CUP-SHAPED COTYLEDON* (*CUC*) family play an important role in boundary formation, certainly by repressing growth in this area but also by activating *STM* expression (Aida et al. 1997; Takada et al. 2001; Vroemen et al. 2003). Other actors from the *LATERAL ORGAN BOUNDARIES DOMAIN* (*LBD*) family have also more recently been shown to play important roles in the formation of the boundary (Husbands et al. 2007; Rast and Simon 2012; Shuai et al. 2002).

## 2 Phyllotaxis: The Dynamics of Organogenesis

Although the fine dissection of the gene regulatory networks underlying meristem functions has pushed forward our understanding of the mechanisms involved in meristem maintenance and organ emergence, this knowledge mostly provides a linear picture of how cell identity changes from the stem niche to the emergence of a new lateral organ and do not explain the spatial positioning of the organs in the meristem. A striking characteristic of the SAM is that organ initiations follow very robust and stereotypical phyllotactic patterns. One the most common pattern of phyllotaxis is the Fibonacci spiral where successive organs appear one by one (the time between two organ initiation being called the plastochrone) at a relative angle (the divergence angle) close to the 137.5° golden angle and either in a clockwise or counter-clockwise direction (Fig. 10.1d). This phyllotaxis is observed in many plants such as the sunflower or *Arabidopsis*, but other patterns can also be found in nature such as the characteristic whorled pattern of the angiosperm flower.

One of the most important early hypotheses to explain phyllotaxis was postulated by Hofmeister (1868) who proposed that new organs appear periodically at the periphery of the meristem in the largest available space between the preexisting organs. This hypothesis was tested experimentally using microsurgery in the middle of the twentieth century by Snow and Snow in *Lupinus albus* (Snow and Snow 1932) and Wardlaw in the fern *Dryopteris* (Wardlaw 1949). The results suggested that the emergence of an organ is influenced by the preexisting organs close to the site of initiation. This led Wardlaw to propose that chemical inhibitory fields prevent the initiation of organs in their vicinity, following the proposition by Schoute that the position of organ initiation is controlled by a chemical inhibitor produced by the existing organs (Schoute 1913). The idea of a key role of a chemical inhibitor was also supported by the work of Richards that proposed the concept of inhibitory fields (Richards 1948; Fig. 10.1c).

In their seminal work, Douady and Couder (1992) used first a physical model (see Chap. 14) and then modeling to demonstrate that both the periodicity of organ initiation and the relative position of the organs can emerge solely from a system where organ produces inhibitory fields on a growing apex (Douady and Couder 1996a, b, c). This theoretical work suggests that a mechanism inhibiting organ initiation in the vicinity of existing organs could be sufficient to drive the spatio-temporal dynamics of phyllotaxis. This prediction holds indeed whatever the nature of the inhibitory mechanism and biophysical theories of phyllotaxis were also developed (Green 1999; Hernandez and Green 1993; Dumais and Steele 2000). However, as we discuss in the following sections, biological data strongly support the existence of auxin-based chemical inhibitory fields, but we will see also that mechanical stress may have a contributing role in stabilizing the patterns of auxin distribution that control phyllotaxis.

# 3 A Central Role for Dynamic Auxin Distribution in Meristem Patterning and Phyllotaxis

## *3.1 Polar Auxin Transport Regulates Organogenesis*

Indole-3-acetic acid (IAA) is auxin main chemical form in most higher plants, including *Arabidopsis*. IAA is a weak acid that can be either in a protonated or deprotonated state depending on the local pH. This chemical property of auxin is at the basis of an active polar transport of auxin (Raven 1975; see Chap. 5) that is able to generate complex distribution of auxin in the SAM. When present in the extracellular space with a pH close to 5.5, auxin is present in its acidic uncharged form. In this neutral form, auxin is in theory able to freely diffuse through the plasma membrane and enter inside the cell. However, influx of auxin has been shown to also be facilitated by the AUX1/LAX family of auxin influx carriers (Bennett et al. 1996; Peret et al. 2012; Yang et al. 2006). When present in the cytoplasm whose pH is close to 7, auxin loses its proton and become charged. In its basic form, auxin cannot diffuse anymore through the plasma membrane. The efflux of auxin from the cell therefore requires the activity of members of the *PIN-FORMED* (*PIN*) family of auxin efflux carriers. The polar localization of the PIN pump in cells then establishes and controls the direction of polar auxin transport within tissues (Petrasek et al. 2006; Benkova et al. 2003; Wisniewska et al. 2006; see Chap. 8). In the meristem, *PIN1* is one of the founding members of this family (Galweiler et al. 1998) and is essential for organogenesis at the SAM. Indeed, while *pin1* mutants are able to generate leaves (Guenot et al. 2012), the inflorescence meristems of *pin1* mostly fail to generate lateral organs (Okada et al. 1991). Organ initiation can however be triggered on *pin1* meristems by applying locally high concentration of auxin (Reinhardt et al. 2000, 2003). These evidence were the first to show that auxin efflux is a limiting step for organogenesis in the inflorescence meristem (while the situation is less clear for the vegetative SAM) and that a local maximum of auxin concentration might be necessary and sufficient for organ initiation in the meristem. *PIN1* is more specifically expressed in the L1 of the SAM and the PIN1 pumps show complex polarity patterns with local convergence foci marking the sites of organ initiation and early primordia (Fig. 10.2a; de Reuille et al. 2006; Heisler et al. 2005; Reinhardt et al. 2003). This leads to propose that the *PIN1* pump network generates local accumulation of auxin in organs from the initium to later stages, with these organs acting as auxin sinks (Reinhardt et al. 2003; Wisniewska et al. 2006). The influx carriers are also involved in the regulation of phyllotaxis by auxin. Loss of *AUX1/LAX* function leads to important alterations of the phyllotactic pattern showing that auxin influx in the cell is required for the emergence of patterns in the meristem (Bainbridge et al. 2008). *AUX1* is expressed specifically in the L1 and was proposed to help concentrating auxin in this layer, and facilitating redistribution by PIN1. The other auxin influx pumps are expressed more broadly in the SAM and the way their activity might influence PIN1-mediated auxin distribution remains largely to be explained (Bainbridge et al. 2008).

In our current view, polar auxin transport not only is responsible for triggering organ initiation but also establishes inhibitory fields through depletion of auxin around organs. Indeed, computer simulation of auxin distribution when considering the entire meristematic PIN1 transport network supports this idea (de Reuille et al. 2006). In addition, the synthetic auxin-inducible *DR5* reporter is specifically activated in the lateral organs, from organ initiation onwards, demonstrating an activation of auxin transcriptional responses during organogenesis and supporting a local accumulation of auxin in organs (Fig. 10.2c; Heisler et al. 2005). The recent development of the *DII-VENUS* auxin biosensor allowed to obtain more direct information on auxin distribution in the SAM (Fig. 10.2b; Brunoud et al. 2012; Vernoux et al. 2011). Auxin perception triggers directly the degradation of a class of highly unstable transcriptional repressors, the Aux/IAAs that act as auxin co-receptors (Chapman and Estelle 2009). *DII-VENUS* was built by fusing the auxin-binding domain of an *Aux/IAA* to a fast-maturating *YFP*, *VENUS*, and expressing it under the constitutive *35S* promoter. Analysis of the distribution of DII-VENUS fluorescence in the SAM confirmed accumulation of auxin during organogenesis but also allowed visualizing auxin-depleted regions in the immediate vicinity of organs (Brunoud et al. 2012; Vernoux et al. 2011). These results provide strong support to the role of polar auxin transport in generating inhibitory fields. Considering the theoretical work on inhibitory fields, our knowledge on auxin distribution in the SAM thus suggests that polar auxin transport could be the necessary and sufficient driving force behind phyllotaxis.

The key role of auxin in the dynamics of organogenesis is further highlighted by experiments that have explored the molecular basis of auxin action on the expression of master genes involved in the control of cell identity (see also Sect. 5). The first evidence that auxin might regulate directly cell identities during organogenesis in the SAM came from the analysis of the expression of various meristematic markers in the inflorescence meristem of the *pin1* mutant (Vernoux et al. 2000).

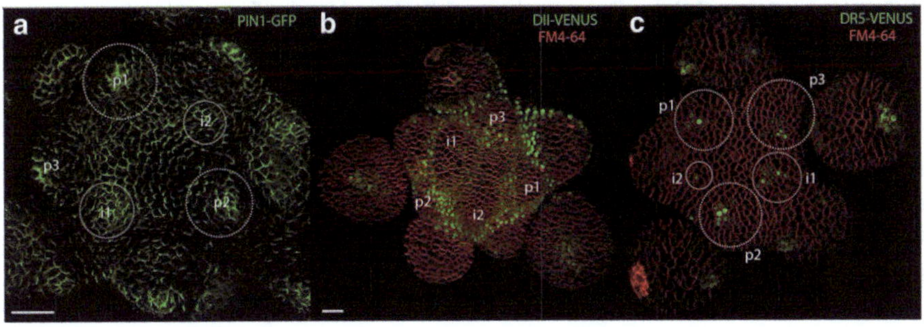


**Fig. 10.2** Polar auxin transport controls auxin distribution. (**a**) PIN1 whole mount immunolocalization in the meristem. (**b**) Meristem expressing the *DII-VENUS* sensor (in *green*) and stained with FM4-64 (in *red*) to label the membranes. (**c**) Meristem expressing the *DR5-VENUS* sensor (in *green*) and stained with FM4-64 (in *red*). Scale bars: 20 μm

Although *WUS* and *STM* are expressed normally, organ markers such as *LFY* and *ANT* and also the boundary marker *CUC2* were found to be expressed in a ring-like domain at the periphery of the SAM. These results indicate that cells in the peripheral zone of *pin1* SAMs have an organ/boundary hybrid identity and that auxin is involved in separating these two identities. Vernoux et al. (2000) also observed that *LFY* was expressed at lower levels in *pin1* meristems further indicating that auxin might regulate positively *LFY* transcription. A role for auxin in repressing *CUC2* expression and the boundary identity was also proposed to explain the expression of *CUC2* throughout the peripheral zone. Live-imaging experiments further allowed to follow concomitantly *PIN1* polarities and auxin signaling activities (using a *DR5::VENUS* marker) and the activation of key morphogenetic regulators during lateral organ initiation (Heisler et al. 2005). This analysis indicates that *LFY*, *FIL* and *REV* are transcriptionally activated after the convergence of PIN1 pumps defining the site of a new initium and leading to activation of the *DR5* marker. In addition, the authors observed that both *STM* and *CUC2* expression are repressed in the same time window in the organ and restricted to the boundary. These results support a causal relationship between auxin distribution and cell identities and are in agreement with the idea that auxin can both activate organ identity markers and repress organ boundary identity. While this remains to be shown for organ boundary genes, a recent study demonstrated using chromatin immunoprecipitation followed by quantitative PCR that both *LFY*, *ANT* and *AIL6/PLT3* are indeed direct targets of the auxin signaling effector MONOPTEROS(MP)/Auxin Response factor 5 (ARF5) during lateral organ initiation (see Sect. 4.1; Yamaguchi et al. 2013). These data clearly show that auxin can directly control cell identities during organogenesis through direct transcriptional regulation of master regulators of organ development and provide a mechanism by which auxin spatiotemporal distribution controls phyllotaxis.

### 3.2 Self-organizing Properties of Polar Auxin Transport: Insights from Conceptual Models

While the data discussed above show that a dynamic reorganization of the pump network controlling polar auxin transport might be driving phyllotaxis, they do not provide an understanding of how polar auxin transport is coordinated in space and time in the SAM. A key question is notably how are PIN1 polarities controlled in order to lead to the emergence of auxin accumulation patterns driving phyllotaxis. The polarity of the different PIN transporters was shown to rely on a complex and rapid mechanism of transcytosis (for a review, see Friml 2010; Chap. 8). Indeed, PIN proteins are constantly undergoing clathrin-dependent endocytosis from the plasma membrane and are incorporated into intracellular vesicles that can be either degraded or retargeted to specific locations on the plasma membrane, which therefore leads to their polar localization. It was shown that auxin itself acts both on the expression level and the polarity of PIN1 through inhibition of its endocytosis (Heisler et al. 2005; Paciorek et al. 2005; Vieten et al. 2005). Therefore, the generation of pattern of phyllotaxis in the meristem could be a self-organizing process relying on local interactions between auxin and its transporter PIN1.

A number of cell-based theoretical models have explored simple but plausible conceptual scenarios that could explain a self-organization of the auxin transport system in the L1 of the SAM (see van Berkel et al. 2013 for an extensive comparison of these different models). A first family of computational models, the concentration-based models, was developed based on a hypothetical mechanism allowing PIN1 polarity in a given cell to be directed toward the neighboring cell with the highest auxin concentration. This hypothetical mechanism could indeed result in part from the inhibitory effect of auxin on PIN1 endocytosis (Jonsson et al. 2006; Paciorek et al. 2005; Smith et al. 2006). Another hypothesis that might explain PIN1 polarization dynamics was proposed by Tsvi Sachs and is used in another family of models, the flux-based or canalization models (Sachs 1969). Working on vascular development, Sachs and others proposed that auxin enhances its own flux across membranes, thus leading to a stabilization of existing fluxes of auxin. This hypothetical feedback mechanism, referred to as the canalization hypothesis, has the ability to generate realistic venation patterns and PIN pattern distribution in vasculature (Mitchison 1980, 1981; Rolland-Lagan and Prusinkiewicz 2005; Scarpella et al. 2006). Models built using either the concentration-based or the canalization hypothesis can explain both the dynamics of PIN1 polarities in the SAM and the patterns of auxin accumulation (Stoma et al. 2008). In addition, a third hybrid model, using both the concentration-based hypothesis in the L1 and flux-based hypothesis in the provasculature, was also shown to reproduce realistic PIN1 polarities patterns during organ initiation in the SAM (Bayer et al. 2009; see also Chap. 15). While these studies leave the question of what is the actual molecular mechanism involved largely unanswered (see below), they however demonstrate that the phyllotactic pattern could emerge from a rather simple PIN1 polarization mechanism controlled by a feedback between auxin and its own transport.

### *3.3 Possible Molecular Mechanisms Controlling Polar Auxin Transport Dynamics and Auxin Distribution in the SAM*

The conceptual models we have discussed above suggest that a PIN1 polarization mechanism relying either on concentration-sensing of auxin in neighboring cells or on flux-sensing of auxin in a given membrane could explain how auxin distribution is controlled and drive phyllotaxis in the SAM. However, so far, modeling approaches specifically designed to understand how auxin drives phyllotaxis do not allow to pinpoint the cellular mechanisms involved. Does our experimental knowledge of PIN1 polarization, auxin perception, and signaling allow us to go further? ABP1, an auxin receptor putatively acting in the extracellular space, has recently been shown to mediate auxin positive feedback on PIN1 membrane localization by inhibiting the clathrin-dependent endocytosis of PIN1 (Nagawa et al. 2012; Xu et al. 2010; Robert et al. 2010; Chen et al. 2012; see Chaps. 6 and 8). If active in the SAM, such a mechanism could provide a concentration-sensing mechanism compatible with the concentration-based model. However, a recent computational study suggests that the existence of an extracellular concentration-sensing mechanism such as the one mediated by ABP1 is actually compatible with the canalization hypothesis and could indeed participate in flux-sensing. Wabnik et al. (2010) built a mechanistic PIN-polarization model considering (1) auxin transport through both influx and efflux carriers; (2) an extracellular auxin receptor which inhibits PIN trafficking upon auxin binding (similarly to ABP1); and (3) the regulation of PIN transcription by auxin through nuclear signaling (see below for details on nuclear signaling), since several other PINs in addition to PIN1 are known to be induced by auxin (Vieten et al. 2005; Peer et al. 2004). The authors demonstrate that vein-like patterns of PIN1 polarities can be obtained with the model in the presence of discrete sources and sink of auxin, as seen in models using the canalization hypothesis (Wabnik et al. 2010). In this model, the patterns of PIN polarities emerge from auxin perception by both extracellular and intracellular concentration sensors and not flux sensors, and still the model appears to have the capacity to simulate canalization of auxin. While it remains to explore whether such a model could simulate phyllotactic patterns, this work by Wabnik et al. (2010) suggests that the molecular components that allow cells to sense auxin fluxes might have been identified.

In this context, it is important to stress that the different PIN1-based models of phyllotaxis are prone to instabilities, independently of the postulated mechanism for PIN1 polarization (Jonsson et al. 2006; Smith et al. 2006; Stoma et al. 2008). This could indicate that other mechanisms essential for auxin homeostasis need to be considered. Auxin biosynthesis is an evident candidate for a mechanism that could be crucial for controlling auxin distribution in the SAM and a positive feedback between auxin transport and biosynthesis is indeed supported by experimental evidence (Ljung et al. 2001; Cheng et al. 2007; see Chap. 2). Auxin biosynthesis is regulated mainly through a two-step pathway implicating enzymes of the TRYPTOPHAN AMINOTRANSFERASE OF ARABIDOPSIS (TAA) family followed by YUCCA (YUC) flavin monooxygenase-like proteins (Won et al. 2011; Mashiguchi et al. 2011). A role for the TAA/YUC pathway is strongly supported by genetic evidence. Several *yuc* mutant combinations display severe defects in the development of leaves and flowers and *pin-like* structures occur in the *yuc1,2,4,6* quadruple mutant (Cheng et al. 2006). In addition, combining *pin1* with the *yuc1* and *yuc4* mutations enhances *pin1* organogenesis defects (Cheng et al. 2007). Since the *YUC* genes are expressed in the SAM, a spatial and temporal control of auxin biosynthesis could play a key role in phyllotaxis. Supporting this idea, the expression of *YUC1* and *YUC4* in the SAM is positively regulated by the transcription factors of the *AIL/PLT* family (Pinon et al. 2013). Phyllotactic defects are observed in the *plt3plt5plt7* mutants (Prasad et al. 2011) and were proposed to result from a lower auxin biosynthesis activity in the SAM which would alter auxin distribution in the SAM (Pinon et al. 2013). A coordinate action between auxin transport and biosynthesis in the SAM, controlled by feedbacks from one to another, could indeed provide robustness to auxin distribution dynamics. Exploring this idea by incorporating auxin biosynthesis in the computational models of phyllotaxis will likely be important to help refining our understanding of auxin-dependent patterning in the SAM.

## 4 The Role of the Gene Network Downstream of the Auxin Signal in Patterning the Shoot Apical Meristem

### 4.1 *The Role of the Spatiotemporal Control of Transcriptional Responses to Auxin in Patterning*

While auxin provides undoubtedly the primary signal driving phyllotaxis, the distribution of this signal needs then to be interpreted by the meristematic cells. 29 *Aux/IAAs* and 23 *ARFs* regulate gene transcription in response to auxin. The properties of this transduction pathway are largely governed by protein–protein interactions involving these two families, downstream of auxin perception by F-box (TIR1 or one the 5 Auxin-related F-Box or AFBs) together with Aux/IAAs that act also as auxin co-receptors (Chapman and Estelle 2009; see Chap. 6). Most *Aux/IAAs* are themselves induced by auxin, thus acting in a negative feedback loop and providing a nonlinear regulation of transcription in response to auxin. The first indication that a spatial control of auxin signaling capacities might be essential to mediate auxin action in the SAM came from the observation that exogenous treatments of *pin1* meristems with auxin could trigger organ initiation and auxin responses throughout the peripheral zone but not in the central zone (Reinhardt et al. 2000). In addition, mutants in the *mp/arf5* develop *pin*-like inflorescence stem and Reinhardt et al. (2003) demonstrated that *mp* meristems are insensitive to exogenous auxin application. This indicates that expression of at least *ARF5* in the SAM regulates the capacity of cells in the peripheral zone to respond to the auxin signal. Vernoux et al. (2011) systematically analyzed the expression patterns of *TIR1/AFBs*, *Aux/IAAs*, and *ARFs* and identified *TIR1*, *AFB1*, *AFB5*, *25 Aux/IAAs*, and *ARFs* (12 and 13, respectively) as the effectors of auxin perception and signaling in the SAM. Moreover, they could show that, similarly to *MP/ARF5*, most *Aux/IAAs* and *ARFs* detected are differentially expressed in the SAM, with a low expression in the central zone and a high expression in the peripheral zone. To understand the functional meaning of this differential distribution of auxin signaling effectors, a mathematical model of the control of gene transcription by auxin was developed using information on the topology of the auxin signaling network obtained through a high-throughput yeast two-hybrid analysis of the Aux/IAA-ARF interactome. The authors could predict that spatial distribution of ARFs establishes a differential sensitivity to auxin between the central zone (low) and the peripheral zone (high) and that the auxin signaling pathway provides buffering capacities to the signaling pathway upon fluctuations of the auxin signal. Confirming the first prediction, analysis of DII-VENUS fluorescence spatial distribution showed high concentration of auxin (high degradation) both at the center of the SAM and in lateral organs. On the contrary, the auxin-inducible *DR5* reporter is expressed only in the lateral organs in the peripheral zone, thus demonstrating the differential sensitivity. Also, the dynamics of DII-VENUS in live imaging experiments demonstrate important changes in fluorescence intensity that are not translated into changes in transcriptional responses, thus supporting buffering capacities for the auxin signaling pathway. These data demonstrate that the expression pattern of the *ARFs*, which is for most of these genes independent of auxin, provides a supplementary layer of regulation of the SAM patterning downstream of the auxin signal distribution. The expression of *ARFs* allows restricting auxin-induced genes (including *Aux/IAAs*) in the peripheral zone and thus participates to the functional zonation of the SAM. The buffering properties of the Aux/IAA-ARF signaling pathway are also likely essential to give robustness to the patterning of the SAM. This regulatory system is therefore a striking example where the integration of both the spatial distribution of a signal and the local signaling capacities controls the dynamics of morphogenesis.

## 4.2 Feedbacks from Cell Identity Genes or How Patterning Might Influence Auxin Distribution

While the first evidence for a direct regulation of the expression of masters genes involved in organogenesis by the auxin signaling pathway is emerging (see Sect. 3), auxin signaling effectors as well as polar auxin transport and auxin biosynthesis regulators have been identified as putative or confirmed direct targets of regulators of organ development (Brandt et al. 2012; Winter et al. 2011; Yamaguchi et al. 2013; Moyroud et al. 2011). PINOID (PID), a kinase regulating PIN polarities (Benjamins et al. 2001; Christensen et al. 2000; Friml et al. 2004), was identified as one of *LFY* putative target. It was demonstrated recently that *PID* is activated transcriptionally downstream of LFY (Yamaguchi et al. 2013). In addition, not only *PID* but also *PIN1* and *DR5* expression levels are reduced in the SAM of *lfy* mutants. Moreover, *LFY* overexpression increases *DR5* expression in the peripheral zone of the SAM and also in the root meristem (Li et al. 2013). These data strongly support a direct positive feedback regulation of the master regulator of flowering *LFY* on auxin transport and signaling that might help in ensuring an irreversible commitment to a floral identity (Yamaguchi et al. 2013). Another positive feedback loop involving *ANT* and several other members of the *AIL/PLT* family has also been identified. Both *ANT* and *AIL6/PLT3* are directly activated by MP during organogenesis and the expression of an *AGH3-2* auxin-inducible reporter is reduced in the *ant plt3* double mutant (Krizek 2009). A reduced expression of *PIN1*, *YUC1*, and *YUC4* is also observed in the *plt3,5,7* triple mutant and, inversely, inducible expression of *PLT5* induces the transcription of these three genes. The *plt3,5,7* triple mutants tend to show an unstable spiral phyllotaxis with frequent shifts to a distichous phyllotaxis, suggesting that the feedback from the *AIL/PLT* genes (this remains to be demonstrated for *ANT*) on auxin transport and biosynthesis is required to stabilize the phyllotactic pattern. Feedback on auxin distribution from cell identity regulators is probably not restricted to the organ per se, and *JAGGED LATERAL ORGANS* (*JLO*), one member of the *LBD* family expressed in organ boundaries in the SAM, was shown to regulate *PIN1* expression (Borghi et al. 2007; Rast and Simon 2012). The establishment of the depletion of auxin in the boundary region, that correspond at least in part to the inhibitory field surrounding the organ, has been proposed to result from an inversion of the polarity of the PIN1 pumps, triggering a redirection of the flux both toward the organ in the boundary cells closest to the organ and toward the meristem in the boundary cells closest to the meristem (Heisler et al. 2005; Jonsson et al. 2006). Feedback from boundary identity genes on auxin transport could thus also participate in establishing and stabilizing the boundary. Further research will be necessary to fully appreciate how feedbacks from cell identity genes on auxin distribution and signaling contribute to the emergence of the spatiotemporal patterns of cell identities in the SAM and thus of phyllotaxis.

# 5 Monitoring the Shape: The Role of Mechanical Signals in Auxin-Induced Morphogenesis

## *5.1 Mechanics of the Shoot Apical Meristem*

Like any other biological system, the SAM is a complex physical structure with specific mechanical properties and which is submitted to mechanical stress both at the cellular level and at the tissue level. At the cellular level, plant cells diverge from animal cells by their high turgor pressure, which results from the differential in osmotic potential between the symplastic and the apoplastic compartments and that is compensated by the stiffness of the cell wall (Cosgrove 1986). Growth of an isolated cell can then be achieved by two means: either by increasing turgor pressure through modulation of the osmotic potential of the cell or by decreasing the yielding properties of the cell wall, that is, the pressure that needs to be applied to induce an irreversible deformation of the wall (Proseus et al. 1999; Schopfer 2006; see Chap. 14). During growth but also at a steady state, each cell is submitted to mechanical stress as turgor pressure is constantly putting the cell walls under tension. At the organ level, both the geometry and the heterogeneity in the mechanical properties between the different tissues impact on cell growth and are believed to generate specific patterns of mechanical stress. In the stem, it was proposed based notably on peeling experiments and on measure of the thickness of cell walls, that the epidermis is under tension and mechanically restricts the growth of the inner tissues under compression (Kutschera and Niklas 2007). This model of growth controlled by the epidermis is thought to be applicable to the shoot apex where the outer cell wall of the epidermis is much thicker than other cell walls as seen in the stem (Kierzkowski et al. 2012). It has thus been proposed that the inner tissues of the meristem are under compression and that the epidermis, which is limiting for growth, is submitted to high tensile stresses, whose intensity and directions would largely depend on the geometry of the organ (Hamant et al. 2008). Supporting the key role proposed for the epidermis in controlling growth, it has also been demonstrated that modulation of BR signaling in the epidermis (including the L1 of the SAM) is sufficient to control shoot growth (Savaldi-Goldstein et al. 2007).

## *5.2 Mechanics Is Involved in the Control of PIN1 Polarity*

The idea that mechanical loads would influence biological processes such as growth and patterning has long been postulated and was notably theorized by d'Arcy Thompson at the beginning of the twentieth century (D'Arcy Thompson 1917). Although some of his assumptions were proven to be wrong, recent observations in animals and plants are strongly supporting the idea that cells are indeed able to perceive their mechanical environment and adapt their behavior to this physical cue

(Chehab et al. 2009; Engler et al. 2006; Farge 2003). In the SAM, it has been shown that mechanical stress can impact on the anisotropy and coordination of growth between cells in the epidermis by influencing the orientation of the cellular microtubules (Hamant et al. 2008; Uyttewaal et al. 2012). Interestingly, several experiments also point out a possible role of mechanical forces in developmental patterning in several plant tissues in addition to their influence on growth. It was shown that the emergence of a lateral root can be induced mechanically, either by gravi-stimulation or by direct bending of the root (Ditengou et al. 2008; Richter et al. 2009). Upon mechanical stimulation, emergence of the root is correlated with a relocation of PIN1 in specific cells of the cortex at the site of induction, suggesting that auxin distribution can be influenced by mechanics. In the SAM, the development of ectopic leaves or flowers can be induced either by changing the mechanical properties of the cells at specific places using enzymes that modify the cell wall properties or directly by compressing the meristem (Green 1999; Peaucelle et al. 2011; Pien et al. 2001). This suggests that mechanical stress can also influence patterning in the SAM. Although the molecular mechanisms by which mechanical stress can influence organ initiation in the SAM or any other tissues are still poorly understood, recent studies have demonstrated an influence of mechanical cue on the polarity of the PIN1 transporter (Heisler et al. 2010; Nakayama et al. 2012). Through various chemical and mechanical treatments modifying the patterns of mechanical stress (Heisler et al. 2010; Nakayama et al. 2012), these two studies were able to show that both the polar localization and the degree of polarity of PIN1 at the membrane in the SAM can be influenced by the directions and the intensity of mechanical stress that the cells are submitted to.

The idea that PIN1 can be influenced by mechanics lead Heisler et al. (2010) to develop a computer model of PIN1 polarization in the SAM based on local mechanical stress. To do so they postulated that higher concentration of auxin in a cell wall leads to higher cell wall relaxation, thus inducing a local increase in the mechanical stress that localizes PIN1 preferentially to this cell wall. This model is very similar to the concentration-based model we discussed earlier (see Sect. 4.2), in the way that it provides an hypothetical mechanisms by which a cell can sense a higher concentration of auxin in neighboring cell and polarize PIN1 toward this cell. However, as we have seen just above, the mechanical stress in a cell wall is also dependent on tissue geometry and on tissue growth. Such a mechanism thus also allows the cell to integrate tissue-level mechanical information and was shown to allow simulating realistic phyllotactic patterns. While this modeling work supports a vision where mechanics could be driving phyllotaxis, the biological data available so far rather indicates that mechanical stress does not play a major role in the generation of these phyllotactic patterns. Notably, applications of mechanical stress have been shown to have only a minor effect on auxin accumulation in primordia (Nakayama et al. 2012) and a modification of the mechanical properties of the meristem using long-term treatments with oryzalin (which depolymerizes microtubules) does not impact phyllotaxis (Hamant et al. 2008). Taken together, this rather suggests a model where an additional level of control of PIN1 polarity by mechanical stress could improve the robustness of the patterning. Through its influence on PIN1 polarity but also on microtubule dynamics, mechanical stress could be used by cells as an integrator of auxin-mediated growth. First, at the cellular level, differential growth between neighboring cells (which could be caused by local variations in auxin concentration between these cells) could locally impact on the distribution of mechanical stress as cells tend to resist to the changes in growth of their neighbors (Uyttewaal et al. 2012). These variations of the mechanical loads could then affect the polarity of PIN1, thus providing a feedback on auxin concentrations and fluxes between neighboring cells (Nakayama et al. 2012). Such a mechanism has not yet been clearly described for PIN1; however, it was recently shown that the mechanical control of microtubule orientation increases local heterogeneities of growth in the meristem (Uyttewaal et al. 2012). Second, at the tissue scale, the emergence of organs in auxin maxima is modifying the overall structure of the meristem and is thus likely influencing at larger scale the patterns of tensions applied on the L1. This information could be used by the cells to assess their position within the SAM and could provide an extra level of control of PIN1 polarity. Such a mechanism could notably be important during the establishment of the boundary between the organs and the meristem where the specific mechanical properties of this crease-like structure (Hamant et al. 2008) are correlated with specific pattern of expression and polarity of PIN1 transporters (Heisler et al. 2005), as we have discussed earlier.

## 6 Conclusions

The SAM is central to the indeterminate growth of the plant and the patterns of organ initiation at the SAM are also the primary determinant of plant architecture, allowing to establish their relative position on the stem. As it is controlling not only the positioning but is also involved in establishing the identity of the organs, the plant hormone auxin is thus a master regulator of plant architecture and therefore likely of the diversity of shape amongst Angiosperms. During the two last decades, the dissection of the mechanisms of auxin-mediated morphogenesis has significantly pushed forward our understanding of how auxin achieves this regulatory function. But important questions still remain to be answered primarily on the molecular mechanisms controlling auxin polar transport but also on the interactions between the auxin pathway and mechanical cues that may be essential for the robustness of the patterning. The study of these interactions together with an exploration of the mechanisms controlling integration of information provided by other hormonal pathways is certainly a major challenge for the future.

**Acknowledgments** We thank Olivier Hamant for critical reading of the manuscript and Geraldine Brunoud for providing pictures. The work in the authors' laboratory is supported by ANR and HFSP grants.

# References

Aichinger E, Kornet N, Friedrich T, Laux T (2012) Plant stem cell niches. Annu Rev Plant Biol 63 (63):615–636

Aida M, Ishida T, Fukaki H, Fujisawa H, Tasaka M (1997) Genes involved in organ separation in Arabidopsis: an analysis of the cup-shaped cotyledon mutant. Plant Cell 9(6):841–857. doi:10.1105/tpc.9.6.841

Bainbridge K, Guyomarc'h S, Bayer E, Swarup R, Bennett M, Mandel T, Kuhlemeier C (2008) Auxin influx carriers stabilize phyllotactic patterning. Genes Dev 22(6):810–823. doi:10.1101/gad.462608

Barton MK (2010) Twenty years on: the inner workings of the shoot apical meristem, a developmental dynamo. Dev Biol 341(1):95–113. doi:10.1016/j.ydbio.2009.11.029

Bayer EM, Smith RS, Mandel T, Nakayama N, Sauer M, Prusinkiewicz P, Kuhlemeier C (2009) Integration of transport-based models for phyllotaxis and midvein formation. Genes Dev 23 (3):373–384. doi:10.1101/gad.497009, 23/3/373 [pii]

Benjamins R, Quint A, Weijers D, Hooykaas P, Offringa R (2001) The PINOID protein kinase regulates organ development in Arabidopsis by enhancing polar auxin transport. Development 128(20):4057–4067

Benkova E, Michniewicz M, Sauer M, Teichmann T, Seifertova D, Jurgens G, Friml J (2003) Local, efflux-dependent auxin gradients as a common module for plant organ formation. Cell 115(5):591–602

Bennett MJ, Marchant A, Green HG, May ST, Ward SP, Millner PA, Walker AR, Schulz B, Feldmann KA (1996) Arabidopsis AUX1 gene: a permease-like regulator of root gravitropism. Science 273(5277):948–950

Borghi L, Bureau M, Simon R (2007) Arabidopsis JAGGED LATERAL ORGANS is expressed in boundaries and coordinates KNOX and PIN activity. Plant Cell 19(6):1795–1808. doi:10.1105/tpc.106.047159

Bowman JL, Eshed Y (2000) Formation and maintenance of the shoot apical meristem. Trends Plant Sci 5(3):110–115

Bowman JL, Floyd SK (2008) Patterning and polarity in seed plant shoots. Annu Rev Plant Biol 59:67–88. doi:10.1146/annurev.arplant.57.032905.105356

Brand U, Fletcher JC, Hobe M, Meyerowitz EM, Simon R (2000) Dependence of stem cell fate in Arabidopsis on a feedback loop regulated by CLV3 activity. Science 289(5479):617–619

Brand U, Grunewald M, Hobe M, Simon R (2002) Regulation of CLV3 expression by two homeobox genes in Arabidopsis. Plant Physiol 129(2):565–575. doi:10.1104/pp. 001867

Brandt R, Salla-Martret M, Bou-Torrent J, Musielak T, Stahl M, Lanz C, Ott F, Schmid M, Greb T, Schwarz M, Choi SB, Barton MK, Reinhart BJ, Liu T, Quint M, Palauqui JC, Martinez-Garcia JF, Wenkel S (2012) Genome-wide binding-site analysis of REVOLUTA reveals a link between leaf patterning and light-mediated growth responses. Plant J 72(1):31–42. doi:10.1111/j.1365-313X.2012.05049.x

Brunoud G, Wells DM, Oliva M, Larrieu A, Mirabet V, Burrow AH, Beeckman T, Kepinski S, Traas J, Bennett MJ, Vernoux T (2012) A novel sensor to map auxin response and distribution at high spatio-temporal resolution. Nature 482(7383):103–106. doi:10.1038/nature10791

Byrne ME, Barley R, Curtis M, Arroyo JM, Dunham M, Hudson A, Martienssen RA (2000) Asymmetric leaves1 mediates leaf patterning and stem cell function in Arabidopsis. Nature 408(6815):967–971. doi:10.1038/35050091

Chandler JW (2012) Floral meristem initiation and emergence in plants. Cell Mol Life Sci 69 (22):3807–3818. doi:10.1007/s00018-012-0999-0

Chapman EJ, Estelle M (2009) Mechanism of auxin-regulated gene expression in plants. Annu Rev Genet 43:265–285. doi:10.1146/annurev-genet-102108-134148

Chehab EW, Eich E, Braam J (2009) Thigmomorphogenesis: a complex plant response to mechano-stimulation. J Exp Bot 60(1):43–56. doi:10.1093/jxb/ern315

Chen X, Naramoto S, Robert S, Tejos R, Lofke C, Lin D, Yang Z, Friml J (2012) ABP1 and ROP6 GTPase signaling regulate clathrin-mediated endocytosis in Arabidopsis roots. Curr Biol 22 (14):1326–1332. doi:10.1016/j.cub.2012.05.020

Cheng YF, Dai XH, Zhao YD (2006) Auxin biosynthesis by the YUCCA flavin monooxygenases controls the formation of floral organs and vascular tissues in Arabidopsis. Genes Dev 20 (13):1790–1799. doi:10.1101/Gad.1415106

Cheng Y, Dai X, Zhao Y (2007) Auxin synthesized by the YUCCA flavin monooxygenases is essential for embryogenesis and leaf formation in Arabidopsis. Plant Cell 19(8):2430–2439. doi:10.1105/tpc.107.053009

Christensen SK, Dagenais N, Chory J, Weigel D (2000) Regulation of auxin response by the protein kinase PINOID. Cell 100(4):469–478

Cosgrove D (1986) Biophysical control of plant cell growth. Annu Rev Plant Physiol 37:377–405. doi:10.1146/annurev.pp. 37.060186.002113

D'Arcy Thompson W (1917) On growth and form. Cambridge University Press, Cambridge

de Reuille PB, Bohn-Courseau I, Ljung K, Morin H, Carraro N, Godin C, Traas J (2006) Computer simulations reveal properties of the cell-cell signaling network at the shoot apex in Arabidopsis. Proc Natl Acad Sci U S A 103(5):1627–1632. doi:10.1073/pnas.0510130103, 0510130103 [pii]

Ditengou FA, Teale WD, Kochersperger P, Flittner KA, Kneuper I, van der Graaff E, Nziengui H, Pinosa F, Li X, Nitschke R, Laux T, Palme K (2008) Mechanical induction of lateral root initiation in Arabidopsis thaliana. Proc Natl Acad Sci U S A 105(48):18818–18823. doi:10.1073/pnas.0807814105

Douady S, Couder Y (1992) Phyllotaxis as a physical self-organized growth-process. Phys Rev Lett 68(13):2098–2101. doi:10.1103/PhysRevLett.68.2098

Douady S, Couder Y (1996a) Phyllotaxis as a dynamical self organizing process.1. The spiral modes resulting from time-periodic iterations. J Theor Biol 178(3):255–274. doi:10.1006/jtbi.1996.0024

Douady S, Couder Y (1996b) Phyllotaxis as a dynamical self organizing process. 2. The spontaneous formation of a periodicity and the coexistence of spiral and whorled patterns. J Theor Biol 178(3):275–294. doi:10.1006/jtbi.1996.0025

Douady S, Couder Y (1996c) Phyllotaxis as a dynamical self organizing process. 3. The simulation of the transient regimes of ontogeny. J Theor Biol 178(3):295–312. doi:10.1006/jtbi.1996.0026

Dumais J, Steele CR (2000) New evidence for the role of mechanical forces in the shoot apical meristem. J Plant Growth Regul 19(1):7–18

Endrizzi K, Moussian B, Haecker A, Levin JZ, Laux T (1996) The SHOOT MERISTEMLESS gene is required for maintenance of undifferentiated cells in Arabidopsis shoot and floral meristems and acts at a different regulatory level than the meristem genes WUSCHEL and ZWILLE. Plant J 10(6):967–979

Engler AJ, Sen S, Sweeney HL, Discher DE (2006) Matrix elasticity directs stem cell lineage specification. Cell 126(4):677–689. doi:10.1016/j.cell.2006.06.044

Farge E (2003) Mechanical induction of Twist in the Drosophila foregut/stomodeal primordium. Curr Biol 13(16):1365–1377

Fletcher JC, Brand U, Running MP, Simon R, Meyerowitz EM (1999) Signaling of cell fate decisions by CLAVATA3 in Arabidopsis shoot meristems. Science 283(5409):1911–1914

Friml J (2010) Subcellular trafficking of PIN auxin efflux carriers in auxin transport. Eur J Cell Biol 89(2–3):231–235. doi:10.1016/j.ejcb.2009.11.003

Friml J, Yang X, Michniewicz M, Weijers D, Quint A, Tietz O, Benjamins R, Ouwerkerk PB, Ljung K, Sandberg G, Hooykaas PJ, Palme K, Offringa R (2004) A PINOID-dependent binary switch in apical-basal PIN polar targeting directs auxin efflux. Science 306(5697):862–865. doi:10.1126/science.1100618

Gallois JL, Woodward C, Reddy GV, Sablowski R (2002) Combined SHOOT MERISTEMLESS and WUSCHEL trigger ectopic organogenesis in Arabidopsis. Development 129(13):3207–3217

Galweiler L, Guan C, Muller A, Wisman E, Mendgen K, Yephremov A, Palme K (1998) Regulation of polar auxin transport by AtPIN1 in Arabidopsis vascular tissue. Science 282 (5397):2226–2230

Green PB (1999) Expression of pattern in plants: combining molecular and calculus-based biophysical paradigms. Am J Bot 86(8):1059–1076

Guenot B, Bayer E, Kierzkowski D, Smith RS, Mandel T, Zadnikova P, Benkova E, Kuhlemeier C (2012) Pin1-independent leaf initiation in Arabidopsis. Plant Physiol 159(4):1501–1510. doi:10.1104/pp. 112.200402

Hamant O, Heisler MG, Jonsson H, Krupinski P, Uyttewaal M, Bokov P, Corson F, Sahlin P, Boudaoud A, Meyerowitz EM, Couder Y, Traas J (2008) Developmental patterning by mechanical signals in Arabidopsis. Science 322(5908):1650–1655. doi:10.1126/science. 1165594, 322/5908/1650 [pii]

Heisler MG, Ohno C, Das P, Sieber P, Reddy GV, Long JA, Meyerowitz EM (2005) Patterns of auxin transport and gene expression during primordium development revealed by live imaging of the Arabidopsis inflorescence meristem. Curr Biol 15(21):1899–1911. doi:10.1016/j.cub. 2005.09.052, S0960-9822(05)01213-3 [pii]

Heisler MG, Hamant O, Krupinski P, Uyttewaal M, Ohno C, Jonsson H, Traas J, Meyerowitz EM (2010) Alignment between PIN1 polarity and microtubule orientation in the shoot apical meristem reveals a tight coupling between morphogenesis and auxin transport. PLoS Biol 8 (10):e1000516. doi:10.1371/journal.pbio.1000516

Hernandez LF, Green PB (1993) Transductions for the expression of structural pattern: analysis in sunflower. Plant Cell 5(12):1725–1738. doi:10.1105/tpc.5.12.1725

Hofmeister W (1868) Allgemeine Morphologie des Gewächse. Handbuch des physiologischeb Botanik, vol 1. Engelmann, Leipzig

Husbands A, Bell EM, Shuai B, Smith HM, Springer PS (2007) LATERAL ORGAN BOUNDARIES defines a new family of DNA-binding transcription factors and can interact with specific bHLH proteins. Nucleic Acids Res 35(19):6663–6671. doi:10.1093/nar/gkm775

Jonsson H, Heisler MG, Shapiro BE, Meyerowitz EM, Mjolsness E (2006) An auxin-driven polarized transport model for phyllotaxis. Proc Natl Acad Sci USA 103(5):1633–1638. doi:10.1073/pnas.0509839103, 0509839103 [pii]

Kierzkowski D, Nakayama N, Routier-Kierzkowska AL, Weber A, Bayer E, Schorderet M, Reinhardt D, Kuhlemeier C, Smith RS (2012) Elastic domains regulate growth and organogenesis in the plant shoot apical meristem. Science 335(6072):1096–1099. doi:10.1126/science.1213100

Kondo T, Sawa S, Kinoshita A, Mizuno S, Kakimoto T, Fukuda H, Sakagami Y (2006) A plant peptide encoded by CLV3 identified by in situ MALDI-TOF MS analysis. Science 313 (5788):845–848. doi:10.1126/science.1128439

Krizek B (2009) AINTEGUMENTA and AINTEGUMENTA-LIKE6 act redundantly to regulate Arabidopsis floral growth and patterning. Plant Physiol 150(4):1916–1929. doi:10.1104/pp. 109.141119

Kutschera U, Niklas KJ (2007) The epidermal-growth-control theory of stem elongation: an old and a new perspective. J Plant Physiol 164(11):1395–1409. doi:10.1016/j.jplph.2007.08.002

Lenhard M, Jurgens G, Laux T (2002) The WUSCHEL and SHOOTMERISTEMLESS genes fulfil complementary roles in Arabidopsis shoot meristem regulation. Development 129(13):3195–3206

Li W, Zhou Y, Liu X, Yu P, Cohen JD, Meyerowitz EM (2013) LEAFY Controls Auxin Response Pathways in Floral Primordium Formation. Sci Signal 6(270):ra23. doi:10.1126/scisignal. 2003937

Ljung K, Bhalerao RP, Sandberg G (2001) Sites and homeostatic control of auxin biosynthesis in Arabidopsis during vegetative growth. Plant J 28(4):465–474

Long JA, Moan EI, Medford JI, Barton MK (1996) A member of the KNOTTED class of homeodomain proteins encoded by the STM gene of Arabidopsis. Nature 379(6560):66–69. doi:10.1038/379066a0

Mashiguchi K, Tanaka K, Sakai T, Sugawara S, Kawaide H, Natsume M, Hanada A, Yaeno T, Shirasu K, Yao H, McSteen P, Zhao YD, Hayashi K, Kamiya Y, Kasahara H (2011) The main auxin biosynthesis pathway in Arabidopsis. Proc Natl Acad Sci USA 108(45):18512–18517. doi:10.1073/pnas.1108434108

Mayer KF, Schoof H, Haecker A, Lenhard M, Jurgens G, Laux T (1998) Role of WUSCHEL in regulating stem cell fate in the Arabidopsis shoot meristem. Cell 95(6):805–815

Mitchison GJ (1980) A model for vein formation in higher plants. Proc R Soc Lond Biol Sci 207:79–109

Mitchison GJ (1981) The polar transport of auxin and vein patterns in plants. Philos Trans R Soc B 295

Mizukami Y, Fischer RL (2000) Plant organ size control: AINTEGUMENTA regulates growth and cell numbers during organogenesis. Proc Natl Acad Sci USA 97(2):942–947

Moyroud E, Minguet EG, Ott F, Yant L, Pose D, Monniaux M, Blanchet S, Bastien O, Thevenon E, Weigel D, Schmid M, Parcy F (2011) Prediction of regulatory interactions from genome sequences using a biophysical model for the Arabidopsis LEAFY transcription factor. Plant Cell 23(4):1293–1306. doi:10.1105/tpc.111.083329

Nagawa S, Xu T, Lin D, Dhonukshe P, Zhang X, Friml J, Scheres B, Fu Y, Yang Z (2012) ROP GTPase-dependent actin microfilaments promote PIN1 polarization by localized inhibition of clathrin-dependent endocytosis. PLoS Biol 10(4):e1001299. doi:10.1371/journal.pbio.1001299

Nakayama N, Smith RS, Mandel T, Robinson S, Kimura S, Boudaoud A, Kuhlemeier C (2012) Mechanical regulation of auxin-mediated growth. Curr Biol 22(16):1468–1476. doi:10.1016/j.cub.2012.06.050

Nimchuk ZL, Tarr PT, Ohno C, Qu X, Meyerowitz EM (2011) Plant stem cell signaling involves ligand-dependent trafficking of the CLAVATA1 receptor kinase. Curr Biol 21(5):345–352. doi:10.1016/j.cub.2011.01.039

Ogawa M, Shinohara H, Sakagami Y, Matsubayashi Y (2008) Arabidopsis CLV3 peptide directly binds CLV1 ectodomain. Science 319(5861):294. doi:10.1126/science.1150083

Okada K, Ueda J, Komaki MK, Bell CJ, Shimura Y (1991) Requirement of the auxin polar transport system in early stages of Arabidopsis floral bud formation. Plant Cell 3(7):677–684. doi:10.1105/tpc.3.7.677

Paciorek T, Zazimalova E, Ruthardt N, Petrasek J, Stierhof YD, Kleine-Vehn J, Morris DA, Emans N, Jurgens G, Geldner N, Friml J (2005) Auxin inhibits endocytosis and promotes its own efflux from cells. Nature 435(7046):1251–1256. doi:10.1038/nature03633

Parcy F, Nilsson O, Busch MA, Lee I, Weigel D (1998) A genetic framework for floral patterning. Nature 395(6702):561–566. doi:10.1038/26903

Peaucelle A, Braybrook SA, Le Guillou L, Bron E, Kuhlemeier C, Hofte H (2011) Pectin-induced changes in cell wall mechanics underlie organ initiation in Arabidopsis. Curr Biol 21 (20):1720–1726. doi:10.1016/j.cub.2011.08.057

Peer WA, Bandyopadhyay A, Blakeslee JJ, Makam SN, Chen RJ, Masson PH, Murphy AS (2004) Variation in expression and protein localization of the PIN family of auxin efflux facilitator proteins in flavonoid mutants with altered auxin transport in Arabidopsis thaliana. Plant Cell 16 (7):1898–1911. doi:10.1105/tpc.021501

Peret B, Swarup K, Ferguson A, Seth M, Yang Y, Dhondt S, James N, Casimiro I, Perry P, Syed A, Yang H, Reemmer J, Venison E, Howells C, Perez-Amador MA, Yun J, Alonso J, Beemster GT, Laplaze L, Murphy A, Bennett MJ, Nielsen E, Swarup R (2012) AUX/LAX genes encode a family of auxin influx transporters that perform distinct functions during Arabidopsis development. Plant Cell 24(7):2874–2885. doi:10.1105/tpc.112.097766

Petrasek J, Mravec J, Bouchard R, Blakeslee JJ, Abas M, Seifertova D, Wisniewska J, Tadele Z, Kubes M, Covanova M, Dhonukshe P, Skupa P, Benkova E, Perry L, Krecek P, Lee OR, Fink GR, Geisler M, Murphy AS, Luschnig C, Zazimalova E, Friml J (2006) PIN proteins perform a rate-limiting function in cellular auxin efflux. Science 312(5775):914–918. doi:10.1126/science.1123542

Phelps-Durr TL, Thomas J, Vahab P, Timmermans MC (2005) Maize rough sheath2 and its Arabidopsis orthologue ASYMMETRIC LEAVES1 interact with HIRA, a predicted histone chaperone, to maintain knox gene silencing and determinacy during organogenesis. Plant Cell 17(11):2886–2898. doi:10.1105/tpc.105.035477

Pien S, Wyrzykowska J, McQueen-Mason S, Smart C, Fleming A (2001) Local expression of expansin induces the entire process of leaf development and modifies leaf shape. Proc Natl Acad Sci USA 98(20):11812–11817. doi:10.1073/pnas.191380498

Pinon V, Prasad K, Grigg SP, Sanchez-Perez GF, Scheres B (2013) Local auxin biosynthesis regulation by PLETHORA transcription factors controls phyllotaxis in Arabidopsis. Proc Natl Acad Sci USA 110(3):1107–1112. doi:10.1073/pnas.1213497110

Prasad K, Grigg SP, Barkoulas M, Yadav RK, Sanchez-Perez GF, Pinon V, Blilou I, Hofhuis H, Dhonukshe P, Galinha C, Mahonen AP, Muller WH, Raman S, Verkleij AJ, Snel B, Reddy GV, Tsiantis M, Scheres B (2011) Arabidopsis PLETHORA transcription factors control phyllotaxis. Curr Biol 21(13):1123–1128. doi:10.1016/j.cub.2011.05.009

Proseus TE, Ortega JK, Boyer JS (1999) Separating growth from elastic deformation during cell enlargement. Plant Physiol 119(2):775–784

Rast MI, Simon R (2012) Arabidopsis JAGGED LATERAL ORGANS acts with ASYMMETRIC LEAVES2 to coordinate KNOX and PIN expression in shoot and root meristems. Plant Cell 24 (7):2917–2933. doi:10.1105/tpc.112.099978

Raven JA (1975) Transport of indoleacetic-acid in plant-cells in relation to pH and electrical potential gradients, and its significance for polar IAA transport. New Phytol 74(2):163–172. doi:10.1111/j.1469-8137.1975.tb02602.x

Reinhardt D, Mandel T, Kuhlemeier C (2000) Auxin regulates the initiation and radial position of plant lateral organs. Plant Cell 12(4):507–518

Reinhardt D, Pesce ER, Stieger P, Mandel T, Baltensperger K, Bennett M, Traas J, Friml J, Kuhlemeier C (2003) Regulation of phyllotaxis by polar auxin transport. Nature 426 (6964):255–260. doi:10.1038/nature02081, nature02081 [pii]

Richards FJ (1948) The geometry of phyllotaxis and its origin. Symp Soc Exp Biol 2:217–245

Richter GL, Monshausen GB, Krol A, Gilroy S (2009) Mechanical stimuli modulate lateral root organogenesis. Plant Physiol 151(4):1855–1866. doi:10.1104/pp. 109.142448

Robert S, Kleine-Vehn J, Barbez E, Sauer M, Paciorek T, Baster P, Vanneste S, Zhang J, Simon S, Covanova M, Hayashi K, Dhonukshe P, Yang Z, Bednarek SY, Jones AM, Luschnig C, Aniento F, Zazimalova E, Friml J (2010) ABP1 mediates auxin inhibition of clathrin-dependent endocytosis in Arabidopsis. Cell 143(1):111–121. doi:10.1016/j.cell.2010.09.027

Rolland-Lagan AG, Prusinkiewicz P (2005) Reviewing models of auxin canalization in the context of leaf vein pattern formation in Arabidopsis. Plant J 44(5):854–865. doi:10.1111/j.1365-313X.2005.02581.x

Sachs T (1969) Polarity and induction of organized vascular tissues. Ann Bot Lond 33(130):263

Savaldi-Goldstein S, Peto C, Chory J (2007) The epidermis both drives and restricts plant shoot growth. Nature 446(7132):199–202. doi:10.1038/Nature05618

Scarpella E, Marcos D, Friml J, Berleth T (2006) Control of leaf vascular patterning by polar auxin transport. Genes Dev 20(8):1015–1027. doi:10.1101/gad.1402406

Schoof H, Lenhard M, Haecker A, Mayer KF, Jurgens G, Laux T (2000) The stem cell population of Arabidopsis shoot meristems in maintained by a regulatory loop between the CLAVATA and WUSCHEL genes. Cell 100(6):635–644

Schopfer P (2006) Biomechanics of plant growth. Am J Bot 93(10):1415–1425. doi:10.3732/ajb. 93.10.1415

Schoute JC (1913) Beitrage zur Blattstellunglehre. I. Die Theorie. Recueil de Travaux Botaniques Neerlandais 10:153–339

Shuai B, Reynaga-Pena CG, Springer PS (2002) The lateral organ boundaries gene defines a novel, plant-specific gene family. Plant Physiol 129(2):747–761. doi:10.1104/pp. 010926

Smith RS, Guyomarc'h S, Mandel T, Reinhardt D, Kuhlemeier C, Prusinkiewicz P (2006) A plausible model of phyllotaxis. Proc Natl Acad Sci USA 103(5):1301–1306. doi:10.1073/pnas.0510457103, 0510457103 [pii]
Snow M, Snow R (1932) Experiments on phyllotaxis I—the effect of isolating a pininordium. Philos Trans R Soc Lond B 221:1–43. doi:10.1098/rstb.1932.0001
Stoma S, Lucas M, Chopard J, Schaedel M, Traas J, Godin C (2008) Flux-based transport enhancement as a plausible unifying mechanism for auxin transport in meristem development. PLoS Comput Biol 4(10):e1000207. doi:10.1371/journal.pcbi.1000207
Takada S, Hibara K, Ishida T, Tasaka M (2001) The CUP-SHAPED COTYLEDON1 gene of Arabidopsis regulates shoot apical meristem formation. Development 128(7):1127–1135
Uyttewaal M, Burian A, Alim K, Landrein B, Borowska-Wykret D, Dedieu A, Peaucelle A, Ludynia M, Traas J, Boudaoud A, Kwiatkowska D, Hamant O (2012) Mechanical stress acts via katanin to amplify differences in growth rate between adjacent cells in Arabidopsis. Cell 149(2):439–451. doi:10.1016/j.cell.2012.02.048
van Berkel K, de Boer RJ, Scheres B, Ten Tusscher K (2013) Polar auxin transport: models and mechanisms. Development 140(11):2253–2268. doi:10.1242/dev.079111
Vernoux T, Kronenberger J, Grandjean O, Laufs P, Traas J (2000) PIN-FORMED 1 regulates cell fate at the periphery of the shoot apical meristem. Development 127(23):5157–5165
Vernoux T, Brunoud G, Farcot E, Morin V, Van den Daele H, Legrand J, Oliva M, Das P, Larrieu A, Wells D, Guedon Y, Armitage L, Picard F, Guyomarc'h S, Cellier C, Parry G, Koumproglou R, Doonan JH, Estelle M, Godin C, Kepinski S, Bennett M, De Veylder L, Traas J (2011) The auxin signalling network translates dynamic input into robust patterning at the shoot apex. Mol Syst Biol 7:508. doi:10.1038/msb.2011.39
Vieten A, Vanneste S, Wisniewska J, Benkova E, Benjamins R, Beeckman T, Luschnig C, Friml J (2005) Functional redundancy of PIN proteins is accompanied by auxin-dependent cross-regulation of PIN expression. Development 132(20):4521–4531. doi:10.1242/dev.02027
Vroemen CW, Mordhorst AP, Albrecht C, Kwaaitaal MA, de Vries SC (2003) The CUP-SHAPED COTYLEDON3 gene is required for boundary and shoot meristem formation in Arabidopsis. Plant Cell 15(7):1563–1577
Wabnik K, Kleine-Vehn J, Balla J, Sauer M, Naramoto S, Reinohl V, Merks RMH, Govaerts W, Friml J (2010) Emergence of tissue polarization from synergy of intracellular and extracellular auxin signaling. Mol Syst Biol 6:447. doi:10.1038/Msb.2010.103
Wardlaw CW (1949) Phyllotaxis and organogenesis in ferns. Nature 164(4161):167–169. doi:10.1038/164167a0
Weigel D, Alvarez J, Smyth DR, Yanofsky MF, Meyerowitz EM (1992) LEAFY controls floral meristem identity in Arabidopsis. Cell 69(5):843–859
Winter CM, Austin RS, Blanvillain-Baufume S, Reback MA, Monniaux M, Wu MF, Sang Y, Yamaguchi A, Yamaguchi N, Parker JE, Parcy F, Jensen ST, Li H, Wagner D (2011) LEAFY target genes reveal floral regulatory logic, cis motifs, and a link to biotic stimulus response. Dev Cell 20(4):430–443. doi:10.1016/j.devcel.2011.03.019
Wisniewska J, Xu J, Seifertova D, Brewer PB, Ruzicka K, Blilou I, Rouquie D, Benkova E, Scheres B, Friml J (2006) Polar PIN localization directs auxin flow in plants. Science 312 (5775):883. doi:10.1126/science.1121356
Won C, Shen XL, Mashiguchi K, Zheng ZY, Dai XH, Cheng YF, Kasahara H, Kamiya Y, Chory J, Zhao YD (2011) Conversion of tryptophan to indole-3-acetic acid by TRYPTOPHAN AMINOTRANSFERASES OF ARABIDOPSIS and YUCCAs in Arabidopsis. Proc Natl Acad Sci USA 108(45):18518–18523. doi:10.1073/pnas.1108436108
Xu T, Wen M, Nagawa S, Fu Y, Chen JG, Wu MJ, Perrot-Rechenmann C, Friml J, Jones AM, Yang Z (2010) Cell surface- and rho GTPase-based auxin signaling controls cellular interdigitation in Arabidopsis. Cell 143(1):99–110. doi:10.1016/j.cell.2010.09.003
Yadav RK, Perales M, Gruel J, Girke T, Jonsson H, Reddy GV (2011) WUSCHEL protein movement mediates stem cell homeostasis in the Arabidopsis shoot apex. Genes Dev 25 (19):2025–2030. doi:10.1101/gad.17258511

Yamaguchi N, Wu MF, Winter CM, Berns MC, Nole-Wilson S, Yamaguchi A, Coupland G, Krizek BA, Wagner D (2013) A molecular framework for auxin-mediated initiation of flower primordia. Dev Cell 24(3):271–282. doi:10.1016/j.devcel.2012.12.017

Yang Y, Hammes UZ, Taylor CG, Schachtman DP, Nielsen E (2006) High-affinity auxin transport by the AUX1 influx carrier protein. Curr Biol 16(11):1123–1127. doi:10.1016/j.cub.2006.04.029

# Chapter 11
# The Role of Auxin for Reproductive Organ Patterning and Development

**Thomas Dresselhaus and Kay Schneitz**

**Abstract** The reproductive structures of flowering plants consist of the gynoecium harboring the female gametophyte (embryo sac) inside one or many ovules as well as stamen and anthers containing the male gametophytes (pollen). The coordinative development and patterning of these structures from undifferentiated flower meristems into gametophyte-containing reproductive organs constitutes the most complex developmental process in plants. Auxin represents a master player as it acts not only as a local morphogenetic trigger in flower organ primordia formation, but also in concert with other hormones during further development, patterning, and function of both reproductive organs. Most of our knowledge about the role of auxin for plant reproduction was obtained from the study of mutants in the Brassicaceae model plant *Arabidopsis thaliana*. Especially mutants defective in biosynthesis and perception of the hormone as well as in auxin-regulated transcriptional response have been analyzed. Although very little is known about the role of auxin during reproduction in crop plants, we will briefly report on reproductive structures in the economically important grass family and highlight auxin signaling data in other species when available.

T. Dresselhaus (✉)
Cell Biology and Plant Biochemistry, Biochemie-Zentrum Regensburg, University of Regensburg, Universitätsstrasse 31, 93053, Regensburg, Germany
e-mail: thomas.dresselhaus@ur.de

K. Schneitz
Entwicklungsbiologie der Pflanzen, Wissenschaftszentrum Weihenstephan, Technische Universität München, Emil-Ramann-Strasse 4, 85354 Freising, Germany

E. Zažímalová et al. (eds.), *Auxin and Its Role in Plant Development*,
DOI 10.1007/978-3-7091-1526-8_11, © Springer-Verlag Wien 2014 

# 1 Introduction

Flowering plants (angiosperms) comprise an extraordinary diversity of species dominating today's terrestrial life. Their mode of reproduction involving the development of highly specialized male and female reproductive organs protecting the gametophytes, the evolution of species-specific fertilization mechanisms, as well as the generation of seeds protecting and nourishing the embryo largely contributes to their success. Angiosperms are characterized by the double fertilization process involving fusion of two sperm cells (male gametes) with the egg and central cell, respectively, representing the female gametes. Sperm cells of angiosperms are immobile and are transported via the pollen tube toward the female gametes to execute fertilization. During their path pollen tubes use the reproductive tract, a unique structure consisting of the stigma, style, transmitting tract, and the sporophytic tissues of the ovule. Initially pollen grains land on the surface of elongated stigma papillae cells (e.g., in the Brassicaceae) or multicellular stigmatic hairs (e.g., in the Poaceae) germinate and invade the stigma during compatible interactions to reach the intercellular space between transmitting tract cells. During stigma invasion and growth through the transmitting tract the pollen tube extensively communicates along its path with the sporophytic tissues of the reproductive tract of the gynoecium in order to reach the ovule and embryo sac, respectively (for review, see Dresselhaus and Franklin-Tong 2013; Palanivelu and Tsukamoto 2012).

The gynoecium represents the female reproductive structure of flowering plants. Depending on the plant family it is composed of one to multiple carpels harboring one (Gramineae) up to hundreds of ovules (e.g., in Solanaceous species). The model plant *Arabidopsis thaliana* develops about 50 ovules in each ovary arising from two fused carpels (for review, see Larsson et al. 2013). In addition to its protective role the gynoecium controls pollen tube germination, invasion at the apical stigma, and growth toward and through the transmitting tract until the final destination of the tube, the female gametophyte containing two female gametes (egg and central cell) (for review, see Palanivelu and Tsukamoto 2012; Dresselhaus and Franklin-Tong 2013). The male reproductive structures (stamen) consist of the filament and the anther. At maturity anthers contain the male gametes (sperm cells) enclosed by the vegetative pollen tube cell inside the pollen grain. This male gametophyte develops during a process called microgametogenesis from microspores that undergo two mitotic divisions forming tricellular pollen grains in many angiosperm genera including the Brassicaceae and Poaceae. After completion of pollen maturation filaments elongate rapidly and anthers dehisce to release pollen grains. Disruption of reproductive organ development, pollen maturation, anther dehiscence, or filament elongation results in male sterility and loss of reproductive success.

Effective reproduction thus depends not only on a timely and coordinated development of reproductive structures and their successful interplay, but is also strongly influenced by environmental stimuli such as heat and drought (Sakata et al. 2010).

The last decade has shown that auxin plays a central role during the development and patterning of both male and female reproductive structures. However, compared to our knowledge about the role of auxin for vegetative development and post-fertilization processes including embryogenesis (see Chaps. 9 and 10), relatively little is known about the role of auxin to provide positional information during reproduction. Moreover, almost all knowledge generated was obtained from studies using the model plant Arabidopsis. It was shown that already the initiation of male and female reproductive structure from the peripheral regions of the floral meristems is associated with auxin signaling. In this chapter we will detail our current knowledge about the role of auxin for gynoecium, stamen, ovule, and gametophyte development and briefly discuss how auxin acts in concert with other hormones during reproduction in angiosperms.

## 2 The Role of Auxin for Gynoecium Patterning and Development

The angiosperm gynoecium represents the female reproductive structure. Its major components are the stigma, style, transmitting tract, and ovary harboring and protecting the ovules (Figs. 11.1 and 11.2). It also regulates and aids pollen tube growth to reach the female gametophyte and controls successful fertilization to maximize reproductive success (Hamamura et al. 2012; Dresselhaus and Sprunck 2012; Larsson et al. 2013). After fertilization the gynoecium of most eudicotyledonous angiosperms develops into a fruit containing numerous seeds or into a single fruit named as caryopsis in the Poaceae (Sundberg and Ostergaard 2009; Kennedy 1899). In the model plant Arabidopsis the gynoecium originates from two fused carpels, which are thought to have evolved from ancestral leaflike structures (e.g., Honma and Goto 2001), while they are formed from a single carpel in the grasses (Kennedy 1899). There is no doubt that auxin plays a major role in gynoecium development and patterning.

Direct measurement of auxin distribution in reproductive organs is technically extremely challenging. However, auxin levels can be indirectly approximated by monitoring reporter signal in a transgenic plant that harbors a construct where the promoter of a gene encoding a key step in auxin biosynthesis, such as members of the *TRYPTOPHAN AMINOTRANSFERASE OF ARABIDOPSIS* (*TAA*) and *YUCCA* (*YUC*) gene families (Cheng et al. 2006; Mashiguchi et al. 2011; Zhao et al. 2001), drives expression of a reporter gene encoding, for example, glucuronidase (GUS) or green fluorescent protein (GFP). In addition, the response to auxin is often assessed by using reporter constructs that carry the artificial auxin-sensitive promoters *DR5* (Ulmasov et al. 1997; Friml et al. 2003) or *DR5rev* (Friml et al. 2002). While these reporters are tremendously useful it should be kept in mind that they may not always reflect endogenous auxin levels. Expression of the auxin response reporter *DR5* (Ulmasov et al. 1997) in Arabidopsis was detectable already in

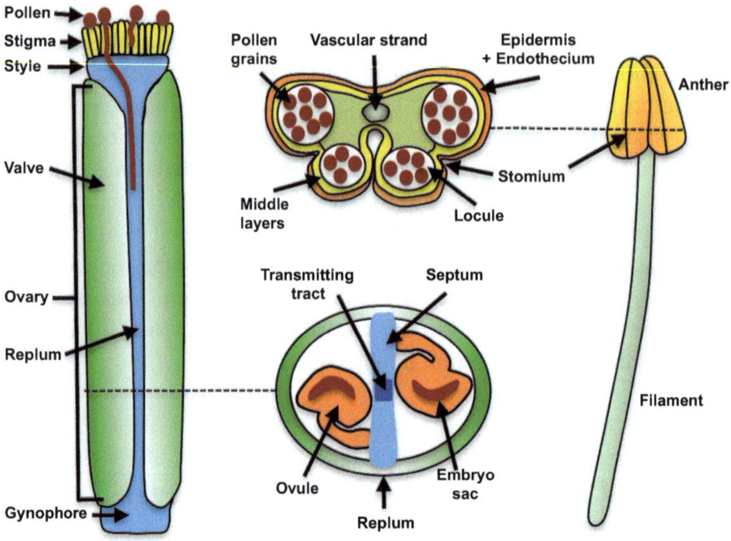


**Fig. 11.1** Scheme of reproductive organ morphology of the Brassicaceae. A mature gynoecium is shown at the *left*. A cross section displays two ovules each containing a female gametophyte (embryo sac). The transmitting tract is indicated in the *center* of the gynoecium. At maturity auxin responses occur in the stigmatic papillae cells and the vasculature associated with the transmitting tract. A stamen consisting of the anther and elongated filament at maturity is shown at the *right*. The cross section displays the *butterfly structure* of a typical angiosperm anther. The tapetum cells are already degenerated and the locules contain mature pollen grains. Auxin responses during late stages of stamen development have been detected in tapetum cells, pollen grains, and the vasculature stand as well as during filament elongation

gynoecium primordia cells with strongest signals towards the primordia tip (Benkova et al. 2003). During further gynoecium development *DR5* activity was detectable apically in two lateral foci of young gynoecia (stage 7), and expression extends to the developing vasculature at stages 8 to 9 and is almost restricted to stigmatic papillae cells at maturity (stage 12) (Larsson et al. 2013). Various auxin efflux facilitator mutants have indicated that the hormone is required during gynoecium development. Of the eight *PIN* (*PIN-FORMED*) genes in Arabidopsis (Grunewald and Friml 2010) *pin1* (Okada et al. 1991) and *pin3 pin7* double mutants (Benkova et al. 2003) display reduced ovary length with enlarged stigma and gynophore (Fig. 11.1). It is likely that these phenotypes are caused by defects of polar auxin transport (PAT) as the auxin-inducible protein kinase PINOID regulating PIN1 localization shows a similar phenotype (Christensen et al. 2000; Friml et al. 2004). Moreover, it has been shown recently that the basic helix-loop-helix (bHLH) proteins SPATULA (SPT) and INDEHISCENT (IND) regulate the maintenance of high auxin levels at the apical pole of maturating gynoecia likely by inhibiting PAT via repressing PINOID activity (Girin et al. 2011).

As mentioned above auxin biosynthesis requires enzymes of the TAA1/TAR and YUC families (for review, Mashiguchi et al. 2011; see also Chap. 2). Expression of

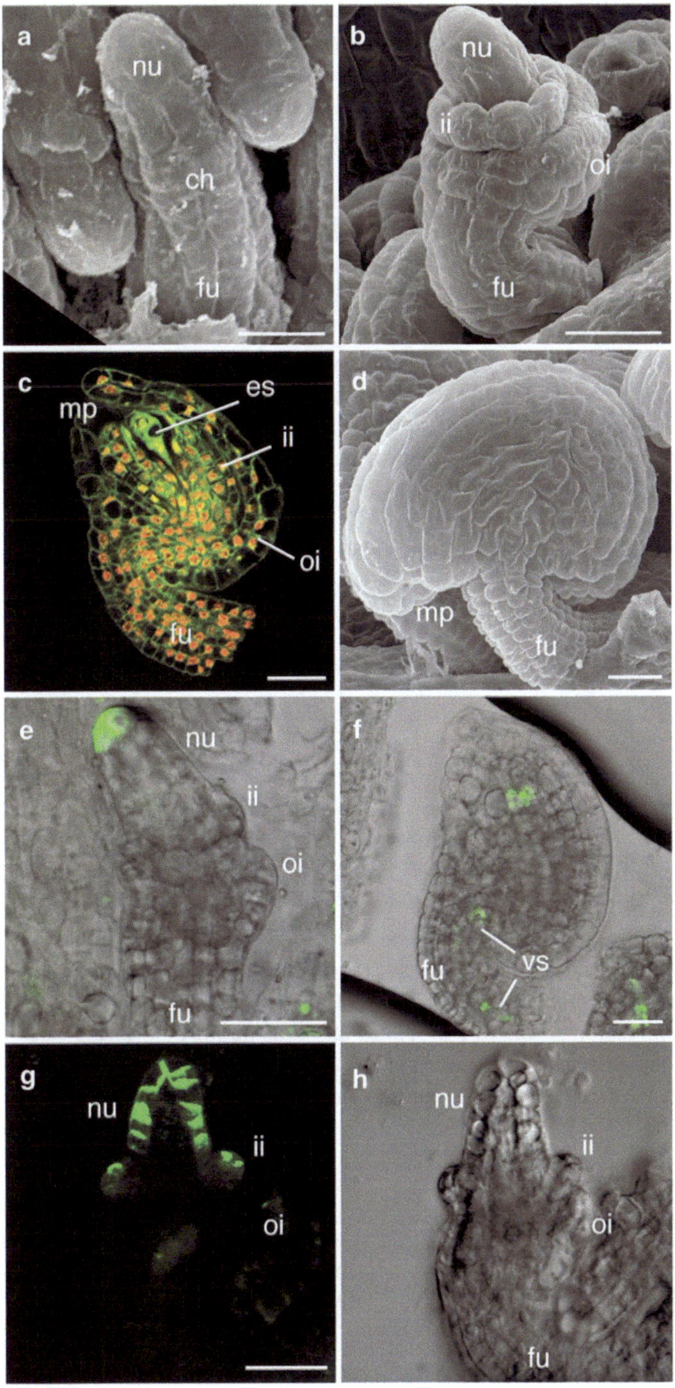


**Fig. 11.2** Wild-type ovule development in Arabidopsis. (**a**) Scanning electron micrograph (SEM) of an early stage 2-II ovule. (**b**) SEM of a stage 2-IV ovule. (**c**) Confocal micrograph of an early stage 3-III ovule. The two-nuclear embryo sac is visible. (**d**) SEM of a near-mature ovule (adapted

members of both families was detected throughout gynoecium development. The whole complexity of gene expression pattern involving several members (*TAA1*, *TAR2*, *YUC1*, *YUC2*, and *YUC4*) and their regulation during gynoecium development is described by Larsson et al. (2013) and will not be further detailed here. Briefly, as shown by in situ hybridization und promoter::GUS marker studies *YUC2* and *YUC4* are expressed in the apical region of the developing gynoecium (Cheng et al. 2006). Regulators of auxin biosynthesis such as the STYLISH (STY)/SHORT INTERNODES (SHI) family proteins targeting *YUC4* or *NGATHA* (*NGA*) genes are predominately expressed in the apical region of the developing gynoecia and seem to regulate style development by directing auxin biosynthesis in the apex of the gynoecium (Kuusk et al. 2002, 2006; Sohlberg et al. 2006; Alvarez et al. 2009; Eklund et al. 2010). These findings may explain the observation that strongest auxin response(s) are detectable in this region (Larsson et al. 2013).

Combinations of various *taa*/*tar yuc* or *sty1* mutants, affecting biosynthesis or the regulation of biosynthesis, respectively, lead to the development of defective gynoecia in Arabidopsis that may be devoid of any ovary tissues (Cheng et al. 2006; Staldal et al. 2008; Eklund et al. 2010). In summary the expression pattern and analyses of corresponding biosynthesis mutants indicate that auxin is generated and required at different regions during gynoecia development, but is largely transported toward the apical region, displaying simultaneously auxin biosynthesis maxima in this region.

The auxin-regulated transcriptional response is mediated by the AUXIN RESPONSE FACTOR (ARF) family (see also Chap. 6). ARF3/ETTIN (ETT) and ARF5/MONOPTEROS (MP) represent the major auxin response regulators during gynoecium development as shown by *arf3*/*ett* and *arf5*/*mp* mutants displaying smaller gynoecia with a reduction of the stigmatic, stylar, and ovary tissues (see, for example, Sessions and Zambryski 1995; Hardtke and Berleth 1998). The phenotypes are even more dramatic in *arf3*/*ett* and *arf5*/*mp* double mutants (Pekker et al. 2005) with a near complete loss of ovary tissues. A very similar phenotype was described for *KANADI* (*KAN*) *kan1 kan2* double mutants (Eshed et al. 2001). A recent report confirmed physical interaction of ARF3/ETT with KAN1 and KAN4 transcriptional regulators (Kelley et al. 2012). Moreover, *KAN* gene activity, which reduces expression levels of *PIN* genes and thus auxin flux during embryogenesis and vasculature development (Izhaki and Bowman 2007; Ilegems et al. 2010), may also affect PAT during gynoecium development in a similar manner. In summary

**Fig. 11.2** (continued) from Enugutti et al. 2012). (**e**, **f**) Confocal micrographs of plants carrying the *pDR5rev::GFP* reporter [adapted from Enugutti and Schneitz (2013)]. Overlays of GFP and bright-field channels. (**e**) Stage 2-III ovule. Note the GFP signal at the tip of the nucellus. (**f**) Stage 3-III ovule. GFP signal can be seen at the tip of the nucellus and the vascular strand. (**g**) Confocal micrograph depicting the PIN1:GFP signal in a stage 2-III ovule. (**h**) Bright-field channel of the ovule shown in (**g**). *ch* chalaza, *es* embryo sac, *fu* funiculus, *ii* inner integument, *mp* micropyle, *nu* nucellus, *oi* outer integument, *vs* vascular strand. Scale bars: 20 μm

these findings suggest that auxin flux, accumulation, and further signaling is required for stigma, style, and ovary development and function.

## 2.1 Auxin and the Transmitting Tract in the Center of the Gynoecium

An important component of the gynoecium is the transmitting tract. Its major function is to connect the apex of the gynoecium (stigma) representing the pollen germination region with the ovules containing the female gametes. It also assists, guides, and nourishes the growing pollen tube during its journey. In the Brassicaceae, the transmitting tract develops between two fused carpels and initially contains two meristematic ridges, which give rise to placenta and ovules, respectively (for review, see Crawford and Yanofsky 2008). Moreover, the transmitting tract cells produce an extracellular matrix (ECM) in many plant species and may undergo programmed cell death before, during, or after pollen tube passage. Development and function of the transmitting tract largely depends on genes related to auxin biosynthesis and response. The above-mentioned transcriptional regulators SPT and STY1, for example, not only regulate stigma and style development at the apical pole of maturating gynoecia, but moreover play a key role during transmitting tract formation (Gremski et al. 2007; Heisler et al. 2001). This structure is strongly reduced in *sty1* mutants (Eklund et al. 2010).

A similar phenotype was recently described by the analysis of the *SPL8* and miR156-targeted *SPL* genes during gynoecium patterning: *spl8* single mutants containing additionally miR156-targeted *SPL* genes showed a reduced style lacking a transmitting tract. The ovary was swollen and the gynophore elongated (Xing et al. 2013). Enhanced sensitivity to PAT inhibition and altered expression of *YUC4* let the authors suggest that *SPL* genes regulate transmitting tract and gynoecium patterning through interference with auxin signaling pathway(s).

## 2.2 Auxin–BR–Cytokinin Cross Talk During Gynoecium Development

The first model for the role of auxin in patterning the Arabidopsis gynoecium along its apical–basal axis was suggested by Nemhauser et al. (2000). According to the model high auxin levels would determine style and stigma formation, intermediate auxin levels are required for ovary and ovule formation, and low levels lead to gynophore formation. The finding that PAT inhibition leads to reduced or almost complete loss of ovules (for review, see Sundberg and Ostergaard 2009) seemed to confirm the model. However, this model was questioned in recent years (Ostergaard 2009; Larsson et al. 2013) due to various observations including the finding that auxin biosynthesis and response are mainly visualized in the stigma and apex of the style (see above), that gynophore elongation was observed in a number of auxin pathway mutants, and that cytokinin signaling takes place in the gynophore (for review, see Sundberg and Ostergaard 2009). It was therefore proposed that high auxin levels are required for stigma and style development, low auxin and cytokinin induce ovary and ovule development, and high cytokinin levels regulate gynophore formation. Especially the role of cytokinin and auxin–cytokinin cross talk has to be verified by further experimentation involving also more sensitive methods (see also 5. Conclusions).

Recently, auxin signaling during gynoecium development was also connected with brassinosteroid (BR) signaling. The major auxin response regulators of gynoecium development ARF6 and ARF8 have been shown to positively regulate *HALF FILLED* (*HAF*) expression in the transmitting tract (Crawford and Yanofsky 2011). *HAF* encoding a bHLH transcription factor acts together with its related sister proteins BRASSINOSTEROID ENHANCED EXPRESSION1/3 (BEE1 and BEE3). The three genes display a similar expression pattern and act together to generate the ECM and cell death during gynoecium maturation in Arabidopsis (Crawford and Yanofsky 2011). Moreover, their corresponding triple mutant significantly reduces pollen tube growth. Whether this effect results, for example, from a malformed transmitting tract or reduced ECM formation remains to be shown. However, a genetic network was elucidated showing that the auxin pathway genes (*STY1*, *ARF6*, and *ARF8*) control the expression of *HAF*, *BEE1*, and *BEE3*, which altogether encode key regulators of stigma, style, and transmitting tract development (Crawford and Yanofsky 2011). Thus although multilevel interactions have been reported about the interaction of auxin with BRs (for review, see Chap. 12), the cross talk among these hormones during reproductive development is still largely unclear.

### 2.3 The Role of Auxin During Gynoecium Development in Other Plants

Auxin pathway genes have been identified in many other plant species, but compared with Arabidopsis almost nothing is known about its role during reproductive development (see also Chap. 13). Most data are available from transcript profiling or promoter::marker experiments that identified, for example, expression of members of the *ARF*, *Aux/IAA*, *GH3*, and *SAUR* gene families during various stages of reproductive development in rice (for example, Jain and Khurana 2009). In maize 14 *PIN/PIN-like* genes have been identified of which some display overlapping expression pattern during both male and female inflorescence development (Forestan et al. 2012). Moreover, PIN1a fused to YFP and studies involving the synthetic auxin-responsive marker *DR5rev:mRFPer* showed that auxin response maxima are created at the flanks of apical inflorescence meristems and thus seem to promote branching during primordia formation (Gallavotti et al. 2008b). Auxin transporters including PIN3, PIN9, AUX/LAX, and PGPs were also shown to be expressed in flowers of sorghum (*Sorghum bicolor*) (Shen et al. 2010). A total of 31 and 36 *ARF* genes, respectively, were identified in the genome of maize (Xing et al. 2011; Wang et al. 2012). Thus the gene family is slightly expanded in this species compared to Arabidopsis (23 genes) and rice (25 genes). Expression studies revealed that members of this gene family display a dynamic post-fertilization expression pattern during embryogenesis and germination. Expression pattern during reproductive organ development has not yet been reported.

Almost nothing is known to date about gynoecia development in other eudicotyledonous plant species. In *Medicago truncatula*, for example, two carpels develop in some *pin10* mutant flowers suggesting that PAT and auxin accumulation might be required to determine carpel number (Peng and Chen 2011). *DR5* promoter activity during gynoecium development in the related Fabaceae pea is similar to the pattern observed in Arabidopsis (DeMason and Polowick 2009), indicating that the role of auxin for gynoecium development may be conserved among eudicots.

## 3 The Role of Auxin for Male Reproductive Organ and Pollen Development

In most hermaphroditic flowers stamen primordia emerge before gynoecium primordia and differentiate into the anther and filament (Fig. 11.1). The filament mainly contains the vascular system and provides water and nutrients for the anthers that contain the male gametophytes at maturity. During morphogenesis of the butterfly-shaped anther archesporial cells of the L2 layer are differentiated in the four anther corners to form the microsporangia containing microsporophytes (for review, see Feng and Dickinson 2007). Microsporocytes undergo meiosis generating microspore tetrads, which are released in the four anther locules (two pairs of neighboring theca) after callose degradation. During progression of gametogenesis pollen grains are formed after two mitotic divisions consisting of the vegetative tube cell enclosing two sperm cells in species generating tricellular pollen (most Brassicaceae and Poaceae species). The second mitosis occurs inside the growing pollen tube of species generating bicellular pollen grains (e.g., many Solanaceous species). During anther maturation, two of the four sporophytic cell layers surrounding the developing pollen grains undergo programmed cell death. From the epidermis, endothecium, middle layers, and tapetum, the middle cell layers start to degenerate while meiosis takes place inside microsporangia. The tapetum degenerates during progression of microgametogenesis and forms the exine of mature pollen grains. Finally, the enthothecium enlarges and lignifies, while the stomium differentiates and ruptures as the final event of anther dehiscence (Feng and Dickinson 2007). The processes of filament elongation, development of sporophytic and gametophytic anther tissues, as well as programmed cell death occurring during anther dehiscence and preanthesis filament elongation require precisely coordinated developmental events, where auxin acts as a masterplayer.

## 3.1 Auxin Is Required Throughout All Stages of Stamen Formation

Although our knowledge about the role of auxin for stamen development is still limited, first mutant approaches have clearly shown that the hormone is required together with jasmonic acid (JA) and giberellic acid (GA) to coordinate and regulate stamen primordia formation, development of filament and anthers, preanthesis filament elongation, pollen maturation, and anther dehiscence (for review, see Song et al. 2013). In Arabidopsis high auxin responses were detected in emerging stamen primordia and mutant approaches indeed confirmed that the hormone is required to regulate stamen number and outgrowth (Heisler et al. 2005). Auxin levels remain high during early stamen development (Cecchetti et al. 2008) and require the formation of auxin gradients during early stamen development as the auxin transport mutant *pin1* and its regulator *pinoid* contain fewer stamen and less than four microsporangia (Reinhardt et al. 2000). Stamen development is even completely arrested in some *pin1* and *pin3 pin7* flowers (Benkova et al. 2003; for review, see Paponov et al. 2005) as well as in the auxin biosynthesis double mutant *yuc1 yuc4* (Cheng et al. 2006). Auxin biosynthesis is also required during anther morphogenesis when *YUC2* and *YUC6* are expressed in microsporangia, surrounding tapetum layer, endothecium, and procambium (Cecchetti et al. 2008; Feng et al. 2006). The *yuc2 yuc6* double mutants but also mutants of genes required for auxin transportation, such as the *MultiDrug Resistance protein 1/P-GlycoProtein 1* (*MDR1/PGP1*) as well as *PIN1*, *PIN3*, and *PIN7*, are defective in filament elongation, anther dehiscence, and pollen maturation generate anthers that rarely contain mature pollen (Geisler et al. 2005; Cheng et al. 2006; Cecchetti et al. 2008; Nagpal et al. 2005).

Mutant approaches have also shown that auxin plays a major role for filament elongation: biosynthesis mutants *yuc2 yuc6*, receptor mutants *tir afb*, as well as auxin response mutants *arf6 arf8* fail to elongate filaments at anthesis (preanthesis filament elongation) and either lack functional pollen grains or release defective pollen grains before completion of filament elongation (Cheng et al. 2006; Cecchetti et al. 2008; Nagpal et al. 2005). The microRNA miR167 restricts *ARF6* and *ARF8* accumulation, thereby showing a similar phenotype than *arf 6 arf8* double mutants (Wu et al. 2006; Ru et al. 2006). This finding further supports the hypothesis that auxin signaling is required for filament elongation.

Feng et al. (2006) suggested that PAT from the filament to the anther may also be required for pollen development. Reduction of auxin levels in the filament not only resulted in short filaments, but also in degeneration of pollen after mitosis arrest.

Moreover, the presence of auxin receptors (TIR1, AFB1-3) at the end of meiosis and *DR5* activation at this stage further indicate a role of auxin for anther morphogenesis and gametophyte development (Cecchetti et al. 2008). Finally, auxin affects anther dehiscence, which takes place precociously in the *tir afb* mutants (Cecchetti et al. 2008). Dehiscence is initiated in parallel to meiosis and results in the degeneration of the tapetum cell layer, degeneration of middle cell layers, thereby forming two locules from each two neighboring thecas, and degeneration of the connective tissue and culminates in the rupture of the stomium, thus releasing mature pollen grains. *DR5*-driven auxin responses are detected in the tapetum layer, the theca, vasculature, and stomium (for review, see Sundberg and Ostergaard 2009). More detailed mutant analyses showed that initiation of meiosis, endothecium lignification, and anther dehiscence occurs earlier in auxin perception mutants *afb1-3* as well as *tir1 afb2 afb3* (Cecchetti et al. 2013). Recently another report demonstrated that auxin acts through MYB26 regulating endothecium lignification, while it regulates stomium opening via the control of JA biosynthesis (see below).

Little is known about the role of auxin during stamen development in other plants. During stamen primordia formation in maize expression of the *YUCCA* family gene *SPI1* becomes visible in proximity to stamen primordia formation (Gallavotti et al. 2008a, b). The corresponding *spi1* mutant displays significantly reduced stamen, indicating that the role of auxin for stamen development is conserved in angiosperms.

In summary auxin accumulation is required for stamen primordia formation and outgrowth and proper auxin transport is essential for filament elongation. However, its major role seems to be the coordination of the timing and progression of anther maturation and dehiscence as well as the generation of functional male gametophytes.

## 3.2 Auxin–JA–GA–BR Cross Talk During Stamen Development

Similar to late auxin mutants, mutations in JA-biosynthesis or -signaling genes fail to elongate filaments, display delayed anther dehiscence, and rarely generate viable pollen (Ishiguro et al. 2001; Nagpal et al. 2005; for review, Song et al. 2013). Exogenous JA can restore anther development, but not filament elongation (Nagpal et al. 2005). ARF6 and ARF8 have been discussed above to represent key auxin response regulators mediating the promotion of stamen filament elongation, anther dehiscence, gynoecium development, and patterning. During stamen development it was further shown that *arf6 arf8* double mutants contain strongly reduced JA levels (Nagpal et al. 2005; Tabata et al. 2010; Reeves et al. 2012). Consequently, the indehiscent phenotype of the *arf6 arf8* double mutant can be rescued by JA treatment, suggesting that auxin signaling acts upstream of JA signaling (Cecchetti et al. 2008). JA triggers expression of the R2R3 MYB transcription factors MYB21 and MYB24 that promote stamen as well as gynoecium growth (Reeves et al. 2012). Furthermore MYB21 feeds back negatively on the expression of JA-biosynthesis genes, thus reducing its levels in flower organs.

Stamen development is also regulated by gibberellic acids (GAs) in Arabidopsis. The *ga1-3*, *ga3*, *ga3ox*, and *ga20ox* mutants defective in GA biosynthesis display late stamen development defects similar to auxin and JA mutants (see above) as well as arrested anther development and short filaments (Cheng et al. 2004; Hu et al. 2008; Rieu et al. 2008; Plackett et al. 2012). These findings suggest that de novo synthesis of GAs is necessary for stamen development. It will be interesting to elucidate whether exogenously applied auxin and/or JA can rescue GA phenotypes. Mutations in the GIBBERELLIN INSENSITIVE DWARF 1a/b/c receptors (GID1a/b/c) show early stamen developmental defects (Griffiths et al. 2006) and likely act by labeling stamen expressed DELLA proteins RGA and RGL2 for degradation (Cheng et al. 2004; Peng 2009). Similar to auxin it is thought that GA may act through JA signaling during late stamen development (Peng 2009), but not during early processes. The molecular mechanisms underlying GA–auxin cross talk during stamen development are quite unclear at the moment.

Brassinosteroids (BRs) represent another hormone group that controls male fertility. Besides cell expansion defects, BR-biosynthetic and -signaling mutants showed reduced pollen number, viability, and release. These phenotypes are correlated with abnormal tapetum and microspore development (Ye et al. 2010). Many key genes required for anther and pollen development are suppressed in BR mutants. Moreover, BES1, a regulator of BR signaling, was shown to directly bind to the promoter regions of genes encoding transcription factors essential for anther and pollen development. Auxin pathway genes have not yet been shown to be regulated by BES1 or BR signaling during stamen development, but the auxin-induced bHLH transcription factor HAF or CESTA is also regulated by the BR-signaling BIN2 kinase (Poppenberger et al. 2011) indicating cross talk among the two hormones.

In summary auxin is required for stamen initiation, auxin and GA are required for early stamen development and preanthesis filament elongation, and auxin and GA seem to act through JA signaling and may involve also BR signaling during late stages regulating anther dehiscence and pollen maturation.

### 3.3 Stress Affects Auxin Signaling During Stamen Development

Stress experiments confirmed the important role of auxin signaling for anther and pollen development. These processes are highly susceptible to various stresses including high temperatures. Initial experiments showed that auxin accumulates upon stress treatment in cotton anthers (*Gossypium hirsutum*) and thus resulted in delayed anther dehiscence and male sterility (Yasuor et al. 2006). Sakata et al. (2010) showed the opposite using Arabidopsis and barley as model plants. Endogenous auxin levels significantly decreased in developing anthers at high temperature conditions. Moreover expression of *YUC* auxin biosynthesis genes was repressed and exogenous application of auxin completely reversed male sterility in both plant species. This important finding indicates that induction or repression (likely depending on the plant species) of auxin biosynthesis at high temperatures represents a major cause of male sterility and emphasizes the importance of balanced and well-coordinated auxin signaling for male reproductive organ and gametophyte development.

## 3.4 Role of Auxin During Pollen Tube Growth

As discussed above high auxin levels exist in papillae cells of mature stigmata and the apical region of the style in Arabidopsis (Larsson et al. 2013). After pollination auxin responses are repressed. The auxin response factor ARF3/ETT was now identified as a mediator of self-incompatibility (SI) and gynoecium maturation in this species. *ARF3/ETT*, which is expressed in the vasculature of the style, downregulates auxin responses in stigma papillae cells likely by regulating a mobile ligand generated in the style (Tantikanjana and Nasrallah 2012). It was suggested that repression of auxin signaling in the stigma epidermis cells promotes inhibition of "self-pollen" in self-incompatible reactions. Auxin dynamics during pollen germination and growth was also investigated in tobacco (*Nicotiana tabacum*). Similar to Arabidopsis high auxin levels were detected in the stigma and style. It was further shown that auxin levels and response (determined by immunolocalization and *DR5* activity, respectively) were highest in the transmitting tract before pollen tube entrance and declined in the regions penetrated by pollen tubes (Chen and Zhao 2008).

High auxin levels were also reported in pollen grains, germinating tubes, and during further growth in Arabidopsis (Chen and Zhao 2008; Aloni et al. 2006), but its role is quite unclear. We show here that also the maize growing pollen tube comprises the whole machinery to perceive, transport, and respond to auxin. Auxin response in Arabidopsis involves at least one of six auxin receptors TIR1/AFB, one of 23 ARF transcriptional regulators, likely a co-repressor (one of 29 Aux/IAA genes) and usually results in the activation of auxin response genes such as members of the *Small Auxin Up RNA* (*SAUR*) family and/or the *Gretchen Hagen3* (*GH3*) family (for review, see Chap. 6). The GH3 family maintains auxin and JA homeostasis by conjugating excess hormones to amino acids, while the function of small SAUR is largely unknown.

As shown in Table 11.1 auxin transporters are expressed at very low levels in growing pollen tubes of maize. However, the auxin receptor gene *AFB5* displays a strong expression level and it is very likely that it acts via *ARF27*, which is the only auxin response factor showing a significant expression pattern. *Aux/IAA* genes are

**Table 11.1** Auxin response of in vitro grown pollen tubes of maize

| Chr no. | Gene_transcript identity | Pollen tube | Leaf | Root | Encoded protein (description according to Zm5b.60) |
|---|---|---|---|---|---|
| **Auxin transport** | | | | | |
| 3 | GRMZM2G127949_T08 | 1.84[a] | 0 | 0 | Auxin transporter-like protein 1 |
| 7 | GRMZM2G461586_T01 | 1.32 | 32.49 | 54.58 | Auxin transport protein |
| 3 | GRMZM2G050089_T02 | 0.5 | 2.33 | 9.45 | Auxin efflux carrier family protein-like (ZmPINY) |
| 2 | GRMZM2G070563_T01 | 0.1[a] | 0 | 0 | Auxin efflux carrier family protein |
| **Auxin receptors** | | | | | |
| 1 | GRMZM2G079112_T01 | 13.77 | 14.27 | 25.63 | Auxin F-box protein 5 |
| 2 | GRMZM2G155849_T01 | 2.71 | 25.13 | 31.39 | Auxin signaling F-box 2 |
| 10 | GRMZM2G078508_T01 | 1.73 | 25.07 | 29.38 | Auxin-binding protein 4 |
| **ARFs** | | | | | |
| 9 | GRMZM2G160005_T01 | 3.08 | 1.02 | 4.81 | Auxin response factor 27 |
| 1 | GRMZM2G179121_T01 | 0.15 | 0 | 0.02 | Auxin response factor 31a |
| 10 | GRMZM2G023813_T01 | 0.12 | 0 | 0.04 | Auxin response factor 31b |
| 3 | GRMZM2G475263_T01 | 0.12 | 4.55 | 13.92 | Auxin response factor 7 |
| **IAAs** | | | | | |
| 1 | GRMZM2G386209_T01 | 0.15 | 2.18 | 16.14 | Auxin induced-like protein |
| 4 | GRMZM2G065244_T01 | 0.09 | 28.82 | 6.65 | Auxin induced-like protein |
| 1 | GRMZM2G137367_T01 | 0.06 | 0.04 | 0.08 | Auxin-responsive protein IAA27 |
| **SAURs** | | | | | |
| 1 | GRMZM2G123896_T04 | 18.9 | 588.56 | 1,260.99 | Dormancy/auxin associated protein |
| 1 | GRMZM2G073755_T01 | 7.06 | 38.28 | 45.86 | Dormancy/auxin associated protein |
| 9 | GRMZM2G043338_T01 | 4.72 | 33.7 | 40.74 | Dormancy/auxin associated protein |
| 1 | GRMZM2G366373_T02 | 3.22 | 0.11 | 84.27 | Auxin responsive protein |
| 6 | GRMZM2G466229_T01 | 1.73 | 3.28 | 9.68 | SAUR-like auxin-responsive protein |
| 3 | GRMZM2G102047_T01 | 0.92 | 0.19 | 13.83 | SAUR-like auxin-responsive protein |
| 2 | GRMZM2G365166_T01 | 0.72 | 0 | 4.54 | SAUR-like auxin-responsive protein |
| 4 | GRMZM2G118717_T01 | 0.69 | 0 | 0 | SAUR-like auxin-responsive protein |

(continued)

**Table 11.1** (continued)

| Chr no. | Gene_transcript identity | Pollen tube | Leaf | Root | Encoded protein (description according to Zm5b.60) |
|---|---|---|---|---|---|
| **GH3** | | | | | |
| 3 | GRMZM2G033359_T01 | 177.89 | 0.06 | 9 | GH3 family protein |
| 8 | GRMZM2G001421_T04 | 1.6 | 0 | 0 | GH3 family protein |
| **Control: Actin** | | | | | |
| 2 | GRMZM2G006765_T02 | 547.32 | 9.15 | 226.25 | Actin-7 |
| 5 | GRMZM2G017847_T01 | 57.18 | 1.28 | 3.49 | Actin |
| 7 | GRMZM2G342386_T02 | 1.45 | 10.72 | 17.27 | Actin |
| 8 | GRMZM2G067985_T01 | 1.4[a] | 0 | 0 | Actin |
| 1 | GRMZM2G082484_T01 | 0.74 | 23.24 | 146.28 | Actin-2 |

RNAseq data was generated by Dr. Mayada Woriedh (University of Regensburg) and filtered gene sets annotated according to release 5b.60 of the maize inbred line B73. Numbers represent log of FPKM values (expected fragments per kilobase of transcript per million fragments sequenced). Note that genes encoding transporters, receptors, SAURs, GH3 and actin with a value <0.5 are not listed (exception: pollen tube-specific transporters). ARFs with a value <0.1 and IAAs with a value <0.05 are not listed. Expression pattern of corresponding genes in leafs and roots as well as *actin* genes is included for comparison

[a]Pollen tube-specific transcript

expressed at extremely low levels, but a number of downstream genes of the dormancy/auxin associated proteins of the SAUR superfamily are expressed, indicating that auxin signaling takes place. A very strong expression was found for a gene encoding a GH3 family member showing that auxin signaling likely takes place during pollen tube growth in maize.

An interesting observation that differs between Arabidopsis and maize is the finding that one of eight Arabidopsis PIN auxin efflux carriers (Grunewald and Friml 2010) is specifically expressed in the male gametophyte. *PIN8* is expressed throughout male gametogenesis and the encoded transporter localizes to the endomembrane system (Dal Bosco et al. 2012; Ding et al. 2012). The transporter remains internally localized during pollen germination and growth and was detected primarily at the endoplasmic reticulum. Its precise role during gametogenesis and pollen tube growth is unclear, but it seems to be required to fine-tune cellular free auxin levels (Ding et al. 2012). Dal Bosco et al. (2012) suggest that PIN8 might control intracellular thresholds and access of auxin to the nucleus and thereby regulating auxin-dependent transcriptional responses (see also Chap. 4).

## 4 Auxin and Ovule Development

The ovule is the central component of the female reproductive system and located within the ovary (Esau 1977). It carries the egg and central cells, respectively, and upon successful fertilization will eventually develop into the seed harboring the embryo and endosperm. The systematics, morphology, and evolution of ovules are well studied in many species (Bouman 1984; Willemse and van Went 1984; Endress 2011), but the genetic and molecular basis of ovule development is best understood in *Arabidopsis thaliana* (Kelley and Gasser 2009; Shi and Yang 2011). However, despite rapidly growing molecular insight into ovule development knowledge regarding the role of auxin in this process is still very limited.

## *4.1 Ovule Development in* Arabidopsis thaliana

The morphology of Arabidopsis ovule development has been described in great detail (Christensen et al. 1997; Robinson-Beers et al. 1992; Schneitz et al. 1995; Webb and Gunning 1990) (Fig. 11.2). During stage 1 (staging according to Schneitz et al. 1995) ovule primordia initiate and outgrow from the placenta, a tissue formed by the adaxial portions of the medial domains of the ovary (Larsson et al. 2013). Eventually, an individual ovule contains three distinct pattern elements along the proximal–distal axis: the distal nucellus that eventually carries the embryo sac or female gametophyte, the central chalaza that initiates the integuments, and the proximal funiculus that harbors the vascular strand and connects the ovule to the placenta (Esau 1977; Schneitz et al. 1995). During stage 2 these elements become morphologically recognizable in a sequential fashion. The nucellus develops first. A large megaspore mother cell (MMC) is singled out from cells of the first subepidermal cell layer, undergoes meiosis, and develops into four haploid megaspores three of which will abort. The surviving chalazal megaspore will develop further into the embryo sac during stage 3. Following the appearance of the nucellus an inner and an outer integument develop from epidermal cells of the chalaza. During stages 2 and 3 cells within a growing integument expand and divide in a regular fashion along the plane of the integument (planar growth). Each integument is characterized by individual adaxial (upper) or abaxial (lower) cell layers and thus exhibits distinct tissue polarity. The outer integument grows asymmetrically and develops into a hoodlike structure eventually enveloping the nucellus and the inner integument. A small cleft or micropyle remains through which the pollen tube enters the ovule and effects fertilization.

## *4.2 Auxin Distribution During Ovule Development*

Spatial distribution of auxin biosynthesis was assayed by analyzing *TAA1* expression patterns using in situ hybridization and by monitoring signals of *TAA1::GFP* and several *YUC::GUS* reporters in wild-type ovules. In the ovule primordium *TAA1* is initially broadly expressed and subsequently becomes restricted to the boundary between nucellus and chalaza (Ceccato et al. 2013; Nole-Wilson et al. 2010). Slightly later, reporter signal could be monitored in the inner integument and the developing funiculus. *YUC1-3::GUS* signal was detected around the distal nucellus starting at early stage 3 (Ceccato et al. 2013; Pagnussat et al. 2009).

In wild-type plants carrying *pDR5rev::GFP* or *pDR5::GFP* reporter signals could be detected in a few epidermal cells at the tip of a late stage 1 ovule primordium (Benkova et al. 2003). During the early phase of integument development a similar but weak signal was seen at the tips of both integuments (Benkova et al. 2003). The nucellar signal continues to be detectable until mid-stage 3 (see also below) (Bencivenga et al. 2012; Ceccato et al. 2013; Enugutti and Schneitz 2013; Pagnussat et al. 2009). In addition, *DR5*-based signal is also detected in the vascular strand during stage 3 (Ceccato et al. 2013; Enugutti and Schneitz 2013; Pagnussat et al. 2009) (Fig. 11.2e, f). These results indicate that higher levels of localized responses to auxin exist at the tips of the nucellus and young integuments, respectively. In addition, a response to auxin appears to occur along the vascular strand.

The localized signal distribution of reporters for auxin synthesis and auxin response indicates that auxin is not present broadly throughout the developing ovule. It further suggests that auxin is synthesized at one position and may be transported toward another location, possibly by members of the PIN family of auxin efflux carriers (Gälweiler et al. 1998; Krecek et al. 2009; see also Chap. 5). Interestingly, analysis of ovules from wild-type plants carrying a *pPIN1::PIN1:GFP* (*PIN1:GFP*) reporter suggested that PIN1 is localized in the nucellar epidermis toward the end of stage 1 (Benkova et al. 2003). At later stages, PIN1:GFP was also observed at the tip of the young inner integument and in the developing vascular strand (Benkova et al. 2003; Ceccato et al. 2013; Pagnussat et al. 2009) (Fig. 11.2g). The subcellular polarity of PIN1:GFP localization suggested a transport of auxin from more proximal parts of the ovule epidermis to the distal tip of the nucellus. Rapidly thereafter PIN1:GFP becomes undetectable in ovules and other PIN family auxin exporters, except PIN3, seem to be absent during prefertilization ovule development altogether (Ceccato et al. 2013; Pagnussat et al. 2009). However, *PIN3* does not seem to play a major role during ovule development (Ceccato et al. 2013).

## 4.3 The Control of Ovule Primordium Outgrowth

Taken together, the results outlined above suggest that auxin plays an important role during early ovule development. Localized auxin maxima regulate the initiation and specification of lateral organs at the shoot apical meristem (Reinhardt et al. 2000, 2003; Yamaguchi et al. 2013). Given the early local *DR5*-based signal at the distal tip of the nucellus one could speculate that auxin plays a role in the outgrowth, specification, and proximal–distal patterning of the ovule primordium as well. Several lines of evidence are compatible with the view that auxin affects the initiation and outgrowth of the primordium. For example, mutations in *PIN1* or the

ARF5 gene *MP* result in malformed gynoecia with either absent or fewer ovules (Okada et al. 1991; Bencivenga et al. 2012; Benkova et al. 2003; Przemeck et al. 1996). In addition, young gynoecia treated with polar auxin transport inhibitors also develop into ovule-less structures (Nemhauser et al. 2000; Nole-Wilson et al. 2010) as do for example gynoecia of plants carrying defects in auxin biosynthesis genes (Cheng et al. 2006; Stepanova et al. 2008). However, these results need to be interpreted with caution as it is often difficult to discriminate between primary effects on ovule outgrowth or secondary effects due to altered gynoecium development.

Additional evidence supporting a role for auxin in ovule primordium outgrowth is derived from studies involving *AINTEGUMENTA* (*ANT*). *ANT* encoding an AP2-type transcription factor (Elliott et al. 1996; Klucher et al. 1996) is thought to act downstream of auxin in the regulation of meristematic competence during organogenesis (Hu et al. 2003; Mizukami and Fischer 2000) and appears to affect auxin homeostasis in developing pistils (Nole-Wilson et al. 2010). Gynoecia of *ant* mutants form fewer but more widely spaced ovules indicating a role of *ANT* in ovule initiation (Elliott et al. 1996). In addition, *SEUSS* (*SEU*), encoding a LIM domain binding putative transcriptional co-regulator (Franks et al. 2002), affects ovule initiation in conjunction with *ANT*. Pistils of *ant seu* double mutant plants exhibit relatively minor defects except for a complete lack of ovules (Azhakanandam et al. 2008). Furthermore, plants defective in *ANT* and *HUELLENLOS* (*HLL*), a gene encoding a mitochondrial ribosomal protein (Skinner et al. 2001), specifically fail to form the proximal-most part of the ovule also supporting a role for *ANT* in the outgrowth of the ovule primordium (Schneitz et al. 1998). Taken together, the available evidence supports the notion that auxin plays an important role in the initiation and outgrowth of the ovule primordium. How auxin controls this process via ANT remains to be investigated.

## 4.4 Pattern Formation in the Ovule Primordium

Outgrowth and proximal–distal pattern formation are coordinated during early ovule development (Schneitz et al. 1995, 1998). It is unclear how this coordination is controlled, but auxin could conceivably be part of such a mechanism. Loss-of-function mutations in *NOZZLE* (*NZZ*) (Schiefthaler et al. 1999), also known as *SPOROCYTELESS* (*SPL*) (Yang et al. 1999), result in a reduced nucellus and the absence of the MMC (Schneitz et al. 1997; Schiefthaler et al. 1999; Yang et al. 1999; Balasubramanian and Schneitz 2000). *NZZ* encodes a putative transcription factor but how it regulates nucellus formation is only partially understood (Balasubramanian and Schneitz 2002; Balasubramanian and Schneitz 2000; Sieber et al. 2004). It was suggested that *NZZ* plays a role in auxin homeostasis (Li et al. 2008) and recent work provided evidence that *NZZ* is a positive regulator of auxin responses and PIN1-mediated PAT in the ovule. The nucellar *DR5*-based GFP signal as well as PIN1:GFP expression was reduced in *nzz* ovules (Bencivenga et al. 2012). However, *pin1-5* mutant flowers carry ovules with a nucellus and an

MMC arguing against a role for *NZZ* in early nucellus development through a primary effect on *PIN1* (Bencivenga et al. 2012; Ceccato et al. 2013). Further work is required to determine whether *NZZ* mediates its effects on ovule development through the regulation of an auxin-related process.

Recent evidence suggests that control of PAT contributes to the development of the chalaza. Chalaza formation requires *BELL1* (*BEL1*) and *NZZ* function. *BEL1* encodes a homeodomain transcription factor (Reiser et al. 1995) and the central region of *bel1* ovules develops outgrowths of unclear identity in place of integuments (Robinson-Beers et al. 1992; Modrusan et al. 1994; Reiser et al. 1995; Schneitz et al. 1997; Brambilla et al. 2007; Bencivenga et al. 2012). In *nzz bel1* double mutants the chalaza is replaced by funiculus-like tissue (Balasubramanian and Schneitz 2000), indicating that *NZZ* and *BEL1* synergistically interact during the control of chalaza identity.

The *bel1* phenotype can be phenocopied by exogenous application of cytokinin to wild-type ovules which is also accompanied by ectopic expression of *pDR5rev::GFP* and PIN1:GFP (Bencivenga et al. 2012). Moreover, the outgrowths of *bel1* ovules show similar ectopic expression of these two reporters and the *bel1* phenotype is suppressed upon exogenous addition of the PAT inhibitor *N*-1-naphthylphthalamic acid (NPA) (Bencivenga et al. 2012). These results indicate that *bel1* outgrowth formation is the result of ectopic auxin accumulation in the developing chalaza. In addition, it was suggested that *BEL1* contributes to the control of PAT in the ovule by repressing *PIN1* in the proximal parts of the chalaza.

## 4.5 Integument Development

Auxin synthesis appears to be present in the inner integument as *TAA1::GFP* signal is present in this tissue during initiation and outgrowth (Ceccato et al. 2013). In addition, *pDR5rev::GFP* signal can be observed at the tips of both young integuments in Arabidopsis (Benkova et al. 2003) and mature integuments in maize (Lituiev et al. 2013), although in Arabidopsis the signal is very weak and frequently undetectable (Ceccato et al. 2013; Enugutti and Schneitz 2013) (Fig. 11.2e). Finally, PIN1:GFP reporter signal can be detected in the adaxial half of the tip of the initiating inner integument (Benkova et al. 2003; Ceccato et al. 2013) (Fig. 11.2g). These findings indicate that local accumulation of auxin and auxin responses may influence integument initiation and/or outgrowth.

Whether or not *PIN1* contributes to integument development is unclear. In *nzz* mutants initiation of the inner integument is delayed (Schiefthaler et al. 1999) and PIN1:GFP expression is strongly reduced (Bencivenga et al. 2012) raising the possibility that *PIN1*-mediated PAT influences the timing of inner integument initiation. Other data do not favor a central role for *PIN1* in integument development. For example, the cytokinin receptor genes *CRE1*, *AHK2*, and *AHK3* (Higuchi et al. 2004) affect the outgrowth of the primordium and development of the female gametophyte (Kinoshita-Tsujimura and Kakimoto 2011; Bencivenga et al. 2012) (see also below). Interestingly, PIN1:GFP expression was undetectable in fully developed young ovules of the strong *cre1 ahk2 ahk3* triple mutant (Bencivenga et al. 2012). However, integument development was either described as unaffected (Kinoshita-Tsujimura and Kakimoto 2011) or it was found that about 10 % of ovules failed to form integuments while the other ovules exhibited normal development (Bencivenga et al. 2012). Regarding integuments a similar result was found for *pin1-5* (Bencivenga et al. 2012) and in plants where *PIN1* was silenced in an ovule-specific manner (Ceccato et al. 2013). Several interpretations of these results are conceivable. For example, *PIN1* could play a minor role in integument development. Alternatively, *PIN1* does not play a role in this process and the 10 % ovules lacking integuments are an indirect result of aberrant ovule primordium outgrowth that can occur in *cre1 ahk2 ahk3* and in *pin1-5* mutants (Bencivenga et al. 2012). In this context it is notable that none of the other *PIN*-family members seem to be expressed in integuments of prefertilization ovules (Ceccato et al. 2013; Pagnussat et al. 2009; B. Enugutti and K. Schneitz, unpublished observations), indicating that integument initiation and/or outgrowth may not depend on *PIN*-mediated PAT.

Auxin appears to affect integument development in a complex fashion. For example *ANT* promotes integument initiation and outgrowth (Elliott et al. 1996; Klucher et al. 1996; Krizek 1999; Mizukami and Fischer 2000; Schneitz et al. 1997) while *ARF2* restricts cell division in integuments (Schruff et al. 2006). Ectopic expression of other *ARF* genes, including *ARF6* and *ARF8*, in integuments, however, results in reduced integument outgrowth (Wu et al. 2006). Yet other auxin-related genes, such as *ARF3/ETT* (Sessions et al. 1997) or the *Aux/IAA* family member *IAA17/AXR3* (Rouse et al. 1998), are expressed in integuments as well (Kelley et al. 2012; Skinner and Gasser 2009). In this context there is an interesting link between *ARF3/ETT* and *ARF4* and the function of the *KANADI* (*KAN*) genes. As discussed already above, members of the *KAN* family of transcription factor genes are required for the establishment of adaxial–abaxial polarity during leaf development (Eshed et al. 2001; Kerstetter et al. 2001). During this process they genetically cooperate with the *ARF* genes *ARF3/ETT* and *ARF4* and it was proposed that ARF3/ETT and ARF4 proteins might function as cofactors of KAN (Pekker et al. 2005). Physical interactions of ARF3/ETT with KAN1 and KAN4 were reported recently (Kelley et al. 2012). *ARF3/ETT* and *ARF4* also affect integument development as for example ovules of *arf3/ett arf4* double mutants exhibit variably reduced integuments (Enugutti and Schneitz 2013; Pekker et al. 2005). Moreover, integument formation is also severely affected in several *kan* mutants (Eshed et al. 2001; McAbee et al. 2006).

Further insight into the close association between *ARF* and *KAN* genes has come from studies on the *KAN* family member *ABERRANT TESTA SHAPE* (*ATS*) and its role in integument development. *ATS* is required for the formation of integument boundary, inner integument outgrowth, and adaxial–abaxial polarity (Balasubramanian and Schneitz 2002; Léon-Kloosterziel et al. 1994; Kelley et al. 2009; McAbee et al. 2006). In ovules of *ats* mutants spacing between the two integuments is reduced resulting in the appearance of "fused" integuments. Interestingly, ovules of single *arf3/ett* mutants exhibit a similar phenotype and recent evidence suggests that ARF3/ETT and ATS proteins form a complex that is part of an auxin-controlled module regulating the spacing between the two initiating integuments (Kelley et al. 2012). This postulated ARF3/ETT-ATS complex appears to be required only transiently during early integument development (Enugutti and Schneitz 2013).

A role for auxin could principally be imagined in the control of planar growth during the outgrowth of the integuments. In this process cells expand and divide along the plane of the integument layer (Schneitz et al. 1995; Truernit and Haseloff 2008). Thus, an interesting question relates to how this planar growth is orchestrated (Enugutti et al. 2013). We are presently only beginning to understand the control of these processes. Planar integument growth involves *UNICORN* (*UCN*) encoding an AGC type kinase (Enugutti et al. 2012). Integuments of loss-of-function *ucn* mutants exhibit localized multicellular outgrowths. The current model states that *ucn* protrusions are the result of hyperactive ATS and that UCN normally suppresses this ectopic growth by repressing ATS in the nucleus, likely through a direct physical interaction of the two proteins (Enugutti et al. 2012; Enugutti and Schneitz 2013). If hyperactive ATS interferes with auxin-related processes remains to be determined. Apical–basal positioning of root hairs, another example of planar polarity, is coordinated relative to an auxin concentration gradient with the maximum concentration of auxin located at the root tip (Grebe et al. 2002; Grebe 2004; Fischer et al. 2006; Ikeda et al. 2009; Masucci and Schiefelbein 1994; Sabatini et al. 1999). It will be interesting to investigate whether a similar mechanism acts during planar integument growth as well and whether auxin plays a fundamental role in both processes.

## 4.6 Does Auxin Pattern the Female Gametophyte?

The haploid female gametophyte (embryo sac) develops during stage 3 in Arabidopsis. After MMC formation and meiosis the surviving megaspore turns into the mononuclear embryo sac which continues syncytial development with three rounds of nuclear divisions and the formation of a large central vacuole. Eventually cellularization takes place resulting in a typical *Polygonum*-type embryo sac containing seven cells and eight nuclei. The maturating embryo sac exhibits a prominent micropylar–chalazal polarity. Two accessory synergid cells and the egg cell form the egg apparatus and locate to the distal or micropylar pole of the embryo sac. A large di-haploid central cell containing two nuclei that eventually fuse and a large vacuole occupies the central part of the embryo sac. At the proximal or chalazal pole three accessory antipodal cells form that eventually will degenerate shortly before or around the time of fertilization (Sprunck and Gross-Hardt 2011). Upon double fertilization the egg cell develops into the diploid zygote and embryo while the central cell will originate the triploid endosperm.

On the basis of auxin synthesis and auxin response reporter studies a dynamic model of auxin flux was proposed in which auxin generated in more proximal regions is transported through the nucellar epidermis to the tip of the nucellus where it could affect MMC and embryo sac development (Ceccato et al. 2013; Pagnussat et al. 2009). Thus, it is believed that auxin contributes to the sporophyte–gametophyte cross talk. However, at present there appears to exist some uncertainties as to the specific part auxin plays in gametophyte development. One set of experiments suggests that PAT is essential for the control of the first nuclear divisions of early embryo sac development. A second set of data indicates that auxin does not play such an early essential role but is a central regulator of pattern formation along the micropylar–chalazal axis of the embryo sac.

For example, plants homozygous for the *pin1-5* mutants develop ovule-bearing pistils (Sohlberg et al. 2006; Bencivenga et al. 2012). While *pin1-5* ovules usually carry normal integuments they exhibit an early arrest of embryo sac development (Bencivenga et al. 2012) as do plants in which *PIN1* was silenced in an ovule-specific manner (Ceccato et al. 2013). This implies that PIN1-mediated PAT is involved in early steps of female gametophyte development. Thus, a sporophytic source of auxin would directly or indirectly regulate early development of the female gametophyte. In line with this notion are results from the study of the role of the cytokinin receptor genes *CRE1*, *AHK2*, and *AHK3* as ovules of a strong *cre1 ahk2 ahk3* triple mutant resemble *pin1-5* ovules with usually normal integuments but an early block in embryo sac development (Bencivenga et al. 2012; Kinoshita-Tsujimura and Kakimoto 2011). Moreover, the absence of detectable PIN1:GFP expression in ovules of the *cre1 ahk2 ahk3* triple mutant (Bencivenga et al. 2012) suggests that the early block in embryo sac development is due to the absence of *PIN1* expression and aberrant PAT. Thus, these experiments are compatible with the view that PIN1-mediated local accumulation of auxin at the distal tip of the nucellus is essential for the control of the initial steps of female gametophyte development.

There is also evidence that auxin plays a central role in patterning the Arabidopsis embryo sac (Pagnussat et al. 2009). The authors found that signals from *DR5*-based reporters eventually accumulated at the micropylar pole inside the early and still syncytial embryo sac. As PIN1:GFP signal was already absent at these stages it was assumed that the *DR5*-based signal was related to local auxin biosynthesis. Indeed *pYUC1-2::GUS* reporters yielded signal at the micropylar pole of the embryo sac. However, recent microscopic evidence does not support the notion of auxin synthesis or auxin response taking place in the developing Arabidopsis embryo sac but suggests dynamic control of these activities in nucellar cells neighboring the developing gametophyte (Ceccato et al. 2013; Lituiev et al. 2013). It is unclear whether significant auxin biosynthesis or response takes place inside the developing female gametophyte as *DR5* reporter signal is absent from embryo sacs of *Hieracium pilosella* (Tucker et al. 2012) or Arabidopsis (Ceccato et al. 2013; Lituiev et al. 2013) and occurs only at maturity in proliferating antipodal cells of maize embryo sacs, but not in Arabidopsis (Lituiev et al. 2013). Nevertheless, embryo sac-specific downregulation of *ARF* genes or upregulation of *YUC* genes did not lead to early gametophyte abortion but resulted in cell identity defects that could be observed in late Arabidopsis embryo sacs (Pagnussat et al. 2009). Morphology and marker gene expression studies indicated that specification defects were particularly evident for egg apparatus cells that are normally located at the micropylar pole of the embryo sac. In one attractive model the authors proposed that PIN1-mediated PAT leads to a first auxin maximum at the distal pole of the nucellus up to the beginning of embryo sac development. Such a sporophytic auxin source could subsequently trigger gametophyte-specific auxin biosynthesis at the micropylar pole of the female gametophyte. This localized auxin biosynthesis would be fundamental for the establishment of an auxin gradient along the micropylar–chalazal (distal–proximal) axis within the syncytium of the developing embryo sac. Auxin might then act as a morphogen in the uncellularized embryo sac with highest auxin concentrations specifying synergids, followed by egg cell and central cell, while low levels of auxin result in the development of antipodal cells (Pagnussat et al. 2009; Sundaresan and Alandete-Saez 2010).

The recent data described above regarding the absence of auxin synthesis and response in the embryo sac and the role of *PIN1* in ovule development pose some challenges to this original model. For example, it is difficult to rationalize the early embryo sac abortion in ovules lacking PIN1:GFP signal, such as in ovules of the *cre1 ahk2 ahk3* triple mutants, or in ovules of *pin1-5* plants if PIN1-mediated PAT would result in the initial trigger leading to the auxin gradient involved in patterning the female gametophyte. Rather, one would expect patterning defects in late embryo sacs as observed in ovules with embryo sac-specific up- or downregulation of *YUC* and *ARF* activities, respectively. In addition, results from mathematical modeling, making use of published auxin degradation rates and diffusion coefficients, are incompatible with a gametophytic auxin gradient patterning the embryo sac (Lituiev et al. 2013). Since there is a dynamic auxin response in sporophytic cells neighboring the embryo sac it was postulated that a sporophytic, auxin-responsive, diffusible signal patterns the embryo sac in a non-cell-autonomous fashion. Obviously, additional work is necessary and it will be of great interest to investigate further the molecular mechanism underlying the control of auxin-dependent pattern formation in the developing embryo sac.

# 5 Conclusions

The past 10 years have seen a tremendous increase in knowledge about the central role of auxin for the timely and coordinated regulation of reproductive organ development in angiosperms. Some findings were perhaps not unexpected while other results, for example, the apparent absence of PIN-mediated PAT during later steps of integument development, are surprising. Moreover, early models such as the auxin gradient theory (Nemhauser et al. 2000) explaining gynoecium patterning turned out to represent oversimplifications as auxin signaling is strongly interconnected with other hormonal pathways (see also Chap. 6). Most of our current knowledge was generated by the analysis of auxin pathway mutants as well as fusion protein and *DR5* reporter studies in the Brassicaceae model

*Arabidopsis thaliana*. Thus, it is not only urgently required to study auxin processes during reproductive development using more advanced tools, for example, more precise Aux/IAA-based DII-VENUS sensors (Brunoud et al. 2012), but to include other plant species, for example, the economically important grasses, in functional studies as well. Considering the existing opportunities and the accelerated pace of research the next 10 years will certainly witness exciting and new insights into the role of auxin during reproductive development.

**Acknowledgments** We thank Mayada Woriedh for the generation of RNAseq data from maize pollen tubes and acknowledge Sapna Sharma for annotation and clustering of the data. Work on reproductive development in the Dresselhaus and Schneitz labs is funded by SFB 924 grants (TP A2/TP A3) as well as by grant SCHN 723/6-1 from the German Research Council (DFG) and supported by the Free State of Bavaria.

## References

Aloni R, Aloni E, Langhans M, Ullrich CI (2006) Role of auxin in regulating Arabidopsis flower development. Planta 223:315–328

Alvarez JP, Goldshmidt A, Efroni I, Bowman JL, Eshed Y (2009) The *NGATHA* distal organ development genes are essential for style specification in Arabidopsis. Plant Cell 21:1373–1393

Azhakanandam S, Nole-Wilson S, Bao F, Franks RG (2008) *SEUSS* and *AINTEGUMENTA* mediate patterning and ovule initiation during gynoecium medial domain development. Plant Physiol 146:1165–1181

Balasubramanian S, Schneitz K (2000) *NOZZLE* regulates proximal-distal pattern formation, cell proliferation and early sporogenesis in *Arabidopsis thaliana*. Development 127:4227–4238

Balasubramanian S, Schneitz K (2002) *NOZZLE* links proximal-distal and adaxial-abaxial pattern formation during ovule development in *Arabidopsis thaliana*. Development 129:4291–4300

Bencivenga S, Simonini S, Benkova E, Colombo L (2012) The transcription factors BEL1 and SPL are required for cytokinin and auxin signaling during ovule development in *Arabidopsis*. Plant Cell 24:2886–2897

Benkova E, Michniewicz M, Sauer M, Teichmann T, Seifertova D, Jürgens G, Friml J (2003) Local, efflux-dependent auxin gradients as a common module for plant organ formation. Cell 115:591–602

Bouman F (1984) The ovule. In: Johri BM (ed) Embryology of angiosperms. Springer, New York, pp 123–157

Brambilla V, Battaglia R, Colombo M, Masiero S, Bencivenga S, Kater MM, Colombo L (2007) Genetic and molecular interactions between BELL1 and MADS box factors support ovule development in Arabidopsis. Plant Cell 19:2544–2556

Brunoud G, Wells DM, Oliva M, Larrieu A, Mirabet V, Burrow AH, Beeckman T, Kepinski S, Traas J, Bennett MJ, Vernoux T (2012) A novel sensor to map auxin response and distribution at high spatio-temporal resolution. Nature 482:103–106

Ceccato L, Masiero S, Sinha RD, Bencivenga S, Roig-Villanova I, Ditengou FA, Palme K, Simon R, Colombo L (2013) Maternal control of PIN1 is required for female gametophyte development in Arabidopsis. PLoS One 8:e66148

Cecchetti V, Altamura MM, Falasca G, Costantino P, Cardarelli M (2008) Auxin regulates Arabidopsis anther dehiscence, pollen maturation, and filament elongation. Plant Cell 20:1760–1774

Cecchetti V, Altamura MM, Brunetti P, Petrocelli V, Falasca G, Ljung K, Costantino P, Cardarelli M (2013) Auxin controls Arabidopsis anther dehiscence by regulating endothecium lignification and jasmonic acid biosynthesis. Plant J 74:411–422

Chen D, Zhao J (2008) Free IAA in stigmas and styles during pollen germination and pollen tube growth of *Nicotiana tabacum*. Physiol Plant 134:202–215

Cheng H, Qin L, Lee S, Fu X, Richards DE, Cao D, Luo D, Harberd NP, Peng J (2004) Gibberellin regulates Arabidopsis floral development via suppression of DELLA protein function. Development 131:1055–1064

Cheng Y, Dai X, Zhao Y (2006) Auxin biosynthesis by the YUCCA flavin monooxygenases controls the formation of floral organs and vascular tissues in Arabidopsis. Genes Dev 20:1790–1799

Christensen CA, King EJ, Jordan JR, Drews GN (1997) Megagametogenesis in *Arabidopsis* wild type and the *Gf* mutant. Sex Plant Reprod 10:49–64

Christensen SK, Dagenais N, Chory J, Weigel D (2000) Regulation of auxin response by the protein kinase PINOID. Cell 100:469–478

Crawford BC, Yanofsky MF (2008) The formation and function of the female reproductive tract in flowering plants. Curr Biol 18:R972–R978

Crawford BC, Yanofsky MF (2011) HALF FILLED promotes reproductive tract development and fertilization efficiency in *Arabidopsis thaliana*. Development 138:2999–3009

Dal Bosco C, Dovzhenko A, Liu X, Woerner N, Rensch T, Eismann M, Eimer S, Hegermann J, Paponov IA, Ruperti B, Heberle-Bors E, Touraev A, Cohen JD, Palme K (2012) The endoplasmic reticulum localized PIN8 is a pollen-specific auxin carrier involved in intracellular auxin homeostasis. Plant J 71:860–870

DeMason DA, Polowick PL (2009) Patterns of DR5:GUS expression in organs of pea (*Pisum sativum*). Int J Plant Sci 170:1–11

Ding Z, Wang B, Moreno I, Duplakova N, Simon S, Carraro N, Reemmer J, Pencik A, Chen X, Tejos R, Skupa P, Pollmann S, Mravec J, Petrasek J, Zazimalova E, Honys D, Rolcik J, Murphy A, Orellana A, Geisler M, Friml J (2012) ER-localized auxin transporter PIN8 regulates auxin homeostasis and male gametophyte development in Arabidopsis. Nat Commun 3:941

Dresselhaus T, Franklin-Tong N (2013) Male-female cross-talk during pollen germination, tube growth and guidance, and double fertilization. Mol Plant 6:1018–1036. doi:10.1093/mp/sst061

Dresselhaus T, Sprunck S (2012) Plant fertilization: maximizing reproductive success. Curr Biol 22:R487–R489

Eklund DM, Staldal V, Valsecchi I, Cierlik I, Eriksson C, Hiratsu K, Ohme-Takagi M, Sundstrom JF, Thelander M, Ezcurra I, Sundberg E (2010) The *Arabidopsis thaliana* STYLISH1 protein acts as a transcriptional activator regulating auxin biosynthesis. Plant Cell 22:349–363

Elliott RC, Betzner AS, Huttner E, Oakes MP, Tucker WQ, Gerentes D, Perez P, Smyth DR (1996) *AINTEGUMENTA*, an *APETALA2*-like gene of Arabidopsis with pleiotropic roles in ovule development and floral organ growth. Plant Cell 8:155–168

Endress PK (2011) Angiosperm ovules: diversity, development, evolution. Ann Bot 107:1465–1489

Enugutti B, Schneitz K (2013) Genetic analysis of ectopic growth suppression during planar growth of integuments mediated by the Arabidopsis AGC protein kinase UNICORN. BMC Plant Biol 13:2

Enugutti B, Kirchhelle C, Oelschner M, Torres Ruiz RA, Schliebner I, Leister D, Schneitz K (2012) Regulation of planar growth by the *Arabidopsis* AGC protein kinase UNICORN. Proc Natl Acad Sci USA 109:15060–15065

Enugutti B, Kirchhelle C, Schneitz K (2013) On the genetic control of planar growth during tissue morphogenesis in plants. Protoplasma 250:651–661

Esau K (1977) Anatomy of seed plants, 2nd edn. Wiley, New York

Eshed Y, Baum SF, Perea JV, Bowman JL (2001) Establishment of polarity in lateral organs of plants. Curr Biol 11:1251–1260

Feng X, Dickinson HG (2007) Packaging the male germline in plants. Trends Genet 23:503–510
Feng XL, Ni WM, Elge S, Mueller-Roeber B, Xu ZH, Xue HW (2006) Auxin flow in anther filaments is critical for pollen grain development through regulating pollen mitosis. Plant Mol Biol 61:215–226
Fischer U, Ikeda Y, Ljung K, Serralbo O, Singh M, Heidstra R, Palme K, Scheres B, Grebe M (2006) Vectorial information for Arabidopsis planar polarity is mediated by combined AUX1, EIN2, and GNOM activity. Curr Biol 16:2143–2149
Forestan C, Farinati S, Varotto S (2012) The maize *PIN* gene family of auxin transporters. Front Plant Sci 3:16
Franks RG, Wang C, Levin JZ, Liu Z (2002) *SEUSS*, a member of a novel family of plant regulatory proteins, represses floral homeotic gene expression with *LEUNIG*. Development 129:253–263
Friml J, Benkova E, Blilou I, Wisniewska J, Hamann T, Ljung K, Woody S, Sandberg G, Scheres B, Jurgens G, Palme K (2002) AtPIN4 mediates sink-driven auxin gradients and root patterning in *Arabidopsis*. Cell 108:661–673
Friml J, Vieten A, Sauer M, Weijers D, Schwarz H, Hamann T, Offringa R, Jürgens G (2003) Efflux-dependent auxin gradients establish the apical-basal axis of Arabidopsis. Nature 426:147–153
Friml J, Yang X, Michniewicz M, Weijers D, Quint A, Tietz O, Benjamins R, Ouwerkerk PB, Ljung K, Sandberg G, Hooykaas PJ, Palme K, Offringa R (2004) A PINOID-dependent binary switch in apical-basal PIN polar targeting directs auxin efflux. Science 306:862–865
Gallavotti A, Barazesh S, Malcomber S, Hall D, Jackson D, Schmidt RJ, McSteen P (2008a) Sparse inflorescence1 encodes a monocot-specific *YUCCA*-like gene required for vegetative and reproductive development in maize. Proc Natl Acad Sci USA 105:15196–15201
Gallavotti A, Yang Y, Schmidt RJ, Jackson D (2008b) The relationship between auxin transport and maize branching. Plant Physiol 147:1913–1923
Gälweiler L, Guan C, Müller A, Wisman E, Mendgen K, Yephremov A, Palme K (1998) Regulation of polar auxin transport by AtPIN1 in *Arabidopsis* vascular tissue. Science 282:2226–2230
Geisler M, Blakeslee JJ, Bouchard R, Lee OR, Vincenzetti V, Bandyopadhyay A, Titapiwatanakun B, Peer WA, Bailly A, Richards EL, Ejendal KF, Smith AP, Baroux C, Grossniklaus U, Muller A, Hrycyna CA, Dudler R, Murphy AS, Martinoia E (2005) Cellular efflux of auxin catalyzed by the Arabidopsis MDR/PGP transporter AtPGP1. Plant J 44:179–194
Girin T, Paicu T, Stephenson P, Fuentes S, Korner E, O'Brien M, Sorefan K, Wood TA, Balanza V, Ferrandiz C, Smyth DR, Ostergaard L (2011) INDEHISCENT and SPATULA interact to specify carpel and valve margin tissue and thus promote seed dispersal in Arabidopsis. Plant Cell 23:3641–3653
Grebe M (2004) Ups and downs of tissue and planar polarity in plants. Bioessays 26:719–729
Grebe M, Friml J, Swarup R, Ljung K, Sandberg G, Terlou M, Palme K, Bennett MJ, Scheres B (2002) Cell polarity signaling in Arabidopsis involves a BFA-sensitive auxin influx pathway. Curr Biol 12:329–334
Gremski K, Ditta G, Yanofsky MF (2007) The *HECATE* genes regulate female reproductive tract development in *Arabidopsis thaliana*. Development 134:3593–3601
Griffiths J, Murase K, Rieu I, Zentella R, Zhang ZL, Powers SJ, Gong F, Phillips AL, Hedden P, Sun TP, Thomas SG (2006) Genetic characterization and functional analysis of the GID1 gibberellin receptors in Arabidopsis. Plant Cell 18:3399–3414
Grunewald W, Friml J (2010) The march of the PINs: developmental plasticity by dynamic polar targeting in plant cells. EMBO J 29:2700–2714
Hamamura Y, Nagahara S, Higashiyama T (2012) Double fertilization on the move. Curr Opin Plant Biol 15:70–77
Hardtke CS, Berleth T (1998) The Arabidopsis gene *MONOPTEROS* encodes a transcription factor mediating embryo axis formation and vascular development. EMBO J 17:1405–1411

Heisler MG, Atkinson A, Bylstra YH, Walsh R, Smyth DR (2001) *SPATULA*, a gene that controls development of carpel margin tissues in Arabidopsis, encodes a bHLH protein. Development 128:1089–1098

Heisler MG, Ohno C, Das P, Sieber P, Reddy GV, Long JA, Meyerowitz EM (2005) Patterns of auxin transport and gene expression during primordium development revealed by live imaging of the Arabidopsis inflorescence meristem. Curr Biol 15:1899–1911

Higuchi M, Pischke MS, Mahonen AP, Miyawaki K, Hashimoto Y, Seki M, Kobayashi M, Shinozaki K, Kato T, Tabata S, Helariutta Y, Sussman MR, Kakimoto T (2004) *In planta* functions of the *Arabidopsis* cytokinin receptor family. Proc Natl Acad Sci USA 101:8821–8826

Honma T, Goto K (2001) Complexes of MADS-box proteins are sufficient to convert leaves into floral organs. Nature 409:525–529

Hu Y, Xie Q, Chua NH (2003) The Arabidopsis auxin-inducible gene *ARGOS* controls lateral organ size. Plant Cell 15:1951–1961

Hu J, Mitchum MG, Barnaby N, Ayele BT, Ogawa M, Nam E, Lai WC, Hanada A, Alonso JM, Ecker JR, Swain SM, Yamaguchi S, Kamiya Y, Sun TP (2008) Potential sites of bioactive gibberellin production during reproductive growth in Arabidopsis. Plant Cell 20:320–336

Ikeda Y, Men S, Fischer U, Stepanova AN, Alonso JM, Ljung K, Grebe M (2009) Local auxin biosynthesis modulates gradient-directed planar polarity in Arabidopsis. Nat Cell Biol 11:731–738

Ilegems M, Douet V, Meylan-Bettex M, Uyttewaal M, Brand L, Bowman JL, Stieger PA (2010) Interplay of auxin, KANADI and class III HD-ZIP transcription factors in vascular tissue formation. Development 137:975–984

Ishiguro S, Kawai-Oda A, Ueda J, Nishida I, Okada K (2001) The DEFECTIVE IN ANTHER DEHISCIENCE gene encodes a novel phospholipase A1 catalyzing the initial step of jasmonic acid biosynthesis, which synchronizes pollen maturation, anther dehiscence, and flower opening in Arabidopsis. Plant Cell 13:2191–2209

Izhaki A, Bowman JL (2007) KANADI and class III HD-Zip gene families regulate embryo patterning and modulate auxin flow during embryogenesis in Arabidopsis. Plant Cell 19:495–508

Jain M, Khurana JP (2009) Transcript profiling reveals diverse roles of auxin-responsive genes during reproductive development and abiotic stress in rice. FEBS J 276:3148–3162

Kelley DR, Gasser CS (2009) Ovule development: genetic trends and evolutionary considerations. Sex Plant Reprod 22:229–234

Kelley DR, Skinner DJ, Gasser CS (2009) Roles of polarity determinants in ovule development. Plant J 57:1054–1064

Kelley DR, Arreola A, Gallagher TL, Gasser CS (2012) ETTIN (ARF3) physically interacts with KANADI proteins to form a functional complex essential for integument development and polarity determination in *Arabidopsis*. Development 139:1105–1109

Kennedy PB (1899) The structure of the caryopsis of grasses with reference to their morphology and classification. USDA Div Agrostol Bull 19:1–44

Kerstetter RA, Bollman K, Taylor RA, Bomblies K, Poethig RS (2001) *KANADI* regulates organ polarity in *Arabidopsis*. Nature 411:706–709

Kinoshita-Tsujimura K, Kakimoto T (2011) Cytokinin receptors in sporophytes are essential for male and female functions in *Arabidopsis thaliana*. Plant Signal Behav 6:66–71

Klucher KM, Chow H, Reiser L, Fischer RL (1996) The *AINTEGUMENTA* gene of Arabidopsis required for ovule and female gametophyte development is related to the floral homeotic gene *APETALA2*. Plant Cell 8:137–153

Krecek P, Skupa P, Libus J, Naramoto S, Tejos R, Friml J, Zazimalova E (2009) The PIN-FORMED (PIN) protein family of auxin transporters. Genome Biol 10:249

Krizek BA (1999) Ectopic expression of *AINTEGUMENTA* in *Arabidopsis* plants results in increased growth of floral organs. Dev Genet 25:224–236

Kuusk S, Sohlberg JJ, Long JA, Fridborg I, Sundberg E (2002) STY1 and STY2 promote the formation of apical tissues during Arabidopsis gynoecium development. Development 129:4707–4717

Kuusk S, Sohlberg JJ, Magnus Eklund D, Sundberg E (2006) Functionally redundant SHI family genes regulate Arabidopsis gynoecium development in a dose-dependent manner. Plant J 47:99–111

Larsson E, Franks RG, Sundberg E (2013) Auxin and the *Arabidopsis thaliana* gynoecium. J Exp Bot 64:2619–2627

Léon-Kloosterziel KM, Keijzer CJ, Koornneef M (1994) A seed shape mutant of *Arabidopsis* that is affected in integument development. Plant Cell 6:385–392

Li LC, Qin GJ, Tsuge T, Hou XH, Ding MY, Aoyama T, Oka A, Chen Z, Gu H, Zhao Y, Qu LJ (2008) *SPOROCYTELESS* modulates *YUCCA* expression to regulate the development of lateral organs in Arabidopsis. New Phytol 179:751–764

Lituiev D, Krohn NG, Müller B, Jackson D, Hellriegel B, Dresselhaus T, Grossniklaus U (2013) Theoretical and experimental evidence indicates that there is no auxin gradient patterning the angiosperm female gametophyte. Development 14022:4544–4553

Mashiguchi K, Tanaka K, Sakai T, Sugawara S, Kawaide H, Natsume M, Hanada A, Yaeno T, Shirasu K, Yao H, McSteen P, Zhao Y, Hayashi K, Kamiya Y, Kasahara H (2011) The main auxin biosynthesis pathway in Arabidopsis. Proc Natl Acad Sci USA 108:18512–18517

Masucci JD, Schiefelbein JW (1994) The *rhd6* mutation of *Arabidopsis thaliana* alters root-hair initiation through an auxin- and ethylene-associated process. Plant Physiol 106:1335–1346

McAbee JM, Hill TA, Skinner DJ, Izhaki A, Hauser BA, Meister RJ, Venugopala Reddy G, Meyerowitz EM, Bowman JL, Gasser CS (2006) *ABERRANT TESTA SHAPE* encodes a KANADI family member, linking polarity determination to separation and growth of Arabidopsis ovule integuments. Plant J 46:522–531

Mizukami Y, Fischer RL (2000) Plant organ size control: *AINTEGUMENTA* regulates growth and cell numbers during organogenesis. Proc Natl Acad Sci USA 97:942–947

Modrusan Z, Reiser L, Feldmann KA, Fischer RL, Haughn GW (1994) Homeotic transformation of ovules into carpel-like structures in Arabidopsis. Plant Cell 6:333–349

Nagpal P, Ellis CM, Weber H, Ploense SE, Barkawi LS, Guilfoyle TJ, Hagen G, Alonso JM, Cohen JD, Farmer EE, Ecker JR, Reed JW (2005) Auxin response factors ARF6 and ARF8 promote jasmonic acid production and flower maturation. Development 132:4107–4118

Nemhauser JL, Feldman LJ, Zambryski PC (2000) Auxin and *ETTIN* in *Arabidopsis* gynoecium morphogenesis. Development 127:3877–3888

Nole-Wilson S, Azhakanandam S, Franks RG (2010) Polar auxin transport together with aintegumenta and revoluta coordinate early Arabidopsis gynoecium development. Dev Biol 346:181–195

Okada K, Ueda J, Komaki MK, Bell CJ, Shimura Y (1991) Requirement of the auxin polar transport system in early stages of *Arabidopsis* floral bud formation. Plant Cell 3:677–684

Ostergaard L (2009) Don't leaf now. The making of a fruit. Curr Opin Plant Biol 12:36–41

Pagnussat GC, Alandete-Saez M, Bowman JL, Sundaresan V (2009) Auxin-dependent patterning and gamete specification in the *Arabidopsis* female gametophyte. Science 324:1684–1689

Palanivelu R, Tsukamoto T (2012) Pathfinding in angiosperm reproduction: pollen tube guidance by pistils ensures successful double fertilization. Wiley Interdiscip Rev Dev Biol 1:96–113

Paponov IA, Teale WD, Trebar M, Blilou I, Palme K (2005) The PIN auxin efflux facilitators: evolutionary and functional perspectives. Trends Plant Sci 10:170–177

Pekker I, Alvarez JP, Eshed Y (2005) Auxin response factors mediate Arabidopsis organ asymmetry via modulation of KANADI activity. Plant Cell 17:2899–2910

Peng J (2009) Gibberellin and jasmonate crosstalk during stamen development. J Integr Plant Biol 51:1064–1070

Peng J, Chen R (2011) Auxin efflux transporter MtPIN10 regulates compound leaf and flower development in *Medicago truncatula*. Plant Signal Behav 6:1537–1544

Plackett AR, Powers SJ, Fernandez-Garcia N, Urbanova T, Takebayashi Y, Seo M, Jikumaru Y, Benlloch R, Nilsson O, Ruiz-Rivero O, Phillips AL, Wilson ZA, Thomas SG, Hedden P (2012) Analysis of the developmental roles of the Arabidopsis gibberellin 20-oxidases demonstrates that GA20ox1, -2, and -3 are the dominant paralogs. Plant Cell 24:941–960

Poppenberger B, Rozhon W, Khan M, Husar S, Adam G, Luschnig C, Fujioka S, Sieberer T (2011) CESTA, a positive regulator of brassinosteroid biosynthesis. EMBO J 30:1149–1161

Przemeck GKH, Mattson J, Hardtke CS, Sung ZR, Berleth T (1996) Studies on the role of the Arabidopsis gene MONOPTEROS in vascular development and plant cell axialization. Planta 200:229–237

Reeves PH, Ellis CM, Ploense SE, Wu MF, Yadav V, Tholl D, Chetelat A, Haupt I, Kennerley BJ, Hodgens C, Farmer EE, Nagpal P, Reed JW (2012) A regulatory network for coordinated flower maturation. PLoS Genet 8:e1002506

Reinhardt D, Mandel T, Kuhlemeier C (2000) Auxin regulates the initiation and radial position of plant lateral organs. Plant Cell 12:507–518

Reinhardt D, Pesce ER, Stieger P, Mandel T, Baltensperger K, Bennett M, Traas J, Friml J, Kuhlemeier C (2003) Regulation of phyllotaxis by polar auxin transport. Nature 426:255–260

Reiser L, Modrusan Z, Margossian L, Samach A, Ohad N, Haughn GW, Fischer RL (1995) The *BELL1* gene encodes a homeodomain protein involved in pattern formation in the Arabidopsis ovule primordium. Cell 83:735–742

Rieu I, Ruiz-Rivero O, Fernandez-Garcia N, Griffiths J, Powers SJ, Gong F, Linhartova T, Eriksson S, Nilsson O, Thomas SG, Phillips AL, Hedden P (2008) The gibberellin biosynthetic genes AtGA20ox1 and AtGA20ox2 act, partially redundantly, to promote growth and development throughout the Arabidopsis life cycle. Plant J 53:488–504

Robinson-Beers K, Pruitt RE, Gasser CS (1992) Ovule development in wild-type Arabidopsis and two female-sterile mutants. Plant Cell 4:1237–1249

Rouse D, Mackay P, Stirnberg P, Estelle M, Leyser O (1998) Changes in auxin response from mutations in an AUX/IAA gene. Science 279:1371–1373

Ru P, Xu L, Ma H, Huang H (2006) Plant fertility defects induced by the enhanced expression of microRNA167. Cell Res 16:457–465

Sabatini S, Beis D, Wolkenfeldt H, Murfett J, Guilfoyle T, Malamy J, Benfey P, Leyser O, Bechtold N, Weisbeek P, Scheres B (1999) An auxin-dependent distal organizer of pattern and polarity in the *Arabidopsis* root. Cell 99:463–472

Sakata T, Oshino T, Miura S, Tomabechi M, Tsunaga Y, Higashitani N, Miyazawa Y, Takahashi H, Watanabe M, Higashitani A (2010) Auxins reverse plant male sterility caused by high temperatures. Proc Natl Acad Sci USA 107:8569–8574

Schiefthaler U, Balasubramanian S, Sieber P, Chevalier D, Wisman E, Schneitz K (1999) Molecular analysis of *NOZZLE*, a gene involved in pattern formation and early sporogenesis during sex organ development in *Arabidopsis thaliana*. Proc Natl Acad Sci USA 96:11664–11669

Schneitz K, Hülskamp M, Pruitt RE (1995) Wild-type ovule development in *Arabidopsis thaliana*: a light microscope study of cleared whole-mount tissue. Plant J 7:731–749

Schneitz K, Hülskamp M, Kopczak SD, Pruitt RE (1997) Dissection of sexual organ ontogenesis: a genetic analysis of ovule development in *Arabidopsis thaliana*. Development 124:1367–1376

Schneitz K, Baker SC, Gasser CS, Redweik A (1998) Pattern formation and growth during floral organogenesis: *HUELLENLOS* and *AINTEGUMENTA* are required for the formation of the proximal region of the ovule primordium in *Arabidopsis thaliana*. Development 125:2555–2563

Schruff MC, Spielman M, Tiwari S, Adams S, Fenby N, Scott RJ (2006) The *AUXIN RESPONSE FACTOR 2* gene of *Arabidopsis* links auxin signalling, cell division, and the size of seeds and other organs. Development 133:251–261

Sessions RA, Zambryski PC (1995) Arabidopsis gynoecium structure in the wild and in ettin mutants. Development 121:1519–1532

Sessions A, Nemhauser JL, McColl A, Roe JL, Feldmann KA, Zambryski PC (1997) *ETTIN* patterns the *Arabidopsis* floral meristem and reproductive organs. Development 124:4481–4491

Shen C, Bai Y, Wang S, Zhang S, Wu Y, Chen M, Jiang D, Qi Y (2010) Expression profile of PIN, AUX/LAX and PGP auxin transporter gene families in *Sorghum bicolor* under phytohormone and abiotic stress. FEBS J 277:2954–2969

Shi DQ, Yang WC (2011) Ovule development in Arabidopsis: progress and challenge. Curr Opin Plant Biol 14:74–80

Sieber P, Petrascheck M, Barberis A, Schneitz K (2004) Organ polarity in *Arabidopsis*. NOZZLE physically interacts with members of the YABBY protein family. Plant Physiol 135:2172–2185

Skinner DJ, Gasser CS (2009) Expression-based discovery of candidate ovule development regulators through transcriptional profiling of ovule mutants. BMC Plant Biol 9:29

Skinner DJ, Baker SC, Meister RJ, Broadhvest J, Schneitz K, Gasser CS (2001) The Arabidopsis *HUELLENLOS* gene, which is essential for normal ovule development, encodes a mitochondrial ribosomal protein. Plant Cell 13:2719–2730

Sohlberg JJ, Myrenas M, Kuusk S, Lagercrantz U, Kowalczyk M, Sandberg G, Sundberg E (2006) *STY1* regulates auxin homeostasis and affects apical-basal patterning of the Arabidopsis gynoecium. Plant J 47:112–123

Song S, Qi T, Huang H, Xie D (2013) Regulation of stamen development by coordinated actions of jasmonate, auxin and gibberellin in Arabidopsis. Mol Plant 6(4):1065–1073

Sprunck S, Gross-Hardt R (2011) Nuclear behavior, cell polarity, and cell specification in the female gametophyte. Sex Plant Reprod 24:123–136

Staldal V, Sohlberg JJ, Eklund DM, Ljung K, Sundberg E (2008) Auxin can act independently of CRC, LUG, SEU, SPT and STY1 in style development but not apical-basal patterning of the Arabidopsis gynoecium. New Phytol 180:798–808

Stepanova AN, Robertson-Hoyt J, Yun J, Benavente LM, Xie DY, Dolezal K, Schlereth A, Jürgens G, Alonso JM (2008) TAA1-mediated auxin biosynthesis is essential for hormone crosstalk and plant development. Cell 133:177–191

Sundaresan V, Alandete-Saez M (2010) Pattern formation in miniature: the female gametophyte of flowering plants. Development 137:179–189

Sundberg E, Ostergaard L (2009) Distinct and dynamic auxin activities during reproductive development. Cold Spring Harb Perspect Biol 1:a001628

Tabata R, Ikezaki M, Fujibe T, Aida M, Tian CE, Ueno Y, Yamamoto KT, Machida Y, Nakamura K, Ishiguro S (2010) Arabidopsis auxin response factor6 and 8 regulate jasmonic acid biosynthesis and floral organ development via repression of class 1 KNOX genes. Plant Cell Physiol 51:164–175

Tantikanjana T, Nasrallah JB (2012) Non-cell-autonomous regulation of crucifer self-incompatibility by auxin response factor ARF3. Proc Natl Acad Sci USA 109:19468–19473

Truernit E, Haseloff J (2008) *Arabidopsis thaliana* outer ovule integument morphogenesis: ectopic expression of *KNAT1* reveals a compensation mechanism. BMC Plant Biol 8:35

Tucker MR, Okada T, Johnson SD, Takaiwa F, Koltunow AM (2012) Sporophytic ovule tissues modulate the initiation and progression of apomixis in *Hieracium*. J Exp Bot 63:3229–3241

Ulmasov T, Murfett J, Hagen G, Guilfoyle TJ (1997) Aux/IAA proteins repress expression of reporter genes containing natural and highly active synthetic auxin response elements. Plant Cell 9:1963–1971

Wang Y, Deng D, Shi Y, Miao N, Bian Y, Yin Z (2012) Diversification, phylogeny and evolution of auxin response factor (ARF) family: insights gained from analyzing maize ARF genes. Mol Biol Rep 39:2401–2415

Webb MC, Gunning BES (1990) Embryo sac development in *Arabidopsis thaliana*. I. Megasporogenesis, including the microtubular cytoskeleton. Sex Plant Reprod 3:244–256

Willemse MTM, van Went JL (1984) The female gametophyte. In: Johri BM (ed) Embryology of angiosperms. Springer, New York, pp 159–196

Wu MF, Tian Q, Reed JW (2006) Arabidopsis microRNA167 controls patterns of *ARF6* and *ARF8* expression, and regulates both female and male reproduction. Development 133:4211–4218

Xing H, Pudake RN, Guo G, Xing G, Hu Z, Zhang Y, Sun Q, Ni Z (2011) Genome-wide identification and expression profiling of auxin response factor (ARF) gene family in maize. BMC Genomics 12:178

Xing S, Salinas M, Garcia-Molina A, Hohmann S, Berndtgen R, Huijser P (2013) SPL8 and miR156-targeted SPL genes redundantly regulate Arabidopsis gynoecium differential patterning. Plant J 75(4):566–577

Yamaguchi N, Wu MF, Winter CM, Berns MC, Nole-Wilson S, Yamaguchi A, Coupland G, Krizek BA, Wagner D (2013) A molecular framework for auxin-mediated initiation of flower primordia. Dev Cell 24:271–282

Yang W-C, Ye D, Xu J, Sundaresan V (1999) The *SPOROCYTELESS* gene of *Arabidopsis* is required for initiation of sporogenesis and encodes a novel nuclear protein. Genes Dev 13:2108–2117

Yasuor H, Abu-Abied M, Belausov E, Madmony A, Sadot E, Riov J, Rubin B (2006) Glyphosate-induced anther indehiscence in cotton is partially temperature dependent and involves cytoskeleton and secondary wall modifications and auxin accumulation. Plant Physiol 141:1306–1315

Ye Q, Zhu W, Li L, Zhang S, Yin Y, Ma H, Wang X (2010) Brassinosteroids control male fertility by regulating the expression of key genes involved in Arabidopsis anther and pollen development. Proc Natl Acad Sci USA 107:6100–6105

Zhao Y, Christensen SK, Fankhauser C, Cashman JR, Cohen JD, Weigel D, Chory J (2001) A role for flavin monooxygenase-like enzymes in auxin biosynthesis. Science 291:306–309

# Chapter 12
# Auxin and Its Henchmen: Hormonal Cross Talk in Root Growth and Development

**Antia Rodriguez-Villalon and Christian S. Hardtke**

**Abstract** Among the intrinsic growth regulators, auxin occupies a central role in plant growth and development. Auxin has been implicated in the regulation of many developmental processes at different scales, including cell division, cell differentiation, organogenesis, and morphogenesis. This is for instance evident in the sometimes similar and sometimes divergent phenotypes of auxin pathway mutants with regard to root system architecture, root meristem maintenance and lateral root organogenesis. Interestingly, other hormone pathways often similarly affect root system development, which has given rise to the notion of hormone pathway cross talk and/or synergy. Indeed, in recent years a few examples of mechanistically defined hormone pathway interactions have been identified. Most of these inputs appear to ultimately converge on the modulation of auxin levels, transport, or signaling and can explain aspects of context-specific auxin action. In this chapter, we will discuss examples of interaction between the pathways of auxin and other hormones, notably brassinosteroid, cytokinin, ethylene, and gibberellin, with a focus on root growth and development.

## 1 Introduction

A broad range of growth processes in root system development, ranging from initiation and organization of root apical meristems to initiation of lateral roots, have been shown to be highly dependent on auxin signaling and homeostasis (Benkova and Hejatko 2009; Bennett and Scheres 2010). This pivotal role of auxin is conserved across the root systems of monocotyledons and dicotyledons,

A. Rodriguez-Villalon • C.S. Hardtke (✉)
Department of Plant Molecular Biology, University of Lausanne, Biophore Building, 1015 Lausanne, Switzerland
e-mail: christian.hardtke@unil.ch

E. Zažímalová et al. (eds.), *Auxin and Its Role in Plant Development*,
DOI 10.1007/978-3-7091-1526-8_12, © Springer-Verlag Wien 2014

despite their variable architectures (Pacheco-Villalobos and Hardtke 2012; de Dorlodot et al. 2007; Hochholdinger and Zimmermann 2008; see Chap. 13). In both mono- and dicotyledons, post-embryonic root growth is driven by cell production of self-renewing apical meristems. This process has been best characterized in Arabidopsis, where the stem cells localized at the apex of the root meristem generate daughter cells that undergo several additional divisions in the so-called meristematic zone before rapid elongation and differentiation into their destined cell type in the so-called transition/elongation zone. For meristem maintenance, and therefore continuous root growth, the rate of cell differentiation must equal the rate of generation of new cells (Dello Ioio et al. 2007; Moubayidin et al. 2009; Scacchi et al. 2010). Moreover, stem cell differentiation must be suppressed, which requires the rarely dividing quiescent center cells at the heart of the stem cell niche. In addition to the embryogenic primary root, the Arabidopsis root system also comprises lateral roots, which are formed throughout the plant life cycle. Lateral roots are essentially similar to primary roots in overall morphology and are initiated from pericycle founder cells located adjacent to the protoxylem (Casimiro et al. 2001; Dubrovsky et al. 2008), which undergo several division and differentiation steps to form lateral root primordia (Malamy and Benfey 1997). The Arabidopsis root system displays all the essential features observed in other species, although the roots of the latter generally tend to be larger and more complex. For example, instead of the single cortex cell layer found in Arabidopsis, frequently 10–15 layers are formed, whereas the Arabidopsis quiescent center of four cells can comprise in the hundreds in other species. Variation is also observed at the level of the root systems, notably in monocotyledons, where it frequently comprises additional root types, such as shoot-borne adventitious roots (Smith and De Smet 2012; Pacheco-Villalobos and Hardtke 2012; Osmont et al. 2007). Finally, the variation extends to the arrangement and number of xylem and phloem tissue poles and their parenchyma within the root vasculature, which is surrounded by the pericycle.

The classic notion of auxin as a key hormonal regulator of root organogenesis has been confirmed through the identification of auxin signaling, transport, and biosynthesis mutants, which display impaired root growth and meristem organization. For instance, impaired polar auxin transport through loss of function in multiple redundant *PIN-FORMED (PIN)* genes, which encode auxin efflux carriers, severely affects root development (Blilou et al. 2005; see Chap. 5), eventually leading to total loss of embryonic root formation, thereby resembling corresponding mutants with severely reduced local auxin biosynthesis or signaling (Stepanova et al. 2008; Hardtke and Berleth 1998; see Chap. 2). Beyond the essential function of auxin, tissue-specific regulation of its biosynthesis, transport, and response in interaction with other hormone pathways is emerging as a key element in the specific roles of auxin during development. Here, we will discuss examples of such signal integration processes in the context of root system growth, with a focus on the Arabidopsis primary root, where these interactions have been characterized best. It is noteworthy however that conceptually similar interactions are frequently observed in other contexts, notably lateral root formation and vascular differentiation (Fig. 12.1). We will focus on those hormones for which clear evidence of

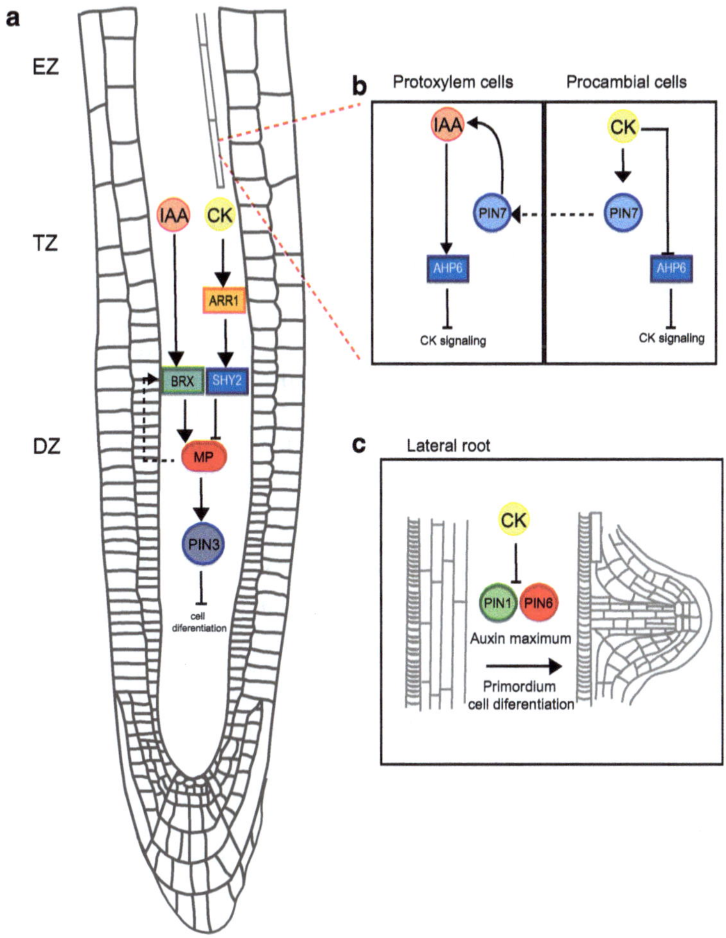


**Fig. 12.1** Molecular mechanisms of auxin and cytokinin interaction in the regulation of the Arabidopsis primary root meristem (**a**), vascular (**b**), and lateral root (**c**) development. (**a**) In the root meristem, a complex network of regulatory interactions balances cell division and differentiation/elongation. Auxin promotes cell division through both degradation of SHY2 and induction of *BRX* expression, which converge on the antagonistica regulation of *PIN3* expression. (**b**) Within the root meristem vasculature, procambial cells display high levels of cytokinin, which promotes the expression of *PIN7*, thereby forcing auxin transport out of the cells and its accumulation in protoxylem cells. High auxin signaling at this position in turn promotes expression of the negative cytokinin regulator *AHP6*, thereby promoting protoxylem specification. (**c**) Early in lateral root development, creation of a local auxin concentration maximum is essential for lateral root primordium formation, whereas cytokinin negatively regulates lateral root initiation by inhibiting the expression of PIN proteins

interaction with the auxin pathway exists, i.e., cytokinin, brassinosteroids, gibberellins, and ethylene.

# 2 Auxin–Cytokinin Cross talk

The maybe most important interaction of the auxin pathway in root development occurs with the cytokinin pathway. Cytokinins are adenine derivatives whose levels are maintained by a complex equilibrium between their synthesis and catabolism (Sakakibara 2006). In Arabidopsis, cytokinins are perceived by three transmembrane histidine kinase receptors, ARABIDOPSIS HISTIDINE KINASE (AHK) 2, AHK3, and WOODEN LEG (WOL; a.k.a. AHK4 or CYTOKININ RECEPTOR 1) (Riefler et al. 2006; Higuchi et al. 2004; Nishimura et al. 2004; Yamada et al. 2001; Inoue et al. 2001). Sensing of cytokinin by their extracellular domain initiates receptor autophosphorylation in the cytosolic kinase domain, followed by transfer of the phosphate to cytosolic Arabidopsis histidine phosphotransfer proteins (AHPs), which can cycle between the cytosol and the nucleus. In the nucleus, AHPs transfer their phosphates to type-B ARABIDOPSIS RESPONSE REGULATORs (B-ARRs), which thus initiate transcription of primary cytokinin-responsive genes (Sakai et al. 2001; Hwang et al. 2001). Among them, the type-A ARRs (A-ARRs), which are stabilized by AHP-mediated phosphorylation (To et al. 2007), mediate negative feedback to fine-tune cytokinin signaling amplitude by repressing the activity of B-ARRs (Hwang and Sheen 2001) (Fig. 12.2).

## 2.1 *Cytokinin Controls Different Aspects of Root System Development*

A role of cytokinin in antagonizing auxin action has initially been realized in classic callus regeneration experiments, in which a high auxin to cytokinin ratio induces root organogenesis. This is also reflected in the finding that contrary to auxin, cytokinin is a negative regulator of lateral root formation; i.e., reduced cytokinin levels result in higher density of lateral roots (Werner et al. 2003; Riefler et al. 2006; Mason et al. 2005; Hutchison et al. 2006), whilst exogenous cytokinin treatment inhibits lateral root formation (Kuderova et al. 2008; Laplaze et al. 2007; Li et al. 2006). A fine-tuned balance in the activity of both hormones is also required to sustain root meristem growth, where simply spoken auxin keeps cells dividing whereas cytokinin promotes cell differentiation (Moubayidin et al. 2009; Bishopp et al. 2011). This antagonism between the two hormones plays out by complex interactions on different levels, which can trigger different downstream responses based on tissue specificity and developmental context.

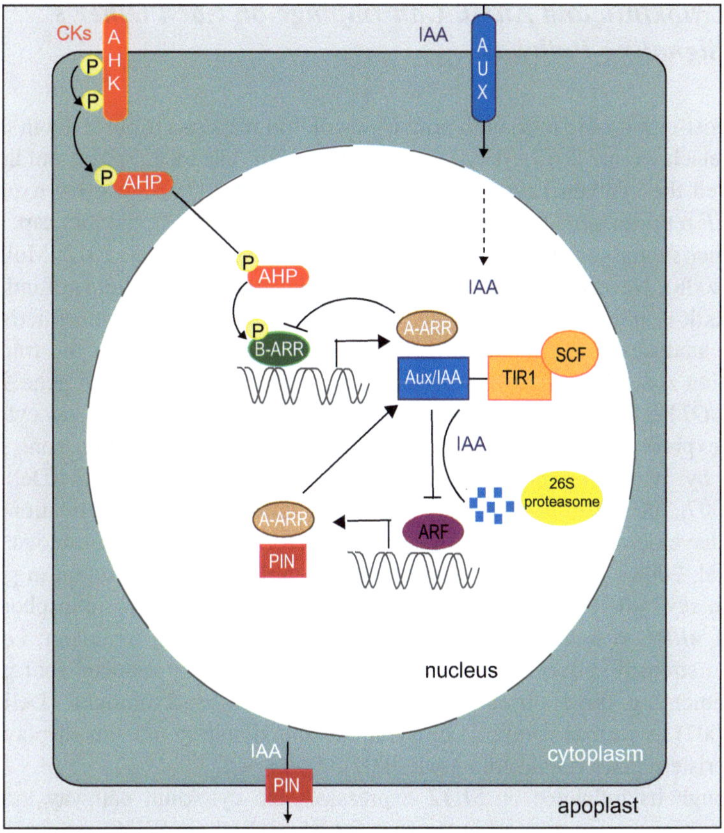


**Fig. 12.2** Schematic representation of the auxin (IAA) and cytokinin (CK) signaling pathways with an emphasis on the nodes where cross talk occurs. CKs are perceived by the AHK receptors, thus initiating receptor autophosphorylation, followed by transfer of the phosphate group to a cytosolic AHP. Translocation of AHP to the nucleus triggers phosphorylation of B-ARRs, promoting the transcription of primary CK-responsive genes, including the negative feedback regulators, the A-ARRs genes. Phosphorylated A-ARRs activate negative regulation of CK signaling through a yet unknown mechanism. Auxin is taken up from the apoplast by passive diffusion as well as the action of influx transporters [(AUXIN-RESISTANT MUTATION 1/LIKE (AUX1 AUX/LAX)] and actively transported out of the cell by the efflux carriers, the PIN proteins. In the nucleus, high auxin concentration promotes interaction of AUX/IAA proteins, which inhibit ARF transcription factors, with SCF-type E3 ubiquitin ligase complexes that contain auxin receptor F-box proteins, such as TRANSPORT INHIBITOR RESPONSE 1 (TIR1). This interaction leads to AUX/IAA degradation and thus releases ARF transcription activation to induce expression of auxin-responsive genes. B-ARRs can activate transcription of certain *AUX/IAA* genes and thereby affect ARF-controlled *PIN* expression, whereas auxin signaling suppresses CK response by the activation of *A-ARR*s

## 2.2 *Cytokinin and Auxin Can Impinge on Each Other's Signaling Pathway*

In the post-embryonic root, high primary cytokinin response is observed in the root tip (Bielach et al. 2012; Aloni et al. 2004). For instance, recent studies that employed the synthetic cytokinin-responsive promoter, *TCS*, to drive expression of a *GFP* reporter gene have shown high cytokinin response in the root cap, but not in the meristematic or transition/elongation zones (Bielach et al. 2012; Muller and Sheen 2008). Nevertheless, due to the differential expression patterns of individual A-ARRs it remains unclear in which cells cytokinin response is most active, and genetic analyses suggest indeed a role for cytokinin signaling in the transition/elongation zone. This has been revealed by analysis of the *AUX/IAA* gene *SHORT HYPOCOTYL 2* (*SHY2*), a mediator of cross talk between auxin and cytokinin, whose expression in the vascular tissue of the transition/elongation zone is controlled by two cytokinin-inducible A-ARRs, *ARR1* and *ARR12* (Dello Ioio et al. 2007, 2008; Moubayidin et al. 2009). Chromatin immunoprecipitation experiments have demonstrated that ARR1 binds directly to the *SHY2* promoter (Dello Ioio et al. 2008). Interestingly, expression of stabilized SHY2 protein in gain-of-function *shy2* alleles or overexpression of a constitutively active (phosphomimic) form of *ARR1* in transgenic lines both phenocopy cytokinin treatment; i.e., they result in strongly reduced meristem size and consequently reduced root growth. Complementing these observations, loss-of-function *shy2* mutants (Dello Ioio et al. 2007) as well as cytokinin biosynthetic and signaling mutants display larger root meristem sizes (Dello Ioio et al. 2007).

Through its influence on *SHY2* expression, the cytokinin pathway indirectly controls polar auxin transport in the root meristem, because SHY2 activity inhibits the expression of *PIN* genes that are under control of auxin response factors (ARFs) (Dello Ioio et al. 2008) (Fig. 12.2). Importantly, this cross talk occurs in a distinct spatiotemporal setting to control root meristem growth and thereby mature meristem size in the early seedling (Fig. 12.1). The *BREVIS RADIX* (*BRX*) gene, a positive regulator of auxin signaling and an antagonist of *SHY2*, is essential for the meristem growth process (Mouchel et al. 2006; Scacchi et al. 2009). BRX is thought to potentiate the activity of the ARF MONOPTEROS (MP), thereby transiently boosting meristematic *PIN* expression and thus polar auxin transport (Scacchi et al. 2010). High polar auxin transport promotes cell proliferation over differentiation and thereby meristem growth. Meristematic *SHY2* expression depends on transcriptional feedback through combined BRX-ARF activity and therefore also on the differential plasma membrane to nucleus transfer of BRX (Scacchi et al. 2009), which is a function of increasing endocytosis in the developing protophloem toward the transition/elongation zone (Santuari et al. 2011). Thus, cell proliferation only ceases once nuclear BRX activity increases in the transition/elongation zone. Eventually, meristem growth is brought to a halt as cytokinin activity in the transition/elongation zone boosts *SHY2* expression independent of BRX-ARF activity, thereby enabling SHY2 to take over the autoregulatory network and locally dampen polar auxin transport to promote cell differentiation. Subsequently, these interactions between the different components result in a dynamic equilibrium that spatially separates and synchronizes cell proliferation and differentiation.

## 2.3 *Homeostasis Feedback Loops Regulate Auxin and Cytokinin Levels*

Interestingly, auxin and cytokinin do not only interact at the level of signaling components and gene expression, but also impinge on each other's biosynthesis. For instance, direct hormone measurements have shown that auxin rapidly downregulates cytokinin biosynthesis (Nordstrom et al. 2004). This result is somewhat counterintuitive given that auxin appears to promote cytokinin biosynthesis via specific activation of genes encoding the rate-limiting step in cytokinin biosynthesis, *ISOPENTENYL TRANSFERASE (IPT) 5* and *IPT7* (Miyawaki et al. 2004; Nordstrom et al. 2004). However, it could reflect the spatiotemporal fluctuation in the interactions of the two hormones, because it has been shown that cytokinin can reciprocally regulate auxin biosynthesis, suggesting the existence of an active homeostatic feedback loop. For example, in experiments combining inducible cytokinin overproduction with stable isotope labeling, an elevation in cytokinin levels led to a rapid increase in auxin biosynthesis rates (Jones et al. 2010). Conversely, cytokinin reduction by induction of catabolic enzymes resulted in lower levels of auxin biosynthesis. The feedbacks between both hormone pathways require their intact signaling pathways. Thus, the specificity of this cross-regulation is likely highly complex and plays out to different levels depending on the root tissue (Jones et al. 2010). This complexity is underlined by interactions with additional pathways. For instance, in early root meristem growth, promotion of *SHY2* expression by cytokinin is indirectly repressed through the gibberellin pathway (see below). This is because high levels of gibberellin signaling in very young meristems repress expression of *ARR1*, which is needed in combination with *ARR12* to fully induce *SHY2* expression in response to cytokinin once gibberellin activity goes down (Moubayidin et al. 2009). Finally, the interactions also involve posttranslational regulation, since cytokinin can for instance promote the degradation of PIN proteins (Marhavy et al. 2011).

In summary, tight control and balance of the antagonistic activities of auxin and cytokinin are particularly important during early phases of primary root growth as well as lateral root organogenesis. A general theme emerging in all contexts is that auxin signaling inhibits the activity of negative regulators of cytokinin signaling, whereas cytokinin signaling in turn promotes the expression of inhibitors of auxin signaling. The two hormones also reciprocally regulate each other's biosynthesis and cytokinin also directs auxin transport, with the generic effect that cytokinin activity can establish zones of high auxin signaling.

# 3 Auxin and Brassinosteroid Cross talk

Another important hormone pathway in root development is the brassinosteroid pathway, which has been shown to modulate auxin activity in various processes, cooperatively in some cases and antagonistically in others (Hardtke et al. 2007). The close relationship between auxin and brassinosteroid effects likely reflects several levels of cross-regulation, which could occur through the regulation of common target genes by auxin- and brassinosteroid-controlled transcription factors. Moreover, brassinosteroid and auxin effects also converge at the level of hormone transport and biosynthesis (Li et al. 2005; Mouchel et al. 2006). Compared with auxin, the brassinosteroid signaling pathway is complex and represents a more classical signaling paradigm. Active brassinosteroids, i.e., in Arabidopsis mainly brassinolide and castasterone, are perceived at the cell surface by direct binding to the extracellular domain of their leucine-rich repeat serine/threonine kinase receptors, which triggers a signal transduction cascade that via several intermediates ultimately results in dephosphorylation of the BRASSINAZOLE RESISTANT 1 (BZR1) and bri1-EMS-SUPPRESSOR 1 (BES1) transcription factors. This event promotes their accumulation in the nucleus and their DNA-binding activity, resulting in gene activation (Vert and Chory 2006). Although BES1 and BZR1 are considered to be redundant transcriptional regulators, their respective mutant phenotypes suggest some level of sub-functionalization (Sun et al. 2010; Yu et al. 2011).

## *3.1 The Role of Brassinosteroids in Root Development*

Pharmacological, genetic, and transcriptomic evidence, for instance, the short-root phenotype of brassinosteroid pathway mutants (Mouchel et al. 2004; Szekeres et al. 1996; Gonzalez-Garcia et al. 2011), suggests an important role of brassinosteroid activity in root growth. Growth stimulation by brassinosteroids has been mainly related to cell elongation, in line with transcriptional and ChIP-chip data for BZR1 and BES1, which suggest that their target genes are mainly involved in cell growth and cell wall organization (Sun et al. 2010; Yu et al. 2011). However, brassinosteroids are also essential to sustain root meristem activity (Mouchel et al. 2006; Gonzalez-Garcia et al. 2011; Gujas et al. 2012), because the brassinosteroid receptor mutants and the gain-of-function *bes1-D* mutant display reduced meristematic cell number and thus reduced root meristem size. Thus, brassinosteroids are needed for cell proliferation in the root meristem, possibly through controlling the cell cycle (Gonzalez-Garcia et al. 2011; Hacham et al. 2011; Hu et al. 2000). Interestingly, cell cycle genes were identified as target genes of BZR1/BES1, although their expression was neither affected in brassinosteroid mutants nor by brassinosteroid treatment (Sun et al. 2010; Yu et al. 2011). It appears possible however that in experiments monitoring expression at the organ level this effect could be masked because it is very localized, and indeed recent data suggest that in the root meristem, brassinosteroids do not generically drive cell proliferation, but rather specifically promote the division of stem cell daughter cells (Gujas et al. 2012).

## 3.2 Auxin–Brassinosteroid Interactions in Modulating Gene Expression

Several transcriptomic studies suggest that auxin and brassinosteroids converge on shared target genes, many of which are synergistically induced when auxin and brassinosteroid are applied simultaneously (Goda et al. 2004; Nemhauser et al. 2006; Mouchel et al. 2006; Vert et al. 2008; Chung et al. 2011). Moreover, expression changes in response to one hormone require an intact biosynthetic and signaling pathway of the other (Chung et al. 2011; Hardtke et al. 2007; Nakamura et al. 2006; Nemhauser et al. 2006). The target genes typically react strongly to auxin and weakly to brassinosteroid, which could reflect differences in the signal transduction kinetics or indirect effects, such as a rate-limiting role of brassinosteroids in auxin action (Mouchel et al. 2006). However, although no direct interactions between ARFs and BES1/BZR1 have been demonstrated to date, a recent study suggests that auxin and brassinosteroid response are mediated by a combination of respective cis-regulatory elements (Walcher and Nemhauser 2012). For example, a Hormone Up at Dawn (HUD)-type E-box in combination with a nearby auxin-responsive element variant has been identified as a target for BES1 and MP, respectively. Moreover, their binding can be enhanced by treatment with either auxin or brassinosteroids (Walcher and Nemhauser 2012; Chandler et al. 2009).

It has also been suggested that brassinosteroids could promote polar auxin transport (Bao et al. 2004; Li et al. 2005), although it remains unclear whether this effect is direct and/or at the transcriptional or posttranslational level. The slow kinetics suggests that at least in part it could be indirect, because brassinosteroid can enhance physiological responses in auxin-saturated background (Nemhauser et al. 2004) and because polar auxin transport is under feedback control through auxin signaling. Consistent with the idea that brassinosteroid limits auxin action at the signaling rather than the biosynthesis level (Mouchel et al. 2006; Nemhauser et al. 2004), auxin biosynthesis does not change dramatically in response to brassinosteroid treatment (Kim et al. 2007; Nakaya et al. 2002; Bao et al. 2004) and genes coding for rate-limiting enzymes in the major developmentally regulated auxin biosynthesis pathway are not among the BZR1/BES1 targets (Sun et al. 2010). However, feedback of the auxin pathway on brassinosteroid homeostasis has been reported. For instance, *BRX* as well as ARFs are likely involved in promoting the expression of rate-limiting enzymes in the brassinosteroid biosynthesis pathway (Yoshimitsu et al. 2011; Mouchel et al. 2006; Chung et al. 2011).

Thus, auxin could promote brassinosteroid biosynthesis to maintain brassinosteroid levels that enable a saturated auxin response. Consistent with this idea, genes encoding rate-limiting enzymes in brassinosteroid biosynthesis are auxin-inducible in suboptimal auxin conditions (Yamamoto et al. 2007). In summary, the data suggest that brassinosteroids limit auxin response and that feedback control by the auxin pathway contributes to the homeostasis of brassinosteroid levels, which might be the primary cause of reduced root growth of brassinosteroid pathway mutants.

## 4 Auxin and Gibberellin Cross talk

Gibberellins are a family of diterpenoid plant hormones that regulate various developmental processes throughout the plant life cycle, from seed germination through leaf expansion, stem elongation, flower induction, and transition from skotomorphogenesis to photomorphogenesis (Weiss and Ori 2007; Teale et al. 2008). Both auxin and gibberellins regulate cell elongation and differentiation through mechanisms that, for a long time, were thought to be independent. However, evidence accumulated over the last years has revealed interaction of at least two types: (1) auxin affects gibberellin biosynthesis and (2) auxin controls the stability of downstream gibberellin signaling components, the DELLA factors, thereby modulating gibberellin response (Fu and Harberd 2003; Ubeda-Tomas et al. 2008; Desgagne-Penix and Sponsel 2008).

### *4.1 The Gibberellin Signaling Pathway*

The DELLA proteins are the most extensively characterized downstream components in the gibberellin signaling pathway. Five largely redundant but also context-specific *DELLA* genes have been identified in Arabidopsis, including the prototypical *REPRESSOR OF ga1-3* (*RGA*) (Peng et al. 1997, 1999; Silverstone et al. 1998). Perception of gibberellin by its predominantly nuclear localized receptors (Griffiths et al. 2007; Willige et al. 2007) triggers their conformational change and promotes their interaction with the DELLA transcriptional co-repressors (Peng et al. 1997; Griffiths et al. 2007). This in turn stabilizes interaction of DELLAs with a specific SCF E3 ubiquitin ligase (Dill et al. 2004; McGinnis et al. 2003), thus resulting in their poly-ubiquitination and degradation by the 26S proteasome (Dill et al. 2004; Griffiths et al. 2007). Beyond this relatively simple gibberellin signaling cascade, other studies have identified additional factors that affect gibberellin response (Feng et al. 2008; de Lucas et al. 2008; Bai et al. 2012). For example, BZR1 can interact with DELLA proteins as well as with the PHYTOCHROME-INTERACTING FACTOR 4 (PIF4) (Bai et al. 2012; Oh et al. 2012) to impinge on a common transcriptome.

## *4.2 Role of Gibberellin in Root Development*

Gibberellins have long been recognized for their pivotal role in shoot elongation; however they also control root growth in conjunction with auxin. Various gibberellin-deficient mutants display a short root phenotype (Fu and Harberd 2003; Ubeda-Tomas et al. 2009), at least in part because of suspended root meristem growth. A set of elegant experiments (in which expression of a dominant negative, because stabilized and gibberellin-insensitive form of a DELLA protein was targeted to select root tissues) has demonstrated that gibberellin is required in the endodermis to reach and maintain full root meristem size after germination (Ubeda-Tomas et al. 2008, 2009). However, since DELLA proteins are expressed throughout the meristem, the biological significance of this observation remains unclear. In fact, local, cell layer-specific inhibition of other hormone signaling pathways similarly affects cell proliferation in the root meristem in a non-cell-autonomous manner, which could mean that asynchronous cellular growth processes in one layer generically limit those processes in neighboring layers, for instance, through mechanically triggered signals.

As pointed out above, in the early meristem gibberellin represses *ARR1* expression, which is likely conferred by destabilization of the activator of *ARR1* expression, RGA. Thus, the root meristem phenotype of gibberellin mutants could simply reflect premature *SHY2* hyperactivity. Consistently, gibberellin promotes root elongation, through increasing meristematic cell proliferation as well as cell elongation, but only in older seedlings, i.e., by prolonging meristem growth (Moubayidin et al. 2010). This requires shoot-derived auxin, because gibberellin-induced root elongation is inhibited when the shoot apex (the main auxin source at the seedling stage) is removed and this effect can be reversed by exogenous auxin treatment (Fu and Harberd 2003). Complicating this scenario, an intact auxin signaling pathway is needed for gibberellin-induced RGA degradation (Fu and Harberd 2003). Thus, RGA seems to be a key point in the cross talk between auxin and gibberellins. Moreover, a recent study established that gibberellins are required for proper polar auxin transport in Arabidopsis, because gibberellin-deficient mutants display reduced PIN protein levels, implying that gibberellin feeds back on auxin response by regulating PIN turnover (Willige et al. 2011). The solution to the first part of this interdependency is the observation that auxin might positively regulate gibberellin biosynthesis (Ross et al. 2000; Wolbang and Ross 2001; Frigerio et al. 2006). However, auxin treatment also induces the expression of genes that encode gibberellin catabolic enzymes (Frigerio et al. 2006). Therefore, it remains difficult to ultimately decide the final effect of auxin on gibberellin metabolism, although the most parsimonious explanation would be that auxin is needed for RGA destabilization in the meristem because it promotes local gibberellin biosynthesis.

## 5 Auxin and Ethylene Cross talk

The final important hormone interactor for auxin is ethylene, a small volatile compound that regulates many aspects of plant development in response to environmental stimuli and has been identified as a general modulator of root development (Ruzicka et al. 2007; Stepanova et al. 2005; Swarup et al. 2002). Indeed, various effects of ethylene can be explained by cross talk with the auxin pathway at different levels. Physiological and genetic characterization of ethylene mutants has revealed a linear signaling pathway, which starts by ethylene binding to its receptors, thereby suppressing their activity. In Arabidopsis, five ethylene receptors have been described (Hua and Meyerowitz 1998) that may act cooperatively rather than independently (Wang et al. 2003; Klee 2004; O'malley et al. 2005; Liu et al. 2010). Their signaling affects the activity of a protein kinase (Huang et al. 2003; Clark et al. 1998), which negatively regulates the pathway. Thus, in the absence of ethylene, the receptors, such as ETHYLENE RESPONSE 1 (ETR1), keep the downstream signaling components ETHYLENE-INSENSITIVE 2 (EIN2) and EIN3 inactive (Chao et al. 1997; Alonso et al. 1999). Ethylene binding causes inactivation of the receptor-kinase complex and accumulation of EIN3 and EIN3-like (EIL) transcription factors in the nucleus (Guo and Ecker 2003). Their activity in turn depends on simultaneous nucleus translocation of the EIN2 C-terminus, after ethylene-triggered dephosphorylation and proteolytic cleavage of EIN2 in the endoplasmic reticulum (Qiao et al. 2012; Ju et al. 2012; Wen et al. 2012). Together, the EIN2 C-terminus, EIN3, and EIL1 are needed to activate the expression of target genes such as *ETHYLENE RESPONSE FACTOR 1 (ERF1)* (Solano et al. 1998), thereby initiating a transcriptional cascade that results in the activation or repression of hundreds of genes (Alonso et al. 1999; Olmedo et al. 2006).

### *5.1 The Role of Ethylene in Root Development*

The best studied effect of the application of ethylene, or the rate-limiting ethylene precursor 1-aminocyclopropane carboxylic acid (ACC), on primary root development is the inhibition of root elongation (Markakis et al. 2012; Rahman et al. 2001; Swarup et al. 2002). This mainly reflects the fact that ethylene reduces the expansion rate of cells in the central elongation zone of the primary root (Swarup et al. 2007; Rahman et al. 2007), which likely reflects an increase in auxin levels to a degree that is eventually inhibitory for cell elongation (see below). A detailed examination of auxin-inducible reporter genes has shown that ethylene treatment increases auxin response throughout the tissues of the transition/elongation zone (Ruzicka et al. 2007; Stepanova et al. 2007; Negi et al. 2008). While a functional auxin signaling network is required for this response (Stepanova et al. 2007; Swarup et al. 2007), it is not clear whether conversely a functional ethylene response is needed for root growth inhibition by auxin application (Ruzicka et al. 2007). Transcriptomic analyses of ACC-treated roots have shown that ethylene inhibition of cell elongation is concomitant with the induction of several auxin-dependent genes (Markakis et al. 2012). However, studies employing mutants in the auxin and ethylene receptors suggest that the primary transcriptional response of the two hormones might be regulated through independent signaling pathways, and secondary cross talk could occur through the expression of genes that are either auxin- or ethylene-responsive (Lewis et al. 2011a, b).

## 5.2 *Ethylene Enhances Auxin Biosynthesis and Vice Versa to Modify Root System Architecture*

The isolation of mutants in polar auxin transport components in screens for reduced ethylene-mediated growth inhibition was a first evidence that ethylene could modulate auxin transport (Pickett et al. 1990; Luschnig et al. 1998). However, whether ethylene effects on polar auxin transport are direct or indirect remains unclear. Again, they could be obscured by the observed reciprocal interaction in the regulation of auxin and ethylene biosynthesis. For instance, ACC synthase catalyzes the rate-limiting step in the ethylene biosynthetic pathway (Wang et al. 2002), and auxin stimulates the transcription of genes encoding this enzyme, thereby enhancing ethylene production (Abel et al. 1995; Stepanova et al. 2007). Eight out of the nine Arabidopsis *ACC SYNTHASE* genes are thus upregulated by auxin, and consistently several of them contain canonical auxin response elements in their promoter, suggesting that the effect is direct (Tsuchisaka and Theologis 2004). Relatively more important however, ethylene was found to promote auxin biosynthesis in Arabidopsis roots (Ruzicka et al. 2007; Swarup et al. 2007). Evidence for ethylene-regulated auxin biosynthesis initially emerged from a screen for weak ethylene-insensitive mutants, surprisingly leading to the identification of several enzymes involved in a developmentally regulated, rather short auxin biosynthesis pathway (Stepanova et al. 2005, 2008). These include the alpha and beta subunits of anthranilate synthase, which catalyzes the first steps of tryptophan biosynthesis, and tryptophan aminotransferases, which catalyze the intermediate step between tryptophan and auxin (Stepanova et al. 2005, 2008). The latter, such as TRYPTOPHAN AMINOTRANSFERASE OF ARRABIDOPSIS 1 (TAA1), maintain proper auxin levels in the root. Loss-of-function mutations in *TAA1* result in root-specific ethylene insensitivity, which can be restored when roots are treated with low levels of exogenous auxin. The ethylene defects of *taa1* mutants are dramatically enhanced in the *taa1 taa1-related 2 (tar2)* double mutants, which display a complete lack of response to ACC in roots and low auxin levels, uncovering functional redundancy between *TAA1* and *TAR2* (Stepanova et al. 2008). Since ACC-dependent upregulation of auxin response is also strongly impaired in *taa1 tar2* roots, ethylene appears to act mainly through regulating auxin biosynthesis in Arabidopsis roots. Indeed, expression of *TAA1* and *TAR* genes, as well as of the

*YUCCA* genes, which encode the second, rate-limiting step in this auxin biosynthesis pathway, is ethylene-responsive (Stepanova et al. 2008; Liang et al. 2012; see Chap. 2), suggesting that upregulation of auxin biosynthesis is the key point in auxin–ethylene cross talk.

# 6 Concluding Remarks

In summary, multiple hormones impinge on the activity of auxin, as best characterized and illustrated in root growth and development. While some of these interactions might be rather indirect and difficult to dissect, for instance, altered polar auxin transport in response to another hormone, examples for direct interactions exist. The latter include evidence for cross-regulation of hormone biosynthesis pathways, but also for a few interactions at the signaling level, mostly converging at the level of transcriptional response. What is lacking for most of these interactions is their detailed characterization with respect to the spatiotemporal dimension and posttranslational regulation. Many of the experiments that have uncovered cross talk between auxin and other hormones one way or another are somewhat artificial, e.g., they report marker responses after flooding plants of various stages with hormones. They do not necessarily tell us if and when the observed regulations are biologically relevant during development. The example of root meristem growth and the sequential interactions between the auxin, cytokinin, and gibberellin pathways highlights this problematic. Thus, in the next step, future studies should aim to verify and quantify the identified, possible regulatory interactions by live imaging of markers at high spatiotemporal resolution.

# References


Abel S, Nguyen MD, Chow W, Theologis A (1995) ACS4, a primary indoleacetic acid-responsive gene encoding 1-aminocyclopropane-1-carboxylate synthase in Arabidopsis thaliana. Structural characterization, expression in Escherichia coli, and expression characteristics in response to auxin [corrected]. J Biol Chem 270:19093–19099

Aloni R, Langhans M, Aloni E, Ullrich CI (2004) Role of cytokinin in the regulation of root gravitropism. Planta 220:177–182

Alonso JM, Hirayama T, Roman G, Nourizadeh S, Ecker JR (1999) EIN2, a bifunctional transducer of ethylene and stress responses in Arabidopsis. Science 284:2148–2152

Bai MY, Shang JX, Oh E, Fan M, Bai Y, Zentella R, Sun TP, Wang ZY (2012) Brassinosteroid, gibberellin and phytochrome impinge on a common transcription module in Arabidopsis. Nat Cell Biol 14:810–817

Bao F, Shen J, Brady SR, Muday GK, Asami T, Yang Z (2004) Brassinosteroids interact with auxin to promote lateral root development in Arabidopsis. Plant Physiol 134:1624–1631

Benkova E, Hejatko J (2009) Hormone interactions at the root apical meristem. Plant Mol Biol 69:383–396

Bennett T, Scheres B (2010) Root development-two meristems for the price of one? Curr Top Dev Biol 91:67–102
Bielach A, Podlesakova K, Marhavy P, Duclercq J, Cuesta C, Muller B, Grunewald W, Tarkowski P, Benkova E (2012) Spatiotemporal regulation of lateral root organogenesis in Arabidopsis by cytokinin. Plant Cell 24:3967–3981
Bishopp A, Benkova E, Helariutta Y (2011) Sending mixed messages: auxin-cytokinin crosstalk in roots. Curr Opin Plant Biol 14:10–16
Blilou I, Xu J, Wildwater M, Willemsen V, Paponov I, Friml J, Heidstra R, Aida M, Palme K, Scheres B (2005) The PIN auxin efflux facilitator network controls growth and patterning in Arabidopsis roots. Nature 433:39–44
Casimiro I, Marchant A, Bhalerao RP, Beeckman T, Dhooge S, Swarup R, Graham N, Inze D, Sandberg G, Casero PJ, Bennett M (2001) Auxin transport promotes Arabidopsis lateral root initiation. Plant Cell 13:843–852
Chandler JW, Cole M, Flier A, Werr W (2009) BIM1, a bHLH protein involved in brassinosteroid signalling, controls Arabidopsis embryonic patterning via interaction with DORNROSCHEN and DORNROSCHEN-LIKE. Plant Mol Biol 69:57–68
Chao Q, Rothenberg M, Solano R, Roman G, Terzaghi W, Ecker JR (1997) Activation of the ethylene gas response pathway in Arabidopsis by the nuclear protein ETHYLENE-INSENSITIVE3 and related proteins. Cell 89:1133–1144
Chung Y, Maharjan PM, Lee O, Fujioka S, Jang S, Kim B, Takatsuto S, Tsujimoto M, Kim H, Cho S, Park T, Cho H, Hwang I, Choe S (2011) Auxin stimulates DWARF4 expression and brassinosteroid biosynthesis in Arabidopsis. Plant J 66:564–578
Clark KL, Larsen PB, Wang X, Chang C (1998) Association of the Arabidopsis CTR1 Raf-like kinase with the ETR1 and ERS ethylene receptors. Proc Natl Acad Sci U S A 95:5401–5406
De Dorlodot S, Forster B, Pages L, Price A, Tuberosa R, Draye X (2007) Root system architecture: opportunities and constraints for genetic improvement of crops. Trends Plant Sci 12:474–481
De Lucas M, Daviere JM, Rodriguez-Falcon M, Pontin M, Iglesias-Pedraz JM, Lorrain S, Fankhauser C, Blazquez MA, Titarenko E, Prat S (2008) A molecular framework for light and gibberellin control of cell elongation. Nature 451:480–484
Dello Ioio R, Linhares FS, Scacchi E, Casamitjana-Martinez E, Heidstra R, Costantino P, Sabatini S (2007) Cytokinins determine Arabidopsis root-meristem size by controlling cell differentiation. Curr Biol 17:678–682
Dello Ioio R, Nakamura K, Moubayidin L, Perilli S, Taniguchi M, Morita MT, Aoyama T, Costantino P, Sabatini S (2008) A genetic framework for the control of cell division and differentiation in the root meristem. Science 322:1380–1384
Desgagne-Penix I, Sponsel VM (2008) Expression of gibberellin 20-oxidase1 (AtGA20ox1) in Arabidopsis seedlings with altered auxin status is regulated at multiple levels. J Exp Bot 59:2057–2070
Dill A, Thomas SG, Hu J, Steber CM, Sun TP (2004) The Arabidopsis F-box protein SLEEPY1 targets gibberellin signaling repressors for gibberellin-induced degradation. Plant Cell 16:1392–1405
Dubrovsky JG, Sauer M, Napsucialy-Mendivil S, Ivanchenko MG, Friml J, Shishkova S, Celenza J, Benkova E (2008) Auxin acts as a local morphogenetic trigger to specify lateral root founder cells. Proc Natl Acad Sci U S A 105:8790–8794
Feng S, Martinez C, Gusmaroli G, Wang Y, Zhou J, Wang F, Chen L, Yu L, Iglesias-Pedraz JM, Kircher S, Schafer E, Fu X, Fan LM, Deng XW (2008) Coordinated regulation of Arabidopsis thaliana development by light and gibberellins. Nature 451:475–479
Frigerio M, Alabadi D, Perez-Gomez J, Garcia-Carcel L, Phillips AL, Hedden P, Blazquez MA (2006) Transcriptional regulation of gibberellin metabolism genes by auxin signaling in Arabidopsis. Plant Physiol 142:553–563
Fu X, Harberd NP (2003) Auxin promotes Arabidopsis root growth by modulating gibberellin response. Nature 421:740–743

Goda H, Sawa S, Asami T, Fujioka S, Shimada Y, Yoshida S (2004) Comprehensive comparison of auxin-regulated and brassinosteroid-regulated genes in Arabidopsis. Plant Physiol 134:1555–1573
Gonzalez-Garcia MP, Vilarrasa-Blasi J, Zhiponova M, Divol F, Mora-Garcia S, Russinova E, Cano-Delgado AI (2011) Brassinosteroids control meristem size by promoting cell cycle progression in Arabidopsis roots. Development 138:849–859
Griffiths J, Murase K, Rieu I, Zentella R, Zhang ZL, Powers SJ, Gong F, Phillips AL, Hedden P, Sun TP, Thomas SG (2007) Genetic characterization and functional analysis of the GID1 gibberellin receptors in Arabidopsis (vol. 18, pg 3399, 2006). Plant Cell 19:726
Gujas B, Alonso-Blanco C, Hardtke CS (2012) Natural Arabidopsis brx loss-of-function alleles confer root adaptation to acidic soil. Curr Biol 22:1962–1968
Guo H, Ecker JR (2003) Plant responses to ethylene gas are mediated by SCF(EBF1/EBF2)-dependent proteolysis of EIN3 transcription factor. Cell 115:667–677
Hacham Y, Holland N, Butterfield C, Ubeda-Tomas S, Bennett MJ, Chory J, Savaldi-Goldstein S (2011) Brassinosteroid perception in the epidermis controls root meristem size. Development 138:839–848
Hardtke CS, Berleth T (1998) The Arabidopsis gene MONOPTEROS encodes a transcription factor mediating embryo axis formation and vascular development. EMBO J 17:1405–1411
Hardtke CS, Dorcey E, Osmont KS, Sibout R (2007) Phytohormone collaboration: zooming in on auxin-brassinosteroid interactions. Trends Cell Biol 17:485–492
Higuchi M, Pischke MS, Mahonen AP, Miyawaki K, Hashimoto Y, Seki M, Kobayashi M, Shinozaki K, Kato T, Tabata S, Helariutta Y, Sussman MR, Kakimoto T (2004) In planta functions of the Arabidopsis cytokinin receptor family. Proc Natl Acad Sci U S A 101:8821–8826
Hochholdinger F, Zimmermann R (2008) Conserved and diverse mechanisms in root development. Curr Opin Plant Biol 11:70–74
Hu Y, Bao F, Li J (2000) Promotive effect of brassinosteroids on cell division involves a distinct CycD3-induction pathway in Arabidopsis. Plant J 24:693–701
Hua J, Meyerowitz EM (1998) Ethylene responses are negatively regulated by a receptor gene family in Arabidopsis thaliana. Cell 94:261–271
Huang Y, Li H, Hutchison CE, Laskey J, Kieber JJ (2003) Biochemical and functional analysis of CTR1, a protein kinase that negatively regulates ethylene signaling in Arabidopsis. Plant J 33:221–233
Hutchison CE, Li J, Argueso C, Gonzalez M, Lee E, Lewis MW, Maxwell BB, Perdue TD, Schaller GE, Alonso JM, Ecker JR, Kieber JJ (2006) The Arabidopsis histidine phosphotransfer proteins are redundant positive regulators of cytokinin signaling. Plant Cell 18:3073–3087
Hwang I, Sheen J (2001) Two-component circuitry in Arabidopsis cytokinin signal transduction. Nature 413:383–389
Hwang UW, Friedrich M, Tautz D, Park CJ, Kim W (2001) Mitochondrial protein phylogeny joins myriapods with chelicerates. Nature 413:154–157
Inoue T, Higuchi M, Hashimoto Y, Seki M, Kobayashi M, Kato T, Tabata S, Shinozaki K, Kakimoto T (2001) Identification of CRE1 as a cytokinin receptor from Arabidopsis. Nature 409:1060–1063
Jones B, Gunneras SA, Petersson SV, Tarkowski P, Graham N, May S, Dolezal K, Sandberg G, Ljung K (2010) Cytokinin regulation of auxin synthesis in Arabidopsis involves a homeostatic feedback loop regulated via auxin and cytokinin signal transduction. Plant Cell 22:2956–2969
Ju C, Yoon GM, Shemansky JM, Lin DY, Ying ZI, Chang J, Garrett WM, Kessenbrock M, Groth G, Tucker ML, Cooper B, Kieber JJ, Chang C (2012) CTR1 phosphorylates the central regulator EIN2 to control ethylene hormone signaling from the ER membrane to the nucleus in Arabidopsis. Proc Natl Acad Sci U S A 109:19486–19491

Kim TW, Lee SM, Joo SH, Yun HS, Lee Y, Kaufman PB, Kirakosyan A, Kim SH, Nam KH, Lee JS, Chang SC, Kim SK (2007) Elongation and gravitropic responses of Arabidopsis roots are regulated by brassinolide and IAA. Plant Cell Environ 30:679–689

Klee HJ (2004) Ethylene signal transduction. Moving beyond Arabidopsis. Plant Physiol 135:660–667

Kuderova A, Urbankova I, Valkova M, Malbeck J, Brzobohaty B, Nemethova D, Hejatko J (2008) Effects of conditional IPT-dependent cytokinin overproduction on root architecture of Arabidopsis seedlings. Plant Cell Physiol 49:570–582

Laplaze L, Benkova E, Casimiro I, Maes L, Vanneste S, Swarup R, Weijers D, Calvo V, Parizot B, Herrera-Rodriguez MB, Offringa R, Graham N, Doumas P, Friml J, Bogusz D, Beeckman T, Bennett M (2007) Cytokinins act directly on lateral root founder cells to inhibit root initiation. Plant Cell 19:3889–3900

Lewis DR, Negi S, Sukumar P, Muday GK (2011a) Ethylene inhibits lateral root development, increases IAA transport and expression of PIN3 and PIN7 auxin efflux carriers. Development 138:3485–3495

Lewis DR, Ramirez MV, Miller ND, Vallabhaneni P, Ray WK, Helm RF, Winkel BS, Muday GK (2011b) Auxin and ethylene induce flavonol accumulation through distinct transcriptional networks. Plant Physiol 156:144–164

Li L, Xu J, Xu ZH, Xue HW (2005) Brassinosteroids stimulate plant tropisms through modulation of polar auxin transport in Brassica and Arabidopsis. Plant Cell 17:2738–2753

Li X, Mo X, Shou H, Wu P (2006) Cytokinin-mediated cell cycling arrest of pericycle founder cells in lateral root initiation of Arabidopsis. Plant Cell Physiol 47:1112–1123

Liang X, Wang H, Mao L, Hu Y, Dong T, Zhang Y, Wang X, Bi Y (2012) Involvement of COP1 in ethylene- and light-regulated hypocotyl elongation. Planta 236:1791–1802

Liu Q, Xu C, Wen CK (2010) Genetic and transformation studies reveal negative regulation of ERS1 ethylene receptor signaling in Arabidopsis. BMC Plant Biol 10:60

Luschnig C, Gaxiola RA, Grisafi P, Fink GR (1998) EIR1, a root-specific protein involved in auxin transport, is required for gravitropism in Arabidopsis thaliana. Genes Dev 12:2175–2187

Malamy JE, Benfey PN (1997) Organization and cell differentiation in lateral roots of Arabidopsis thaliana. Development 124:33–44

Marhavy P, Bielach A, Abas L, Abuzeineh A, Duclercq J, Tanaka H, Parezova M, Petrasek J, Friml J, Kleine-Vehn J, Benkova E (2011) Cytokinin modulates endocytic trafficking of PIN1 auxin efflux carrier to control plant organogenesis. Dev Cell 21:796–804

Markakis MN, De Cnodder T, Lewandowski M, Simon D, Boron A, Balcerowicz D, Doubbo T, Taconnat L, Renou JP, Hofte H, Verbelen JP, Vissenberg K (2012) Identification of genes involved in the ACC-mediated control of root cell elongation in Arabidopsis thaliana. BMC Plant Biol 12:208

Mason MG, Mathews DE, Argyros DA, Maxwell BB, Kieber JJ, Alonso JM, Ecker JR, Schaller GE (2005) Multiple type-B response regulators mediate cytokinin signal transduction in Arabidopsis. Plant Cell 17:3007–3018

Mcginnis KM, Thomas SG, Soule JD, Strader LC, Zale JM, Sun TP, Steber CM (2003) The Arabidopsis SLEEPY1 gene encodes a putative F-box subunit of an SCF E3 ubiquitin ligase. Plant Cell 15:1120–1130

Miyawaki K, Matsumoto-Kitano M, Kakimoto T (2004) Expression of cytokinin biosynthetic isopentenyltransferase genes in Arabidopsis: tissue specificity and regulation by auxin, cytokinin, and nitrate. Plant J 37:128–138

Moubayidin L, Di Mambro R, Sabatini S (2009) Cytokinin-auxin crosstalk. Trends Plant Sci 14:557–562

Moubayidin L, Perilli S, Dello Ioio R, Di Mambro R, Costantino P, Sabatini S (2010) The rate of cell differentiation controls the Arabidopsis root meristem growth phase. Curr Biol 20:1138–1143

Mouchel CF, Briggs GC, Hardtke CS (2004) Natural genetic variation in Arabidopsis identifies BREVIS RADIX, a novel regulator of cell proliferation and elongation in the root. Genes Dev 18:700–714

Mouchel CF, Osmont KS, Hardtke CS (2006) BRX mediates feedback between brassinosteroid levels and auxin signalling in root growth. Nature 443:458–461

Muller B, Sheen J (2008) Cytokinin and auxin interaction in root stem-cell specification during early embryogenesis. Nature 453:1094–1097

Nakamura A, Nakajima N, Goda H, Shimada Y, Hayashi K, Nozaki H, Asami T, Yoshida S, Fujioka S (2006) Arabidopsis Aux/IAA genes are involved in brassinosteroid-mediated growth responses in a manner dependent on organ type. Plant J 45:193–205

Nakaya M, Tsukaya H, Murakami N, Kato M (2002) Brassinosteroids control the proliferation of leaf cells of Arabidopsis thaliana. Plant Cell Physiol 43:239–244

Negi S, Ivanchenko MG, Muday GK (2008) Ethylene regulates lateral root formation and auxin transport in Arabidopsis thaliana. Plant J 55:175–187

Nemhauser JL, Mockler TC, Chory J (2004) Interdependency of brassinosteroid and auxin signaling in Arabidopsis. PLoS Biol 2:E258

Nemhauser JL, Hong F, Chory J (2006) Different plant hormones regulate similar processes through largely nonoverlapping transcriptional responses. Cell 126:467–475

Nishimura C, Ohashi Y, Sato S, Kato T, Tabata S, Ueguchi C (2004) Histidine kinase homologs that act as cytokinin receptors possess overlapping functions in the regulation of shoot and root growth in Arabidopsis. Plant Cell 16:1365–1377

Nordstrom A, Tarkowski P, Tarkowska D, Norbaek R, Astot C, Dolezal K, Sandberg G (2004) Auxin regulation of cytokinin biosynthesis in Arabidopsis thaliana: a factor of potential importance for auxin-cytokinin-regulated development. Proc Natl Acad Sci U S A 101:8039–8044

O'malley RC, Rodriguez FI, Esch JJ, Binder BM, O'donnell P, Klee HJ, Bleecker AB (2005) Ethylene-binding activity, gene expression levels, and receptor system output for ethylene receptor family members from Arabidopsis and tomato. Plant J 41:651–659

Oh E, Zhu JY, Wang ZY (2012) Interaction between BZR1 and PIF4 integrates brassinosteroid and environmental responses. Nat Cell Biol 14:802–809

Olmedo G, Guo H, Gregory BD, Nourizadeh SD, Aguilar-Henonin L, Li H, An F, Guzman P, Ecker JR (2006) ETHYLENE-INSENSITIVE5 encodes a 5′–>3′ exoribonuclease required for regulation of the EIN3-targeting F-box proteins EBF1/2. Proc Natl Acad Sci U S A 103:13286–13293

Osmont KS, Sibout R, Hardtke CS (2007) Hidden branches: developments in root system architecture. Annu Rev Plant Biol 58:93–113

Pacheco-Villalobos D, Hardtke CS (2012) Natural genetic variation of root system architecture from Arabidopsis to Brachypodium: towards adaptive value. Philos Trans R Soc Lond B Biol Sci 367:1552–1558

Peng J, Carol P, Richards DE, King KE, Cowling RJ, Murphy GP, Harberd NP (1997) The Arabidopsis GAI gene defines a signaling pathway that negatively regulates gibberellin responses. Genes Dev 11:3194–3205

Peng J, Richards DE, Hartley NM, Murphy GP, Devos KM, Flintham JE, Beales J, Fish LJ, Worland AJ, Pelica F, Sudhakar D, Christou P, Snape JW, Gale MD, Harberd NP (1999) 'Green revolution' genes encode mutant gibberellin response modulators. Nature 400:256–261

Pickett FB, Wilson AK, Estelle M (1990) The aux1 mutation of Arabidopsis confers both auxin and ethylene resistance. Plant Physiol 94:1462–1466

Qiao H, Shen Z, Huang SS, Schmitz RJ, Urich MA, Briggs SP, Ecker JR (2012) Processing and subcellular trafficking of ER-tethered EIN2 control response to ethylene gas. Science 338:390–393

Rahman A, Amakawa T, Goto N, Tsurumi S (2001) Auxin is a positive regulator for ethylene-mediated response in the growth of Arabidopsis roots. Plant Cell Physiol 42:301–307

Rahman A, Bannigan A, Sulaman W, Pechter P, Blancaflor EB, Baskin TI (2007) Auxin, actin and growth of the Arabidopsis thaliana primary root. Plant J 50:514–528
Riefler M, Novak O, Strnad M, Schmulling T (2006) Arabidopsis cytokinin receptor mutants reveal functions in shoot growth, leaf senescence, seed size, germination, root development, and cytokinin metabolism. Plant Cell 18:40–54
Ross JJ, O'neill DP, Smith JJ, Kerckhoffs LH, Elliott RC (2000) Evidence that auxin promotes gibberellin A1 biosynthesis in pea. Plant J 21:547–552
Ruzicka K, Ljung K, Vanneste S, Podhorska R, Beeckman T, Friml J, Benkova E (2007) Ethylene regulates root growth through effects on auxin biosynthesis and transport-dependent auxin distribution. Plant Cell 19:2197–2212
Sakai H, Honma T, Aoyama T, Sato S, Kato T, Tabata S, Oka A (2001) ARR1, a transcription factor for genes immediately responsive to cytokinins. Science 294:1519–1521
Sakakibara H (2006) Cytokinins: activity, biosynthesis, and translocation. Annu Rev Plant Biol 57:431–449
Santuari L, Scacchi E, Rodriguez-Villalon A, Salinas P, Dohmann EM, Brunoud G, Vernoux T, Smith RS, Hardtke CS (2011) Positional information by differential endocytosis splits auxin response to drive Arabidopsis root meristem growth. Curr Biol 21:1918–1923
Scacchi E, Osmont KS, Beuchat J, Salinas P, Navarrete-Gomez M, Trigueros M, Ferrandiz C, Hardtke CS (2009) Dynamic, auxin-responsive plasma membrane-to-nucleus movement of Arabidopsis BRX. Development 136:2059–2067
Scacchi E, Salinas P, Gujas B, Santuari L, Krogan N, Ragni L, Berleth T, Hardtke CS (2010) Spatio-temporal sequence of cross-regulatory events in root meristem growth. Proc Natl Acad Sci U S A 107:22734–22739
Silverstone AL, Ciampaglio CN, Sun T (1998) The Arabidopsis RGA gene encodes a transcriptional regulator repressing the gibberellin signal transduction pathway. Plant Cell 10:155–169
Smith S, De Smet I (2012) Root system architecture: insights from Arabidopsis and cereal crops. Philos Trans R Soc Lond B Biol Sci 367:1441–1452
Solano R, Stepanova A, Chao Q, Ecker JR (1998) Nuclear events in ethylene signaling: a transcriptional cascade mediated by ETHYLENE-INSENSITIVE3 and ETHYLENE-RESPONSE-FACTOR1. Genes Dev 12:3703–3714
Stepanova AN, Hoyt JM, Hamilton AA, Alonso JM (2005) A Link between ethylene and auxin uncovered by the characterization of two root-specific ethylene-insensitive mutants in Arabidopsis. Plant Cell 17:2230–2242
Stepanova AN, Yun J, Likhacheva AV, Alonso JM (2007) Multilevel interactions between ethylene and auxin in Arabidopsis roots. Plant Cell 19:2169–2185
Stepanova AN, Robertson-Hoyt J, Yun J, Benavente LM, Xie DY, Dolezal K, Schlereth A, Jurgens G, Alonso JM (2008) TAA1-mediated auxin biosynthesis is essential for hormone crosstalk and plant development. Cell 133:177–191
Sun Y, Fan XY, Cao DM, Tang W, He K, Zhu JY, He JX, Bai MY, Zhu S, Oh E, Patil S, Kim TW, Ji H, Wong WH, Rhee SY, Wang ZY (2010) Integration of brassinosteroid signal transduction with the transcription network for plant growth regulation in Arabidopsis. Dev Cell 19:765–777
Swarup R, Parry G, Graham N, Allen T, Bennett M (2002) Auxin cross-talk: integration of signalling pathways to control plant development. Plant Mol Biol 49:411–426
Swarup R, Perry P, Hagenbeek D, Van Der Straeten D, Beemster GT, Sandberg G, Bhalerao R, Ljung K, Bennett MJ (2007) Ethylene upregulates auxin biosynthesis in Arabidopsis seedlings to enhance inhibition of root cell elongation. Plant Cell 19:2186–2196
Szekeres M, Nemeth K, Koncz-Kalman Z, Mathur J, Kauschmann A, Altmann T, Redei GP, Nagy F, Schell J, Koncz C (1996) Brassinosteroids rescue the deficiency of CYP90, a cytochrome P450, controlling cell elongation and de-etiolation in Arabidopsis. Cell 85:171–182
Teale WD, Ditengou FA, Dovzhenko AD, Li X, Molendijk AM, Ruperti B, Paponov I, Palme K (2008) Auxin as a model for the integration of hormonal signal processing and transduction. Mol Plant 1:229–237

To JP, Deruere J, Maxwell BB, Morris VF, Hutchison CE, Ferreira FJ, Schaller GE, Kieber JJ (2007) Cytokinin regulates type-A Arabidopsis Response Regulator activity and protein stability via two-component phosphorelay. Plant Cell 19:3901–3914

Tsuchisaka A, Theologis A (2004) Unique and overlapping expression patterns among the Arabidopsis 1-amino-cyclopropane-1-carboxylate synthase gene family members. Plant Physiol 136:2982–3000

Ubeda-Tomas S, Swarup R, Coates J, Swarup K, Laplaze L, Beemster GT, Hedden P, Bhalerao R, Bennett MJ (2008) Root growth in Arabidopsis requires gibberellin/DELLA signalling in the endodermis. Nat Cell Biol 10:625–628

Ubeda-Tomas S, Federici F, Casimiro I, Beemster GT, Bhalerao R, Swarup R, Doerner P, Haseloff J, Bennett MJ (2009) Gibberellin signaling in the endodermis controls Arabidopsis root meristem size. Curr Biol 19:1194–1199

Vert G, Chory J (2006) Downstream nuclear events in brassinosteroid signalling. Nature 441:96–100

Vert G, Walcher CL, Chory J, Nemhauser JL (2008) Integration of auxin and brassinosteroid pathways by Auxin Response Factor 2. Proc Natl Acad Sci U S A 105:9829–9834

Walcher CL, Nemhauser JL (2012) Bipartite promoter element required for auxin response. Plant Physiol 158:273–282

Wang KL, Li H, Ecker JR (2002) Ethylene biosynthesis and signaling networks. Plant Cell 14 (Suppl):S131–S151

Wang W, Hall AE, O'malley R, Bleecker AB (2003) Canonical histidine kinase activity of the transmitter domain of the ETR1 ethylene receptor from Arabidopsis is not required for signal transmission. Proc Natl Acad Sci U S A 100:352–357

Weiss D, Ori N (2007) Mechanisms of cross talk between gibberellin and other hormones. Plant Physiol 144:1240–1246

Wen X, Zhang C, Ji Y, Zhao Q, He W, An F, Jiang L, Guo H (2012) Activation of ethylene signaling is mediated by nuclear translocation of the cleaved EIN2 carboxyl terminus. Cell Res 22:1613–1616

Werner T, Motyka V, Laucou V, Smets R, Van Onckelen H, Schmulling T (2003) Cytokinin-deficient transgenic Arabidopsis plants show multiple developmental alterations indicating opposite functions of cytokinins in the regulation of shoot and root meristem activity. Plant Cell 15:2532–2550

Willige BC, Ghosh S, Nill C, Zourelidou M, Dohmann EM, Maier A, Schwechheimer C (2007) The DELLA domain of GA INSENSITIVE mediates the interaction with the GA INSENSITIVE DWARF1A gibberellin receptor of Arabidopsis. Plant Cell 19:1209–1220

Willige BC, Isono E, Richter R, Zourelidou M, Schwechheimer C (2011) Gibberellin regulates PIN-FORMED abundance and is required for auxin transport-dependent growth and development in Arabidopsis thaliana. Plant Cell 23:2184–2195

Wolbang CM, Ross JJ (2001) Auxin promotes gibberellin biosynthesis in decapitated tobacco plants. Planta 214:153–157

Yamada H, Suzuki T, Terada K, Takei K, Ishikawa K, Miwa K, Yamashino T, Mizuno T (2001) The Arabidopsis AHK4 histidine kinase is a cytokinin-binding receptor that transduces cytokinin signals across the membrane. Plant Cell Physiol 42:1017–1023

Yamamoto R, Fujioka S, Iwamoto K, Demura T, Takatsuto S, Yoshida S, Fukuda H (2007) Co-regulation of brassinosteroid biosynthesis-related genes during xylem cell differentiation. Plant Cell Physiol 48:74–83

Yoshimitsu Y, Tanaka K, Fukuda W, Asami T, Yoshida S, Hayashi K, Kamiya Y, Jikumaru Y, Shigeta T, Nakamura Y, Matsuo T, Okamoto S (2011) Transcription of DWARF4 plays a crucial role in auxin-regulated root elongation in addition to brassinosteroid homeostasis in Arabidopsis thaliana. PLoS One 6:e23851

Yu X, Li L, Zola J, Aluru M, Ye H, Foudree A, Guo H, Anderson S, Aluru S, Liu P, Rodermel S, Yin Y (2011) A brassinosteroid transcriptional network revealed by genome-wide identification of BESI target genes in Arabidopsis thaliana. Plant J 65:634–646

# Chapter 13
# Evolutionary Aspects of Auxin Signalling

**Priya Ramakrishna and Ive De Smet**

**Abstract** Auxin is one of the first phytohormones to be discovered in plants. It plays a key role in plant growth and development and in the evolution of land plants. The presence of auxin has been reported from microalgae to higher seed plants. However, tracing the origin of auxin response and of the associated proteins has proven to be more difficult. This chapter will summarize recent molecular developments on the origin of auxin metabolism, transport and signalling in green, red and brown algae, mosses and spikemosses.

## 1 Introduction

Auxin, the first phytohormone to be discovered in plants, is a key player in the regulation of plant growth and development and in response to environmental changes. The term "auxin" is derived from the Greek word "*auxein*", which means "to grow" or "to increase". Auxin is important for various developmental processes, including gamete specification in the female gametophyte, local patterning during embryogenesis and, post-embryonically, for the iterative production of tissues and organs and in the development of vascular strands (Reinhardt et al. 2003; Benkova et al. 2003; Heisler et al. 2005; Swarup et al. 2008; Dubrovsky

P. Ramakrishna
Division of Plant and Crop Sciences, School of Biosciences, University of Nottingham, Sutton Bonington Campus, Loughborough LE12 5RD, UK

I. De Smet (✉)
Division of Plant and Crop Sciences, School of Biosciences, University of Nottingham, Sutton Bonington Campus, Loughborough LE12 5RD, UK

Centre for Plant Integrative Biology, University of Nottingham, Nottingham LE12 5RD, UK
e-mail: Ive.De_Smet@nottingham.ac.uk

E. Zažímalová et al. (eds.), *Auxin and Its Role in Plant Development*,
DOI 10.1007/978-3-7091-1526-8_13,  

et al. 2008; Pagnussat et al. 2009; Péret et al. 2009; De Smet et al. 2010; Bennett and Scheres 2010; Peris et al. 2010; Ha et al. 2010; Scarpella and Helariutta 2010). Auxin acts by moving over long distances (Petrášek and Friml 2009), from shoot to root, influencing growth and development along its transport path (Thimann and Skoog 1933; Marchant et al. 2002; Fu and Harberd 2003). However, the specific expression patterns of some auxin biosynthesis genes suggest that localized auxin biosynthesis may also have a role in plant development (Tao et al. 2008; Cheng et al. 2006; Stepanova et al. 2005; Stepanova 2008; Ikeda et al. 2009; Pinon et al. 2013).

Indole-3-acetic acid (IAA) is the most common auxin involved in many of the physiological processes in plants and is broadly found from bacteria to seed plants, including, algae, fungi and even animals (Arteca 1995; De Smet et al. 2011; Cooke et al. 2002; Gruen 1959; Spaepen et al. 2007; Ichimura and Yamaki 1975). IAA was discovered in the first half of the twentieth century (Went and Thimann 1937; Abel and Theologis 2010), though the concept of plant hormones and their role in plant development had been around since the late 1880s (Went and Thimann 1937). IAA is a very dynamic molecule, which can move from the place of synthesis to the regions where it will exert its function. This regulated polar auxin transport (PAT) within plant tissues appears to be unique to IAA, as it has not been detected for any other signalling molecule (Zažímalová et al. 2007; Benjamins and Scheres 2008; Petrášek and Friml 2009; Boot et al. 2012; Went and Thimann 1937). In addition to IAA, plants synthesize other compounds referred to as "endogenous auxins", including indole-3-butyric acid (IBA) (Zimmerman and Wilcoxon 1935), phenylacetic acid (PAA) (Koepfli et al. 1938) and 4-chloroindole-3-acetic acid (4-Cl-IAA) (Porter and Thimann 1965). However, their roles and mechanisms of action have not been entirely determined (Simon and Petrášek 2011).

The plant hormone auxin is possibly one of the most extensively studied molecules in plants as it impacts on nearly every aspect of a plant's life cycle. However, very little is known about the evolutionary origin of its signalling activity. This chapter will summarize recent developments on the origin of auxin metabolism, transport and signalling in green, red and brown algae, mosses and spikemosses.

## 2 A Brief History of Auxin Biology

Auxin was one of the first plant hormones to be discovered. Charles Darwin, one of the first scientists to perform plant hormone research, described the effects of light on movement of canary grass (*Phalaris canariensis*) coleoptiles (Darwin 1880). Darwin's experiments expanded upon Theophil Ciesielski's research examining roots bending towards gravity (Ciesielski 1871). In one of his experiments, Darwin covered the tip of the coleoptile with aluminium foil and observed that no bending would occur towards the unidirectional light. However, when the tip of the coleoptile was left uncovered, but the portion just below the tip was covered, exposure to unidirectional light resulted in curvature towards the light. Darwin's experiment suggested that the tip of the coleoptile was the tissue responsible for perceiving the light and producing some signal which was transported to the lower part of the coleoptile where the physiological response of bending occurred. He then cut off the tip of the coleoptile and exposed the rest to unidirectional light to observe if curvature occurred. Curvature did not occur, confirming the results of his first experiment (Darwin 1880). Building on Darwin's observations, Peter Boysen-Jensen showed that the phototropic stimulus is transmitted from an excised tip across a gelatin barrier into the lower part of an *Avena* coleoptile where it still elicits the curvature response (Boysen Jensen and Nielsen 1925). Arpad Paál further developed this line of experimentation and concluded that the tip must produce and release a chemical substance that travels toward the coleoptile base to promote growth and that unilateral light causes an asymmetric transmission of this substance. This later became a cornerstone of the Cholodny–Went hypothesis (Went and Thimann 1937). For a direct demonstration of the postulated growth-promoting substance, Frits Went adapted Stark's agar block method (Stark 1921) in which he placed excised *Avena* coleoptile tips on agar blocks that received the growth-promoting substance by diffusion and could then serve as an artificial source for inducing curvature of decapitated coleoptiles (Went 1927). Between 1926 and 1928, Went made the definitive discovery of the "hormone", which was later in 1934–1935 called "auxin". Around a decade later, IAA was discovered in *Zea mays* (Haagen-Smit et al. 1946), and it soon became clear that IAA is the principal auxin in all plant species (Woodward and Bartel 2005; Davies 2010). Soon after the chemical identification of IAA, a number of structurally related compounds with auxin activity were reported which led to systematic study of structural activity relationships. The large body of experimental data led to predictions of structural requirements for auxin activity and spatial features of a hypothetical receptor site (Thimann 1969).

Recent evidence has provided further support for the Cholodny–Went hypothesis and its applicability to plant tropisms and development (see Chap. 16; Orbovik 1993; Litwack 2005). Several experiments that establish the plant signalling molecule auxin as a key player in organogenesis and vascular tissue formation (Reinhardt 2005), and the unique property of auxin being polarly transported from cell to cell through whole tissues, led to the formulation of the "canalization hypothesis". It proposes a feedback effect of the phytohormone auxin on tissues based on a feedback effect that auxin exerts on the polarity of its transport at the single cell level (Sauer et al. 2006). However, the critical requirement for these models of an auxin flux sensor was not addressed (Sauer et al. 2006) and aspects like the signalling mechanisms responsible for accumulation of auxin on, for example, the shaded side of the stem remain poorly defined (Christie and Murphy 2013).

New biologically plausible mathematical models of auxin transport during shoot development have been proposed (see Chap. 15; de Reuille et al. 2006; Heisler and Jönsson 2006; Jonsson et al. 2006; Smith et al. 2006), each of them being able to reproduce some aspects of phyllotaxis in simulations. Other auxin-related mathematical models were also proposed for plant mechanics (Mjolsness 2006) and for the interaction between mechanics and biochemistry (Shipman and Newell 2005;

Newell et al. 2008). The idea of canalization without flux sensors was derived from the model of phyllotaxis (Jonsson et al. 2006) and a travelling-wave hypothesis for the formation of polar auxin transport channels was formulated (Merks et al. 2007). It was assumed that auxin is transported actively using auxin transporter proteins, and possibly passively, at a very low rate, by leaking away from auxin maxima. The model neglects apoplastic (intercellular) compartments, so auxin moves directly from one cell into the next, and the effect of this has not fully been studied (Merks et al. 2007). A model proposed by Wabnik et al. (2011) integrates feedback circuits utilizing the conserved nuclear auxin signalling for the regulation of PIN transcription and a hypothetical endoplasmic reticulum (ER)-based signalling for the regulation of PIN-dependent transport activity at the ER. Hošek et al. (2012) used a single-cell-based systems and their proposed mathematical model of 2,4-D transport at a single-cell level successfully predicts the course of 2,4-D accumulation and confirms the consistency of the current concept of cellular auxin transport. Feller et al. (2013) studied equilibrium configurations of auxin and have come up with a model to explain the pattern formation in auxin flux. Recently, multiscale modelling of auxin transport has been carried out for the plant-root elongation zone (Band and King 2012) and with respect to protoxylem and protophloem of *Arabidopsis thaliana* root tips (Novoselova et al. 2013).

Over the years, various components of auxin biosynthesis (see Chaps. 2 and 3), transport (see Chaps. 4 and 5) and signalling (see Chap. 6) have been identified (Dharmasiri et al. 2005; Kepinski and Leyser 2005; Zažímalová et al. 2007; Zhao 2010; Normanly 2010; Leyser 2011). One of the most ground-breaking discoveries was the role of TRANSPORT INHIBITOR RESPONSE 1 (TIR1)-AUXIN SIGNALING F-BOX PROTEINs (AFBs) and their mode of action (Dharmasiri et al. 2005; Kepinski and Leyser 2005; Tan et al. 2007). In *Arabidopsis*, TIR1 and its closest paralogues AFB1 to AFB5 belong to the C3 subfamily of leucine-rich-repeat-containing F-box proteins (Gagne et al. 2002; Napier 2005). Auxin is sensed by a co-receptor complex consisting of a TIR1/AFB protein and an AUXIN/INDOLE-3-ACETIC ACID (AUX/IAA) protein, collectively mediating auxin-regulated responses throughout plant growth and development (Dharmasiri et al. 2005; Calderon Villalobos et al. 2012). Auxin promotes SKP1/CULLIN/F-BOX PROTEIN$^{TIR1}$–substrate binding by acting as a "molecular glue" rather than an allosteric switch (Tan et al. 2007).

## 3 Basics of Auxin Biology

### 3.1 *Auxin Metabolism*

Auxin metabolism involves the processes of biosynthesis, conjugation and degradation (see Chap. 2). Auxin biosynthesis in plants is extremely complex and multiple pathways contribute to de novo auxin production (Zhao 2010). Five pathways of auxin (IAA) biosynthesis have been proposed so far (Ljung 2013): one tryptophan (Trp) independent and four Trp dependent. Due to the unknown identities of some key enzymes and extensive functional redundancy, the importance of each of these pathways in auxin biosynthesis has been difficult to assess. In the indole-3-pyruvic acid (IPyA) pathway, the TRYPTOPHAN AMINOTRANSFERASE OF ARABIDOPSIS 1 (TAA1) and its close homologues TRYPTOPHAN AMINOTRANSFERASE-RELATED 1 (TAR1) and TAR2 convert L-Trp to IPyA, and the YUCCA (YUC) enzymes subsequently synthesize IAA from IPyA (Zhao 2012). The GH3 enzyme is responsible for the homeostatic feedback regulatory loop that controls the intracellular IAA level (Park et al. 2007). The indole-3-acetamide (IAM) pathway is well studied in bacteria and is present in many plant species, including *Arabidopsis*, maize, rice and tobacco (Sugawara et al. 2009; Novák et al. 2012). The tryptamine (TRA) pathway was originally thought to be an intermediate in the YUCCA pathway, but this has recently been questioned (Tivendale et al. 2010; Mano and Nemoto 2012). It is possible that TRA could function both as a precursor for IAA and in indole alkaloid and serotonin biosynthesis in different plant species (Mano and Nemoto 2012). More details on other pathways can be found in recent reviews on this topic (Lehmann et al. 2010; Zhao 2012; Mano and Nemoto 2012; Ljung 2013).

De novo synthesis of IAA is one important homeostatic mechanism operating in plant cells, but IAA levels can also be attenuated by conjugation (mainly to amino acids and sugars) and by degradation (Ruiz Rosquete et al. 2012). Proteins involved in IAA conjugation and IAA conjugate hydrolysis have been identified (Woodward and Bartel 2005; Ludwig-Mueller 2011). The metabolites 2-oxoindole-3-acetic acid (oxIAA) and oxIAA-glucose (oxIAA-Glc) are the major degradation products of IAA (Östin et al. 1998; Kai et al. 2007; Novák et al. 2012), but the genes involved in IAA catabolism have not been identified so far (Ljung 2013).

## 3.2 *Auxin Transport*

Auxin is a weak organic acid that does not move from cell to cell just by plain diffusion alone and requires specific transporters (Zažímalová et al. 2010; Ruiz Rosquete et al. 2012). Passive auxin uptake is aided by the amino acid permease-like proteins (influx carriers) of the AUXIN RESISTANT 1 (AUX1)-LIKE AUX1 (LAX) family (Bennett et al. 1996; Yang et al. 2006; Swarup et al. 2008; Jones et al. 2009; Ugartechea-Chirino et al. 2009; Péret et al. 2012). The *Arabidopsis thaliana* genome encodes four types of these proteins in total (AUX1 and LAX1-3). They have been shown to be involved in almost any auxin-dependent process for example, lateral root formation, gravitropism, embryonic development or phyllotaxis (Bennett et al. 1996; Swarup et al. 2008; Bainbridge et al. 2008; Ugartechea-Chirino et al. 2009). AUX1, being asymmetrically localized to the plasma membrane of root protophloem cells, promotes the acropetal, post-phloem movement of auxin to the root apex (Bennett et al. 1996; Swarup et al. 2001).

The PIN-FORMED (PIN) proteins are secondary transporters acting in the efflux of auxin from cells. The PIN family proteins can be classified into two groups: (1) the PIN1-type proteins (PIN1, 2, 3, 4, and 7) that are plasma membrane (PM) localized and function as auxin transporters at the PM for intercellular transport (long PINs), and (2) the PIN5-type proteins (PIN5, 6, and 8) that localize to the endoplasmic reticulum (ER) and carry out intracellular regulation of auxin homeostasis (short PINs) (Petrasek et al. 2006; Mravec et al. 2009). *Arabidopsis* has eight annotated *PIN* genes, which have been functionally characterized: *PIN1* (Gälweiler et al. 1998), *PIN2* (Ludwig-Mueller 2011; Muller et al. 1998; Luschnig et al. 1998; Chen et al. 1998; Utsuno et al. 1998), *PIN3* (Friml et al. 2002b), *PIN4* (Friml et al. 2002a), *PIN5* (Mravec et al. 2009), *PIN6* (Krecek et al. 2009; Sawchuk et al. 2013), *PIN7* (Friml et al. 2003) and *PIN8* (Ding et al. 2012; Bosco et al. 2012). PINs play an important role in regulating asymmetric auxin distribution in multiple developmental processes, including embryogenesis, organogenesis, tissue differentiation and tropic responses (Vieten et al. 2007; Petrášek and Friml 2009; Grunewald and Friml 2010).

The ATP-BINDING CASSETTE PROTEIN SUB-FAMILY B (ABCB)–MULTI-DRUG RESISTANCE (MDR)–P-GLYCOPROTEIN (PGP) transporters also facilitate the transport of IAA to and from the cell (Terasaka et al. 2005; Blakeslee et al. 2007; Mravec et al. 2008; Cho and Cho 2012; Kubeš et al. 2012). They are a sub-family of the ancient ABC family which is shown to have the most conserved sequences coding for their nucleotide-binding domains in living organisms (Isenbarger et al. 2008). At least five members of this sub-family have been reported to mediate cellular transport of auxin or auxin derivatives (Cho and Cho 2012). For example, ABCB19 was shown to work in coordination with PIN1 in specific PM domains, thereby at least taking part in directional auxin transport (Titapiwatanakun et al. 2009) and ABCB21 was found to function as a facultative importer/exporter controlling auxin concentrations in plant cells (Kamimoto et al. 2012).

PIN-LIKES (PILS) are a putative auxin carrier family of seven members in *Arabidopsis* (Barbez et al. 2012). These proteins are named PIN-LIKES since their predicted protein topology is highly similar to the topology of the PIN proteins. In addition, PILS contain the so-called Interpro auxin carrier domain, an in silico-defined domain to predict auxin transport function. These putative auxin carriers localize to the endoplasmic reticulum (ER) and contribute to cellular auxin homeostasis. PILS proteins regulate intracellular auxin accumulation, the rate of auxin conjugation and, subsequently, affect nuclear auxin signalling. Consequently, these proteins are important for auxin-regulated developmental such as de novo organ formation and growth regulation (Barbez et al. 2012).

### 3.3 Auxin Signalling

Auxin signalling (see Chap. 6) is central to the growth and development of higher plants (Mockaitis and Estelle 2008). Intracellular auxin is perceived by the TIR1–AFB1–3 receptors (Dharmasiri et al. 2005; Kepinski and Leyser 2005), or more correctly co-receptors, because for high affinity binding, a complex with proteins of the AUX/IAA family is needed (Calderon-Villalobos et al. 2010). TIR1 is an integral component of the SKP1/CULLIN/F-BOX PROTEIN$^{TIR1}$ complex that mediates the auxin-dependent ubiquitination of AUXIN/INDOLE-3-ACETIC ACID (AUX/IAA) repressor proteins and thereby destines them for 26S proteasome-dependent degradation (Maraschin et al. 2009). The AUX/IAA family of proteins plays a central role in the auxin signal transduction that regulates auxin-responsive gene expression. AUX/IAAs at low auxin concentrations together with co-repressor proteins, such as TOPLESS (Long et al. 2006; Szemenyei et al. 2008), form dimers with AUXIN RESPONSE FACTOR (ARF) transcription factors, thereby blocking the activity of at least the activating ARFs. When freed from the AUX/IAAs, these ARFs regulate the expression of auxin-responsive genes and control many aspects of plant growth and development (Lau et al. 2008; Mockaitis and Estelle 2008).

Regulation of gene expression by the SCF$^{TIR1/AFB}$ pathway does not account for early auxin responses, which do not necessarily require primary modifications of gene expression (Tromas et al. 2010). The potential receptor AUXIN-BINDING PROTEIN 1 (ABP1) is involved in early responses at the plasma membrane with the regulation of rapid cytosol-based signalling mechanisms. ABP1 contains a KDEL-type endoplasmic reticulum (ER) retention sequence at the C-terminus, which suggests that the ABP1 protein would be predominantly localized in the ER (Jones and Herman 1993). ABP1 can sense the transported auxin concentration in the apoplast and regulate PIN activity and is also involved in other wide variety of auxin-dependent responses, including regulation of gene expression, cell division and cell expansion (Braun et al. 2008; Robert et al. 2010; Chen et al. 2012).

Finally, INDOLE-3-BUTYRIC ACID RESPONSE5 (IBR5) is a phosphatase protein that acts as a positive regulator of auxin responses, including auxin-inducible transcription, without affecting TIR1-mediated destabilisation of AUX/IAAs. IBR5 acts downstream of auxin recognition by the SCF$^{TIR1/AFB}$-AUX/IAA complexes (Monroe-Augustus et al. 2003; Strader et al. 2008).

## 4 Auxin Biology Throughout Evolution

### 4.1 A Few Basics of Algae and Land Plant Evolution

Land plants (Embryophytes) are believed to have evolved from freshwater multicellular algae and are possibly related to the existing Charophyte groups

Charales or Coleochaetales. Embryophytes and Charophytes collectively form a monophyletic group, the Streptophytes, which is sister to the other green algae, i.e. the Chlorophytes (Karol et al. 2001; Lewis and McCourt 2004; Becker and Marin 2009; Pires and Dolan 2012; Kenrick et al. 2012). The Streptophytes themselves likely split from *Chlorophyta* (all other green algae) in between 725 and 1,200 Mya (Becker and Marin 2009). Brown algae are a group of (photosynthetic) organisms or "plant systems" belonging to the Heterokonts (Van den Hoek et al. 1996), an extremely diverse kingdom. Brown algae are among the eukaryotes that acquired complex multicellularity (Bogaert et al. 2013). They share a common ancestry with land plants well over 1,500 Mya (Yoon et al. 2004).

The two monophyletic lineages, the *Chlorophyta* and the *Streptophyta* comprise a clade *Viridiplantae* (Latin for "green plants") (Fig. 13.1) (Simon et al. 2006). Many important cellular structures (phragmoplast, plasmodesmata, hexameric cellulose synthase, sporopollenin) and physiological characteristics (photorespiration, phytochrome system) originated within the *Streptophyta* (Becker and Marin 2009). Zygnematales, the conjugating algae, have been identified as the closest relatives of land plants (Turmel 2006; Finet et al. 2010). The transition of Streptophytes to terrestrial environments was allied with the evolution of the key features that define land plants, for instance a multicellular sporophyte, retention of the zygote and embryo within the female gametophyte, and apical cells with three cutting faces that allow the generation of three-dimensional parenchymatous tissues (Graham et al. 2000; Niklas and Kutschera 2009; Pires and Dolan 2012).

The first proto-land plants, represented by the sporophytes of *Cooksonia* and similar forms, appear on older Mid-Late Silurian strata, around 425 Mya. Since then, *Bryophyta, Pteridophyta* and *Spermatophyta* have emerged (Wodniok et al. 2011). The Bryophytes (mosses, hornworts and liverworts) represent the most basal lineage of land plants. They supposedly separated about 400 Mya, only 50 Mya after the actual separation of land plants themselves (Sanderson et al. 2004; Becker and Marin 2009). They differ a great deal from the rest of the group, as seen from the domination of the gametophyte in their life cycle and the absence of vascular tissues, genuine roots, stalks and leaves. Since the available fossils suggest that the first land plants were moss-like organisms (Rensing et al. 2008), they are an excellent model to uncover the aspects of the transition to land. This more "recent" evolution of land plants appears to mainly utilize expansion of genes and their differential expression rather than sequence variations. Horizontal gene transfer is extremely rare, except for mitochondrial genes (Becker and Marin 2009). Plants have been challenged to coordinate the patterning of their multicellular body plans during embryogenesis and for postembryonic development. Genes involved in cell-to-cell communication, cell adhesion and cell differentiation had to evolve predating or accompanying the emergence of multicellularity (Ruiz-Trillo et al. 2007).

*Selaginellaceae* (belonging to the *Lycopsida* class) have their origins dating to the late Silurian/early Devonian (Banks 2009) and are heterosporous (producing megaspores and microspores). Homospory is a character known to be shared by the earliest vascular plants and hence heterospory is thought to have evolved multiple

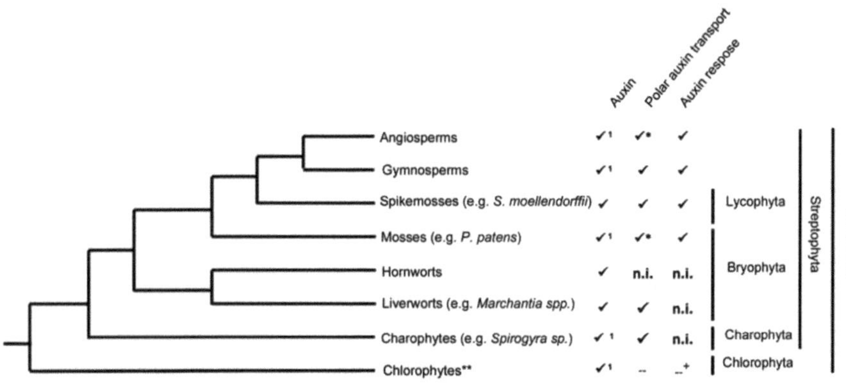


**Fig. 13.1** Schematic evolutionary tree highlighting some key aspects of auxin biology. The symbols are explained as follows: **, Chlorophyte species studied are *Ostreococcus tauri*, *Ostreococcus lucimarinus*, *Micromonas pusilla* strain CCMP1545 and strain RCC299, *Chlamydomonas reinhardtii*, *Volvox carteri*, *Chlorella vulgaris* and *Chlorella variabilis* NC64A; [1], presence of auxin confirmed via experimental evidence; ✓, indicative of presence; *, presence only in sporophyte; n.i., not investigated in detail; [+], no response to auxin induction in the Chlorophyte *Chlamydomonas reinhardtii* based on DR5 promoter activity. Figure mainly based on Finet and Jaillais (2012), De Smet et al. (2011) and Lau et al. (2009)

times in different vascular plant lineages (including the lycophytes) and this is considered a key innovation in land plant evolution (Bateman and DiMichele 1994). The spikemoss *Selaginella moellendorffii* is an extant member of the lycopsid lineage, which diverged from the *Euphyllophyta* (ferns and seed plants) ca. 400 Mya (Banks 2009).

## 4.2 *Where Can We Find Auxin?*

Auxin is ubiquitous in heterotrophic and photoautotrophic organisms (Evans and Trewavas 1991). Auxin has been shown to be present in several kingdoms and in a wide range of organisms (Fig. 13.1), ranging from bacteria, some plant parasitic nematodes to several single-celled and multicellular green, red and brown algae, bryophytes and flowering plants (Overbeek 1940; Jacobs 1950; Abe et al. 1972; Jacobs et al. 1985; Cooke et al. 2002; De Meutter et al. 2005; Decker et al. 2006; Tsavkelova et al. 2007; Spaepen et al. 2007; Lau et al. 2008, 2009; Hayashi et al. 2008; Banks 2009; Ross and Reid 2010; Yokoya et al. 2010; Bradley 1991; Evans and Trewavas 1991; Basu 2002; Rensing et al. 2008; De Smet et al. 2011).

## 4.3 The Role of Auxin in Algae

Auxin and its action have been studied for a long time in algae (Bradley 1991). There is evidence that auxin may play a role in various aspects of algal development, such as the establishment of polarity, embryo development and rhizoid elongation.

In green algae, auxin formation and distribution have been shown to control rhizoid formation in *Bryopsis* (Jacobs 1950), and in *Ulothrix* auxin was shown to have a striking effect with a dramatic increase in stalk elongation (Conrad et al. 1959). Polar auxin transport was shown to be present in *Charophyta* and probably auxin may play a role in directing growth and development (Boot et al. 2012). With respect to brown algae, *Fucus distichus* embryos accumulate IAA in early stages and this acts on the formation of apical basal patterns in *Fucus* embryo development (Basu 2002). There are also suggested interactions between the actin cytoskeleton and auxin transport in *Fucus* (Sun et al. 2004). IAA was also detected in mature *Ectocarpus siliculosus* and was found to be present mainly at the apices of the filaments in the early stages of development (Hayashi et al. 2008). It has been suggested that IAA is used by *Ectocarpus* to relay cell–cell positional information and induces a signalling pathway different from that known in land plants (Le Bail et al. 2010). In red algae (*Rhodophyta*), there has been very little research performed with respect to the role of auxin so far. There are a few reports on IAA having a positive effect on growth of *Gracilaria* tissue cultures (Yokoya et al. 2004) and playing a role in regulating physiological processes (Yokoya et al. 2010).

## 4.4 The Role of Auxin in Bryophytes

Bryophytes are the earliest diverging group of land plants where polar auxin transport has been unequivocally detected and where auxin-signalling response was found to be functional (Hayashi et al. 2008; Poli et al. 2003; Fujita et al. 2008; Fujita and Hasebe 2009; Viaene et al. 2013; Prigge et al. 2010; Rensing et al. 2008). Auxin-driven mechanisms are important for Bryophyte growth, reproduction and development. In *Physcomitrella. patens*, the development of rhizoids (muticellular filaments of epidermal origin) was found to be auxin dependent (Ashton et al. 1979; Sakakibara 2003; Bierfreund et al. 2003). Auxin treatment was also found to stimulate stem elongation (Fujita et al. 2008), and elongation of gametophore phyllids (leaves) (Decker et al. 2006) and triggers physiological responses such as the chloronema-to-caulonema transition by the auxin-dependent positive regulation of ROOT HAIR DEFECTIVE SIX-LIKE1 (PpRSL1) and PpRSL2 transcription factors (Johri and Desai 1973; Jang and Dolan 2011).

# 5 Molecular Components for Auxin Biology Throughout Evolution

The conquest of terrestrial habitats was associated with an expansion of gene families in several signalling pathways including that of signalling mediated by the growth hormone auxin (De Smet et al. 2011; Rensing et al. 2008). Surveys of green algae and plants with special emphasis on available genomes of chlorophyte algae, such as *Ostreococcus tauri*, *Ostreococcus lucimarinus*, *Micromonas pusilla*, *Chlamydomonas reinhardtii*, *Volvox carteri*, *Chlorella vulgaris* and *Chlorella variabilis* (De Smet et al. 2011), brown algae *Ectocarpus siliculosus* (Hayashi et al. 2008), the mosses *Physcomitrella patens* (Nishiyama et al. 2003; Fujita et al. 2008) and *Funaria hygrometrica* (Panigrahi et al. 2009; Rensing et al. 2008) and the spikemoss *Sellaginella* (Sztein et al. 2000; Paponov et al. 2009; Banks 2009), revealed a potential early origin for proteins related to auxin (see Fig. 13.2 and below for more details). The completion of the genome sequence for the moss *Physcomitrella patens* and *Sellaginella moellendorffii* has established them as a useful experimental tool to study the ancestral role of auxin (Banks et al. 2011; Rensing et al. 2008).

## 5.1 *Auxin Metabolism Components*

The Trp-independent pathway is poorly characterized in plants, which limits evolutionary genomic approaches. *Chlorophyta* generally seem to possess many orthologues of auxin metabolism genes (Finet and Jaillais 2012; De Smet et al. 2011). The key enzymes that convert the important precursor chorismate into indole and Trp in land plants had clear orthologues in *Chlorophyta*. Although not all genetic pathways known from land plants for auxin biosynthesis are represented in the Chlorophytes, based on the genome evidence, they very likely may produce auxin by the IAM pathway (Patten and Glick 1996; De Smet et al. 2011). Finally, in silico surveys showed that IAA biosynthesis genes from land plants also have orthologues in the brown alga *Ectocarpus siliculosus* (Cock et al. 2010; Hayashi et al. 2008).

In *P. patens* GH3 proteins were shown to be involved in auxin homeostasis by conjugating excess of physiologically active free auxin to inactive IAA-amide conjugates (Ludwig-Muller et al. 2009). With regard to the Trp-dependent pathway of auxin biosynthesis, the YUCCA gene family is ancient in land plants (Finet and Jaillais 2012). The green algal species *Chlorella vulgaris* encodes a putative functional YUC since the putative FAD-binding and NADPH-binding motifs of the protein are largely conserved (De Smet et al. 2011). Homologues have also been identified in the moss *P. patens* (Pires and Dolan 2012; Rensing et al. 2008) and in *Selaginella* (Banks 2009). Finally, orthologues of *Arabidopsis* SHI/STY family

| | | Chlorophyta | Charophyta | Bryophyta | Lycophyta | Gymnosperms | Angiosperms |
|---|---|---|---|---|---|---|---|
| METABOLISM | YUCCA | ✓+ | n.i. | ✓ | ✓ | ✓ | ✓ |
| | SUR1 | ✓ | n.i. | ✓ | ✓ | ✓ | ✓ |
| | IAN | -- | ✓ | ✓ | ✓ | ✓ | ✓ |
| | TAA1-TAR | -- | n.i. | ✓ | ✓ | ✓ | ✓ |
| | GH3 | ✓+ | n.i. | ✓ | ✓ | ✓ | ✓ |
| TRANSPORT | PIN | -- | ✓ | ✓ | ✓ | ✓ | ✓ |
| | PILS | ✓ | n.i. | ✓ | ✓ | ✓ | ✓ |
| | PGP | ✓ | ✓ | ✓ | ✓ | ✓ | ✓ |
| | AUX1-LAX | ✓* | -- | ✓ | ✓ | ✓ | ✓ |
| SIGNALLING | TIR1-AFB | -- | n.i. | ✓ | ✓ | ✓ | ✓ |
| | ARF | -- | n.i. | ✓ | ✓ | ✓ | ✓ |
| | AUX/IAA | -- | n.i. | ✓ | ✓ | ✓ | ✓ |
| | ABP1 | ✓ | n.i. | ✓ | ✓ | ✓ | ✓ |
| | IBR5 | ✓ | n.i. | ✓ | ✓ | ✓ | ✓ |

**Fig. 13.2** Summary of molecular components throughout plant evolution. The upper panel displays the evolutionary clades. The symbols are explained as follows: ✓, indicative of presence; –, absence; n.i., not investigated; ✓+, unclear putative orthologue for most Chlorophytes except *Chlorella vulgaris*; ✓*, no orthologues identified except in *Chlorella sp.* Figure mainly based on Finet and Jaillais (2012) and De Smet et al. (2011)

proteins, which are positive regulators of auxin biosynthesis, are also present in *P. patens* (Eklund et al. 2010).

## 5.2 *Auxin Transport Components*

Auxin transport mechanisms appear to have already been present, at least in part (Fig. 13.2), before plants invaded the land (De Smet et al. 2011).

The ABCB/PGP/MDR family of ABC transporters have been shown to be involved in auxin transport (Geisler and Murphy 2006; Verrier et al. 2008; Yang and Murphy 2009), and putative homologues of the ABCB efflux IAA transporter family have been identified in brown algae *Ectocarpus siliculosus* (Hayashi et al. 2008) and in *Chlorophyta* and *Streptophyta* species (De Smet et al. 2011).

AUX1–LAX homologues have been reported to be present throughout the plant kingdom (Hochholdinger et al. 2000; de Billy et al. 2001; Péret et al. 2007; Pattison and Catala 2012) and are present in mosses (Rensing et al. 2008). However, AUX1–LAX influx carriers were not identified in all *Chlorophyta*, except for one orthologue in *Chlorella* (De Smet et al. 2011).

There is no evidence of presence of PINs in *Chlorophyta* and neither the precise origin of PIN proteins in the evolutionary history of plants is known (Krecek et al. 2009; Viaene et al. 2013). All green algae with genomes sequenced so far (*Chlamydomonas*, *Volvox*, *Ostreococcus* and *Micromonas*) belong to the clade *Chlorophyta* and none of these organisms contains a *PIN* gene (Viaene et al. 2013; De Smet et al. 2011). However, a putative *PIN* gene orthologue was found in the genome of *Spirogyra pratensis* (De Smet et al. 2011). Sequence data from *P. patens* and *S. moellendorffi* have revealed the presence of both group 1 and group 2 PIN genes (Krecek et al. 2009; Fujita et al. 2008). The most ancient PIN proteins currently known from mosses are localized to the ER, which suggests that intracellular function is evolutionarily ancestral (Paponov et al. 2005). *Physcomitrella* PIN proteins are functionally related to the PIN5-type proteins that regulate subcellular homeostasis of auxin and not to the PIN1-type proteins that are responsible for auxin efflux from cell to cell in angiosperms (Mravec et al. 2009). Within the group of multicellular plants the PINs have undergone changes during evolution, both in structure and in number.

Finally, the PILS family was shown to be present already in algae and to have diversified in the different plant lineages. Ancient species, such as algae, mosses and spike mosses, have 1–8 *PILS* genes, while seed plants, such as *Oryza*, *Zea*, *Medicago* or *Populus* have 6–18 *PILS* genes. PILS proteins based on their evolution diversified into three clades. Clade I consists of PILS algae orthologues clustered together. In *Chlorophyta*, relatively low number of only one or two *PILS* genes per species were identified (Feraru et al. 2012). The evolutionary Clade II and III already emerged early during non-vascular plant evolution and both contain PILS sequences from land plants. Clade II includes the well-conserved PILS2- and PILS6-like subclades and includes the orthologues of PILS2 and PILS6 from *Physcomitrella*, *Selaginella*, *Brachypodium* and/or *Oryza*. Clade III encompasses the PILS1/PILS3/PILS4- and PILS5/PILS7-like subclades and displays particular expansion in higher seed plants (Feraru et al. 2012).

In conclusion, while sequence data suggests the presence of auxin transporters in green algae, further detailed functional characterisation will be required to demonstrate auxin transport. For example, in *Physcomitrella* the reduced sporophyte seemingly exhibits polarized auxin flow and sensitivity to its inhibitor (NPA), while the dominant gametophyte shows apolar auxin distribution (Fujita et al. 2008).

## 5.3 Auxin Signalling

The auxin signalling pathway has undergone substantial functional diversification and specialization within vascular plants since they diverged from bryophytes (Fig. 13.2). Although the auxin signalling pathways in land plants are nowadays relatively well understood and the respective signalling mechanisms are also present in basal lineages such as bryophytes (Paponov et al. 2005; Rensing et al. 2008), the evolution of the associated molecular components is much less clear. According to Cooke et al. (2002), auxin signalling mechanisms in both brown algae and green plants originated independently by convergent evolution. It has been suggested that auxin signalling emerged to coordinate multicellular growth, at least in land plants (Rensing et al. 2008), but particular aspects of auxin signalling could of course have an earlier origin. With respect to the readout of auxin signalling, DR5-visualised auxin response occurs in *Physcomitrella* (Bierfreund et al. 2003), but seemingly not in *Chlamydomonas* (De Smet et al. 2011).

In silico analyses of available algal genomes (green, brown and red) for "classical" auxin signalling mechanisms in plants involving TIR1-AFB, ARFs and AUX/IAA proteins revealed no complete AUX/IAAs and ARFs-dependent signalling machinery to be present in the investigated *Chlorophyta* (Lau et al. 2009; De Smet et al. 2011) though potential orthologues of single domains can be found. Orthologues for the transcriptional corepressor TOPLESS (TPL) and related proteins were not present in the investigated *Chlorophyta*, but seemed to be present in *P. patens* and *S. moellendorffii* (De Smet et al. 2011). Comparative analysis of the *P. patens*, *S. moellendorffii* and *A. thaliana* genomes suggested that the well-established rapid transcriptional response to auxin of flowering plants, evolved in vascular plants after their divergence from the last common ancestor shared with mosses (Paponov et al. 2009). However, a phylogenomic analysis of auxin signaling genes indicates that the lower land plants *P. patens* and *S. moellendorffii* contain fewer AUX/IAA and ARF family members (3 and 3 AUX/IAAs and 12 and 7 ARFs in *P. patens* and *S. moellendorffii*, respectively) (Paponov et al. 2009; Hayashi 2012). The *Physcomitrella* genome encodes four closely related TIR1 homologues PpAFB 1–4 and two slightly more distantly related homologues from a lineage not present in angiosperms (Paponov et al. 2009; Rensing et al. 2008).

Orthologues of ABP1 and IBR5 were detected in the green algae (Tromas et al. 2010; Lau et al. 2009; De Smet et al. 2011), suggesting that they evolved in the streptophyte lineage. The presence of ABP1 was also shown in all species from bryophytes to flowering plants by comparative analysis of the *Arabidopsis* ABP1 sequence against available genome sequence databases of land plants (Tromas et al. 2010).

# 6 Conclusion

Auxin appears to play a role already in green algae and even in brown algae. However, these organisms only appear to have some components of the machinery related to auxin biology. In future, more sequence information on algae, and especially multicellular green algae, will provide further insight in how auxin biology components evolved. However, it will be important to subsequently functionally characterise relevant proteins in algae.

# References

Abe H, Uchiyama M, Sato R (1972) Isolation and identification of native auxins in marine algae. Agric Biol Chem 36(12):2259–2260

Abel S, Theologis A (2010) Odyssey of auxin. Cold Spring Harb Perspect Biol 2:a004572. doi:10.1101/cshperspect.a004572

Arteca RN (1995) Plant growth substances: principles and applications. Chapman & Hall, USA

Ashton NW, Grimsley NH, Cove DJ (1979) Analysis of gametophytic development in the moss, Physcomitrella patens, using auxin and cytokinin resistant mutants. Planta 144(5):427–435. doi:10.1007/bf00380118

Bainbridge K, Guyomarc'h S, Bayer E, Swarup R, Bennett M, Mandel T, Kuhlemeier C (2008) Auxin influx carriers stabilize phyllotactic patterning. Genes Dev 22(6):810–823. doi:10.1101/gad.462608

Band LR, King JR (2012) Multiscale modelling of auxin transport in the plant-root elongation zone. J Math Biol 65(4):743–785. doi:10.1007/s00285-011-0472-y

Banks JA (2009) Selaginella and 400 million years of separation. Annu Rev Plant Biol 60:223–238. doi:10.1146/annurev.arplant.59.032607.092851

Banks JA, Nishiyama T, Hasebe M, Bowman JL, Gribskov M, de Pamphilis C, Albert VA, Aono N, Aoyama T, Ambrose BA, Ashton NW, Axtell MJ, Barker E, Barker MS, Bennetzen JL, Bonawitz ND, Chapple C, Cheng C, Correa LGG, Dacre M, DeBarry J, Dreyer I, Elias M, Engstrom EM, Estelle M, Feng L, Finet C, Floyd SK, Frommer WB, Fujita T, Gramzow L, Gutensohn M, Harholt J, Hattori M, Heyl A, Hirai T, Hiwatashi Y, Ishikawa M, Iwata M, Karol KG, Koehler B, Kolukisaoglu U, Kubo M, Kurata T, Lalonde S, Li K, Li Y, Litt A, Lyons E, Manning G, Maruyama T, Michael TP, Mikami K, Miyazaki S, Morinaga SI, Murata T, Mueller-Roeber B, Nelson DR, Obara M, Oguri Y, Olmstead RG, Onodera N, Petersen BL, Pils B, Prigge M, Rensing SA, Riaño-Pachón DM, Roberts AW, Sato Y, Scheller HV, Schulz B, Schulz C, Shakirov EV, Shibagaki N, Shinohara N, Shippen DE, Sørensen I, Sotooka R, Sugimoto N, Sugita M, Sumikawa N, Tanurdzic M, Theißen G, Ulvskov P, Wakazuki S, Weng JK, Willats WWGT, Wipf D, Wolf PG, Yang L, Zimmer AD, Zhu Q, Mitros T, Hellsten U, Loqué D, Otillar R, Salamov A, Schmutz J, Shapiro H, Lindquist E, Lucas S, Rokhsar D, Grigoriev IV (2011) The selaginella genome identifies genetic changes associated with the evolution of vascular plants. Science 332(6032):960–963. doi:10.1126/science.1203810

Barbez E, Kubeš M, Rolčík J, Beziat C, Pěnčík A, Wang B, Rosquete MR, Zhu J, Dobrev PI, Lee Y, Zažímalová E, Petrášek J, Geisler M, Friml J, Kleine-Vehn J (2012) A novel putative auxin carrier family regulates intracellular auxin homeostasis in plants. Nature 485(7396):119–122. doi:10.1038/nature11001

Basu S (2002) Early embryo development in fucus distichus is auxin sensitive. Plant Physiol 130 (1):292–302. doi:10.1104/pp. 004747

Bateman RM, DiMichele WA (1994) Heterospory: the most iterative key innovation in the evolutionary history of the plant kingdom. Biol Rev 69(3):345–417. doi:10.1111/j.1469-185X.1994.tb01276.x
Becker B, Marin B (2009) Streptophyte algae and the origin of embryophytes. Ann Bot 103 (7):999–1004. doi:10.1093/aob/mcp044
Benjamins R, Scheres B (2008) Auxin: the looping star in plant development. Annu Rev Plant Biol 59:443–465. doi:10.1146/annurev.arplant.58.032806.103805
Benkova E, Michniewicz M, Sauer M, Teichmann T, Seifertova D, Jurgens G, Friml J (2003) Local, efflux-dependent auxin gradients as a common module for plant organ formation. Cell 115(5):591–602. doi:10.1016/S0092-8674(03)00924-3
Bennett T, Scheres B (2010) Chapter Three – Root development—two meristems for the price of one? In: Marja CPT (ed) Current topics in developmental biology, vol 91. Academic Press, USA, pp 67–102. doi:10.1016/S0070-2153(10)91003-X
Bennett MJ, Marchant A, Green HG, May ST, Ward SP, Millner PA, Walker AR, Schulz B, Feldmann KA (1996) Arabidopsis AUX1 gene: a permease-like regulator of root gravitropism. Science (New York, NY) 273(5277):948–950. doi:10.1126/science.273.5277.948
Bierfreund NM, Reski R, Decker EL (2003) Use of an inducible reporter gene system for the analysis of auxin distribution in the moss Physcomitrella patens. Plant Cell Rep 21(12):1143–1152. doi:10.1007/s00299-003-0646-1
Blakeslee JJ, Bandyopadhyay A, Lee OR, Mravec J, Titapiwatanakun B, Sauer M, Makam SN, Cheng Y, Bouchard R, Adamec J, Geisler M, Nagashima A, Sakai T, Martinoia E, Friml J, Peer WA, Murphy AS (2007) Interactions among PIN-FORMED and P-glycoprotein auxin transporters in Arabidopsis. Plant Cell 19:131–147. doi:10.1105/tpc.106.040782
Bogaert KA, Arun A, Coelho SM, Clerck O (2013) Brown algae as a model for plant organogenesis. Methods Mol Biol 959:97–125. doi:10.1007/978-1-62703-221-6_6
Boot KJ, Libbenga KR, Hille SC, Offringa R, van Duijn B (2012) Polar auxin transport: an early invention. J Exp Bot 63(11):4213–4218. doi:10.1093/jxb/ers106
Bosco CD, Dovzhenko A, Liu X, Woerner N, Rensch T, Eismann M, Eimer S, Hegermann J, Paponov IA, Ruperti B, Heberle-Bors E, Touraev A, Cohen JD, Palme K (2012) The endoplasmic reticulum localized PIN8 is a pollen-specific auxin carrier involved in intracellular auxin homeostasis. Plant J 71(5):860–870. doi:10.1111/j.1365-313X.2012.05037.x
Boysen Jensen P, Nielsen N (1925) Studien über die hormonalen Beziehungen zwischen Spitze und Basis der Avenacoleoptile. Planta 1(3):321–331. doi:10.1007/bf02039223
Bradley PM (1991) Plant hormones do have a role in controlling growth and development of algae. J Phycol 27(3):317–321. doi:10.1111/j.0022-3646.1991.00317.x
Braun N, Wyrzykowska J, Muller P, David K, Couch D, Perrot-Rechenmann C, Fleming AJ (2008) Conditional repression of AUXIN BINDING PROTEIN1 reveals that it coordinates cell division and cell expansion during postembryonic shoot development in Arabidopsis and tobacco. Plant Cell 20(10):2746–2762. doi:10.1105/tpc.108.059048
Calderon Villalobos LI, Lee S, De Oliveira C, Ivetac A, Brandt W, Armitage L, Sheard LB, Tan X, Parry G, Mao H, Zheng N, Napier R, Kepinski S, Estelle M (2012) A combinatorial TIR1/AFB-Aux/IAA co-receptor system for differential sensing of auxin. Nat Chem Biol 8(5):477–485. doi:10.1038/nchembio.926
Calderon-Villalobos LI, Tan X, Zheng N, Estelle M (2010) Auxin perception—structural insights. Cold Spring Harb Perspect Biol 2(7):a005546. doi:10.1101/cshperspect.a005546
Chen R, Hilson P, Sedbrook J, Rosen E, Caspar T, Masson PH (1998) The Arabidopsis thaliana AGRAVITROPIC 1 gene encodes a component of the polar-auxin-transport efflux carrier. Proc Natl Acad Sci U S A 95(25):15112–15117
Chen X, Naramoto S, Robert S, Tejos R, Lofke C, Lin D, Yang Z, Friml J (2012) ABP1 and ROP6 GTPase signaling regulate clathrin-mediated endocytosis in Arabidopsis roots. Curr Biol 22 (14):1326–1332. doi:10.1016/j.cub.2012.05.020

Cheng Y, Dai X, Zhao Y (2006) Auxin biosynthesis by the YUCCA flavin monooxygenases controls the formation of floral organs and vascular tissues in Arabidopsis. Genes Dev 20:1790–1799. doi:10.1101/gad.1415106
Cho M, Cho HT (2012) The function of ABCB transporters in auxin transport. Plant Signal Behav 8(2):e22990. doi:10.4161/psb.22990
Christie JM, Murphy AS (2013) Shoot phototropism in higher plants: new light through old concepts. Am J Bot 100(1):35–46. doi:10.3732/ajb.1200340
Ciesielski T (1871) Untersuchungen über die Abwärtskrümmung der Wurzel. [s.n.], Breslau
Cock JM, SterckL RP, Scornet D, Allen AE, Amoutzias G, Anthouard V, Artiguenave F, Aury J, Badger JH, Beszteri B, Billiau K, Bonnet E, Bothwell JH, Bowler C, Boyen C, Brownlee C, Carrano CJ, Charrier B, Cho GY, Coelho SM, Collén J, Corre E, Silva CD, Delage L, Delaroque N, Dittami SM, Doulbeau S, Elias M, Farnham G, Gachon CMM, Gschloessl B, Heesch S, Jabbari K, Jubin C, Kawai H, Kimura K, Kloareg B, Küpper FC, Lang D, Bail AL, Leblanc C, Lerouge P, Lohr M, Lopez PJ, Martens C, Maumus F, Michel G, Miranda-Saavedra D, Morales J, Moreau H, Motomura T, Nagasato C, Napoli CA, Nelson DR, Nyvall-Collén P, Peters AF, Pommier C, Potin P, Poulain J, Quesneville H, Read B, Rensing SA, Ritter A, Rousvoal S, Samanta M, Samson G, Schroeder DC, Ségurens B, Strittmatter M, Tonon T, Tregear JW, Valentin K, Dassow PV, Yamagishi T, de Peer YV, Wincker P (2010) The Ectocarpus genome and the independent evolution of multicellularity in brown algae. Nature 465(7298):617–621
Conrad H, Saltman P, Eppley R (1959) Effects of auxin and gibberellic acid on growth of ulothrix. Nature 184(4685):556–557. doi:10.1038/184556a0
Cooke TJ, Poli D, Sztein AE, Cohen JD (2002) Evolutionary patterns in auxin action. Plant Mol Biol 49(3–4):319–338. doi:10.1023/A:1015242627321
Darwin C (1880) The power of movement in plants. John Murray, London
Davies P (2010) The plant hormones: their nature, occurrence, and functions. In: Davies P (ed) Plant hormones. Springer, Netherlands, pp 1–15. doi:10.1007/978-1-4020-2686-7_1
de Billy F, Grosjean C, May S, Bennett M, Cullimore JV (2001) Expression studies on AUX1-like genes in Medicago truncatula suggest that auxin is required at two steps in early nodule development. Mol Plant Microbe Interact 14(3):267–277. doi:10.1094/MPMI.2001.14.3.267
De Meutter J, Tytgat T, Prinsen E, Gheysen G, Van Onckelen H (2005) Production of auxin and related compounds by the plant parasitic nematodes Heterodera schachtii and Meloidogyne incognita. Commun Agric Appl Biol Sci 70(1):51–60
de Reuille PB, Bohn-Courseau I, Ljung K, Morin H, Carraro N, Godin C, Traas J (2006) Computer simulations reveal properties of the cell-cell signaling network at the shoot apex in Arabidopsis. Proc Natl Acad Sci U S A 103(5):1627–1632. doi:10.1073/pnas.0510130103
De Smet I, Lau S, Mayer U, Jürgens G (2010) Embryogenesis – the humble beginnings of plant life. Plant J 61(6):959–970. doi:10.1111/j.1365-313X.2010.04143.x
De Smet I, Voβ U, Lau S, Wilson M, Shao N, Timme RE, Swarup R, Kerr I, Hodgman C, Bock R, Bennett M, Jürgens G, Beeckman T (2011) Unraveling the evolution of auxin signaling. Plant Physiol 155(1):209–221, doi:pp.110.168161 [pii]
Decker EL, Frank W, Sarnighausen E, Reski R (2006) Moss systems biology en route: phytohormones in Physcomitrella development. Plant Biol (Stuttg) 8(3):397–405. doi:10.1055/s-2006-923952
Dharmasiri N, Dharmasiri S, Estelle M (2005) The F-box protein TIR1 is an auxin receptor. Nature 435(7041):441–445. doi:10.1038/nature03543
Ding Z, Wang B, Moreno I, Dupláková N, Simon S, Carraro N, Reemmer J, Pěnčík A, Chen X, Tejos R, Skůpa P, Pollmann S, Mravec J, Petrášek J, Zažímalová E, Honys D, Rolčík J, Murphy A, Orellana A, Geisler M, Friml J (2012) ER-localized auxin transporter PIN8 regulates auxin homeostasis and male gametophyte development in Arabidopsis. Nat Commun 3:941. doi:10.1038/ncomms1941

Dubrovsky JG, Sauer M, Napsucialy-Mendivil S, Ivanchenko MG, Friml J, Shishkova S, Celenza J, Benkova E (2008) Auxin acts as a local morphogenetic trigger to specify lateral root founder cells. Proc Natl Acad Sci U S A 105:8790–8794. doi:10.1073/pnas.0712307105

Eklund DM, Thelander M, Landberg K, Ståldal V, Nilsson A, Johansson M, Valsecchi I, Pederson ERA, Kowalczyk M, Ljung K, Ronne H, Sundberg E (2010) Homologues of the Arabidopsis thaliana SHI/STY/LRP1 genes control auxin biosynthesis and affect growth and development in the moss Physcomitrella patens. Development 137(8):1275–1284. doi:10.1242/dev.039594

Evans LV, Trewavas AJ (1991) Is algal development controlled by plant growth substances? J Phycol 27(3):322–326. doi:10.1111/j.0022-3646.1991.00322.x

Feller C, Gabriel JP, Mazza C, Yerly F (2013) Pattern formation in auxin flux. J Math Biol 23:23. doi:10.1007/s00285-013-0655-9

Feraru E, Vosolsobě S, Feraru MI, Petrášek J, Kleine-Vehn J (2012) Evolution and structural diversification of PILS putative auxin carriers in plants. Front Plant Sci 3:227. doi:10.3389/fpls.2012.00227

Finet C, Jaillais Y (2012) Auxology: when auxin meets plant evo-devo. Dev Biol 369(1):19–31. doi:10.1016/j.ydbio.2012.05.039S0012-1606(12)00309-0

Finet C, Timme RE, Delwiche CF, Marlétaz F (2010) Multigene phylogeny of the green lineage reveals the origin and diversification of land plants. Curr Biol 20(24):2217–2222. doi:10.1016/j.cub.2010.11.035

Friml J, Benková E, Blilou I, Wisniewska J, Hamann T, Ljung K, Woody S, Sandberg G, Scheres B, Jürgens G, Palme K (2002a) AtPIN4 mediates sink-driven auxin gradients and root patterning in arabidopsis. Cell 108(5):661–673. doi:10.1016/S0092-8674(02)00656-6

Friml J, Wisniewska J, Benkova E, Mendgen K, Palme K (2002b) Lateral relocation of auxin efflux regulator PIN3 mediates tropism in Arabidopsis. Nature 415(6873):806–809. doi:10.1038/415806a

Friml J, Vieten A, Sauer M, Weijers D, Schwarz H, Hamann T, Offringa R, Jurgens G (2003) Efflux-dependent auxin gradients establish the apical-basal axis of Arabidopsis. Nature 426(6963):147–153. doi:10.1038/nature02085

Fu X, Harberd NP (2003) Auxin promotes Arabidopsis root growth by modulating gibberellin response. Nature 421(6924):740–743. doi:10.1038/nature01387

Fujita T, Hasebe M (2009) Convergences and divergences in polar auxin transport and shoot development in land plant evolution. Plant Signal Behav 4(4):313–315. doi:10.4161/psb.4.4.8090

Fujita T, Sakaguchi H, Hiwatashi Y, Wagstaff SJ, Ito M, Deguchi H, Sato T, Hasebe M (2008) Convergent evolution of shoots in land plants: lack of auxin polar transport in moss shoots. Evol Dev 10(2):176–186. doi:10.1111/j.1525-142X.2008.00225.x

Gagne JM, Downes BP, Shiu S-H, Durski AM, Vierstra RD (2002) The F-box subunit of the SCF E3 complex is encoded by a diverse superfamily of genes in Arabidopsis. Proc Natl Acad Sci 99(17):11519–11524. doi:10.1073/pnas.162339999

Gälweiler L, Guan C, Müller A, Wisman E, Mendgen K, Yephremov A, Palme K (1998) Regulation of polar auxin transport by AtPIN1 in Arabidopsis vascular tissue. Science 282(5397):2226–2230. doi:10.1126/science.282.5397.2226

Geisler M, Murphy AS (2006) The ABC of auxin transport: the role of p-glycoproteins in plant development. FEBS Lett 580(4):1094–1102. doi:10.1016/j.febslet.2005.11.054

Graham LE, Cook ME, Busse JS (2000) The origin of plants: body plan changes contributing to a major evolutionary radiation. Proc Natl Acad Sci 97(9):4535–4540. doi:10.1073/pnas.97.9.4535

Gruen HE (1959) Auxins and fungi. Annu Rev Plant Physiol Plant Mol Biol 10:405–440. doi:10.1146/annurev.pp. 10.060159.002201

Grunewald W, Friml J (2010) The march of the PINs: developmental plasticity by dynamic polar targeting in plant cells. EMBO J 29(16):2700–2714. doi:10.1038/emboj.2010.181

Ha CM, Jun JH, Fletcher JC (2010) Chapter Four – Shoot apical meristem form and function. In: Marja CPT (ed) Current topics in developmental biology, vol 91. Academic Press, USA, pp 103–140. doi:10.1016/S0070-2153(10)91004-1

Haagen-Smit A, Dandliker W, Wittwer S, Murneek A (1946) Isolation of 3-indoleacetic acid from immature corn kernels. Am J Bot 33:118–120. doi:10.2307/2437327

Hayashi K (2012) The interaction and integration of auxin signaling components. Plant Cell Physiol 53(6):965–975. doi:10.1093/pcp/pcs035

Hayashi K-I, Tan X, Zheng N, Hatate T, Kimura Y, Kepinski S, Nozaki H (2008) Small-molecule agonists and antagonists of F-box protein–substrate interactions in auxin perception and signaling. Proc Natl Acad Sci 105(14):5632–5637. doi:10.1073/pnas.0711146105

Heisler MG, Jönsson H (2006) Modeling auxin transport and plant development. J Plant Growth Regul 25(4):302–312. doi:10.1007/s00344-006-0066-x

Heisler MG, Ohno C, Das P, Sieber P, Reddy GV, Long JA, Meyerowitz EM (2005) Patterns of auxin transport and gene expression during primordium development revealed by live imaging of the Arabidopsis inflorescence meristem. Curr Biol 15(21):1899–1911. doi:10.1016/j.cub.2005.09.052

Hochholdinger F, Wulff D, Reuter K, Park WJ, Feix G (2000) Tissue-specific expression of AUX1 in maize roots. J Plant Physiol 157(3):315–319. doi:10.1016/S0176-1617(00)80053-X

Hošek P, Kubeš M, Laňková M, Dobrev PI, Klima P, Kohoutová M, Petrášek J, Hoyerová K, Jiřina M, Zažímalová E (2012) Auxin transport at cellular level: new insights supported by mathematical modelling. J Exp Bot 63(10):3815–3827. doi:10.1093/jxb/ers074

Ichimura H, Yamaki T (1975) Indole-3-acetic acid in chick embryos. Dev Growth Differ 17 (3):275–279. doi:10.1111/j.1440-169X.1975.00275.x

Ikeda Y, Men S, Fischer U, Stepanova AN, Alonso JM, Ljung K, Grebe M (2009) Local auxin biosynthesis modulates gradient-directed planar polarity in Arabidopsis. Nat Cell Biol 11:731–738. doi:10.1038/ncb1879

Isenbarger TA, Carr CE, Johnson SS, Finney M, Church GM, Gilbert W, Zuber MT, Ruvkun G (2008) The most conserved genome segments for life detection on earth and other planets. Orig Life Evol Biosph 38(6):517–533. doi:10.1007/s11084-008-9148-z

Jacobs WP (1950) Auxin effects in bryopsis. Biol Bull 99(2):369–370

Jacobs WP, Falkenstein K, Hamilton RH (1985) Nature and amount of auxin in algae: IAA from extracts of Caulerpa paspaloides (Siphonales). Plant Physiol 78(4):844–848. doi:10.1104/pp.78.4.844

Jang G, Dolan L (2011) Auxin promotes the transition from chloronema to caulonema in moss protonema by positively regulating PpRSL1and PpRSL2 in Physcomitrella patens. New Phytol 192(2):319–327. doi:10.1111/j.1469-8137.2011.03805.x

Johri MM, Desai S (1973) Auxin regulation of caulonema formation in moss protonema. Nat New Biol 245:1. doi:10.1038/newbio245223a0

Jones AM, Herman EM (1993) KDEL-containing auxin-binding protein is secreted to the plasma membrane and cell wall. Plant Physiol 101(2):595–606. doi:10.1104/pp. 101.2.595

Jones AR, Kramer EM, Knox K, Swarup R, Bennett MJ, Lazarus CM, Leyser HM, Grierson CS (2009) Auxin transport through non-hair cells sustains root-hair development. Nat Cell Biol 11:78–84. doi:10.1038/ncb1815

Jonsson H, Heisler MG, Shapiro BE, Meyerowitz EM, Mjolsness E (2006) An auxin-driven polarized transport model for phyllotaxis. Proc Natl Acad Sci U S A 103:1633–1638. doi:10.1073/pnas.0509839103

Kai K, Horita J, Wakasa K, Miyagawa H (2007) Three oxidative metabolites of indole-3-acetic acid from Arabidopsis thaliana. Phytochemistry 68(12):1651–1663. doi:10.1016/j.phytochem.2007.04.030

Kamimoto Y, Terasaka K, Hamamoto M, Takanashi K, Fukuda S, Shitan N, Sugiyama A, Suzuki H, Shibata D, Wang B, Pollmann S, Geisler M, Yazaki K (2012) Arabidopsis ABCB21 is a facultative auxin importer/exporter regulated by cytoplasmic auxin concentration. Plant Cell Physiol 53(12):2090–2100. doi:10.1093/pcp/pcs149

Karol KG, McCourt RM, Cimino MT, Delwiche CF (2001) The closest living relatives of land plants. Science 294(5550):2351–2353. doi:10.1126/science.1065156

Kenrick P, Wellman CH, Schneider H, Edgecombe GD (2012) A timeline for terrestrialization: consequences for the carbon cycle in the Palaeozoic. Philos Trans R Soc Lond B Biol Sci 367 (1588):519–536. doi:10.1098/rstb.2011.0271

Kepinski S, Leyser O (2005) The Arabidopsis F-box protein TIR1 is an auxin receptor. Nature 435 (7041):446–451. doi:10.1038/nature03542

Koepfli JB, Thimann KV, Went FW (1938) Phytohormones: structure and physiological activity. I. J Biol Chem 122(3):763–780

Krecek P, Skupa P, Libus J, Naramoto S, Tejos R, Friml J, Zazimalova E (2009) The PIN-FORMED (PIN) protein family of auxin transporters. Genome Biol 10(12):249, doi:gb-2009-10-12-249 [pii]10.1186/gb-2009-10-12-249

Kubeš M, Yang H, Richter GL, Cheng Y, Mlodzinska E, Wang X, Blakeslee JJ, Carraro N, Petrášek J, Zažímalová E, Hoyerová K, Peer WA, Murphy AS (2012) The Arabidopsis concentration-dependent influx/efflux transporter ABCB4 regulates cellular auxin levels in the root epidermis. Plant J 69(4):640–654. doi:10.1111/j.1365-313X.2011.04818.x

Lau S, Jurgens G, De Smet I (2008) The evolving complexity of the auxin pathway. Plant Cell 20 (7):1738–1746. doi:10.1105/tpc.108.060418

Lau S, Shao N, Bock R, Jurgens G, De Smet I (2009) Auxin signaling in algal lineages: fact or myth? Trends Plant Sci 14(4):182–188. doi:10.1016/j.tplants.2009.01.004

Le Bail A, Billoud B, Kowalczyk N, Kowalczyk M, Gicquel M, Le Panse S, Stewart S, Scornet D, Cock JM, Ljung K, Charrier B (2010) Auxin metabolism and function in the multicellular brown alga Ectocarpus siliculosus. Plant Physiol 153(1):128–144. doi:10.1104/pp. 109.149708

Lehmann T, Hoffmann M, Hentrich M, Pollmann S (2010) Indole-3-acetamide-dependent auxin biosynthesis: a widely distributed way of indole-3-acetic acid production? Eur J Cell Biol 89 (12):895–905, doi:S0171-9335(10)00135-4 [pii]

Lewis LA, McCourt RM (2004) Green algae and the origin of land plants. Am J Bot 91(10):1535–1556. doi:10.3732/ajb.91.10.1535

Leyser O (2011) Auxin, self-organisation, and the colonial nature of plants. Curr Biol 21(9):R331–R337, doi:S0960-9822(11)00222-3 [pii]

Litwack G (2005) Plant hormones. Elsevier Academic Press Inc, USA

Ljung K (2013) Auxin metabolism and homeostasis during plant development. Development 140 (5):943–950. doi:10.1242/dev.086363

Long JA, Ohno C, Smith ZR, Meyerowitz EM (2006) TOPLESS regulates apical embryonic fate in Arabidopsis. Science 312(5779):1520–1523. doi:10.1126/science.1123841

Ludwig-Mueller J (2011) Auxin conjugates: their role for plant development and in the evolution of land plants. J Exp Bot 62(6):1757–1773. doi:10.1093/jxb/erq412

Ludwig-Muller J, Julke S, Bierfreund NM, Decker EL, Reski R (2009) Moss (Physcomitrella patens) GH3 proteins act in auxin homeostasis. New Phytol 181(2):323–338. doi:10.1111/j. 1469-8137.2008.02677.x

Luschnig C, Gaxiola RA, Grisafi P, Fink GR (1998) EIR1, a root-specific protein involved in auxin transport, is required for gravitropism in Arabidopsis thaliana. Genes Dev 12(14):2175–2187. doi:10.1101/gad.12.14.2175

Mano Y, Nemoto K (2012) The pathway of auxin biosynthesis in plants. J Exp Bot 63(8):2853–2872. doi:10.1093/jxb/ers091

Maraschin Fdos S, Memelink J, Offringa R (2009) Auxin-induced, SCF(TIR1)-mediated poly-ubiquitination marks AUX/IAA proteins for degradation. Plant J 59(1):100–109. doi:10.1111/j.1365-313X.2009.03854.x

Marchant A, Bhalerao R, Casimiro I, Eklöf J, Casero PJ, Bennett M, Sandberg G (2002) AUX1 promotes lateral root formation by facilitating indole-3-acetic acid distribution between sink and source tissues in the Arabidopsis seedling. Plant Cell 14(3):589–597. doi:10.1105/tpc. 010354

Merks RM, Van de Peer Y, Inze D, Beemster GT (2007) Canalization without flux sensors: a traveling-wave hypothesis. Trends Plant Sci 12:384–390. doi:10.1016/j.tplants.2007.08.004

Mjolsness E (2006) The growth and development of some recent plant models: a viewpoint. J Plant Growth Regul 25(4):270–277. doi:10.1007/s00344-006-0069-7

Mockaitis K, Estelle M (2008) Auxin receptors and plant development: a new signaling paradigm. Annu Rev Cell Dev Biol 24:55–80. doi:10.1146/annurev.cellbio.23.090506.123214

Monroe-Augustus M, Zolman BK, Bartel B (2003) IBR5, a dual-specificity phosphatase-like protein modulating auxin and abscisic acid responsiveness in Arabidopsis. Plant Cell 15 (12):2979–2991. doi:10.1105/tpc.017046

Mravec J, Kubes M, Bielach A, Gaykova V, Petrasek J, Skupa P, Chand S, Benkova E, Zazimalova E, Friml J (2008) Interaction of PIN and PGP transport mechanisms in auxin distribution-dependent development. Development 135:3345–3354. doi:10.1242/dev.021071

Mravec J, Skupa P, Bailly A, Hoyerova K, Krecek P, Bielach A, Petrasek J, Zhang J, Gaykova V, Stierhof YD, Dobrev PI, Schwarzerova K, Rolcik J, Seifertova D, Luschnig C, Benkova E, Zazimalova E, Geisler M, Friml J (2009) Subcellular homeostasis of phytohormone auxin is mediated by the ER-localized PIN5 transporter. Nature 459(7250):1136–1140. doi:10.1038/nature08066

Muller A, Guan C, Galweiler L, Tanzler P, Huijser P, Marchant A, Parry G, Bennett M, Wisman E, Palme K (1998) AtPIN2 defines a locus of Arabidopsis for root gravitropism control. EMBO J 17(23):6903–6911. doi:10.1093/emboj/17.23.6903

Napier RM (2005) TIRs of joy: new receptors for auxin. Bioessays 27(12):1213–1217. doi:10.1002/bies.20329

Newell AC, Shipman PD, Sun Z (2008) Phyllotaxis: cooperation and competition between mechanical and biochemical processes. J Theor Biol 251(3):421–439. doi:10.1016/j.jtbi.2007.11.036

Niklas KJ, Kutschera U (2009) The evolutionary development of plant body plans. Funct Plant Biol 36(8):682–695. doi:10.1071/fp09107

Nishiyama T, Fujita T, Shin IT, Seki M, Nishide H, Uchiyama I, Kamiya A, Carninci P, Hayashizaki Y, Shinozaki K, Kohara Y, Hasebe M (2003) Comparative genomics of Physcomitrella patens gametophytic transcriptome and Arabidopsis thaliana: implication for land plant evolution. Proc Natl Acad Sci U S A 100(13):8007–8012. doi:10.1073/pnas.0932694100

Normanly J (2010) Approaching cellular and molecular resolution of auxin biosynthesis and metabolism. Cold Spring Harb Perspect Biol 2(1):a001594. doi:10.1101/cshperspect.a001594

Novák O, Hényková E, Sairanen I, Kowalczyk M, Pospíšil T, Ljung K (2012) Tissue-specific profiling of the Arabidopsis thaliana auxin metabolome. Plant J 72(3):523–536. doi:10.1111/j.1365-313X.2012.05085.x

Novoselova ES, Mironova VV, Omelyanchuk NA, Kazantsev FV, Likhoshvai VA (2013) Mathematical modeling of auxin transport in protoxylem and protophloem of Arabidopsis thaliana root tips. J Bioinform Comput Biol 11(1):6. doi:10.1142/S0219720013400106

Orbovik V, Poff KL (1993) Growth distribution during Phototropism of Arabidopsis thaliana seedlings. Plant Physiology 103(1):157–163. doi:10.1104/pp. 103.1.157

Östin A, Kowalyczk M, Bhalerao RP, Sandberg G (1998) Metabolism of indole-3-acetic acid in Arabidopsis. Plant Physiol 118(1):285–296. doi:10.1104/pp. 118.1.285

Overbeek JV (1940) Auxin in marine algae. Plant Physiol 15(2):291–299. doi:10.1104/pp. 15.2.291

Pagnussat GC, Alandete-Saez M, Bowman JL, Sundaresan V (2009) Auxin-dependent patterning and gamete specification in the Arabidopsis female gametophyte. Science 324(5935):1684–1689, doi:1167324 [pii] 10.1126/science.1167324

Panigrahi KC, Panigrahy M, Vervliet-Scheebaum M, Lang D, Reski R, Johri MM (2009) Auxin-binding proteins without KDEL sequence in the moss Funaria hygrometrica. Plant Cell Rep 28 (11):1747–1758. doi:10.1007/s00299-009-0775-2

Paponov IA, Teale WD, Trebar M, Blilou K, Palme K (2005) The PIN auxin efflux facilitators: evolutionary and functional perspectives. Trends Plant Sci 10(4):170–177. doi:10.1016/j.tplants.2005.02.009

Paponov IA, Teale W, Lang D, Paponov M, Reski R, Rensing SA, Palme K (2009) The evolution of nuclear auxin signalling. BMC Evol Biol 9:126, doi:1471-2148-9-126 [pii] 10.1186/1471-2148-9-126

Park JE, Park JY, Kim YS, Staswick PE, Jeon J, Yun J, Kim SY, Kim J, Lee YH, Park CM (2007) GH3-mediated auxin homeostasis links growth regulation with stress adaptation response in Arabidopsis. J Biol Chem 282(13):10036–10046. doi:10.1074/jbc.M610524200

Patten CL, Glick BR (1996) Bacterial biosynthesis of indole-3-acetic acid. Can J Microbiol 42 (3):207–220. doi:10.1139/m96-032

Pattison RJ, Catala C (2012) Evaluating auxin distribution in tomato (Solanum lycopersicum) through an analysis of the PIN and AUX/LAX gene families. Plant J 70(4):585–598. doi:10.1111/j.1365-313X.2011.04895.x

Péret B, Swarup R, Jansen L, Devos G, Auguy F, Collin M, Santi C, Hocher V, Franche C, Bogusz D, Bennett M, Laplaze L (2007) Auxin influx activity is associated with frankia infection during actinorhizal nodule formation in Casuarina glauca. Plant Physiol 144 (4):1852–1862. doi:10.1104/pp. 107.101337

Péret B, De Rybel B, Casimiro I, Benkova E, Swarup R, Laplaze L, Beeckman T, Bennett MJ (2009) Arabidopsis lateral root development: an emerging story. Trends Plant Sci 14:399–408. doi:10.1016/j.tplants.2009.05.002

Péret B, Swarup K, Ferguson A, Seth M, Yang Y, Dhondt S, James N, Casimiro I, Perry P, Syed A, Yang H, Reemmer J, Venison E, Howells C, Perez-Amador MA, Yun J, Alonso J, Beemster GT, Laplaze L, Murphy A, Bennett MJ, Nielsen E, Swarup R (2012) AUX/LAX genes encode a family of auxin influx transporters that perform distinct functions during Arabidopsis development. Plant Cell 24(7):2874–2885. doi:10.1105/tpc.112.097766

Peris CIL, Rademacher EH, Weijers D (2010) Green beginnings – pattern formation in the early plant embryo. In: Timmermans MCP (ed) Plant development, vol 91, Current topics in developmental biology. Elsevier Academic, San Diego, pp 1–27. doi:10.1016/s0070-2153 (10)91001-6

Petrášek J, Friml J (2009) Auxin transport routes in plant development. Development 136 (16):2675–2688. doi:10.1242/dev.030353

Petrasek J, Mravec J, Bouchard R, Blakeslee JJ, Abas M, Seifertova D, Wisniewska J, Tadele Z, Kubes M, Covanova M, Dhonukshe P, Skupa P, Benkova E, Perry L, Krecek P, Lee OR, Fink GR, Geisler M, Murphy AS, Luschnig C (2006) PIN proteins perform a rate-limiting function in cellular auxin efflux. Science 312:914–918. doi:10.1126/science.1123542

Pinon V, Prasad K, Grigg SP, Sanchez-Perez GF, Scheres B (2013) Local auxin biosynthesis regulation by PLETHORA transcription factors controls phyllotaxis in Arabidopsis. Proc Natl Acad Sci U S A 110(3):1107–1112. doi:10.1073/pnas.1213497110

Pires ND, Dolan L (2012) Morphological evolution in land plants: new designs with old genes. Philos Trans R Soc Lond B Biol Sci 367(1588):508–518. doi:10.1098/rstb.2011.0252

Poli DB, Jacobs M, Cooke TJ (2003) Auxin regulation of axial growth in bryophyte sporophytes: its potential significance for the evolution of early land plants. Am J Bot 90(10):1405–1415. doi:10.3732/ajb.90.10.1405

Porter WL, Thimann KV (1965) Molecular requirements for auxin action-I. Halogenated indoles and indoleacetic acid. Phytochemistry 4(2):229–243. doi:10.1016/S0031-9422(00)86169-5

Prigge MJ, Lavy M, Ashton NW, Estelle M (2010) Physcomitrella patens auxin-resistant mutants affect conserved elements of an auxin-signaling pathway. Curr Biol 20(21):1907–1912. doi:10.1016/j.cub.2010.08.050

Reinhardt D (2005) Regulation of phyllotaxis. Int J Dev Biol 49(5–6):539–546, doi:041922dr [pii] 10.1387/ijdb.041922dr

Reinhardt D, Pesce ER, Stieger P, Mandel T, Baltensperger K, Bennett M, Traas J, Friml J, Kuhlemeier C (2003) Regulation of phyllotaxis by polar auxin transport. Nature 426:255–260. doi:10.1038/nature02081

Rensing SA, Lang D, Zimmer AD, Terry A, Salamov A, Shapiro H, Nishiyama T, Perroud PF, Lindquist EA, Kamisugi Y, Tanahashi T, Sakakibara K, Fujita T, Oishi K, Shin IT, Kuroki Y, Toyoda A, Suzuki Y, Hashimoto S, Yamaguchi K, Sugano S, Kohara Y, Fujiyama A, Anterola A, Aoki S, Ashton N, Barbazuk WB, Barker E, Bennetzen JL, Blankenship R, Cho SH, Dutcher SK, Estelle M, Fawcett JA, Gundlach H, Hanada K, Heyl A, Hicks KA, Hughes J, Lohr M, Mayer K, Melkozernov A, Murata T, Nelson DR, Pils B, Prigge M, Reiss B, Renner T, Rombauts S, Rushton PJ, Sanderfoot A, Schween G, Shiu SH, Stueber K, Theodoulou FL, Tu H, Van de Peer Y, Verrier PJ, Waters E, Wood A, Yang L, Cove D, Cuming AC, Hasebe M, Lucas S, Mishler BD, Reski R, Grigoriev IV, Quatrano RS, Boore JL (2008) The physcomitrella genome reveals evolutionary insights into the conquest of land by plants. Science 319(5859):64–69, doi:1150646 [pii] 10.1126/science.1150646

Robert S, Kleine Vehn J, Barbez E, Sauer M, Paciorek T, Baster P, Vanneste S, Zhang J, Simon S, Covanova M, Hayashi K, Dhonukshe P, Yang Z, Bednarek SY, Jones AM, Luschnig C, Aniento F, Zazimalova E, Friml J (2010) ABP1 mediates auxin inhibition of clathrin-dependent endocytosis in Arabidopsis. Cell 143(1):111–121. doi:10.1016/j.cell.2010.09.027

Ross J, Reid J (2010) Evolution of growth-promoting plant hormones. Funct Plant Biol 37:10. doi:10.1071/FP10063

Ruiz Rosquete M, Barbez E, Kleine-Vehn J (2012) Cellular auxin homeostasis: gatekeeping is housekeeping. Mol Plant 5(4):772–786. doi:10.1093/mp/ssr109

Ruiz-Trillo I, Burger G, Holland PWH, King N, Lang BF, Roger AJ, Gray MW (2007) The origins of multicellularity: a multi-taxon genome initiative. Trends Genet 23(3):113–118. doi:10.1016/j.tig.2007.01.005

Sakakibara K (2003) Involvement of auxin and a homeodomain-leucine zipper I gene in rhizoid development of the moss Physcomitrella patens. Development 130(20):4835–4846. doi:10.1242/dev.00644

Sanderson M, Thorne J, Wikstrom N, Bremer K (2004) Molecular evidence on plant divergence times. Am J Bot 91(10):1656–1665. doi:10.3732/ajb.91.10.1656

Sauer M, Balla J, Luschnig C, Wisniewska J, Reinohl V, Friml J, Benkova E (2006) Canalization of auxin flow by Aux/IAA-ARF-dependent feedback regulation of PIN polarity. Genes Dev 20 (20):2902–2911. doi:10.1101/gad.390806

Sawchuk MG, Edgar A, Scarpella E (2013) Patterning of leaf vein networks by convergent auxin transport pathways. PLoS Genet 9(2):e1003294. doi:10.1371/journal.pgen.1003294

Scarpella E, Helariutta Y (2010) Chapter Eight – Vascular pattern formation in plants. In: Marja CPT (ed) Current topics in developmental biology, vol 91. Academic Press, USA, pp 221–265. doi:10.1016/S0070-2153(10)91008-9

Shipman PD, Newell AC (2005) Polygonal planforms and phyllotaxis on plants. J Theor Biol 236 (2):154–197. doi:10.1016/j.jtbi.2005.03.007

Simon S, Petrášek J (2011) Why plants need more than one type of auxin. Plant Sci 180(3):454–460. doi:10.1016/j.plantsci.2010.12.007

Simon A, Glockner G, Felder M, Melkonian M, Becker B (2006) EST analysis of the scaly green flagellate Mesostigma viride (Streptophyta): implications for the evolution of green plants (Viridiplantae). BMC Plant Biol 6(1):2. doi:10.1186/1471-2229-6-2

Smith RS, Guyomarc'h S, Mandel T, Reinhardt D, Kuhlemeier C, Prusinkiewicz P (2006) A plausible model of phyllotaxis. Proc Natl Acad Sci USA 103:1301–1306. doi:10.1073/pnas.0510457103

Spaepen S, Vanderleyden J, Remans R (2007) Indole-3-acetic acid in microbial and microorganism-plant signaling. FEMS Microbiol Rev 31(4):425–448. doi:10.1111/j.1574-6976.2007.00072.x

Stark P (1921) Studien iiber traumatotrope und haptotrope Reizleitungsvorgange mit besonderer Berücksichtigung der Reizübertragung auf fremde Arten und Gattungen. Jahr biicher fiir wissenschaftliche Botanik 60:67
Stepanova AN (2008) TAA1-mediated auxin biosynthesis is essential for hormone crosstalk and plant development. Cell 133:177–191. doi:10.1016/j.cell.2008.01.047
Stepanova AN, Hoyt JM, Hamilton AA, Alonso JM (2005) A link between ethylene and auxin uncovered by the characterization of two root-specific ethylene-insensitive mutants in Arabidopsis. Plant Cell 17:2230–2242. doi:10.1105/tpc.105.033365
Strader LC, Monroe-Augustus M, Bartel B (2008) The IBR5 phosphatase promotes Arabidopsis auxin responses through a novel mechanism distinct from TIR1-mediated repressor degradation. BMC Plant Biol 8:41. doi:10.1186/1471-2229-8-41
Sugawara S, Hishiyama S, Jikumaru Y, Hanada A, Nishimura T, Koshiba T, Zhao Y, Kamiya Y, Kasahara H (2009) Biochemical analyses of indole-3-acetaldoxime-dependent auxin biosynthesis in Arabidopsis. Proc Natl Acad Sci 106(13):5430–5435. doi:10.1073/pnas.0811226106
Sun H, Basu S, Brady SR, Luciano RL, Muday GK (2004) Interactions between auxin transport and the actin cytoskeleton in developmental polarity of Fucus distichus embryos in response to light and gravity. Plant Physiol 135(1):266–278. doi:10.1104/pp. 103.034900
Swarup R, Friml J, Marchant A, Ljung K, Sandberg G, Palme K, Bennett M (2001) Localization of the auxin permease AUX1 suggests two functionally distinct hormone transport pathways operate in the Arabidopsis root apex. Genes Dev 15:2648–2653. doi:10.1101/gad.210501
Swarup K, Benkova E, Swarup R, Casimiro I, Peret B, Yang Y, Parry G, Nielsen E, De Smet I, Vanneste S, Levesque MP, Carrier D, James N, Calvo V, Ljung K, Kramer E, Roberts R, Graham N, Marillonnet S, Patel K (2008) The auxin influx carrier LAX3 promotes lateral root emergence. Nat Cell Biol 10:946–954. doi:10.1038/ncb1754
Szemenyei H, Hannon M, Long JA (2008) TOPLESS mediates auxin-dependent transcriptional repression during Arabidopsis embryogenesis. Science 319(5868):1384–1386. doi:10.1126/science.1151461
Sztein AE, Cohen JD, Cooke TJ (2000) Evolutionary patterns in the auxin metabolism of green plants. Int J Plant Sci 161(6):849–859. doi:10.1086/317566
Tan X, Calderon-Villalobos LIA, Sharon M, Zheng C, Robinson CV, Estelle M, Zheng N (2007) Mechanism of auxin perception by the TIR1 ubiquitin ligase. Nature 446(7136):640–645. doi:10.1038/nature05731
Tao Y, Ferrer JL, Ljung K, Pojer F, Hong F, Long JA, Li L, Moreno JE, Bowman ME, Ivans LJ, Cheng Y, Lim J, Zhao Y, Ballare CL, Sandberg G, Noel JP, Chory J (2008) Rapid synthesis of auxin via a new tryptophan-dependent pathway is required for shade avoidance in plants. Cell 133:164–176. doi:10.1016/j.cell.2008.01.049
Terasaka K, Blakeslee JJ, Titapiwatanakun B, Peer WA, Bandyopadhyay A, Makam SN, Lee OR, Richards EL, Murphy AS, Sato F, Yazaki K (2005) PGP4, an ATP binding cassette P-glycoprotein, catalyzes auxin transport in Arabidopsis thaliana roots. Plant Cell 17 (11):2922–2939. doi:10.1105/tpc.105.035816
Thimann KV (1969) The auxins. In: Wilkins MB (ed) The physiology of plant growth and development. McGraw-Hill, New York, p 345
Thimann KV, Skoog F (1933) Studies on the growth hormone of plants: III. The inhibiting action of the growth substance on bud development. Proc Natl Acad Sci 19(7):714–716
Titapiwatanakun B, Blakeslee JJ, Bandyopadhyay A, Yang H, Mravec J, Sauer M, Cheng Y, Adamec J, Nagashima A, Geisler M, Sakai T, Friml J, Peer WA, Murphy AS (2009) ABCB19/PGP19 stabilises PIN1 in membrane microdomains in Arabidopsis. Plant J 57(1):27–44. doi:10.1111/j.1365-313X.2008.03668.x
Tivendale ND, Davies NW, Molesworth PP, Davidson SE, Smith JA, Lowe EK, Reid JB, Ross JJ (2010) Reassessing the role of N-hydroxytryptamine in auxin biosynthesis. Plant Physiol 154 (4):1957–1965. doi:10.1104/pp. 110.165803
Tromas A, Paponov I, Perrot-Rechenmann C (2010) AUXIN BINDING PROTEIN 1: functional and evolutionary aspects. Trends Plant Sci 15(8):436–446. doi:10.1016/j.tplants.2010.05.001

Tsavkelova E, Cherdyntseva T, Klimova S, Shestakov A, Botina S, Netrusov A (2007) Orchid-associated bacteria produce indole-3-acetic acid, promote seed germination, and increase their microbial yield in response to exogenous auxin. Arch Microbiol 188(6):655–664. doi:10.1007/s00203-007-0286-x

Turmel M (2006) The chloroplast genome sequence of Chara vulgaris sheds new light into the closest green algal relatives of land plants. Mol Biol Evol 23(6):1324–1338. doi:10.1093/molbev/msk018

Ugartechea-Chirino Y, Swarup R, Swarup K, Peret B, Whitworth M, Bennett M, Bougourd S (2009) The AUX1 LAX family of auxin influx carriers is required for the establishment of embryonic root cell organization in Arabidopsis thaliana. Ann Bot 105(2):277–289. doi:10.1093/aob/mcp287

Utsuno K, Shikanai T, Yamada Y, Hashimoto T (1998) AGR, an agravitropic locus of Arabidopsis thaliana, encodes a novel membrane-protein family member. Plant Cell Physiol 39(10):1111–1118

Van den Hoek C, Mann D, Jahns HM (1996) Algae: an introduction to phycology. Cambridge University Press, Cambridge

Verrier PJ, Bird D, Burla B, Dassa E, Forestier C, Geisler M, Klein M, Kolukisaoglu Ü, Lee Y, Martinoia E, Murphy A, Rea PA, Samuels L, Schulz B, Spalding EP, Yazaki K, Theodoulou FL (2008) Plant ABC proteins – a unified nomenclature and updated inventory. Trends Plant Sci 13(4):151–159. doi:10.1016/j.tplants.2008.02.001

Viaene T, Delwiche CF, Rensing SA, Friml J (2013) Origin and evolution of PIN auxin transporters in the green lineage. Trends Plant Sci 18(1):5–10, doi:S1360-1385(12)00194-X [pii] 10.1016/j.tplants.2012.08.009

Vieten A, Sauer M, Brewer PB, Friml J (2007) Molecular and cellular aspects of auxin-transport-mediated development. Trends Plant Sci 12(4):160–168. doi:10.1016/j.tplants.2007.03.006

Wabnik K, Kleine-Vehn J, Govaerts W, Friml J (2011) Prototype cell-to-cell auxin transport mechanism by intracellular auxin compartmentalization. Trends Plant Sci 16(9):468–475. doi:10.1016/j.tplants.2011.05.002S1360-1385(11)00100-2

Went FW (1927) Wuchsstoff und wachstum. J.H. de Bussy, Amsterdam

Went F, Thimann K (1937) Phytohormones. Macmillan, New York, NY

Wodniok S, Brinkmann H, Glöckner G, Heidel AJ, Philippe H, Melkonian M, Becker B (2011) Origin of land plants: do conjugating green algae hold the key? BMC Evol Biol 11(1):104. doi:10.1186/1471-2148-11-104

Woodward AW, Bartel B (2005) Auxin: regulation, action, and interaction. Ann Bot 95(5):707–735. doi:10.1093/aob/mci083

Yang H, Murphy AS (2009) Functional expression and characterization of Arabidopsis ABCB, AUX1 and PIN auxin transporters in Schizosaccharomyces pombe. Plant J 59(1):179–191. doi:10.1111/j.1365-313X.2009.03856.x

Yang Y, Hammes UZ, Taylor CG, Schachtman DP, Nielsen E (2006) High-affinity auxin transport by the AUX1 influx carrier protein. Curr Biol 16(11):1123–1127. doi:10.1016/j.cub.2006.04.029

Yokoya NS, West JA, Luchi AE (2004) Effects of plant growth regulators on callus formation, growth and regeneration in axenic tissue cultures of *Gracilaria tenuistipitata* and Gracilaria perplexa (Gracilariales, Rhodophyta). Phycol Res 52(3):244–254. doi:10.1111/j.1440-183.2004.00349.x

Yokoya NS, Stirk WA, van Staden J, Novák O, Turečková V, Pěnčík A, Strnad M (2010) Endogenous cytokinins, auxins, and abscisic acid in red algae from Brazil. J Phycol 46(6):1198–1205. doi:10.1111/j.1529-8817.2010.00898.x

Yoon HS, Hackett JD, Ciniglia C, Pinto G, Bhattacharya D (2004) A molecular timeline for the origin of photosynthetic eukaryotes. Mol Biol Evol 21(5):809–818. doi:10.1093/molbev/msh075

Zažímalová E, Křeček P, Skůpa P, Hoyerová K, Petrášek J (2007) Polar transport of the plant hormone auxin – the role of PIN-FORMED (PIN) proteins. Cell Mol Life Sci 64(13):1621–1637. doi:10.1007/s00018-007-6566-4

Zažímalová E, Murphy AS, Yang H, Hoyerová K, Hošek P (2010) Auxin transporters—why so many? Cold Spring Harb Perspect Biol 2(3):a001552. doi:10.1101/cshperspect.a001552

Zhao Y (2010) Auxin biosynthesis and its role in plant development. Annu Rev Plant Biol 61:49–64. doi:10.1146/annurev-arplant-042809-112308

Zhao Y (2012) Auxin biosynthesis: a simple two-step pathway converts tryptophan to indole-3-acetic acid in plants. Mol Plant 5(2):334–338. doi:10.1093/mp/ssr104

Zimmerman PW, Wilcoxon F (1935) Several chemical growth substances which cause initiation of root and other responses in plants. Contrib Boyce Thompson Inst 7:209–229

# Chapter 14
# Auxin and Self-Organisation

**Peter Nick**

**Abstract** This essay develops a conceptual framework to understand the role of auxin for the genesis of plant organisms. This framework has to consider the specificities of the plant lifestyle and underscores the fact that plant organisation is highly modular. The assembly of these modules is controlled through robust self-organisation driven by autocatalytic loops linked to lateral inhibition which can be formally described as reaction–diffusion system in sensu Turing. Instead of actual inhibitory molecules as in the original Turing model, they achieve lateral inhibition by mutual competition for an activator (auxin). This can be demonstrated for phyllotaxis, but also for vascular differentiation. We study self-organisation in cell strains from tobacco and find that individual cell divisions within a file are synchronised through weak coupling based on a directional flow of auxin. We use this system as a simple minimal organism we have identified an oscillatory circuit as central element of self-organisation. This self-referring circuit connects auxin-dependent remodelling of the actin cytoskeleton with actin-dependent remodelling of auxin flux. The essay concludes with the working hypothesis that the contiguity of plant organisms is manifest in time ("rhythm") rather than in space ("body") and describes an experimental model where the induction of cell axis and polarity as base for self-organisation can be studied de novo in regenerating protoplasts.

P. Nick (✉)
Molecular Cell Biology, Botanical Institute, Karlsruhe Institute of Technology, Kaiserstr. 2, Karlsruhe 76128, Germany
e-mail: peter.nick@bio.uni-karlsruhe.de

E. Zažímalová et al. (eds.), *Auxin and Its Role in Plant Development*,
DOI 10.1007/978-3-7091-1526-8_14, © Springer-Verlag Wien 2014 

# 1 Why Plant Organisms Are Really Different

## *1.1 It All Starts with Biophysics*

All life has to balance supply with consumption. Supply has to occur through the surface, whereas consumption is a function of volume. A growing cell will increase surface by the second power of the radius, but volume by the third power. As a consequence, supply (surface, $r^2$) and consumption (volume, $r^3$) will diverge progressively. To bridge this gap, metabolic efficiency has to be elevated—however, this is possible only due to the innate constraints set for instance by protein structure. When this tunable component of metabolic efficiency is enriched, the surface has to be increased by invaginations or protrusions, a phenomenon already observed already in unicellular organisms. Such surface increases confer a selective advantage, because a larger organism acquires buffering against environmental fluctuations and, most important, is less readily devoured by competitors.

As a consequence of their photosynthetic lifestyle, plants have to augment their surface by centrifugal extension, generating a considerable degree of mechanical load (Fig. 14.1a). As long as plants remained aquatic, this load was at least partially relieved by buoyancy, allowing considerable sizes even for fairly simple architectures. However, when plants began to conquer terrestrial habitats, they had to develop flexible, yet robust, mechanical supports. The invention of vasculature-based modules, so-called telomes (Zimmermann 1965), became a decisive factor for the evolutionary success of land plants (Fig. 14.1b). Mechanical load shaped plant architecture down to the cellular level: Plant cells are endowed with a rigid cell wall with specific and fundamental consequences for cell division and cell expansion. These cellular specificities are of tremendous agronomical impact. For instance, the reduction of lodging in cereals is considered as pivotal factor for the success of the so-called Green Revolution (for the cellular details, refer to Nick 2012).

The central point of these considerations is that plants were channelled towards a sessile lifestyle due to biophysical constraints.

## *1.2 A Consequence: Cells Versus Organisms—Why the Plant Approach Is Different*

In addition to plant architecture, the sessile lifestyle has shaped the mode of morphogenesis. In animals, the *Bauplan* is laid down early in development. In some cases, even maternal factors have been found to complement the DNA of the embryo providing a kind of "morphogenetic inheritance". For instance, the anterior–posterior polarity in the *Drosophila* embryo is determined by a gradient of maternal, untranslated mRNA encoding transcription factors such as BICOID or

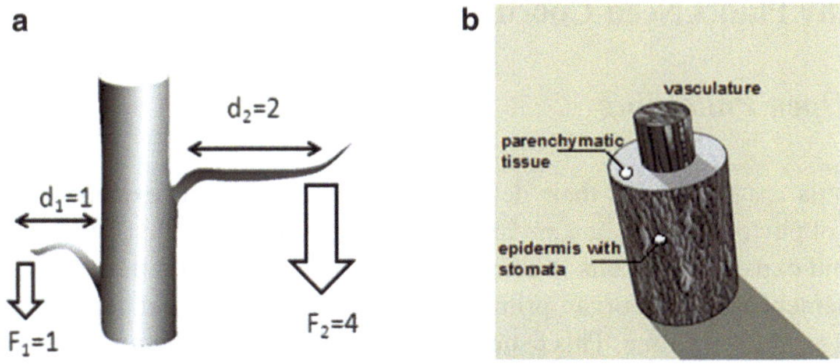


**Fig. 14.1** Plant architecture is shaped by mechanical load. (**a**) The lever force produced when a branch doubles its length (from $d_1 = 1$ to $d_2 = 2$) grows fourfold. (**b**) The development of telomes as load-bearing architectural module in the early Devonian was decisive for the evolutionary success of land plants

NANOS (Nüsslein-Volhard 1995). Even in classical models for epigenetic morphogenesis, such as the amphibian embryo (Spemann 1936), the dorsiventral polarity of frog eggs is established by autocatalytic feedback of polarising signals (gravity, sperm entrance) upon inherited patterns. These inherent patterns include not only preformed morphogenetic movements, but also transport and translation of maternal mRNA coding for cytoskeletal proteins and polar determinants (Elinson and Rowning 1988). In these models, the *Bauplan* is laid down during early development, often prior to cellularisation. Differentiation proceeds from the level of the entire organism down to the level of individual cells.

Again, plants are different: The genetic determination of plant shape is not as stringent as for animal development, but depends strongly on the environment. As central feature of this open morphogenesis, growth is not confined to early development, but continues throughout the entire life cycle. The ability to adjust growth in response to environmental stimuli is central for the adaptation of the individual plant to the challenges of its habitat. As a consequence of the rigid cell walls, cellular movements, a central mechanism in animal development, are not relevant for plant morphogenesis. The basic morphogenetic unit in plant development is the individual cell. Differentiation initiates from the level of individual cells and subsequently proceeds up to the level of the entire organism. This fact is highlighted by the ability to regenerate entire plants from almost any plant cell. In animals, such totipotency is confined to the fertilised egg cell and, sometimes, to its immediate descendants (Spemann 1936).

Thus, the principal difference (although there are definitely transitions that are now ignored for the sake of being clear) between plant and animal morphogenesis can be condensed into the following statement: In animals, the organism produces cells, whereas in plants, cells produce an organism. This means that the potency to form an organism must be enshrined in the individual plant cell.

# 2 Why Plants Need Coordinative Signalling?

## 2.1 *Open Patterning*

Plant cells are flexible in their developmental potency. Cell expansion is under control of phytohormones and environmental factors such as light (Lockhard 1960). The rapid expansion of cells is complemented by a slower addition of morphogenetic elements (cells or organ primordia), which does not occur randomly, but is ordered in space and time. This pattern formation (in sensu Bünning 1965) depends, on the one hand, on intrinsic signals that are obviously defined by genetic factors (otherwise there would be no base for classical plant taxonomy!). On the other hand, plant patterning can integrate signals from the environment. Environmental integration is evident, when a shoot meristem is committed for flowering controlled by day length and subsequently will form floral instead of vegetative organs. In animal patterning, the elements that are organised during pattern formation are generated prior to being differentiated. In a fruit fly embryo, for example, numerous nuclei are produced before they are patterned depending on gradients derived from maternal factors. Plants follow different developmental rules—here, the pattern is perpetuated in an iterative manner when new elements are continuously added *during* the patterning process.

This pattern iteration could be achieved, in principle, by assigning different developmental fates to the daughter cells during cell division. The pattern would then result from an ordered sequence of such formative divisions. Such a mechanism had been proposed for the root meristem of the model plant *Arabidopsis thaliana*, which is characterised by a highly stereotypic cell lineage (Scheres et al. 1994). However, elegant laser ablation experiments (Van den Berg et al. 1995) and functional analysis of mutants with aberrant tissue layers (Nakajima et al. 2001) revealed that even in this stereotypic system, cell fate was defined by signals (transcription factors) from adjacent cells and not by cellular genealogy.

Generally, the principal totipotency of plant cells is difficult to reconcile with a strong impact of cell lineage. Patterning in plants must result from coordinative signals between the already defined (older) regions of the pattern and the newly formed elements of the field that still have to acquire a specific identity.

## 2.2 *Coordinative Signalling During Patterning Is Evolutionary Ancient*

Plants acquired photosynthesis by sustainable symbiosis with autotrophic cyanobacteria. Functional multicellularity is already present in this class of prokaryotes. Filamentous cyanobacteria are capable of a simple cell differentiation yielding so-called heterocysts that can convert atmospheric nitrogen into ammonium and thus overcome the limitations of nitrogen bioavailability. Since filamentous cyanobacteria combine open patterning with developmental flexibility, coordinative signalling would be expected already in these prokaryotic precursors of the plant lifestyle. The nitrogenase required for the fixation of nitrogen dates back to the earliest anoxic phases of life on this planet and is therefore highly sensitive to oxygen. To safeguard the functionality of nitrogenase, any photosynthetic activity (releasing oxygen) has to be excluded from heterocysts. Thus, the heterocysts must be supplied with assimilates from their photosynthetic neighbours. Nitrogen export and assimilate import have to be balanced even though the total number of cells grows continuously, which represents a classical problem of open patterning. This balance is regulated by iterative algorithms, whereby preexisting heterocysts suppress the differentiation of new heterocysts over a range of around ten cells. When, as a consequence of cell division, the distance between the heterocysts exceeds this threshold, a new heterocyst will differentiate between them. Using patterning mutants in *Anabaena*, the factor responsible for this lateral inhibition could be identified as the diffusible peptide patS (Yoon and Golden 1998). Differentiation (including the synthesis of patS) will begin in clusters of neighbouring cells. However, one of these cells will excel the others and then immediately start to suppress further differentiation in its neighbourhood (Yoon and Golden 2001).

Thus, already the photosynthetic pluricellular prokaryotes do not use a predetermined cell fate but "negotiate" differentiation by signalling between neighbouring cells.

## 3 The LEGO Principle of Plant Morphogenesis

Plants use a modular version of "organism" and "identity". Due to their centrifugal architecture, plant organisms lack contiguous borders, a hierarchy of the "body" over its parts (which is linked with a strong cell autonomy). Moreover, there is not any impact of cellular genealogy on the set-up of the *Bauplan*, nor a predefined developmental programme. Nevertheless, they are able to defend their identity against the fluctuations of their environment. Their buffering capacity even excels that of animals by orders of magnitude. Despite strong variations in the details of individual development (which is tuned with the respective environmental conditions), the characteristics of each plant species emerge as a specific way to develop, respond and propagate. Without this specificity, no classical plant taxonomy would be possible.

We encounter here a seemingly paradox combination of flexible and species characteristic development. This paradox can be resolved considering the pronounced modularity of plant development. To use a metaphor: plant development resembles a play of LEGO bricks. Each brick is simple in shape and robust enough to survive most if not all challenges posed by a young architect. The assembly of these bricks is extremely flexible, though, and allows for almost any conceivable

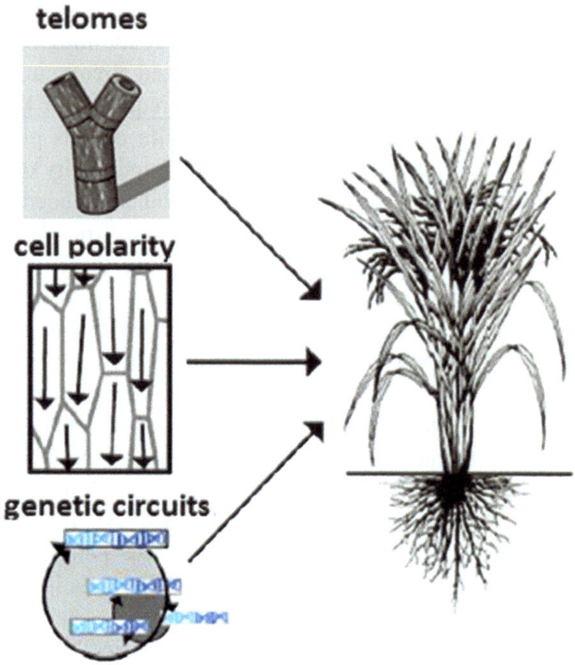


**Fig. 14.2** Modular organisation of plant development. Plant architecture is based on morphological modules (telomes) that are combined in a flexible manner depending on environmental conditions. Individual plant cells are endowed with innate directionality (cell polarity) that is dynamically aligned by signal flow through the morphological modules. Self-referring robust genetic circuits guide the differentiation of the cellular and the morphological modules and are recombined in temporal patterns

variation of architecture. Where would be genetic information be placed in this metaphor? Probably less in the inspiration of the young architect, but rather in the way the bricks are produced in the factory.

What are these "LEGO bricks" of plant development? There are principally three types of bricks: architectural, cellular, and genetic (Fig. 14.2).

## *3.1 Architectural Modules*

The architectural "LEGO bricks" are the telomic modules arranged in a flexible way integrating mechanical load with a panel of environmental signals (with light as central component). The arrangement of telomic modules is under control of auxin flux from the aerial organs towards the roots driving the differentiation of ground tissue into vasculature (as core element of the developing telome). This architectural principle is simple, robust, and flexible. Cellular details of recently discovered fossil finds of the progymnosperm *Archaeopteris* (Rothwell and Lev-Yadun 2005) suggest that already in the Upper Devonian, 375 million years before our time, the arrangement of telomes was controlled by auxin flow. The secret of the land plant success story seems to reside in this modular morphology.

## 3.2 Cellular Modules

The polarity of vascular cells is aligned with the shoot–root axis and represents the cellular correlate of the "LEGO bricks" forming the base for the telomic principle. The directional transport of auxin (see Chap. 4) is brought about by the combination of multidirectional, "exploratory" influx and directional efflux (due to the polar localisation of auxin efflux carriers). The polar localisation of auxin efflux carriers is a continuous process rather than a fixed structure: These carriers cycle continuously and rapidly (the lifetime of the carriers at the membrane are in the range of a few minutes!) between an endocytic compartment and the site of their activity at the plasma membrane. The intensity of cycling depends on the presence of auxin (Paciorek et al. 2005) and differs between the different poles of the cell establishing such a polar distribution (Dhonukshe et al. 2008) providing the positive amplification loop required for the auxin canalisation mechanism driving vascular patterning. Sensory input on the auxin distribution of surrounding cells continuously "questions" this loop either reinforcing the existing directionality of the cell or leading to a new polarity. Since morphology and cellular architecture are brought about by modular, self-organised processes, the genetic control might be relatively simple.

## 3.3 Genetic Modules

The genetic "LEGO bricks" underlying the *Bauplan* of land plants might be robust regulatory circuits that are launched under control of fairly permissive temporal patterns. By shifting these programmes in time (so called heterochrony), new architectures can be generated that at first sight are very spectacular. The power of heterochrony is illustrated by the case of the "Skye" ecotype of the model plant *Arabidopsis thaliana* (thale cress). Normal thale cress plants produce a leaf rosette but, upon bolting, switch to the formation of small, single leaves protruding from the elongating inflorescence. This switch is impaired in the "Skye" ecotype resulting in a fundamentally altered architecture with aerial rosettes formed from the axillary meristems of the bolting inflorescence (Grbić and Bleecker 1996). It could be shown that this spectacular change of the *Bauplan* was caused by mutations in two relatively inconspicuous genes that modulate, among numerous other factors, the timing of developmental processes. The mutations simply delay the inactivation of the vegetative programming during bolting. The ongoing vegetative development at simultaneous launch of a floral programme accounted for a fundamentally different morphology that at first sight seemed to result from "macroevolution". A comparative approach on plant development rapidly reveals that evolutionary adaptations of plant architecture can often be deduced from heterochronic shifts between fairly simply morphogenetic processes (for review, see Li and Johnston 2000).

### 3.4 *The Secret of Plant Morphogenesis: Robust Modules, Flexible Assembly*

In summary, plant organisms assemble morphological, cellular, and genetic modules to accommodate challenges from the environment with the innate necessities of physiology. These modules stem from fairly robust self-organisation providing a mechanism to maintain the specific quality of the respective plant. The assembly of these modules is rather flexible and can be tuned with the exogenous necessity of environment. Since the modules are relatively robust and autonomous, the signals that regulate modular assembly may be very simple. Complexity is provided by the receiving modular process, not by signal triggering this process. Indole-acetic acid, the natural auxin, is astonishingly small and simple. However, it combines three molecular properties (none of which is spectacular): Auxin is a small organic acid and therefore easily moves through the acidic environment of the apoplast. Auxin carries a lipophilic indole ring and therefore can permeate the cell membrane from any direction, which allows a cell to "explore" the auxin levels in its neighbourhood. Auxin is a weak acid and thus readily trapped in the neutral cytoplasm and has to be actively exported by carriers, which allows to create a directionality of auxin efflux. When the localisation of the efflux transporter is shifted under the control of auxin itself, this will generate a self-regulatory circuit that perfectly meets the criteria of a reaction–diffusion system in sensu Turing (1952).

## 4 Auxin and Coordinative Signalling

### 4.1 *Plant Patterning and Coordinative Signalling: Phyllotaxis*

Phyllotaxis allows to optimise the position of leaves to maximise photosynthetic efficiency. A prospective leaf primordium will form at maximal distance from the older primordia indicating inhibitory fields (Schoute 1913). In fact, when the youngest primordium is isolated from its environment by tangential incisions, this will be the subsequently formed primordia will shift their position (Snow and Snow 1931). This shift was originally interpreted in terms of the additional space created by the incision that would allow the incipient primordia to move to a position where they otherwise were excluded (first available space model). Later, inhibitory fields emanating from the older primordia have been proposed, but the nature of these inhibitory signals has been under debate for a long time. Basically, there were two standpoints in this debate: biophysics versus biochemistry.

## 4.2 Biophysical Model of Phyllotaxis

Buckling from the older primordial would, under conditions of the tissue tension present in a growing meristem, inhibit by mechanical stresses the formation of new primordial in the neighbourhood. In fact, the position of prospective primordial could be perfectly predicted by models of stress–strain patterns (for review, see Green 1980). As expected from a biophysical model, local release of tissue tension by beads coated with the cell wall loosening protein extensin could invert the phyllotactic pattern (Fleming et al. 1997). As early event of incipient primordial commitment, membrane-associated microtubules reorient sharply and subsequently align with the stress–strain pattern (Hardham et al. 1980). By means of GFP-labelled microtubules, this phenomenon could be followed in a non-invasive manner in living shoot apices of *Arabidopsis thaliana* (Hamant et al. 2008). A combination of live cell imaging with mathematical modelling of stress–strain pattern revealed that cortical microtubules align in the direction of maximal mechanical stress in a transcellular pattern. When the outer meristem layer was removed by laser ablation, microtubules reoriented in orientations predicted by the mathematical model, followed by a compensatory bulging of the apex. The impact of cortical microtubules is further corroborated by recent evidence for a role of the microtubule-severing protein katanin for meristem patterning (Uyttewaal et al. 2012).

## 4.3 Biochemical Model of Phyllotaxis

As alternative to the biophysical inhibition, chemical signals from the older primordia were proposed to inhibit the initiation of a new primordium in their proximity. This model was supported by studies in apices that had been freed from primordia by application of auxin transport inhibitors (Reinhardt et al. 2000), an experimental system that allows study of the *de novo* generation of a pattern in the absence of any prepattern. These studies showed that the coordinative signal depends on auxin. Unexpectedly, the preexisting primordia did not act as sources, but as sinks for auxin. This leads to mutual competition for free auxin within the apical belt that is competent for the initiation of leaf primordia. Since pre-existing primordia attract auxin fluxes from the meristem, they drain their neighbourhood from diffusible auxin, such that the initiation of new primordia is inhibited (Reinhardt et al. 2003). When the auxin efflux regulator PIN1 is mutated, this will, as a consequence, strongly perturb phyllotaxis, supporting the importance of directional auxin efflux for the pattern. However, a recent detailed analysis of the *pin1* mutant revealed that the phyllotactic pattern is disturbed but not eliminated, indicating that PIN1 is not the only factor capable of guiding the pattern (Guenot et al. 2012).

## 4.4 Synthesis: The Auxin–Microtubule–Tension Loop

As for most dichotomous debates in biology, reality seems to be a synthesis of the two seemingly exclusive standpoints: By measuring tissue rigidity with Atomic Force Microscopy in the vegetative apex of *Arabidopsis thaliana*, an auxin-dependent local softening of the cell wall could be demonstrated. It was further shown that this relaxation of wall tension was mediated by a specific modification of wall pectins homogalacturonan de-methyl-esterification (Braybrook and Peaucelle 2013). Interestingly, when this modification was administered in the absence of auxin transport, it was not effective. Thus, both functional auxin transport and local reduction of mechanical stress were necessary and sufficient for phyllotactic patterning. The resulting model is based on a regulatory feedback loop between auxin transport and tissue tension and requires that the direction of mechanic stress can be transduced into altered localisation of auxin efflux transporters. Microtubules that can guide the localisation of PIN1 (Heisler et al. 2010) and simultaneously perceive mechanic stress might be the mechanistic link. This auxin–microtubule–tension loop might thus provide the synthetic bridge reconciling the traditional antagonistic viewpoints on phyllotaxis.

## 4.5 Plant Patterning and Coordinative Signalling: Vasculature

All land plants (except the archaic mosses) are made of load-bearing modular elements, the telomes. These telomic modules consist of conductive woody vasculature embedded in a cylinder of parenchymatic tissue with the potency for vascular development, enclosed by an epidermal layer. The flexible arrangement of telomes represents the core process of plant architecture. The vasculature can be adjusted by differentiation of parenchymatic cells to tune mechanic load and transport with the perturbations, for instance, from wounds (Sachs 2000) or from growth as in the venation of developing leaves (Mattsson et al. 1999). All cells of the parenchymatic tissue are competent to differentiate into vasculature. This differentiation depends on the flow of auxin through these cells.

Although auxin can enter the cells of higher plants through specific import channels, for the pattern of vasculature, the non-directional influx through the membrane is important (discussed in more detail in Sect. 6). The major natural auxin, indolyl-3-acetic acid, is a weak acid and relatively small. In the acidic environment at the outer face of a plant cells, it will be uncharged and therefore can permeate through the membrane even without the help of influx carriers. In the more or less neutral cytoplasm, auxin is deprotonated and thus acquires a negative charge that will prevent its spontaneous exit. Due to this ion-trap mechanism, auxin will accumulate in the cell. However, it can exit by means of specific export pumps that are localised asymmetrically, guided by cell polarity. The combination of non-directional influx and directional influx produces a mutual competition of individual cells for free auxin and a directional flow in the direction of cell polarity. A cell with more active or more localised auxin exporters will transport more auxin than its neighbours and therefore cause a drainage of auxin. This mechanism for lateral inhibition of individual elements is now combined by autocatalytic feedback: The differentiation from the ground state into a vascular cell fate is induced by the flux of auxin passing through the respective cell. Conversely, this differentiation promotes cell polarity resulting in a stronger gradient of auxin exporters what, in turn, will further stimulate auxin drainage of the neighbourhood.

This positive feedback, in combination with a lateral inhibition (caused by mutual competition for auxin), drives the pattern of conductive tissue and thus the arrangement of telomes. This "auxin canalisation" model has been extensively studied and modelled mathematically and is capable, for instance, to predict venation patterns in leaves (for review, see Berleth and Sachs 2001).

## 4.6 Plant Patterning: Order Without a "Great Chairman"

Biological patterns are shapes that become manifest on the level of a population of cells or organs. They are holistic in quality and represent classical system properties that emerge when the system is considered as an entity. At first glance, this would call for a strong hierarchy controlling the behaviour of the individual elements. To use a metaphor from human societies: collectivism is usually bound to strong (and often autocratic) leader personalities. This approach will not work for plant development, though. As pointed out above, plant cells maintain a high level of autonomy and are not easily subdued to the rule of a "Great Chairman". In addition, plant cells behave in a highly stochastic manner, a property that ultimately can be attributed to the diffuse organisation of environmental sensing (details are given in Nick 2006): Plants lack specialised sensory organs. In shorthand, each individual cell is able to sense most environmental signals in a monadic way and therefore has to employ extreme amplification of the sensory input resulting in all-or-none type outputs on the level of individual cells. If all cells of a tissue would respond in a homogenous manner, plant responses would be saturated already at very low input. In case of light, the input from the new moon would produce the same output as full sunlight at noon time. Due to the strong variation of sensory thresholds, plants can differentiate between weak and strong stimuli by the frequency of individual cells where sensing is activated. By integrating over the population of activated cells via cell–cell communication, plants can extend their dynamic range of sensing combining high sensitivity with differential responses to different signal input.

What can we generalise from phyllotaxis and vascular patterning? Both patterns resist stochastic fluctuations of the initial situation, they are based on lateral inhibition between the elements of the pattern, and they contain qualitative decisions generated via autocatalytic feedback loops. Both patterns can be described by the mathematics of reaction–diffusion systems that were adapted to biology by

Turing (1952) and have been quite successfully used to model various biological patterns such as foot-head patterning in *Hydra* (Gierer et al. 1972) and segmentation in *Drosophila* (Meinhard 1986), but also leaf venation (Meinhard 1976). In reaction–diffusion systems, a locally constrained, self-amplifying feedback loop of an activator is linked to a far-ranging mutual inhibition (Gierer and Meinhard 1972). Auxin-dependent patterning differs in one aspect from the original model, where the inhibitor is usually described as a positive entity (such as the patS peptide acting in cyanobacterial patterning). In auxin-dependent patterning, lateral inhibition is brought about by mutual competition for the activator.

Self-activation combined to mutual competition also provides *proportional harmony*, a system property of many organisms meaning that a new holistic organisation can emerge independently of size, when the original system is either divided or fused. In plants, this astounding ability becomes manifest as the lack of physical body individuality: the plant body can be subdivided and the parts will readily organise a new independent plantlet that in shape and architecture resembles its progenitor organism.

## 5 Organismic Modules Are Built by "Auxin Resonance"

### 5.1 *"Leaves in the Test Tube": Experimental Reduction of Plant Self-Organisation*

Plants add, during their entire lifetime, new cells to the tip of roots and shoots. As shown for the root meristem by elegant laser ablation experiments (van den Berg et al. 1995), cell differentiation in the mitotically active meristems is controlled by signals from the neighbouring, already differentiated, cells. However, when the meristem becomes accessible to cell biological inspection, differentiation is already channelled. At this stage, it is very difficult, if not impossible, to manipulate the pattern in a fundamental manner. Thus, meristems are beautiful systems to study how patterns are perpetuated, but for the analysis of pattern induction, simpler systems are needed, where determination has not progressed that far. Several years ago, we have introduced cell lines derived from the ground tissue of tobacco shoots as experimental system to study the primordial stages of division patterning (Campanoni et al. 2003). These cell lines can be readily cultivated in suspensions maintained under continuous rotation. Plant suspension cell lines are generally considered as dedifferentiated and have even been termed "HeLa cells of plant biology" (Nagata et al. 1992). However, they often preserve certain features from their source tissue, such as the ability to generate the structured cell wall thickenings characteristic for vascular cells (Nick et al. 2000), the ability to generate, through a series of axial cell divisions, cell files with a clear axis and polarity, and

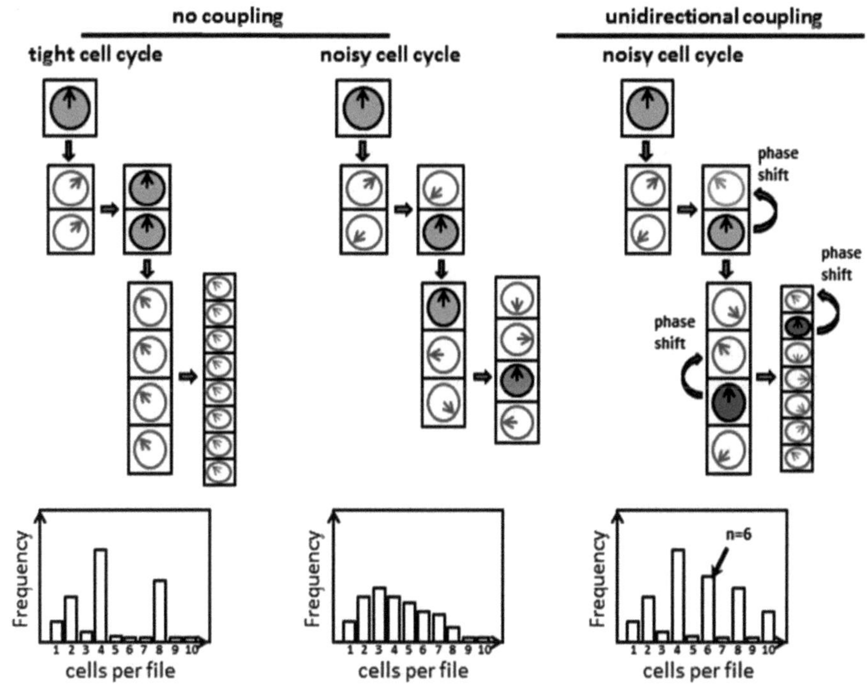


**Fig. 14.3** Models for the synchrony of division patterns in cell lines derived from tobacco parenchyma. There are two concurrent possibilities—either individual cells act autonomously (no coupling of their cell cycles, *left*) or they show temporary, unidirectional coupling of their cell cycles (*right*). The schematic clocks represent the position of the cell cycle for the respective cell. A position at "high noon" stands for the onset of mitosis. The predicted frequency distributions of cell number per file are shown in the *lower panel*. Three principle cases are shown: (1) In the absence of coupling, but under tight control of cell cycle duration would result in a sequence, where frequency peaks are predicted for 2, 4, 8,. . ., $2^n$ cells per file (*left-hand column*). (2) In the absence of coupling and for a noisy cell cycle, there should be no clear frequency peaks, but odd- and even-numbered cell files should occur at the same frequency (*central column*). (3) In case of unidirectional coupling, a cell entering mitosis generates a signal that is conveyed to its downstream neighbour. This signal causes a phase shift by accelerating the cell cycle of the receiver cell. For this unidirectional coupling, even for noisy cell cycles, a partial synchrony is predicted with frequency peaks at 2, 4, 6,. . ., $2n$ cells per file (*right-hand column*). This model is the only that predicts a frequency peak for six cells per file. This frequency peak (*arrow*) is diagnostic for unidirectional coupling (bidirectional coupling would lead to a pattern as observed for uncoupled cells under tight control of the cell cycle; see *left-hand panel*)

they have preserved responsiveness to the controlling signal, auxin. Since these files derive from singular cells, they cannot rely on positional information inherited from the mother tissue. Patterns of competence within a cell file must originate de novo during the culture cycle.

### 5.2 Weak Coupling of Autonomous Oscillators

During the work with these tobacco cell files, we observed that files consisting of even numbers of cells were dominating over files with uneven cell numbers (Campanoni et al. 2003; Maisch and Nick 2007). At first sight, frequency peaks of even-numbered files might occur, when the cell cycle proceeds with a precise timing (Fig. 14.3). This should generate files in a sequence of

$$f(n) = 1, 2, 4, 8, \ldots 2n$$

individual cells (with $n$ representing the number of cell cycles). However, the length of individual cell cycles varies over a broad range, and there is, in addition to the expected peaks at $2n$, a curious frequency peak for files composed of six cells (in some cases accompanied by a smaller peak of ten cells). This observed feature could be simulated using a mathematical model derived from non-linear dynamics, where elementary oscillators (cycling cells) with a high level of noise (variation in the length of individual cell cycles) were weakly coupled, and where the number of these oscillators was not constant, but grows with time (Campanoni et al. 2003). In contrast to concurrent models, this weak coupling algorithm was able to predict the observed frequency peak of hexacellular files. Moreover, this model predicted several non-intuitive properties of the experimental system. A striking feature of the model was the prediction that coupling must be unidirectional, i.e. that the coordinating signal is transported in a polar fashion. The coupling is seen as a phase shift in the cell cycle, i.e. a dividing cell will cause its downstream neighbour to accelerate its cell cycle such that it will also initiate mitosis. Unidirectional signalling is a diagnostic feature of auxin transport. In fact, the predominance of even-numbered cell files could be eliminated by low concentrations of 1-*N*-naphthylphthalamic acid, a specific inhibitor of auxin exporters (and thus of directional auxin transport). Although the noise in this system was considerable, with high variation in the cycling period over the cell population, the division of adjacent cells was synchronised to such a degree that files with uneven cell numbers were rare compared to files with even numbers. Frequency distributions over the cell number per file thus exhibited oscillatory behaviour with characteristic peaks at even cell numbers (Fig. 14.3).

### 5.3 Sensitive Muscles: The Actin–Auxin Oscillator

Auxin efflux carriers are not static, but undergo dynamic cycling between intracellular compartments and the plasma membrane. Treatment with the fungal toxin Brefeldin A (BFA) traps the carriers in the intracellular compartments (Geldner et al. 2001). This trapping is suppressed by cytochalasin D, an inhibitor of actin assembly suggesting that actin is involved in the cycling of auxin efflux carriers. On the other hand, the cargo of these carriers, auxin, controls the conformation of actin, whereby the massive bundles prevalent in the absence of auxin are rapidly detached into finer filaments after addition of auxin (for review, see Nick 2010). Auxin can stimulate its own transport by improving the polar localisation of the auxin efflux carriers at the cell poles (Paciorek et al. 2005), suggesting that these transporters are more efficiently moved along the finer actin filaments in response to auxin. This model was tested in rice seedlings expressing different levels of the actin-binding domain of mouse talin in fusion with the yellow fluorescent protein (Fig. 14.4). By feeding radioactively labelled auxin to the tip of the seedling, the amount of radioactivity recovered in an agar block at the seedling base (quantified by a scintillation counter) could be used as measure for the efficiency of auxin transport. Based on this experimental system, the debundling of actin filaments by exogenous auxin could be shown to precede the concomitant stimulation of transport efficiency (Nick et al. 2009). Upon overexpression of the talin marker, actin filaments were constitutively bundled accompanied by a reduced capacity to transport auxin. However, when exogenous auxin was added, these bundles relaxed into numerous fine strands of actin filaments followed by a promotion of auxin transport. These findings demonstrate that

1. Actin reorganisation into fine strands precedes the stimulation of auxin transport.
2. Fine strands of actin are necessary for efficient auxin transport.
3. Actin reorganisation into fine strands is sufficient to promote auxin transport.

Thus, manipulation of actin can be used as tool to manipulate auxin transport—at least in experimental systems, where polar auxin flux is elevated to an extent that the steady-state level of active transporters at the membrane becomes limiting. We therefore transferred this strategy to further dissect the role of auxin transport for division synchrony in the tobacco cell model. If actin is part of an auxin-driven feedback loop, it should be possible to manipulate auxin-dependent patterning through manipulation of actin. To test this prediction, we had to create a situation, where actin is excessively bundled. For this purpose we employed a transgenic approach, where we expressed the actin-binding domain of mouse talin in fusion with the yellow fluorescent protein. Mouse talin competes with endogenous actin depolymerisation factors for binding sites on actin such that the actin filaments are progressively trapped in a bundled configuration (Ketelaar et al. 2004). In fact, overexpression of the construct in tobacco cells produced constitutively bundled fluorescent actin filaments. As predicted, the synchrony of cell division was impaired in this line, but could be restored by addition of auxins along with a normal organisation of actin. A screen for actin-binding proteins mediating the effect of auxin upon actin organisation identified tobacco actin depolymerisation factor 2 (NtADF2) as central player. A cell line overexpressing this factor is impaired in division synchrony in a highly specific way—in this line, the frequency peak at $n = 6$ diagnostic of unidirectional weak coupling is absent, but can be rescued by addition of PIP2, a phospholipid specifically sequestering ADFs (Durst et al. 2013).

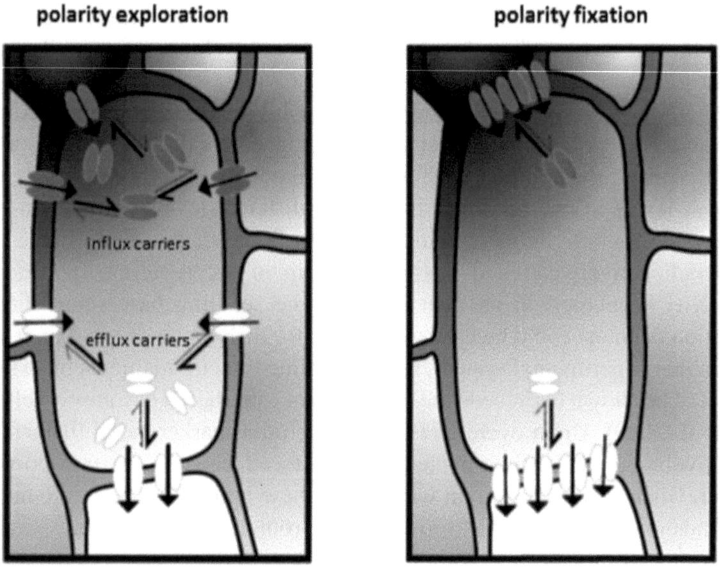


**Fig. 14.4** Two-phase model for the exploration of space by auxin in plant self-organisation. A self-amplifying feedback between non-directional auxin influx through ion-trapping and gradient-dependent cycling of auxin efflux and influx carriers allows to integrate auxin concentration over the environment of a given cell and to generate a polar flux driving self-organisation

We therefore arrive at a model of a self-referring regulatory circuit between polar auxin transport and actin organisation, where auxin promotes its own transport by shaping actin filaments. This circuit seems to contribute to the self-amplification of auxin transport, a central element in current models of auxin-dependent patterning. The implications of this model are to be explored, but already at this stage it can be used to derive characteristic properties of basipetal auxin transport. For instance, the model predicts that the transport of IAA should oscillate. Auxin will induce fine actin strands that will partition auxin efflux carriers more efficiently to the plasma membrane, such that the intracellular auxin concentration will decrease. This decrease will cause bundling of actin filaments and, as a consequence, efflux carriers will be sequestered in intracellular compartments, culminating in a reduced efflux such that auxin received from the adjacent cells will accumulate and trigger a new cycle. The frequency of these oscillations should depend on the dynamics of actin reorganisation (around 20 min) and the speed of carrier cycling which is in the range of 5–10 min (as inferred from the comparison of auxin uptake in control versus BFA-treated cells; Paciorek et al. 2005). From these parameters, auxin transport is predicted to oscillate with a period of about 25–30 min. In fact, such oscillations with a period of 25 min had been observed during classical experiments on basipetal auxin transport in maize coleoptiles (Hertel and Flory 1968).

## 5.4 New Approaches to Morphogenesis: Chemical Engineering

The polarity of ground tissue cells provides the cellular base for the alignment of telomes as architectural modules. A self-referring, oscillatory circuit involving actin remodelling, rapid cycling of auxin efflux carriers, non-directional auxin influx, and directional auxin efflux has been identified as core element of this cell polarity. To dissect this circuit, it is not sufficient to identify molecular players such as ADF2 (Durst et al. 2013), but it is necessary to generate and manipulate the spatial patterns of molecules at subcellular resolution. Although genetic engineering allows to target transgenes to specific compartments using localisation motives, the spatial resolution of this strategy is too coarse-grained. New strategies are warranted to increase spatial resolution. To achieve this goal we used chemical engineering based on caged auxin that can be released by localised irradiation in single cells or even parts of a cell (Kusaka et al. 2009). Caged compounds conventionally use 2-nitrobenzyl-esters as caging group. However, the ester bond was found to be enzymatically hydrolysed in plant cells such that auxin was released prior to photolysis producing high unspecific background activities. By molecular modelling of the active centres of these enzymes, an esterase-resistant caging group, (2,5-dimethoxyphenyl)(2-nitrobenzyl) ester, could be designed and employed successfully. We administered this tool to the actin–auxin oscillator to demonstrate in a proof-of-principle experiment that a biological response can be controlled by light at cellular resolution. By using an auxin-inducible promoter (DR5) driving a GFP reporter, we were able to confirm that auxin was released only in the irradiated cell. Subsequently, we used the cell line overexpressing talin in fusion with the yellow fluorescent protein. In this cell line, actin is constitutively bundled, but can be rescued by addition of exogenous auxin (Maisch and Nick 2007). By feeding caged auxin to this cell line and irradiating individual cells of a file, we could trigger a specific reorganisation of actin filaments that was confined to the irradiated cell (Kusaka et al. 2009). Thus, chemical engineering using light-switchable triggers can now be exploited to steer auxin gradients during self-organisation of the tobacco cell model. At present, we are completing a study, where auxin is released in different cells of a file during specific stages of the culture cycle accompanied by specific changes in division patterns. Recent experiments using protoplasts from fluorescently tagged actin marker lines could demonstrate that even intracellular auxin gradients can be produced that will then be transduced into intracellular gradients of actin organisation (Liu et al. 2013).

## 6 The Influx Issue: Auxin as Exploratory Molecule

The generation of spatial patterns by coordinative signals requires that space can be explored in different directions. This seems to contrast with the pronounced polarity of auxin transport. This directionality has been classically explained by polar efflux of auxin (Rubery and Sheldrake 1974). The exploratory part of coordination would be the non-directional influx of indolyl-acetic acid through the plasma membrane maintained by a chemical gradient, where IAA is stripped from its proton in the more or less neutral cytoplasm. The molecular identification of auxin influx carriers (Bennett et al. 1996) that are localised in a polar fashion opposite to the PIN efflux carriers (Swarup et al. 2001) has shifted focus a bit. These findings led to a model, where not only efflux but also directional influx contributes to the polarity of auxin flow.

A molecule that is easily transported through the acidic environment of the apoplast, but that is readily trapped in the cytoplasm and then has to be actively exported is ideally suited to convey lateral inhibition between neighbouring cells. When the localisation of the efflux transporter is placed under the control of auxin itself (Paciorek et al. 2005), this will establish a perfect reaction–diffusion system in sensu Turing (1952). This system is able to establish a clear cell polarity from even minute and noisy directional cues. However, when auxin *influx were exclusively* directional, due to the polar localisation of the influx carrier AUX1, exploration of space as prerequisite of coordinative signalling for pattern formation, would not work.

This apparent dilemma might be less dramatic as it seems at first sight. The impact of carrier-based auxin influx depends strongly on apoplastic pH: since the $pK_s$ value for indole-acetic acid is 4.75, the proportion of the anionic form that definitely requires a carrier to enter the cell is relatively high for pH 5 (74 % $IAA^-$); for a pH of 5.5 even 95 % of auxin are present in the anionic form (Swarup and Péret 2012). However, is this the relevant pH of the cell wall? To determine the pH of plant cell walls is far from trivial, due to ion exchange at the carbon hydrate matrix. Most measurements of cell wall pH systematically underestimate the acidity of the chemical environment for the apoplastic auxin. Reliable measurements can only be achieved by using a pH-stat approach, because here the metric component is buffered. Using this strategy, the physiological pH of the cell wall has been found to range between 4.0 and 4.5 (Lüthen et al. 1990), i.e. in a range, where the uncharged form of auxin predominates. Thus, “exploratory” ion trapping is a substantial component of auxin influx.

Alternatively, exploration of space might be achieved by the cycling of the efflux carrier as well (see Chap. 8). When the auxin effect on the cycling of PIN proteins (Paciorek et al. 2005) is not homogenous over the auxin-stimulated cell, but depends on the local auxin concentration at the respective flank of the cell, this would provide a mechanism, by which a cell can “explore” gradients of auxin across a tissue. This mechanism has been proposed for phyllotaxis (Jönsson et al. 2006) and has been integrated into models for auxin channelling that are congruent with predictions from the classical auxin canalisation model (Roeland et al. 2007).

A polar transport of auxin can be detected already in several lines of multicellular algae (see Chap. 13; Dibb-Fuller and Morris 1992; Cooke et al. 2002) including *Chara* as close relative of the land plant ancestor (Boot et al. 2012; for review see Raven 2013). In phaeophycean algae, polar auxin transport has been recruited for the establishment of polarity (Basu et al. 2002). Transcellular auxin gradients are necessary for polarity, because when these gradients are overrun by exogenous IAA symmetry break is suppressed in *Fucus* (Whitaker 1942). This indicates that the central role of auxin in cell communication developed from evolutionarily quite ancient preadaptations already present prior to the transition to a terrestrial lifestyle. However, in order to integrate plant architecture, the directional output must be integrated with input that is non-directional. The cell must explore its neighbourhood in different directions, which is possible through the ion-trap mechanism of auxin influx. This does not exclude that the resulting cellular polarity will subsequently reinforce the main route of influx by partitioning auxin influx carriers of the AUX1/LAX family to the sites, where ion trapping was most active. In fact, both mechanisms of auxin influx might act in concert (Fig. 14.4): the ion-trap mechanism would be used in a phase of polarity exploration, whereas repartitioning of influx carriers (along with repartitioning of efflux carriers) would provide a fixation of the initial, still flexible, polarity.

Patterning of a tissue is a complex phenomenon, and at the time that tissues become amenable to experimental manipulation, cell polarity is already laid down. This means that in tissues it is possible to investigate pattern perpetuation. How a pattern is laid down requires experimental systems, where cell polarity is still on the move.

## 7 It Is All Geometry: The *Tabula Rasa* Approach

Polarity induction de novo has been classically studied in the brown alga *Fucus* (Goodner and Quatrano 1993; Hable and Hart 2010). The spherical zygote undergoes asymmetric division yielding progenitor cells for thallus and rhizoid. The orientation of this division can be aligned by unilateral blue light inducing a calcium influx at the shaded flank, where later the rhizoid will emerge (Jaffe 1966). A cap of dynamic actin filaments is formed at this site and attracts vesicles transporting cell wall material driving the outgrowth of a rhizoid. The polarity seen in response to blue light is produced by reorientation of a preformed polarity, but truly generated de novo, demonstrated by induction with strong plane-polarised blue light producing a high fraction of birhizoidal twins. This beautiful system has enabled a wealth of phenomenological, physiological, and cell biological insights into polarity induction, but it suffers from limited molecular accessibility.

Comparable systems, where spherical cells undergo formative divisions, are rare in higher plants. The closest version, developing microspores, are quite different, in that they harbour a distinct preformed polarity that becomes manifest as nuclear movements as well as asymmetric cell fate of the daughter cells: the generative daughter will inherit immortality, whereas the vegetative cell is doomed to death at fertilisation, giving a neat illustration of Weismann's germ line/soma concept. By colchicine or other antimicrotubular drugs, this developmental asymmetry can be eliminated (Twell et al. 1998).

To obtain symmetric, apolar cells in higher plants is possible, however, when the cell is stripped off its wall by cellulase. These protoplasts correspond to a tabula rasa situation and lack any axis and polarity, but retain the ability to regenerate complete plants as shown in spectacular experiments on tobacco (Nagata and Takebe 1970). Thus, protoplasts resemble the zygotes of *Fucus* with respect to de novo generation of polarity. The observation that regenerating protoplasts of the moss *Physcomitrella patens* show a redistribution of calcium channels (visualised by a fluorescent channel antagonist (Bhatla et al. 2002) indicates that the underlying mechanisms might be similar.

We therefore used protoplasts of tobacco BY-2 cells to study how polarity and axis are induced de novo (Zaban et al. 2013). The presence of fluorescently tagged transgenic marker lines allowed us to follow the behaviour and role of the cytoskeleton during this phenomenon. The system could be standardised to such a degree that the temporal pattern of regenerative stages could be investigated on the quantitative level such that functional analysis became possible. Using anti-cytoskeletal compounds and inducible expression of actin-bundling proteins it could be shown that a dynamic population of actin was necessary for polarity. When actin dynamics were suppressed, curious tripolar cells ensued in analogy to the twinned embryos observed in *Fucus* for induction by strong polarised light.

In the meantime, we succeeded to integrate this tabula rasa system into a microfluidics platform, which allows us to provide gradients of auxin through controlling the flux through the system and preformed geometries through rectangular microvessels (Sun et al. 2009). Using this system we currently investigate how the regenerating protoplast induces polarity after having explored the geometry of its environment.

## References

Basu S, Sun H, Brian L, Quatrano RL, Muday GK (2002) Early embryo development in *Fucus distichus* is auxin sensitive. Plant Physiol 130:292–302

Bhatla SC, Kießling J, Reski R (2002) Observation of polarity induction by cytochemical localization of phenylalkylamine-binding receptors in regenerating protoplasts of the moss Physcomitrella patens. Protoplasma 219:99–105

Bennett MJ, Marchant A, Green HG, May ST, Ward SP, Millner PA, Walker AR, Schulz B, Feldmann KA (1996) Arabidopsis AUX1 gene: a permease-like regulator of root gravitropism. Science 273:948–950

Berleth T, Sachs T (2001) Plant morphogenesis: long-distance coordination and local patterning. Curr Opin Plant Biol 4:57–62

Boot KJ, Libbenga KR, Hille SC, Offringa R, van Duijn B (2012) Polar auxin transport: an early invention. J Exp Bot 63:4213–4218

Braybrook SA, Peaucelle A (2013) Mechano-chemical aspects of organ formation in *Arabidopsis thaliana*: the relationship between auxin and pectin. PLoS One 8:e57813

Bünning E (1965) Die Entstehung von Mustern in der Entwicklung von Pflanzen. In: Ruhland W (ed) Handbuch der Pflanzenphysiologie, vol 15/1. Springer, Berlin, pp 383–408

Campanoni P, Blasius B, Nick P (2003) Auxin transport synchronizes the pattern of cell division in a tobacco cell line. Plant Physiol 133:1251–1260

Cooke TJ, Poli DB, Sztein AE, Cohen JD (2002) Evolutionary patterns in auxin action. Plant Mol Biol 49:319–338

Dhonukshe P, Tanaka H, Goh T, Ebine K, Mähonen AP, Kalika Prasad K, Blilou I, Geldner N, Xu J, Uemura T, Chory J, Ueda T, Nakano A, Scheres B, Friml J (2008) Generation of cell polarity in plants links endocytosis, auxin distribution and cell fate decisions. Nature 456:962–967

Dibb-Fuller JB, Morris DA (1992) Studies on the evolution of auxin carriers and phytotropin receptors: transmembrane auxin transport in unicellular and multicellular Chlorophyta. Planta 186:219–226

Durst S, Nick P, Maisch J (2013) Actin-depolymerizing factor 2 is involved in auxin dependent patterning. J Plant Physiol 170(12):1057–1066. doi:10.1016/j.jplph.2013.03.002

Elinson RP, Rowning B (1988) Transient array of parallel microtubules in frog eggs: potential tracks for a cytoplasmic rotation that specifies the dorso-ventral axis. Dev Biol 128:185–197

Fleming AJ, McQueen-Mason S, Mandel T, Kuhlemeier C (1997) Induction of leaf primordia by the cell wall protein expansion. Science 276:1415–1418

Geldner N, Friml J, Stierhof YD, Jürgens G, Palme K (2001) Auxin transport inhibitors block PIN1 cycling and vesicle trafficking. Nature 413:425–428

Gierer A, Meinhard H (1972) A theory of biological pattern formation. Kybernetik 12:30–39

Gierer A, Berking S, Bode H, David CN, Flick K, Hansmann G, Schaller H, Trenkner E (1972) Regeneration of *Hydra* from reaggregated cells. Nature 239:98–101

Goodner B, Quatrano RS (1993) *Fucus* embryogenesis: a model to study the establishment of polarity. Plant Cell 5:1471–1481

Grbić V, Bleecker AB (1996) An altered body plan is conferred on Arabidopsis plants carrying dominant alleles of two genes. Development 122:2395–2403

Green PB (1980) Organogenesis – a biophysical view. Annu Rev Plant Physiol 31:51–82

Guenot B, Bayer E, Kierzkowski D, Smith RS, Mandel T, Zádníková P, Benková E, Kuhlemeier C (2012) PIN1-independent leaf initiation in Arabidopsis. Plant Physiol 159:1501–1510

Hable WE, Hart PE (2010) Signaling mechanisms in the establishment of plant and fucoid algal polarity. Mol Reprod Dev 77:751–758

Hamant O, Heisler MG, Jönsson H, Krupinski P, Uyttewaal M, Bokov P, Corson F, Sahlin P, Boudaoud A, Meyerowitz EM, Couder Y, Traas J (2008) Developmental patterning by mechanical signals in Arabidopsis. Science 322:1650–1655

Hardham AR, Green PB, Lang JM (1980) Reorganizatioonf cortical microtubules and cellulose deposition during leaf formation of *Graptopetala paraguyense*. Planta 149:181–195

Heisler MG, Hamant O, Krupinski P, Uyttewaal M, Ohno C et al (2010) Alignment between PIN1 polarity and microtubule orientation in the shoot apical meristem reveals a tight coupling between morphogenesis and auxin transport. PLoS Biol 8:e1000516

Hertel R, Flory R (1968) Auxin movement in corn coleoptiles. Planta 82:123–144

Jaffe LF (1966) Electrical currents through the developing *Fucus* egg. Proc Natl Acad Sci U S A 56:1102–1109

Jönsson H, Heisler MG, Shapiro BE, Meyerowitz EM, Mjolsness E (2006) An auxin-driven polarized transport model for phyllotaxis. Proc Natl Acad Sci U S A 103:1633–1638

Ketelaar T, Anthony RG, Hussey PJ (2004) Green fluorescent protein-mTalin causes defects in actin organization and cell expansion in Arabidopsis and inhibits actin depolymerizing factor's actin depolymerizing activity *in vitro*. Plant Physiol 136:3990–3998

Kusaka N, Maisch J, Nick P, Hayashi KI, Nozaki H (2009) Manipulation of intercellular auxin in a single cell by light with esterase-resistant caged auxins. ChemBioChem 10:2195–2202
Li P, Johnston MO (2000) Heterochrony in plant evolutionary studies through the twentieth century. Bot Rev 66:57–88
Liu Q, Qiao F, Ismail A, Chang X, Nick P (2013) The plant cytoskeleton controls regulatory volume increase. Biochim Biophys Acta 1828(2111):2120
Lockhard J (1960) Intracellular mechanism of growth inhibition by radiant energy. Plant Physiol 35:129–135
Lüthen H, Bigdon M, Böttger M (1990) Reexamination of the acid growth theory of auxin action. Plant Physiol 93:931–939
Maisch J, Nick P (2007) Actin is involved in auxin-dependent patterning. Plant Physiol 143:1695–1704
Mattsson J, Sung ZR, Berleth T (1999) Responses of plant vascular systems to auxin transport inhibition. Development 126:2979–2991
Meinhard H (1976) Morphogenesis of lines and nets. Differentiation 6:117–123
Meinhard H (1986) The threefold subdivision of segments and the initiation of legs and wings in insects. Trends Genet 3:36–41
Nagata T, Takebe I (1970) Cell wall regeneration and cell division in isolated tobacco mesophyll protoplasts. Planta 92:301–308
Nagata T, Nemoto Y, Hasezawa S (1992) Tobacco BY-2 cell line as the "Hela" cell in the cell biology of higher plants. Int Rev Cytol 132:1–30
Nakajima K, Sena G, Nawy T, Benfey PN (2001) Intercellular movement of the putative transcription factor SHR in root patterning. Nature 413:307–311
Nick P (2006) Noise yields order – auxin, actin, and polar patterning. Plant Biol 8:360–370
Nick P (2010) Probing the actin-auxin oscillator. Plant Signal Behav 5:4–9
Nick P (2012) Microtubules and the tax payer. Protoplasma 249(Suppl 2):S81–S94
Nick P, Heuing A, Ehmann B (2000) Plant chaperonins: a role in microtubule-dependent wall-formation? Protoplasma 211:234–244
Nick P, Han M, An G (2009) Auxin stimulates its own transport by actin reorganization. Plant Physiol 151:155–167
Nüsslein-Volhard C (1995) The identification of genes controlling development of flies and fishes. In: Noble Lecture, held on 8 Dec 1995
Paciorek T, Zažimalová E, Ruthardt N, Petrášek J, Stierhof YD, Kleine-Vehn J, Morris DA, Emans N, Jürgens G, Geldner N, Friml J (2005) Auxin inhibits endocytosis and promotes its own efflux from cells. Nature 435:1251–1256
Raven JA (2013) Polar auxin transport in relation to long-distance transport of nutrients in the Charales. J Exp Bot 64:1–9
Reinhardt D, Mandel T, Kuhlemeier C (2000) Auxin regulates the initiation and radial position of plant lateral organs. Plant Cell 12:507–518
Reinhardt D, Pesce ER, Stieger P, Mandel T, Baltensperger K, Bennett M, Traas J, Friml J, Kuhlemeier C (2003) Regulation of phyllotaxis by polar auxin transport. Nature 426:255–260
Roeland MH, Merks RMH, Van de Peer Y, Inzé D, Beemster GTS (2007) Canalization without flux sensors: a traveling-wave hypothesis. Trends Plant Sci 12:384–390
Rothwell GW, Lev-Yadun S (2005) Evidence of polar auxin flow in 375 million-year-old fossil wood. Am J Bot 92:903–906
Rubery PH, Sheldrake AR (1974) Carrier-mediated auxin transport. Planta 118:101–121
Sachs T (2000) Integrating cellular and organismic aspects of vascular differentiation. Plant Cell Physiol 41:649–656
Scheres B, Wolkenfelt H, Willemsen V, Terlouw M, Lawson E, Dean C, Weisbeek P (1994) Embryonic origin of the Arabidopsis root and root meristem initials. Development 120:2475–2487
Schoute JC (1913) Beiträge zur Blattstellungslehre. I. Die Theorie. Rec Trav Bot Neerl 10:153–325

Snow M, Snow R (1931) Experiments on phyllotaxis. I. The effect of isolating a primordium. Phil Trans Roy Soc Lond B 221:1–43
Spemann H (1936) Experimentelle Beiträge zu einer Theorie der Entwicklung. Springer, Berlin
Sun Y, Liu Y, Qu W, Jiang X (2009) Combining nanosurface chemistry and microfluidics for molecular analysis and cell biology. Anal Chim Acta 650:98–105
Swarup R, Péret B (2012) AUX/LAX family of auxin influx carriers—an overview. Front Plant Sci 3:e225
Swarup R, Friml J, Marchant A, Ljung K, Sandberg G, Palme K, Bennett M (2001) Localization of the auxin permease AUX1 suggests two functionally distinct hormone transport pathways operate in the Arabidopsis root apex. Genes Dev 15:2648–2653
Turing AM (1952) The chemical basis of morphogenesis. Phil Trans Roy Soc Lond B 237:37–72
Twell D, Park SK, Lalanne E (1998) Asymmetric division and cell-fate determination in developing pollen. Trends Plant Sci 3:305–310
Uyttewaal M, Burian A, Alim K, Landrein B, Borowska-Wykręt D, Dedieu A, Peaucelle A, Ludynia M, Traas J, Boudaoud A, Kwiatkowska D, Hamant O (2012) Mechanical stress acts via katanin to amplify differences in growth rate between adjacent cells in Arabidopsis. Cell 149:439–451
Van den Berg C, Willensen V, Hage W, Weisbeek P, Scheres B (1995) Cell fate in the Arabidopsis root meristem is determined by directional signalling. Nature 378:62–65
Whitaker DM (1942) Ultraviolet light and the development of Fucus eggs as affected by auxin and pH. Biol Bull 82:127–137
Yoon HS, Golden JW (1998) Heterocyst pattern formation controlled by a diffusible peptide. Science 282:935–938
Yoon HS, Golden JW (2001) PatS and products of nitrogen fixation control heterocyst pattern. J Bacteriol 183:2605–2613
Zaban B, Maisch J, Nick P (2013) Dynamic actin controls polarity induction de novo in protoplasts. J Int Plant Biol 55:142–159
Zimmermann W (1965) Die Telomtheorie. Gustav Fischer, Stuttgart

# Chapter 15
# Computational Models of Auxin-Driven Development

**Adam Runions, Richard S. Smith, and Przemyslaw Prusinkiewicz**

**Abstract** Auxin plays a key regulatory role in plant development. According to our current understanding, the morphogenetic action of auxin relies on its polar transport and the feedback between this transport and the localization of auxin transporters. Computational models complement experimental data in studies of auxin-driven development: they help understand the self-organizing aspects of auxin patterning, reveal whether hypothetical mechanisms inferred from experiments are plausible, and highlight differences between competing hypotheses that can be used to direct further experimental studies. In this chapter we present the state of the art in the computational modeling of auxin patterning and auxin-driven development in plants. We first discuss the methodological foundations of model construction: computational representations of tissues, cells, and molecular components of the studied systems. On this basis, we present mathematical models of auxin transport and the essential properties of pattern formation mechanisms involving auxin. We then review some of the key areas that have been investigated with the use of models: phyllotactic patterning of lateral organs in the shoot apical meristem, determination of leaf shape and vasculature, long-distance signaling and apical control of development, and auxin patterning in the root. The chapter is concluded with a brief review of current open problems.

A. Runions • P. Prusinkiewicz (✉)
Department of Computer Science, University of Calgary, Calgary, AB, Canada
e-mail: pwp@cpsc.ucalgary.ca

R.S. Smith
Department of Comparative Development and Genetics, Max Planck Institute for Plant Breeding Research, Cologne, Germany

E. Zažímalová et al. (eds.), *Auxin and Its Role in Plant Development*,
DOI 10.1007/978-3-7091-1526-8_15, © Springer-Verlag Wien 2014

# 1 Introduction

A key objective of developmental biology is to understand how molecular processes drive the development of tissues, organs, and entire organisms. In plants, the growth regulator auxin plays a commanding role on which many developmental processes depend. The morphogenetic role of auxin begins in the embryo, where its dynamic, differential distribution establishes the shoot–root polarity (Weijers and Jürgens 2005 ). In post-embryonic development, diverse patterning, signaling, and regulatory functions of auxin are summarized by the reverse/inverse fountain model (Benková et al. 2003) (Fig. 15.1). According to this model, auxin is produced in the vicinity of the shoot apical meristem and is transported in the epidermis towards the peripheral zone of the apex. There it accumulates in emergent convergence points, which determine the phyllotactic pattern of the incipient plant organs: leaves, flowers, and new branches (Reinhardt et al. 2003; Jönsson et al. 2006; Smith et al. 2006a). As a leaf grows and becomes flat, further convergence points appear at the leaf margin (Scarpella et al. 2006; Hay et al. 2006). These points may be correlated with the growth pattern localized near the margin, leading to the formation of serrated (Hay et al. 2006; Bilsborough et al. 2011), lobed, or compound (Barkoulas et al. 2008; Koenig et al. 2009; Ben-Gera et al. 2012; Townsley and Sinha 2012) leaves. From the primordia auxin flows into the subepidermal layers of the apex and, subsequently, into the plant stem. In this process, it is "canalized" into narrow paths (Sachs 1969, 1991; Mitchison 1980, 1981; Rolland-Lagan and Prusinkiewicz 2005; Feugier et al. 2005; Bayer et al. 2009; O'Connor et al. 2014), which, in the case of a leaf, mark the location of the primary vein and its extension into stem vasculature. Within the stem, auxin is involved in the patterning of the vascular system and the activation of lateral buds (Bennett et al. 2006; Prusinkiewicz et al. 2009; Crawford et al. 2010), thus coordinating the development of the branching plant structure (Leyser 2011). From the stem, auxin continues on to the root system, flowing through the root–shoot transition zone towards the apical meristems of the main and lateral roots, and reversing its direction in the root epidermis. In this process, it is involved in the maintenance and growth of sharply bounded meristematic and elongation zones (Grieneisen et al. 2007), initiation of lateral rootlets (Benková et al. 2003; Laskowski et al. 2008; Lucas et al. 2008a, b; Moreno-Risueno et al. 2010), and tropic responses to gravity (Swarup et al. 2005; Zažímalová et al. 2010).

The ability of auxin to perform these diverse functions is related to the pattern of its transport and the feedback between transport and the intercellular distribution of transporters. This includes, in particular, the highly mobile efflux carriers from the PIN protein family. In recent years, the interplay between auxin and further morphogenetic factors, such as other hormones, nutrients, light, and mechanical forces acting on cells, has also been considered (Leyser 2009). Computational models play a significant role in the studies of auxin-related patterning. The importance of these models stems from the self-organization of the patterning processes. Causal links underlying the emergence of patterns through self-

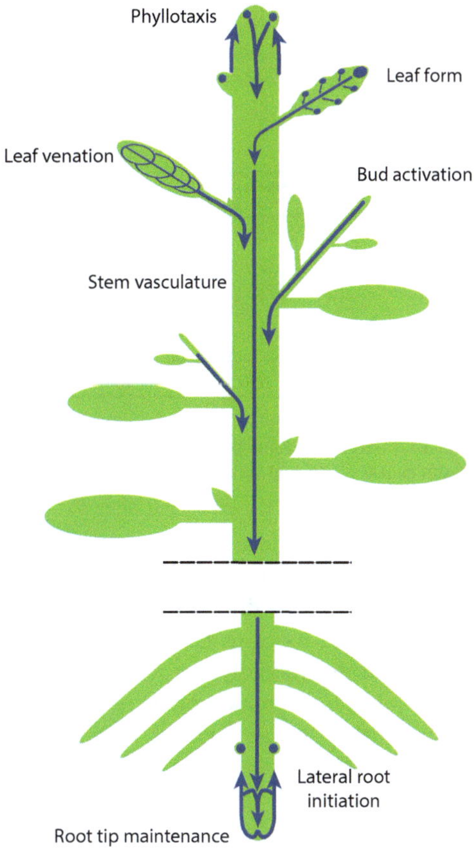


**Fig. 15.1** Processes and patterns regulated by auxin in post-embryonic development according to the reverse (shoot) and inverse (root) fountain model (Benková et al. 2003). *Blue arrows* indicate the paths and directions of auxin flow. *Blue circles* mark points of auxin accumulation. From (Prusinkiewicz and Runions 2012)

organization are generally nonintuitive, and computational models are a valuable tool facilitating their understanding (Camazine et al. 2001; Prusinkiewicz and Runions 2012). Models of auxin-driven patterning range from those directly rooted in biochemistry (Renton et al. 2012; Steinacher et al. 2012; Hošek et al. 2012) to more abstract constructs that aim at deducing morphogenetic characteristics of molecular-level process from the observed patterns and forms. In some cases, several hypotheses have been proposed to explain the same phenomenon, for example, the formation of phyllotactic and vascular patterns (Merks et al. 2007; Stoma et al. 2008; Bayer et al. 2009). While there is no consensus which of these hypotheses, if any, is the right one, the alternative models highlight their logical consequences and help formulate experiments that may support or falsify each hypothesis. Eventually, the models that survive the test of experiments will establish a causal chain linking molecular processes to macroscopic patterns and forms (Fig. 15.2).

The survey presented in this chapter begins with an outline of computational representations of tissues, cells, and cell states used in models of auxin-driven development. This is a fundamental aspect of model construction, as different

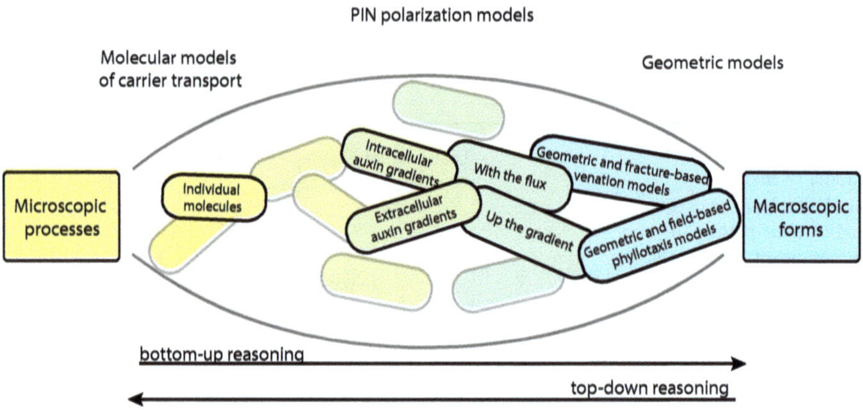


**Fig. 15.2** Models formulated at different scales and levels of abstraction represent a partial view of the causal chain that links molecular-level patterning and macroscopic forms

representations reflect different assumptions concerning the modeled processes. Models of auxin transport and polarization of transporters in cells are presented next, followed by a discussion of fundamental patterning properties of the postulated feedback loops (e.g., capability of forming a pattern of peaks or canals of auxin transport). Finally, the reverse/inverse fountain model is used to organize a review of specific models of auxin-driven patterning and plant development. Parts of this survey are an updated version of an earlier work by Prusinkiewicz and Runions (2012).

## 2 Computational Representations of Cells and Tissues

The choice of computational representations (data structures) of the modeled phenomena affects the range of processes that can be captured by the model, the level of abstraction at which they will be considered, the ease of creating, modifying, and exploring the model, and the computational efficiency of simulations. In the case of auxin-driven patterning, the central question is the relation between processes taking place in individual cells and the patterns emerging at the level of tissues. Consequently, the data structures typically consist of an explicit representation of cells connected into a tissue. Within this general framework, a number of choices exist and have been incorporated into different models.

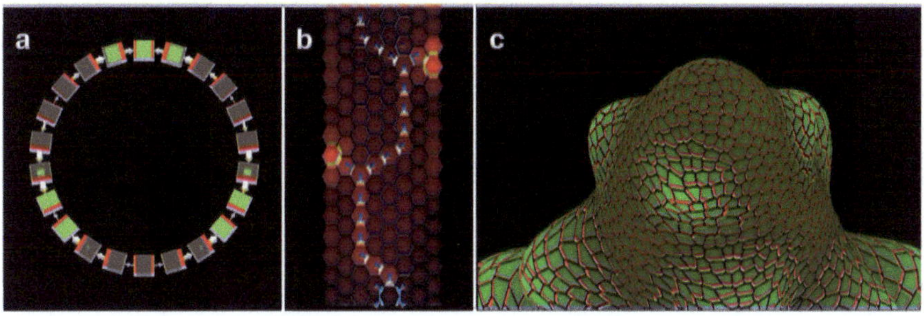

**Fig. 15.3** Some tissue representations used in cellular-level models of auxin-driven development. (**a**) One-dimensional ring representing the peripheral zone of a shoot apical meristem patterning the position of three primordia. The model assumes up-the-gradient polarization as described in Sect. 4.1 (Jönsson et al. 2006; Smith et al. 2006a). Auxin concentration in each cell is indicated by the size of the *green square*, PIN localization by the width of *red bars*, and auxin fluxes by the *white arrows* between cells. (**b**) Hexagonal tissue representing a longitudinal section through the shoot of *Brachypodium distachyon*. The model captures phyllotaxis and vascular development as described in Sect. 5.3 (O'Connor et al. 2014). Auxin concentration is shown in *red* and the localization of different PIN types in *yellow*, *blue*, and *white*. (**c**) A polygonal mesh representing the surface of a shoot apical meristem during a simulation of spiral phyllotaxis described in Sect. 5.1 (Smith et al. 2006a). Auxin concentration is shown in *green* and PIN in cell membranes in *red*

## 2.1 Dimensionality of the Model

Selected basic aspects of patterning, for example, the emergence of auxin concentration peaks (Smith et al. 2006a; Jönsson et al. 2006) or uniformly polarized cell files (Abley et al. 2013), can be explored in one-dimensional models: a sequence, ring (Fig. 15.3a), or branching arrangement of cells. One-dimensional models can be specified, implemented, and analyzed more easily than two- or three-dimensional models, but the scope of phenomena that they can capture is limited. In particular, canalization, i.e., the consolidation of auxin transport into narrow channels that pattern vascular tissues, can only be considered in two or three dimensions.

Two-dimensional models that abstract cells as polygons and cell walls as polygon edges are widely used as a compromise between the limited expressive power of one-dimensional models and the complexity of creating and visualizing fully three-dimensional models. The use of two-dimensional models is often further justified by the nature of the studied processes. For example, phyllotactic patterning in the apical shoot meristem of Arabidopsis is assumed to take place in the single layer of epidermal cells (Reinhardt et al. 2003; Barbier de Reuille et al. 2006; Jönsson et al. 2006; Smith et al. 2006a; Stoma et al. 2008; Kierzkowski et al. 2013), although subepidermal tissues may also play a role ( Larson 1975; Banasiak 2011). Patterning of leaf veins is essentially a two-dimensional process. The radial symmetry of roots suggests that modeling a longitudinal root section may suffice to capture the essential features of root morphogenesis (Grieneisen et al. 2007). Even processes that break this radial symmetry can be modeled in two dimensions if an appropriate section plane is chosen. For instance, the impact of root bending on the distribution of auxin and initiation of lateral roots was successfully modeled in two dimensions (Laskowski et al. 2008). Fully three-dimensional models still present a technical challenge.

## 2.2 *Representation of Cells and Tissues*

The simplest two-dimensional models are constructed assuming identical, square (Mitchison 1981; Rolland-Lagan and Prusinkiewicz 2005), or hexagonal (Feugier et al. 2005; Stoma et al. 2008; O'Connor et al. 2014) cells (Fig. 15.3b) . These cells are arranged in a regular tiling pattern, allowing for a straightforward and computationally efficient representation of the tissue as an array of cells. Artifacts of the regular tilings include directional bias (for example, a tissue made of square cells has different properties in the horizontal and vertical directions than in the diagonal directions) and a limited capacity for simulating growth: divisions of internal cells would break the regularity of the tiling, and thus new cells can only be added at the tissue boundary.

In more realistic models, tissues are represented by two-dimensional assemblies of polygons resembling the shape of cells, which form polygon meshes. For example, positions of cell walls and vertices at which they meet may be specified explicitly (Fig. 15.3c), inferred from the position of cell centroids through the construction of a Voronoi diagram, or result from a physically based simulation (see Prusinkiewicz and Runions (2012) for a review and Shapiro et al. (2013) for the most recent result). Diverse computational representations of polygon meshes are possible and have been widely studied due to their importance to geometric modeling and computer graphics. In particular, the vertex–vertex data structure (Smith et al. 2004; Smith 2006), rooted in the mathematical formalism of graph rotation systems (Edmonds 1960; White 1973), has been used in several models of auxin-driven patterning (Smith et al. 2006a; Bayer et al. 2009; Chitwood et al. 2012; Abley et al. 2013; O'Connor et al. 2014). This is due, in part, to the relatively simple specification of tissue growth and cell divisions in this formalism. The model of cell division by Besson and Dumais (2011) highlights the need for representing cells with curved walls, which likely will be incorporated into future models of auxin-driven morphogenesis. Models of jigsaw-shaped pavement cells in leaf epidermis will require even more flexibility in representing cell geometry.

In the above representations, cells typically partition the tissue without gaps or overlaps (except for the shared walls). Accumulation and diffusion of auxin in the intercellular space is neglected, and auxin leaving a cell is bound to enter the neighboring cell. This implies, in particular, that the relative roles of auxin efflux and influx carriers (e.g., PIN vs. AUX/LAX proteins) are difficult to distinguish. Recognizing these limitations, Kramer (2004) pioneered the incorporation of intercellular space into auxin transport models. He estimated the range of diffusion in the intercellular space to be of the same order as the cell size, which could be interpreted as an argument for both including extracellular auxin in more detailed models and excluding it from less detailed models. More recently, intercellular space was postulated to play a fundamental role in the molecular-level models of auxin-based cell polarization proposed by Wabnik et al. (2010), Roussel and Slingerland (2012) and Abley et al. (2013). In addition to auxin itself, candidate molecules involved in auxin-driven patterning involve ABP1 (Napier et al. 2002) and ROP (Xu et al. 2010).

Increasing the spatial resolution of cell models, Kramer (2004) incorporated vacuoles as a factor affecting the diffusion of auxin in the cells, and Hamant et al. (2008) subdivided the cell wall in order to analyze cell wall mechanics with the finite element method. In addition, Hamant et al. (2008) showed a correlation between the orientation of cortical microtubules, which are thought to be sensitive to stresses, and PIN polarity. The likely significant role of mechanosensing creates the need of representing the cytoskeleton in detailed models of auxin patterning as well.

The above representations of cells and tissues are of the Lagrangian type: they describe *where* in space the cells and their components are located. In contrast, Eulerian representations characterize *what* is located in different points in space. An example of the Eulerian viewpoint is the Cellular Potts model (Merks and Glazier 2005), which was employed to simulate auxin distribution and flow in the root of Arabidopsis by Grieneisen et al. (2007). Changes of shape due to deformations or growth are more easily represented from the Lagrangian viewpoint (Fan et al. 2013), which suggests why it has been used more frequently in tissue modeling.

## 2.3 *Static vs. Dynamic Tissue Models*

Some models of auxin-based patterning operate on tissues with a fixed geometry: cell arrangements generated programmatically (e.g., Mitchison 1981; Rolland-Lagan and Prusinkiewicz 2005; Feugier et al. 2005) or templates obtained by digitizing a microphotograph (e.g., Barbier de Reuille et al. 2006; Stoma et al. 2008; Bayer et al. 2009; Santuari et al. 2011). The underlying assumption is that the modeled patterning processes are fast compared to tissue growth, and thus growth can be neglected (c.f. Bayer et al. 2009). However, patterning may also be driven by growth or coupled with growth in a feedback loop of interactions. Sample models exploring such connections include phyllotactic patterning in a growing shoot apical meristem (Jönsson et al. 2006; Smith et al. 2006a; O'Connor et al. 2014), the sequential production of serrations in a growing leaf (Bilsborough et al. 2011), and the maintenance of the pattern of auxin flow ("reflux") in a growing Arabidopsis root (Grieneisen et al. 2007; Mironova et al. 2012). Tissue growth can be modeled geometrically, as a consequence of the expansion of the surface in which the cells are embedded (e.g., Smith et al. 2006a), or using a physically based model (e.g., Jönsson et al. 2006; Merks et al. 2011). In the latter case, cell expansion is attributed to an imbalance between the internal pressure in the cell and cell wall tension (Lockhart 1965; Prusinkiewicz and Lindenmayer 1990). Compared to the geometric models with prescribed growth, physically based models facilitate the inclusion of the impact of patterning on growth. In addition, they inherently incorporate physical forces, which may be morphogenetically relevant due to mechanosensing (Hamant et al. 2008; Heisler et al. 2010).

Models of tissue growth involve cell divisions, which are frequently simulated using the Errera (1886) rule. In the context of auxin-driven patterning, cell divisions pose a problem, because the impact of divisions on auxin transport and the distribution of the transporters is not sufficiently understood. The assumptions that the daughter cells preserve the polarization of the parent cell (Bilsborough et al. 2011) or that the polarization of the daughter cells is immediately established by the neighboring cells (Smith et al. 2006a) have been used in practice.

## 2.4 The State of the Cell

In most models of auxin-based patterning, each cell is characterized by the concentrations of the relevant substances, e.g., auxin, PIN, AUX/LAX, and CUC. Additional parameters are used to nuance this representation, for example, by indicating the polar allocation of PIN proteins to different sections of the membrane (Sect. 3.4) or by specifying the gradient of auxin concentration within the cell (Mitchison 1981). This level of abstraction is closely related to microscopic observations and is often used in models and their visualizations. Further details can be given by subdividing the cell into compartments and specifying relevant parameters individually for each compartment (Kramer 2004).

It is also possible to account for individual molecules of auxin and other substances (Garnett et al. 2008; Renton et al. 2012), instead of characterizing them summarily as concentrations. Potential advantages of this approach include a more intuitive model of interaction and transport of molecules, and the sustained validity of the model when the numbers of molecules are small and the continuous notion of concentration no longer applies (Gillespie 1976, 1977). At present, the numbers of molecules involved in auxin-driven patterning are not known, and thus it is not clear whether the increased computational cost of simulating individual molecules, compared to solving systems of differential equations used in the continuous case, is justified.

# 3 Auxin Transport

Auxin-induced patterning in plants is intimately related to auxin distribution and transport, in which auxin efflux carriers from the PIN family (Zažímalová et al. 2010) and auxin influx carriers from the AUX/LAX family (Swarup and Péret 2012) have received the closest attention. The currently recognized key processes involved in auxin transport are shown in Fig. 15.4a. The concentration of PIN on each membrane is determined by allocation (exocytosis, 1) and deallocation (endocytosis, 2) from a pool of free PIN in the cell. PINs located at the membrane export auxin from the cell to the extracellular space (3). From there, auxin is transported back into cells with the help of AUX/LAX proteins (4), which are assumed to be uniformly distributed along the cell membranes. Auxin also moves between the cells and the extracellular space by diffusion and background transport due to the residual presence of auxin exporters along the cell membranes (5). Finally, auxin diffuses between neighboring extracellular compartments (6). If the extracellular space is neglected, there is no extracellular diffusion, and any auxin leaving a cell directly enters the adjacent cell, as shown in Fig. 15.4b.

Below we present the typical equations used to model these processes. We first discuss the case when extracellular space is included and then introduce the simplified equations employed when this space is omitted. For simplicity, we assume that each cell has unit volume and each cell wall has unit area. Extensions to nonuniform volumes and lengths are described, for example, by Smith et al. (2006a), Jönsson et al. (2006), Stoma et al. (2008) and Bayer et al. (2009).

## 3.1 Auxin Mass Conservation

In cell $i$, auxin concentration $[\mathrm{IAA}_i]$ changes according to the equation:

$$\frac{d[\mathrm{IAA}_i]}{dt} = \text{Production} - \text{Removal} - \text{Flux}. \tag{15.1}$$

The *Production* term accounts for auxin biosynthesis, the level of which has a qualitative impact on some patterning processes (Pinon et al. 2013). The *Removal* term captures auxin turnover or conversion of auxin to a conjugated form. Both these terms may depend on the auxin concentration $[\mathrm{IAA}_i]$. For example, auxin production may have the form of a polynomial or rational polynomial function (e.g., Smith et al. 2006a, Eq. 5), which are easily derived from the chemically plausible laws of mass action (Shapiro et al. 2013). The *Production* term may also incorporate the effect of exogenous application of auxin in simulated experiments (e.g., Smith et al. 2006a), and both the *Production* and *Removal* terms may include sources or sinks representing tissues not explicitly included in the simulation (e.g., Mitchison 1980).

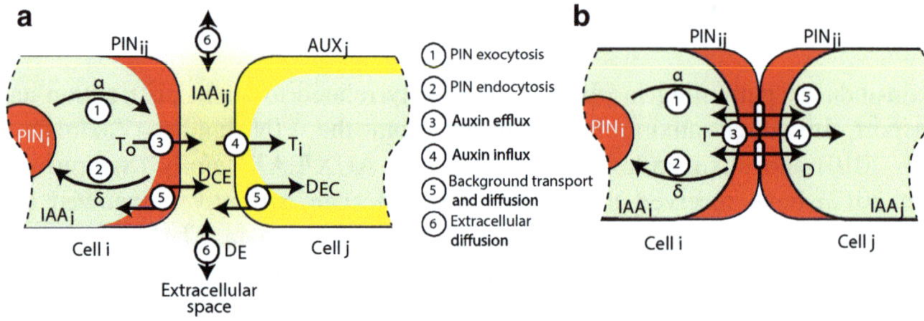


**Fig. 15.4** Processes underlying cellular-level models of polar auxin transport with (**a**) and without (**b**) intercellular space. Variables and parameters denoted by letters are explained in Sect. 3

The last term, *Flux*, represents the net flow (difference between outflux and influx) of auxin from cell $i$. It is the sum of fluxes $\Phi_{ij}$ through the membranes of cell $i$ facing cells $j$ in the neighborhood $N(i)$ of cell $i$:

$$\text{Flux} = \sum_{j \in N(i)} \Phi_{ij}. \tag{15.2}$$

Cells do not exchange fluxes directly, but via the extracellular space. Auxin concentration $[\text{IAA}_{ij}]$ in the extracellular compartment between cells $i$ and $j$ changes according to the fluxes from cells $i$ and $j$ and diffusion to neighboring extracellular regions:

$$\frac{d[\text{IAA}_{ij}]}{dt} = \Phi_{ij} + \Phi_{ji} - D_{\text{E}} \sum_{(k,l) \in N(i,j)} \left([\text{IAA}_{ij}] - [\text{IAA}_{kl}]\right), \tag{15.3}$$

where the sum is taken over all extracellular neighbors $(k, l)$ of the extracellular compartment $(i, j)$. The coefficient $D_{\text{E}}$ represents the rate of diffusion between adjacent compartments. Fluxes $\Phi_{ij}$ through the walls are captured by the chemiosmotic model of auxin transport, described next.

## 3.2 *The Chemiosmotic Model*

Inside a cell and within the intercellular space, auxin is assumed to move by diffusion. However, the transport of auxin into and out of cells is more complicated. The chemiosmotic model (Rubery and Sheldrake 1974; Raven 1975; Goldsmith 1977; Mitchison 1980) was proposed to provide a physicochemical basis for the description of this transport.

Auxin is a weak acid, and in the neutral pH inside cells it is largely dissociated. In this ionic form, auxin is hydrophilic and unable to diffuse across the plasma membrane. In order to leave a cell, auxin requires the activity of carriers located at the plasma membrane (Zažímalová et al. 2010), among which PIN proteins appear to play the most prominent morphogenetic role. Outside the cell, a significant portion of auxin becomes protonated in the lower pH of the extracellular space. This auxin is lipophilic, which makes it able to cross the plasma membrane and diffuse back into cells. Additionally, auxin is transported into cells by the AUX/LAX family of auxin import carriers, located in the plasma membrane.

Equations often used to implement the chemiosmotic model of auxin transport have been presented by Kramer (2009) (see also Kramer 2004; Swarup et al. 2005; Jönsson et al. 2006; Heisler and Jönsson 2006; Sahlin et al. 2009). The flux of auxin from the extracellular space $ij$ into cell $i$ is described as a combination of fluxes due to diffusion—$\Phi_{\text{diff}}$, export—$\Phi_{\text{export}}$, and import—$\Phi_{\text{import}}$:

$$\Phi_{ij} = \Phi_{\text{diff}} + \Phi_{\text{export}} + \Phi_{\text{import}}. \tag{15.4}$$

The flux due to diffusion, $\Phi_{\text{diff}}$, is proportional to the difference in concentration of protonated auxin between the cell $i$ and the extracellular space $ij$. Given pK, the negative log of dissociation constant for auxin, and $\text{pH}_\text{c}$, the pH of a compartment c, the protonated proportion of auxin in this compartment is (Weiss 1996):

$$K_\text{c}^{\text{IAAH}} = \frac{1}{1 + 10^{\text{pH}_\text{c} - \text{pK}}}. \tag{15.5}$$

Flux due to diffusion can thus be calculated as

$$\Phi_{\text{diff}} = P_{\text{diff}}\left(K_i^{\text{IAAH}}[\text{IAA}_i] - K_{ij}^{\text{IAAH}}\left[\text{IAA}_{ij}\right]\right), \tag{15.6}$$

where $P_{\text{diff}}$ is the membrane permeability for auxin diffusion.

The fluxes due to active transport, $\Phi_{\text{export}}$ and $\Phi_{\text{import}}$, are typically modeled using the Goldman–Hodgkin–Katz equation (Weiss 1996), assuming that the membrane is a homogeneous material and that the permeability to auxin depends on membrane potential. If we let

$$f(z) = \frac{\phi z}{e^{\phi z} - 1} \quad \text{with} \quad \phi = \frac{VF}{RT}, \tag{15.7}$$

where $V$ is the membrane potential, $F$ is the Faraday constant, $R$ is the gas constant, $T$ is the absolute temperature, and $z$ is the net valence of the ions being transported, the equations for import and export can be written as

$$\Phi_{\text{export}} = P_{\text{export}}\left(f(z)K_i^{\text{IAA}-}[\text{IAA}_i] - f(-z)K_{ij}^{\text{IAA}-}\left[\text{IAA}_{ij}\right]\right), \tag{15.8}$$

$$\Phi_{\text{import}} = P_{\text{import}}\Big(f(-z)K_i^{\text{IAA}-}[\text{IAA}_i] - f(z)K_{ij}^{\text{IAA}-}\big[\text{IAA}_{ij}\big]\Big). \tag{15.9}$$

In the above equations, we use the notation

$$K_{\text{c}}^{\text{IAA}-} = 1 - K_{\text{c}}^{\text{IAAH}} = \frac{1}{1 + 10^{\text{pK}-\text{pH}_{\text{c}}}}. \tag{15.10}$$

$P_{\text{import}}$ is the membrane permeability for import of auxin by AUX/LAX, and $P_{\text{export}}$ is the membrane permeability for the export of auxin by PIN and other exporters, such as the ABCB proteins (Zažímalová et al. 2010). These equations represent diffusion down the electrochemical gradient. Equations (15.8) and (15.9) are similar to Eq. (15.6) except for the factor $f$ () required to account for the dependence of fluxes on membrane potential. $P_{\text{import}}$ depends on the membrane concentration of importers [AUX$_i$], whereas $P_{\text{export}}$ depends on the membrane concentration [PIN$_{ij}$] of PIN proteins, as well as on ABCB proteins, which are assumed to be present at a background level (Grieneisen et al. 2007; Kramer 2009).

With a few exceptions (Steinacher et al. 2012; Hošek et al. 2012), the terms in Eqs. (15.5), (15.7), and (15.10) have been assumed constant in simulation models. Kramer (2009) calculated the three fluxes given by Eqs. (15.6), (15.8), and (15.9) by setting $z = 1$, using a membrane voltage of 120 mV, pK = 4.8, $\text{pH}_{ij} = 5.3$, and $\text{pH}_i = 7.2$. This yielded:

$$\begin{aligned}
\Phi_{\text{diff}} &= \big(0.004[\text{IAA}_i] - 0.24\big[\text{IAA}_{ij}\big]\big)P_{\text{diff}},\\
\Phi_{\text{export}} &= \big(4.68[\text{IAA}_i] - 0.034\big[\text{IAA}_{ij}\big]\big)P_{\text{export}},\\
\Phi_{\text{import}} &= \big(0.045[\text{IAA}_i] - 3.57\big[\text{IAA}_{ij}\big]\big)P_{\text{import}}.
\end{aligned} \tag{15.11}$$

A comparison of coefficients in the expression for $\Phi_{\text{diff}}$ shows that diffusion into the cell is favored over diffusion out of the cell by almost two orders of magnitude. However, the coefficient 3.57 is much larger than the other influx terms (preceded by the minus sign in the equations), which suggests that carrier-mediated influx dominates in cells where importers are expressed (provided that permeabilities $P_{\text{diff}}$, $P_{\text{export}}$, and $P_{\text{import}}$ have comparable values). Likewise, of the three terms controlling auxin efflux, coefficient 4.68 of the export term is significantly larger than the other two terms, which is consistent with the biological importance of PIN proteins. Note that the model implies a small influx involving exporters of auxin, and efflux involving its importers.

### 3.3 *Auxin Fluxes*

To obtain the typical equations used to model flux through a membrane, $\Phi_{ij}$, we eliminate negligible terms in Eqs. (15.6), (15.8), and (15.9) according to the analysis of Eq. (15.11). Assuming that $P_{\text{export}}$ is proportional to [PIN$_{ij}$] + $\beta$, where

$\beta$ is the background concentration of efflux carriers, and $P_{\text{import}}$ is proportional to $[\text{AUX}_i]$, we then obtain:

$$\begin{aligned} \Phi_{\text{diff}} &= -0.24P_{\text{diff}}\left[\text{IAA}_{ij}\right], \\ \Phi_{\text{export}} &= 4.68P_{\text{export}}\left[\text{IAA}_i\right] &&= 4.68K_{\text{PIN}}\left(\left[\text{PIN}_{ij}\right]+\beta\right)\left[\text{IAA}_i\right], \\ \Phi_{\text{import}} &= -3.57P_{\text{import}}\left[\text{IAA}_{ij}\right] &&= -3.57K_{\text{AUX}}\left[\text{AUX}_i\right]\left[\text{IAA}_{ij}\right], \end{aligned} \tag{15.12}$$

where $K_{\text{PIN}}$ and $K_{\text{AUX}}$ are coefficients of proportionality. By combining constant terms and parameters $P_{\text{diff}}$ , $\beta$, $K_{\text{PIN}}$ , and $K_{\text{AUX}}$ , we can rewrite the fluxes as

$$\begin{aligned} \Phi_{\text{diff}} &= -D_{\text{EC}}\left[\text{IAA}_{ij}\right], \\ \Phi_{\text{export}} &= \left(T_{\text{O}}\left[\text{PIN}_{ij}\right]+D_{\text{CE}}\right)\left[\text{IAA}_i\right], \\ \Phi_{\text{import}} &= -T_{\text{I}}\left[\text{AUX}_i\right]\left[\text{IAA}_{ij}\right], \end{aligned} \tag{15.13}$$

which, when summed, yield the net flux through the membrane equal to

$$\Phi_{ij} = T_{\text{O}}\left[\text{PIN}_{ij}\right]\left[\text{IAA}_i\right] + D_{\text{CE}}\left[\text{IAA}_i\right] - D_{\text{EC}}\left[\text{IAA}_{ij}\right] - T_{\text{I}}\left[\text{AUX}_i\right]\left[\text{IAA}_{ij}\right]. \tag{15.14}$$

All the elements of this equation are illustrated in Fig. 15.4. The first term accounts for the transport of auxin from cell $i$ to the extracellular space between cells $i$ and $j$ by PIN, with efficiency $T_{\text{O}}$ . It is sometimes assumed that this transport is nonlinear and the efficiency of PIN decreases as the concentration of auxin in cell $i$ increases (Jönsson et al. 2006) or as the concentration of auxin in the adjacent compartment increases (Smith et al. 2006a; Bayer et al. 2009; Roussel and Slingerland 2012; Chitwood et al. 2012). The second and third terms account for background auxin transport into the extracellular space with rate $D_{\text{CE}}$ and diffusion from the extracellular space into the cell with rate $D_{\text{EC}}$, respectively. The last term captures active import of auxin from the extracellular space by AUX/LAX proteins, with rate $T_{\text{I}}$ . For AUX/LAX the same concentration $[\text{AUX}_i]$ is used for all segments of the cell membrane, as these proteins are typically uniformly localized throughout the membrane.

When extracellular compartments are included, all communication between cells is mediated by the extracellular space. Explicit representation of extracellular space is particularly useful in models including the action of AUX/LAX proteins (Kramer 2004; Wabnik et al. 2010) and those interrogating the fundamental mechanisms that underlie PIN polarization (Kramer 2009; Wabnik et al. 2010; Roussel and Slingerland 2012; Abley et al. 2013). However, in patterning models the extracellular space is often assumed to play a secondary role and is omitted (Fig. 15.4b). In this case, auxin is transported directly between neighboring cells, i.e., every efflux implies a corresponding influx. Equation (15.14) then takes the form:

$$\Phi_{ij} = T\left[\mathrm{PIN}_{ij}\right][\mathrm{IAA}_i] - \left[\mathrm{PIN}_{ji}\right]\left[\mathrm{IAA}_j\right] + D[\mathrm{IAA}_i] - \left[\mathrm{IAA}_j\right]. \tag{15.15}$$

Equation (15.15) has been used in numerous models, e.g., Mitchison (1981); Smith et al. (2006a); Jönsson et al. (2006); Rolland-Lagan and Prusinkiewicz (2005); Stoma et al. (2008); Feugier et al. (2005); Prusinkiewicz et al. (2009); Bilsborough et al. (2011); O'Connor et al. (2014). The first term accounts for polar transport of auxin from cell $i$ to cell $j$ by PIN located in the membrane of cells $i$ facing $j$, with efficiency given by $T$. The second term accounts for polar transport from cell $j$ to cell $i$ in an analogous way. The last two terms account for nonpolar transport between the cells, with rate given by $D$. They represent the combined effect of phenomena such as diffusion through the extracellular space, background transport of auxin by PIN and other efflux carriers (e.g., ABCB proteins), and diffusion through the plasmodesmata (Rutschow et al. 2011). Note that $\Phi_{ij} = -\Phi_{ji}$, which is not the case when extracellular space is present. Equation (15.15) can be contrasted with that appearing in facilitated diffusion models (Mitchison 1981; Rolland-Lagan and Prusinkiewicz 2005; van Berkel et al. 2013), which postulate regulated permeability of the cell membranes instead of polar auxin transport controlled by the membrane concentration of influx and efflux carriers. In terms of Eq. (15.15), $T$ is then equal to 0, and the values of $D$ change locally as a function of the absolute value of $\Phi_{ij}$ . Unlike polar transport, facilitated diffusion cannot move auxin up a concentration gradient.

## 3.4 PIN Cycling

The concentration of PIN in the membrane of cell $i$ abutting cell $j$ changes due to allocation from (exocytosis) and deallocation to (endocytosis) a pool of unallocated PIN in the cell $i$:

$$\frac{d\left[\mathrm{PIN}_{ij}\right]}{dt} = \alpha[\mathrm{PIN}_i] - \delta\left[\mathrm{PIN}_{ij}\right]. \tag{15.16}$$

Here $[\mathrm{PIN}_i]$ denotes PIN concentration within the cell, $\alpha$ is the rate of exocytosis, and $\delta$ is the rate of endocytosis. These rates may depend on several factors. For $\alpha$, typical examples include auxin concentration in the neighboring cell $j$ (Smith et al. 2006a; Jönsson et al. 2006; Bayer et al. 2009; Bilsborough et al. 2011; Draelants et al. 2012; O'Connor et al. 2014) and auxin flux through the membrane (Feugier et al. 2005; Stoma et al. 2008; Bayer et al. 2009; Farcot and Yuan 2013; O'Connor et al. 2014). In contrast, $\delta$ may be a function of cellular auxin concentration (Paciorek et al. 2005) and also likely depends on cytokinin (Marhavý et al. 2011). Bilsborough et al. (2011) postulated that CUC2 may be required in some instances to modify cellular PIN polarizations, which could be accomplished by acting on $\alpha$ and $\delta$. A broad survey of the various PIN allocation schemes proposed in the literature is provided by van Berkel et al. (2013), who examined properties of these schemes at the level of cell membranes, cells, and one-dimensional files of cells.

Balancing the allocation of PIN proteins to the cell membranes, the change in concentration of PIN in the cytosol is

$$\frac{d[\mathrm{PIN}_i]}{dt} = -\sum_{j\in N(i)} \frac{d[\mathrm{PIN}_{ij}]}{dt}. \tag{15.17}$$

Initial models of polar auxin transport did not employ Eqs. (15.16) and (15.17), and instead assumed independent production of PIN-like efflux carriers at different segments of the cell membrane (Mitchison 1981; Rolland-Lagan and Prusinkiewicz 2005). However, competitive allocation of PIN proteins from a common pool appears to be more justified in view of biological data (Geldner et al. 2001), and readily leads to high auxin concentrations in developing veins (Feugier et al. 2005), consistent with observations (Sect. 4.2). Recent mathematical analysis (van Berkel et al. 2013; Farcot and Yuan 2013) shows that competitive allocation increases the range of parameters for which stable pattern formation may occur.

# 4 Elements of Auxin-Based Patterning

Molecular-level observations suggest that auxin regulates its own transport through a feedback with PIN proteins (Reinhardt et al. 2003; Scarpella et al. 2006; Hay et al. 2006; Bayer et al. 2009) (Fig. 15.5). This feedback likely provides the basis for the self-organized patterning of many elements of plant anatomy (Reinhardt et al. 2003; Scarpella et al. 2006; Hay et al. 2006; Barkoulas et al. 2008; Bayer et al. 2009; Bilsborough et al. 2011; O'Connor et al. 2014). Two different types of feedback between auxin and the cellular localization of PIN have been proposed, not precluding a possibility that they are different manifestations of a common mechanism. On the one hand, leaf primordia, as well as serrations, lobes, and leaflets, are initiated at auxin maxima (as inferred through auxin reporters such as DR5), with PIN1 in surrounding tissues polarized towards these maxima (Reinhardt et al. 2003; Hay et al. 2006; Koenig et al. 2009; Barkoulas et al. 2008; Bayer et al. 2009; Bilsborough et al. 2011; O'Connor et al. 2014). This has led to the hypothesis that PIN polarizes *up the gradient* of auxin concentration to generate convergence points (Jönsson et al. 2006; Smith et al. 2006a). On the other hand, during vascular initiation, PIN1 expression is refined into highly polarized strands (Scarpella et al. 2006; Bayer et al. 2009; O'Connor et al. 2014). The patterning of these strands is generally consistent with the canalization hypothesis proposed by Sachs (1969, 1981), according to which auxin flux through cells increases their capacity to transport auxin. The corresponding computational models thus assume that PIN polarizes *with the flux* of auxin transport (Mitchison 1980, 1981;

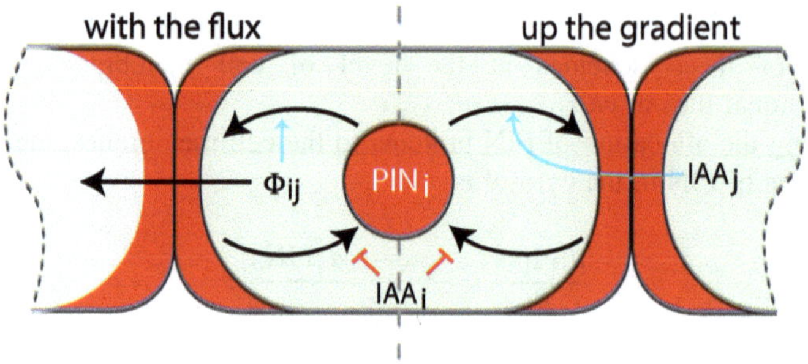


**Fig. 15.5** Hypothetical feedbacks controlling the localization of PIN proteins. With-the-flux models assume that (positive) auxin flux $\Phi_{ij}$ (c.f. Eq. 15.15) through a cell membrane increases exocytosis, whereas up-the-gradient models assume that high auxin concentration $[IAA_j]$ in the adjoining cell increases exocytosis. Some models also assume that cellular auxin concentration $[IAA_i]$ inhibits endocytosis

Rolland-Lagan and Prusinkiewicz 2005; Feugier et al. 2005; Fujita and Mochizuki 2006; Stoma et al. 2008; Bayer et al. 2009; O'Connor et al. 2014). Computational models employing these two types of feedback reproduce a broad range of the observed spatiotemporal patterns of auxin signaling and PIN polarization.

## *4.1 Up-the-Gradient Models*

In up-the-gradient models, PIN is allocated to each cell membrane according to the auxin concentration in the neighboring cell (Fig. 15.5). This causes small differences in cellular auxin concentration to be amplified, leading to the emergence of a stable pattern of periodic auxin maxima (Jönsson et al. 2006; Smith et al. 2006a; Sahlin et al. 2009; Draelants et al. 2012; van Berkel et al. 2013). Formally, up-the-gradient polarization can be enacted by making the rate of exocytosis, $\alpha$ in Eq. (15.16), an increasing function of auxin concentration in the neighboring cell $j$, while keeping the rate of endocytosis $\delta$ constant. As noted by Sahlin et al. (2009, p. 66), the opposite case, where the rate of exocytosis is constant and the rate of endocytosis is regulated, is mathematically equivalent; what matters is the ratio between both processes.

In up-the-gradient models constructed to date, PIN polarization has been assumed to be fast compared to the production and turnover of PINs, as well as changes in cellular auxin concentration. Consequently, the concentrations of PIN at each cell membrane and inside each cell were set to their steady-state values at each simulation step:

$$[\mathrm{PIN}_{ij}] = \frac{\alpha([\mathrm{IAA}_j])}{\sum_{k\in N(i)}\alpha([\mathrm{IAA}_k]) + \delta}, \tag{15.18}$$

$$[\mathrm{PIN}_i] = \frac{\delta}{\sum_{k\in N(i)}\alpha([\mathrm{IAA}_k]) + \delta}. \tag{15.19}$$

These equations can be derived by assuming that the total amount of PIN proteins in the cell, $[\mathrm{PIN}_i] + \sum_{j\in N(i)}[\mathrm{PIN}_{ij}]$, is constant, and setting Eqs. (15.16) and (15.17) to 0 (see Jönsson et al. (2006) for details). A key difference in initial models was the choice of the function $\alpha([\mathrm{IAA}_j])$ relating the rate of PIN allocation to a membrane to the auxin concentration in the abutting cell. Jönsson et al. (2006) employed a Hill function and Smith et al. (2006a) an exponential function. Simulations and mathematical analysis showed that, with either function, up-the-gradient polarization can generate one- and two-dimensional periodic patterns of approximately equidistant auxin maxima (Jönsson et al. 2006; Smith et al. 2006a; Sahlin et al. 2009; Draelants et al. 2012; van Berkel et al. 2013) (Fig. 15.6a, b). Different spacings can be achieved by adjusting model parameters, with the number of cells between peaks depending on the efficiency of polar auxin transport $T$ compared to diffusion rate $D$ (Eq. 15.15) (Fig. 15.6c). Further analysis in two dimensions showed that up-the-gradient models are also capable of creating striped patterns (Sahlin et al. 2009), similar to those emerging in reaction–diffusion models (Meinhardt 1982; Chap. 12). Differentiating between variants of up-the-gradient polarization models, recent mathematical analysis by Draelants et al. (2012) demonstrated that the model of Smith et al. (2006a) can produce oscillating steady states and confirmed the observation by Jönsson et al. (2006) that their model cannot.

Vieten et al. (2005) reported strong upregulation of *PIN1* expression at the sites of primordia initiation, suggesting the dependence of PIN1 production on auxin. Model studies by Smith et al. (2006a) and Heisler and Jönsson (2006) showed that such an upregulation can destabilize auxin peaks. Specifically, if PIN levels increase with auxin concentration, a cell with a high concentration of auxin will also have a high concentration of PIN, resulting in a large outflux of auxin. This may cause the maximum to shift to neighboring cells, which Smith et al. (2006a) and Heisler and Jönsson (2006) found undesirable in the context of phyllotactic patterning. In contrast, Merks et al. (2007) exploited the instability of auxin peaks, motivated by the appeal of a unified model potentially explaining both the formation of convergence points and vascular strands. In their model, the auxin maximum that initiates a leaf primordium subsequently moves into subepidermal tissues. PIN polarity follows this moving peak, leaving behind a strand of polarized PINs patterning a future vein. Unfortunately, predictions of this model are not consistent with the observed spatiotemporal patterns of auxin maxima and PIN polarization in developing leaves. For example, the predicted progression of the auxin maximum from the leaf tip towards the base during midvein formation is not observed in Arabidopsis leaves, where the maximum indicated by the DR5 reporter remains at

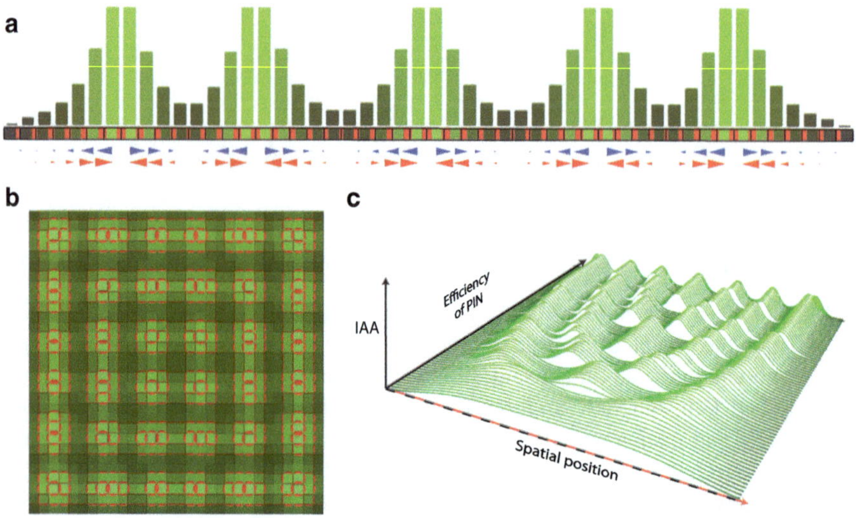


**Fig. 15.6** Up-the-gradient patterning. (**a**) One-dimensional pattern of equidistant auxin peaks that emerge when PINs orient up the gradient of auxin concentration. PIN polarization in each cell is shown in *red* and auxin concentration in *green*. Polar transport up the auxin gradient (*red arrows*) balances diffusion down the gradient (*blue arrows*) in the steady state shown. (**b**) A two-dimensional counterpart of the simulation from (**a**) also produces a pattern of auxin peaks. (**c**) The steady-state auxin concentration in a row of 50 cells plotted as a function of the efficiency of PIN transport $T$ (Eq. 15.15). *Red* and *black dashes* indicate the approximate size and position of each cell. As the efficiency of transport increases, the number of maxima increases as well

the tip as the midvein develops. Consequently, most models of vein patterning assume a different mode of PIN polarization, discussed next.

## 4.2 *With-the-Flux Models*

In with-the-flux models, PIN allocation to a cell membrane is promoted by auxin flux through this membrane. With-the-flux polarization is the cornerstone of the canalization hypothesis formulated by Sachs (1969, 1981, 1991, 2003). Historically, it was the first conceptual model of patterning that involved auxin and postulated the feedback of auxin on its own transport.

Sachs postulated that the export of auxin across a cell membrane promotes further auxin transport in the same direction and hypothesized that this feedback creates canals of auxin flow in a manner analogous to the carving of riverbeds by flowing water (Sachs 2003). Using a computational model operating on a square array of cells, Mitchison (1980, 1981) showed that the with-the-flux polarization model proposed by Sachs can indeed generate canals of high auxin flux. A reimplementation of Mitchison's model by Rolland-Lagan and Prusinkiewicz (2005) (Fig. 15.7a) and its reinterpretation in terms of a feedback between auxin

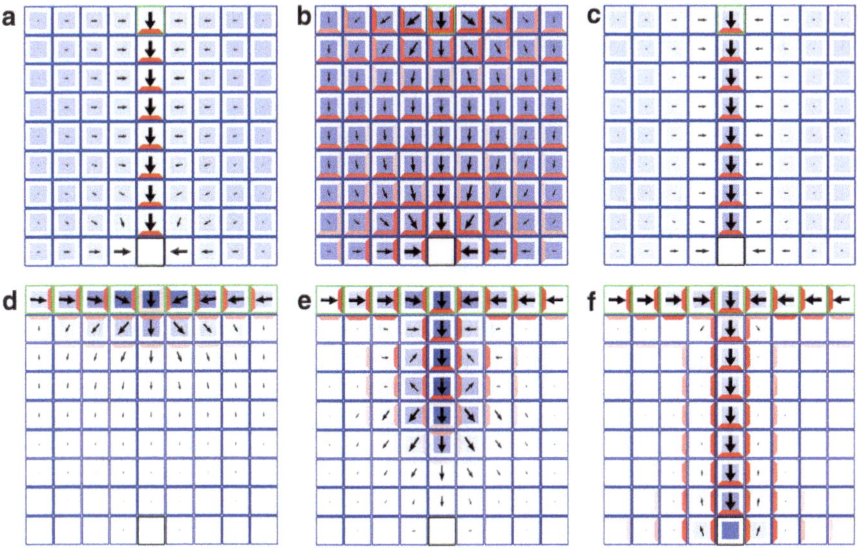

**Fig. 15.7** Patterns generated by with-the-flux (**a–c**) and dual-polarization (**d–f**) models. (**a**) A reimplementation (Rolland-Lagan and Prusinkiewicz 2005) of the model proposed by Mitchison (1980). PINs (*red*) are allocated assuming a quadratic dependence on auxin flux (*black arrows*). A canal of polarized cells is formed, connecting the auxin source at the top of the grid (outlined in *green*) to the sink at the bottom (middle cell, bottom row). The canal is characterized by high flux and low concentration of auxin (*blue*). (**b**) A linear PIN allocation function results in a broad coordination of PIN polarity across the tissue. (**c**) An implementation of the canalization model of Feugier et al. (2005). In contrast to panel (**a**), PINs are drawn from a limited pool, causing transport to saturate and auxin to build up in the strand. (**d–f**) Three frames of a simulation using the dual-polarization model by Bayer et al. (2009). (**d**) Epidermal cells (top row) initially polarize up the gradient, causing a convergence point to form in the center of the top row. (**e**) As auxin levels increase, the peak extends into the inner tissue. (**f**) The resulting strand elongates until it reaches the sink

flow and polarization of PIN1 proteins confirmed that the canalization hypothesis is generally consistent with observations of vein formation in developing leaves.

Mitchison (1980) proposed two main variants of his model: facilitated diffusion and polar transport. Each variant suggested a different molecular mechanism. In the case of facilitated diffusion, transport was affected by passive channels. The diffusion rate between cells was assumed to increase with net auxin flux, irrespective of the flux direction. Mitchison (1980) suggested plasmodesmata as potential candidates for the channels. Although it is likely that auxin can move through plasmodesmata to some extent (Rutschow et al. 2011), experimental support for a feedback based on auxin flux is currently lacking.

Polar transport is more compatible with the chemiosmotic model of auxin transport and molecular data on the localization and polarity of the PIN proteins (Rolland-Lagan and Prusinkiewicz 2005). At the cellular level, the impact of auxin on carrier allocation is captured by making parameter $\alpha$ in Eq. (15.16) a function of the net flux through the cell membrane:

$$\alpha(\Phi_{ij}) = \begin{cases} h(\Phi_{ij}) & \Phi_{ij} \geq 0 \\ 0 & \text{otherwise} \end{cases}, \tag{15.20}$$

where $h(\Phi_{ij})$ is an increasing function of net flux. According to this equation, the export of auxin across a cell membrane promotes further auxin transport in the same direction. Mitchison (1980) used a quadratic allocation function $h(\Phi_{ij})$ (Fig. 15.7a) and reported that it must be supralinear for canalization to occur. This feature was later investigated by Feugier et al. (2005) who found that a variety of supralinear functions for carrier allocation produced strands, including a step function. Feugier et al. (2005) also showed that if allocation is linear or sublinear then broad patterns of coordinated polarity over many cells arise (Fig. 15.7b). Stoma et al. (2008) exploited this regime in a model which, similar to the model of Merks et al. (2007), attempted to encompass phyllotaxis and vein formation using a common mechanism. In this model, linear polarization was assumed in the epidermis of the shoot apical meristem, producing broad patterns of PIN polarization towards primordia, and quadratic polarization was used to model the subepidermal patterning of veins. The produced patterns of PIN polarization closely matched those observed in the shoot apical meristem, but the model predicted a decrease of auxin concentration at the tips of leaf primordia that did not match auxin patterns reported by DR5.

Mitchison's model produces canals with high flux and low concentration of auxin (Fig. 15.7a), whereas experiments suggest that auxin concentration in canals is high (Scarpella et al. 2006). Exploring this discrepancy, Feugier et al. (2005) proposed and analyzed variants of Mitchison's models that operated according to two scenarios: with PINs allocated to different membrane sectors independently, and with PINs allocated to membranes from a fixed pool within each cell (c.f. Sect. 3.4). In the first case, simulations confirmed that the concentration of auxin in canals was lower than in the surrounding tissue, as originally predicted by Mitchison's model. In contrast, when cell membranes competed for the PINs within each cell, the models produced canals with auxin concentration higher than in the surrounding tissue (Fig. 15.7c). This result removed a key inconsistency between the canalization hypothesis and experimental data.

Competitive allocation of PIN qualitatively modifies the results of simulation compared to the noncompetitive allocation for the following reason. Given the fixed pool of PIN proteins in a cell, competitive allocation of PIN to one segment of the membrane (bottom segment of the provascular cells in Fig. 15.7c) reduces PIN allocation to the remaining segments of the membrane in the same cell. Consequently, auxin outflux from the provascular strand is reduced. From the viewpoint of the cells adjacent to this strand, this situation is indistinguishable from the reduction of outflux due to low concentration of auxin in Mitchison's model (Fig. 15.7a). This can be seen by rewriting Eq. (15.15) into the form:

$$\Phi_{ij} = \left(T\left[\mathrm{PIN}_{ij}\right] + D\right)\left[\mathrm{IAA}_i\right] - \left(T\left[\mathrm{PIN}_{ji}\right] + D\right)\left[\mathrm{IAA}_j\right]. \tag{15.21}$$

Reduction in the concentration of $[\mathrm{PIN}_{ji}]$ postulated by Feugier's model, but not by Mitchison's model, has the same effect on the flux $\Phi_{ij}$ as a reduction of auxin concentration $[\mathrm{IAA}_j]$.

## *4.3 The Dual-Polarization Model*

The proposed modes of PIN polarization by auxin, up the gradient and with the flux, involve the same molecular players. This raises the question of how a plant decides where and when to deploy each mode. Addressing this question, Bayer et al. (2009) investigated the development of the midvein in tomato leaf primordia. There the auxin peak that causes leaf initiation in the meristem remains in place while the strand that prepatterns the midvein is formed. To explain these dynamics, Bayer et al. (2009) proposed a dual-polarization model, according to which up-the-gradient and with-the-flux modes operate concurrently, with the weights dependent on the tissue type and auxin concentration. Figure 15.7d–f shows a simulation of this model. At first, auxin levels are low, allowing PINs to polarize up the gradient in the L1 and form a new convergence point (Fig. 15.7d). As the auxin levels increase, cells at the convergence point begin to favor with-the-flux polarization, which directs auxin flow towards inner tissues. This causes the peak to extend into a canal that eventually connects the source to the sink (Fig. 15.7e, f). The model reliably produces canals with high auxin concentration, as any drop in concentration would restore the up-the-gradient polarization mode, replenishing auxin in the canal.

The existence of an auxin-dependent transition between these two modes of PIN1 polarization has recently been supported by Furutani et al. (2014), who showed that genes from the *MAB4* family mediate the transition from up-the-gradient PIN1 polarization at lower auxin concentrations to with-the-flux polarization at higher concentrations. An interesting hypothesis is that PINOID is also involved in the deployment of each mode (van Berkel et al. 2013), as it is known to regulate apical vs. basal polarization of members of the PIN family in the root (Friml et al. 2004) in a manner dependent on auxin (Fozard et al. 2013).

The work of Bayer et al. (2009) suggests that the combined action of the up-the-gradient and with-the-flux polarization modes suffices to explain patterning induced by polar auxin transport in the shoot. Further support for the coordinated operation of up-the-gradient and with-the-flux polarization modes is presented by (O'Connor et al. 2014), who showed that in grasses these modes of polarization may be associated with distinct proteins related to AtPIN1 (c.f. Sect. 5.3).

## 4.4 The Role of Import Carriers

In addition to export carriers, the flow of auxin is affected by the AUX/LAX family of import carriers (Bennett et al. 1996; Parry et al. 2001) (Eqs. 15.11, 15.12, 15.13 and 15.14). These proteins are typically, although not always (Swarup et al. 2001), located uniformly on the cell membranes. Experimental results and models have focused on the role of AUX/LAX in enhancing and maintaining patterns of high auxin concentration in selected cells, vascular strands, and tissues. In contrast, studies of PINs have been focused on their primary role in the self-organization of patterns.

The first computational model by Kramer (2004) showed that AUX/LAX proteins can contribute to the maintenance of high auxin concentrations in vascular strands. A subsequent model by Swarup et al. (2005) pointed to the importance of AUX/LAX proteins in maintaining gradients of auxin concentration that are responsible for gravitropic responses in the root. Heisler and Jönsson (2006) used computational models to support the hypothesis that AUX/LAX proteins play a role concentrating auxin in the epidermis of shoot apical meristems (Reinhardt et al. 2003), although the retention or concentration of auxin in the epidermis also involves PIN1 (Bainbridge et al. 2008; Bayer et al. 2009; Kierzkowski et al. 2013). Heisler and Jönsson (2006) and Sahlin et al. (2009) also showed that auxin-induced AUX/LAX proteins may help to fix auxin maxima at the locations at which they emerged (i.e., the convergence points), and thus stabilize phyllotactic patterns. This role of AUX/LAX is consistent with the observations of irregular phyllotaxis patterns in plants with multiple mutations of these importers (Bainbridge et al. 2008).

Auxin application has been shown to upregulate AUX1 in roots (Laskowski et al. 2006, 2008; Paponov et al. 2008). On this basis, Laskowski et al. (2008) proposed that a positive feedback between auxin and its importers in the pericycle reinforces auxin peaks during lateral root initiation. Smith and Bayer (2009) explored this idea further using a model of a line of cells. They showed that a positive feedback between auxin-dependent importer production and the retention of auxin by importers not only can reinforce preexisting patterns, but can also generate patterns of equidistant peaks de novo (Fig. 15.8). These patterns resemble those generated by up-the-gradient polar transport of auxin by PIN (Fig. 15.6a). In contrast to peak formation by PIN proteins, peak formation by auxin importers does not require polarized transporters.

## 4.5 Molecular Basis of Cell Polarization

Although formulated in molecular terms, neither the up-the-gradient nor with-the-flux model explains the molecular mechanism of PIN polarization. As experimental data remain limited, several computational models have recently been proposed to

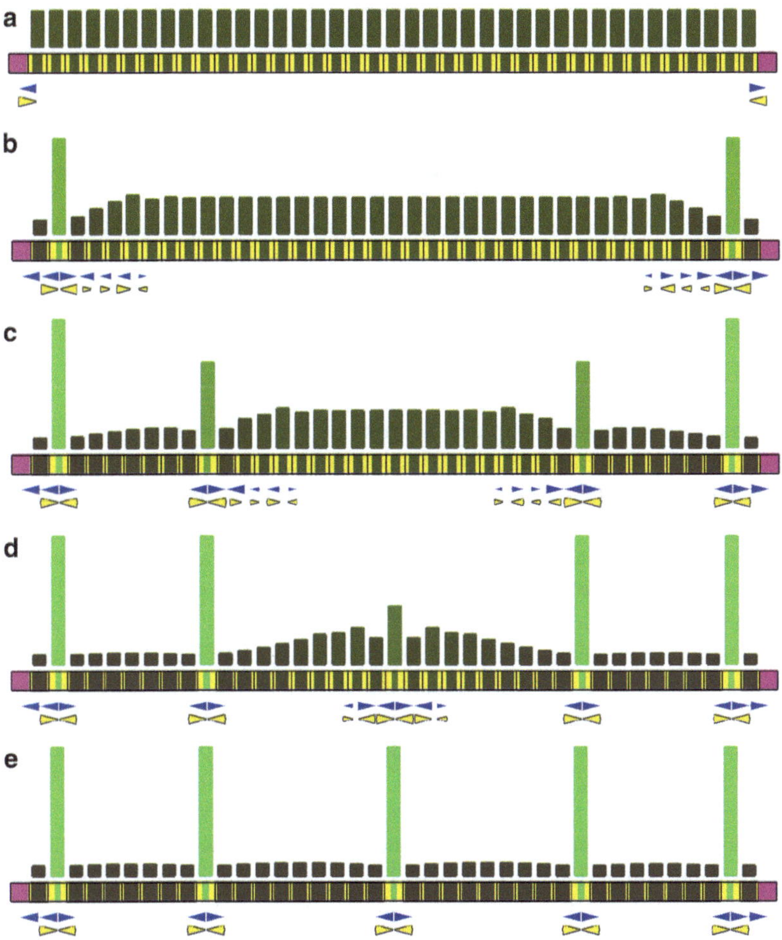


**Fig. 15.8** One-dimensional simulation of a hypothetical pattern formation process driven by AUX/LAX. Panels (**a–e**) represent subsequent stages of the simulation. Auxin concentration in each cell is shown in *green* and AUX/LAX concentration on cell membranes in *yellow*. Auxin is produced at the same rate in each cell. The first and last cells, shown in *purple*, are auxin sinks. The concentration of AUX/LAX is a quadratic function of auxin concentration. As cellular auxin levels increase, influx due to AUX/LAX (*yellow arrows*) begins to exceed efflux due to diffusion or transport by background efflux carriers (*blue arrows*), leading to auxin accumulation in some cells (progression from **a** to **b**). A competition between cells results, where the cells achieving a high auxin concentration deplete auxin from nearby cells. A pattern of approximately equidistant auxin maxima gradually emerges (**c**, **d**, **e**)

explore hypothetical mechanisms. Generally, these models can be divided into two classes: those postulating a purely biochemical polarization mechanism (Kramer 2009; Wabnik et al. 2010; Roussel and Slingerland 2012; Abley et al. 2013) and those incorporating biomechanical factors (Heisler et al. 2010).

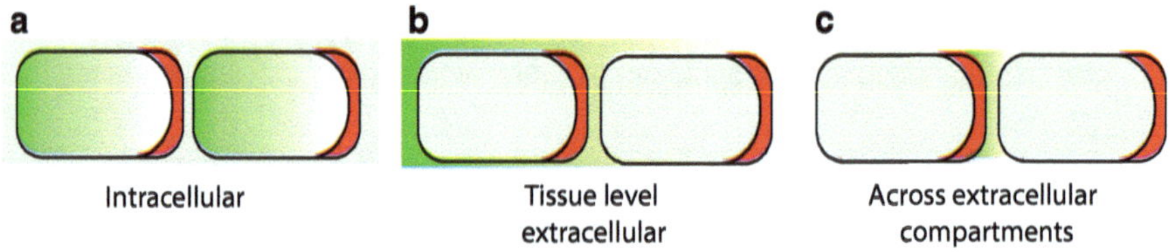


**Fig. 15.9** Auxin concentration gradients assumed in postulated molecular-level models of PIN polarization. High auxin concentrations are shown in *green* and low in *white*. (**a**) Gradients are present in individual cells (Kramer 2009). (**b**) Tissue-level gradient are present in the extracellular space (Roussel and Slingerland 2012; Abley et al. 2013). (**c**) Gradients are present in compartments of the extracellular space (Wabnik et al. 2010)

The biochemical models explore the emergence of a coherent polarization in a set of cells under different assumptions regarding auxin gradients. These assumptions include emergent coordination of auxin gradients in individual cells (Fig. 15.9a), alignment of polarizations governed by a tissue-level gradient in the intercellular space (Fig. 15.9b), and transmission of polarizing information via auxin gradients in the extracellular spaces between adjoining cells (Fig. 15.9c).

Kramer (2009) explored Mitchison's (1981) idea that flux sensing could result from a readout of intracellular auxin gradients (Fig. 15.9a). He suggested a role for the auxin-binding protein ABP1 in sensing these gradients, and showed that the resulting auxin fluxes can become canalized. In the reported simulations, vascular strands were always initiated at auxin sinks. In contrast, experimental observations suggest that the midvein in the leaf is initiated at an auxin source (Scarpella et al. 2006; Bayer et al. 2009; O'Connor et al. 2014). Kramer (2009) did not comment whether his model could reproduce these dynamics as well.

Roussel and Slingerland (2012) investigated another model of PIN polarization. They postulated that auxin in the extracellular space inhibits PIN exocytosis and, consequently, PIN proteins polarize towards regions of low auxin concentration in the extracellular space (Fig. 15.9b). This model produced a tissue-scale gradient of extracellular auxin spanning its source and sink, with PINs in the cells polarized consistently towards the sink. The source and the sink were connected by a path of high auxin flux in a manner resembling a vein, but auxin concentration in this path was not consistently elevated, in contrast to experimental data (c.f. Sect. 4.2).

Abley et al. (2013) systematically explored several hypothetical mechanisms that potentially could underlie cell polarization in both animals and plants. The mechanism they proposed to describe polarization in plants employed a PIN-like transporter molecule and an auxin-like substance that was exported from cells into the extracellular space by the transporter molecule. The extracellular auxin promoted PIN endocytosis, thus decreasing PIN concentration on both membranes abutting the same extracellular compartment. An inherent part of the model is the assumption of two antagonistic membrane-bound substances (possibly the proteins ROP2 and ROP6), one of which correlates positively and another one negatively with the localization of PIN. These substances enforce inherent cell polarization even in the absence of external stimuli. Abley et al. showed that a coordinated polarization of cells in a tissue results, and the steady-state auxin concentration in consecutive cells may either decrease or increase towards the sink, depending on model parameters. They did not apply their model to specific patterning processes, such as the formation of convergence points or veins.

In both the models of Roussel and Slingerland (2012) and Abley et al. (2013), auxin in the extracellular space acted symmetrically on the adjacent cells. In contrast, Wabnik et al. (2010) proposed that auxin in the extracellular compartments forms gradients, and these gradients provide asymmetric cues guiding PIN polarization in the adjacent cells (Fig. 15.9c). Similar to Kramer (2009), Wabnik et al. (2010) postulated that the auxin-binding protein ABP1 plays a role in this process, but they assumed that ABP1 interacts with auxin in the apoplast rather than within cells. PIN polarization would thus emerge from the intercellular competition for the ABP1 proteins that prevent PIN endocytosis. This hypothesis is consistent with experimental data showing that ABP1 is secreted from the cell where it is physiologically active (Napier et al. 2002) and is involved in the inhibition of endocytosis by auxin (Robert et al. 2010). The resulting model reproduced numerous details of vascular patterning and regeneration. Furthermore, bifurcation analysis indicated that it was capable of transitioning between up-the-gradient and with-the-flux transport regimes. Potentially, it could thus also account for phyllotaxis and other up-the-gradient phenomena. Nevertheless, the question remains whether significant auxin gradients are possible in the very narrow spaces between cells in the tissues where patterning occurs.

A model assuming that PINs are polarized by mechanical forces was proposed by Heisler et al. (2010), who built on their earlier model (Hamant et al. 2008) to explain peak formation in the shoot apex. Heisler et al. showed that PIN polarity is correlated with microtubule patterns, which can be modified by mechanical stresses. They proposed that high auxin concentration in a cell causes its wall to loosen, transferring load onto the wall of the adjacent cell (the loads acting on adjacent cell walls, abutting the same extracellular compartment, may thus be different). These stresses were sensed by the cells and used as a cue to polarize PIN proteins. Using a computational model operating on a fixed template of hexagonal cells, Heisler et al. (2010) showed that the feedback between the polarization of PIN proteins and stresses can generate a whorled pattern of auxin maxima.

Mechanical forces have also been involved in models of vascular patterning in leaves (Couder et al. 2002; Laguna et al. 2008; Corson et al. 2009), but links between these models and polar auxin transport are tenuous.

# 5 Review of Specific Models

A tight synergy between laboratory experiments and computational models underlies recent studies of growth regulation and patterning focused on the role of auxin. The fountain model (Fig. 15.1) suggests an exciting possibility of reducing fundamental features of plant development to a small number of general mechanisms. At a more immediate level, it presents a structured set of hypotheses regarding some of the key elements of plant development. Below we discuss these elements in more detail.

## *5.1 Phyllotaxis*

The first morphogenetic process involving auxin, in the order implied by the reverse fountain model, is the generation of a phyllotactic pattern of leaf and flower primordia on the shoot apical meristem (SAM). Microscopic observations of meristems in Arabidopsis by Reinhardt et al. (2003) showed that PIN1 proteins are oriented towards spatially separated convergence points, creating auxin maxima that predict the location of new primordia. Following these observations, they proposed that phyllotactic patterns emerge from a competition for auxin, where existing primordia drain auxin from their neighborhoods. This creates zones of low auxin concentration surrounding each primordium, where new primordia cannot be formed. The conceptual model of Reinhardt et al. can thus be viewed as a molecular implementation of the inhibitory mechanism of phyllotaxis proposed by Hofmeister (1868): the absence of auxin plays the role of an inhibitor. It leaves open, however, the question of what information is used to polarize PINs towards a convergence point, and what biochemical or biomechanical mechanisms effect this polarization. Addressing the first question, Jönsson et al. (2006) and Smith et al. (2006a) postulated a feedback between auxin distribution and PIN localization. According to these models, active auxin transport by PIN proteins creates auxin maxima that localize the incipient primordia. PINs orient themselves preferentially towards these maxima, promoting further auxin flux that reinforces them (up-the-gradient model, c.f. Sect. 4.1). Operating on a growing surface approximating the shoot apical meristem, this basic mechanism creates a relatively irregular pattern of auxin maxima. However, with additional assumptions—the restriction of the initiation of new primordia to the peripheral zone, the immobilization of auxin maxima, and the strengthening of PIN1 polarization towards the incipient primordia after their initiation (Smith et al. 2006a)—the model generates typical, highly regular spiral phyllotactic patterns (Fig. 15.10). Van Mourik et al. (2012) have recently proposed a related model to explain the patterning of floral organ primordia in Arabidopsis.

Motivated by the auxin-driven models of phyllotaxis, Smith et al. (2006b) and Mirabet et al. (2012) analyzed the robustness of phyllotactic patterning using models that abstract inhibitory properties of auxin in geometric terms. Both studies

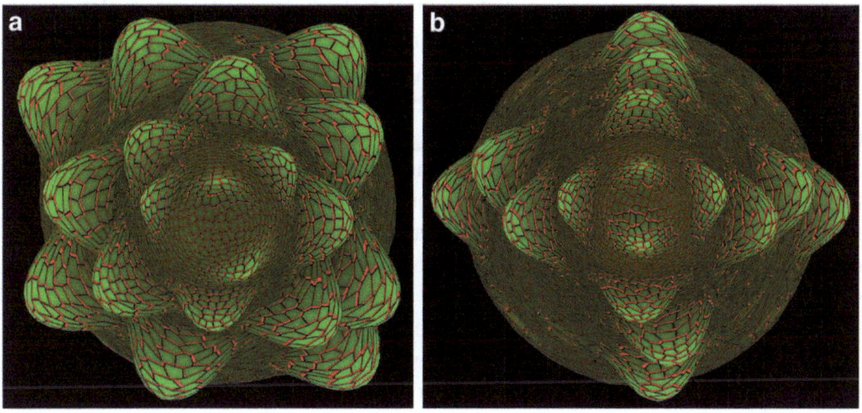


**Fig. 15.10** Simulation model of organ formation in the shoot apical meristem (Smith et al. 2006a). Transport of auxin (*green*) by PIN proteins (*red*) creates a self-organizing pattern of auxin maxima. PINs are polarized up the gradient, resulting in a spacing mechanism that positions auxin peaks as far as possible from previously existing ones. These peaks trigger the formation of new organs that bulge out from the apex surface. Growth of the shoot apex creates space at the tip, giving room for new organs to appear. Depending on model parameters and initial conditions, this can lead to a pattern of spiral (**a**) or decussate (**b**) phyllotaxis

postulated a secondary inhibitory field as a means through which the robustness of phyllotactic pattern formation can be increased. The more detailed analysis by Mirabet et al. (2012) has also considered a form of instability manifested by incorrect order of the initiation of primordia. Besnard et al. (2014) have subsequently shown that cytokinin establishes a secondary field which reduces the frequency of such instabilities in Arabidopsis. In addition to the inherent value of these results, they point to the need and usefulness of extending auxin-driven models with other regulatory processes and substances.

## 5.2 *Leaf Development*

Once positioned, a leaf primordium begins to grow, bulging out of the shoot apical meristem and gradually flattening along the abaxial–adaxial axis. During this growth, new convergence points emerge along the leaf margin, while the convergence point that initiated the leaf remains at the leaf tip (Scarpella et al. 2006; Hay et al. 2006). The formation of convergence points along the leaf margin appears to be governed by a mechanism similar to phyllotactic patterning in the SAM (Berleth et al. 2007; Smith and Bayer 2009; Bilsborough et al. 2011). As in phyllotaxis, existing convergence points locally inhibit the formation of new convergence points by draining auxin. New points thus only emerge when sufficient space is created for them by leaf growth. Similar to their counterparts at the shoot apical meristem, the convergence points at the leaf margin mark locations of increased outgrowth, yielding serrations in the case of Arabidopsis leaves (Bilsborough et al. 2011) and, possibly, lobes in leaves of other species (Barkoulas et al. 2008; Koenig et al. 2009). This similarity is consistent with the "partial shoot theory" (Arber 1950), which emphasizes parallels between the growth of shoots and leaves (Champagne and Sinha 2004). Following this train of thought, the strikingly different appearance of spiral phyllotactic patterns and leaves does not result from fundamentally different morphogenetic processes, but from the different geometries on which they operate: an approximately paraboloid shoot apical meristem that dynamically maintains its form vs. a flattening leaf that changes its shape and size as it grows.

Bilsborough et al. (2011) constructed a computational model of Arabidopsis leaf serration to further explore leaf development (Fig. 15.11). The general features of the observed serration patterns could be explained in terms of a feedback between auxin and PIN proteins, but the model showed that an additional factor was required to stabilize auxin maxima and thus robustly position serrations (Fig. 15.11b, c). This stabilizing role was fulfilled by the CUC2 protein, known to play a major role in leaf serration development (Nikovics et al. 2006; Kawamura et al. 2010). Following experimental data (PIN1 convergence points do not form in *cuc2* mutants), Bilsborough et al. (2011) hypothesized that PIN1 repolarization may only occur in the presence of CUC2. Auxin, however, downregulates *CUC2* expression, thus fixing PIN1 localization at the convergence points. It is an interesting question whether a related mechanism also stabilizes auxin maxima during phyllotactic pattern generation at the SAM [as suggested by Nikovics et al. (2006) and Berger et al. (2009)].

Chitwood et al. (2012) observed that auxin maxima in the model of phyllotaxis by Smith et al. (2006a) have an asymmetric shape and hypothesized that this asymmetry may disrupt the bilateral symmetry of leaf forms. They validated this hypothesis experimentally in tomato. Specifically, they observed the predicted asymmetric DR5 expression due to differences in distances between a given maximum and adjacent primordia in the clockwise and counterclockwise directions around the shoot and confirmed a relation between the direction of phyllotaxis (clockwise or counterclockwise) and the resulting asymmetry of leaves using a statistical analysis of leaf form.

## 5.3 *Vascular Patterning*

The models of phyllotaxis and leaf formation discussed above operate at the boundary of the organs considered: in the epidermis of the shoot apical meristem and at the margin of the leaf. The localization of PIN1 proteins and the activation of the DR5 auxin reporter in emerging leaves indicate that auxin reaching convergence points is redirected there towards the leaf interior. Its flow is then organized into canals: narrow paths that define the position of future veins. Modeling of vein formation is intimately linked with auxin canalization discussed in Sect. 4.2.

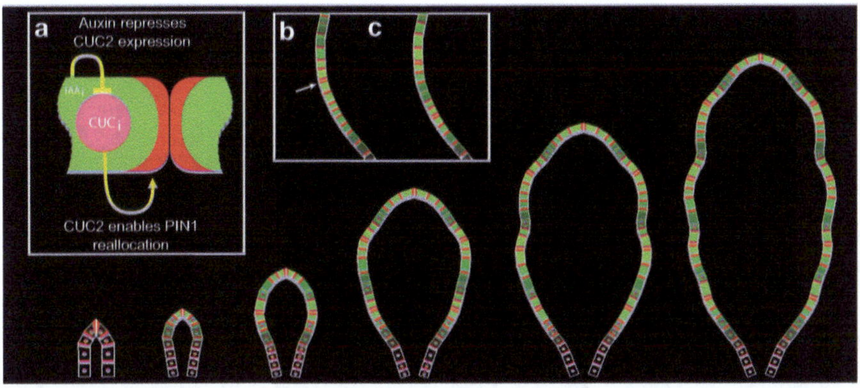


**Fig. 15.11** The model of leaf margin development proposed by Bilsborough et al. (2011). Polarized PIN1 proteins are shown in *red*, cellular auxin concentrations in *green*, and CUC2 expression is indicated by the presence of *pink circles* in the center of the cells. The model assumes that PIN1 polarizes up the gradient of auxin concentration and incorporates CUC2 as the enabling factor (inset **a**). As the leaf grows, the feedback between PIN1, auxin, and CUC2 generates an interspersed pattern of auxin maxima and CUC2 expression. Increased growth at auxin maxima and growth repression at sites of CUC2 expression modulate leaf growth, producing serrations. Inset (**b**)–(**c**): a variant of the model, where PIN1 can (re)polarize in the absence of CUC2. The convergence point marked by an *arrow* in (**b**) is unstable and splits into two in (**c**). The resulting convergence points travel away from each other until a stable spacing is achieved. Figure based on Bilsborough et al. (2011)

Initial models of the initiation of leaf midveins used pure up-the-gradient (Merks et al. 2007) or with-the-flux (Stoma et al. 2008) polarization modes. These models did not fully reproduce the spatiotemporal dynamics of DR5 and PIN1 expression (Sects. 4.1 and 4.2). Bayer et al. (2009) reproduced detailed observations of leaf midvein initiation with the dual-polarization model (Sect. 4.3), which blends between both polarization modes based on auxin concentration and tissue type (Fig. 15.12). In this model, up-the-gradient polarization is dominant in the epidermis and at low auxin concentrations, whereas with-the-flux polarization is dominant in the subepidermal tissues and at high auxin concentrations. The dual-polarization model reproduces the experimentally observed spatiotemporal sequence of PIN polarizations and auxin distribution in a leaf primordium. It shows that up-the-gradient and with-the-flux polarization modes can plausibly coexist in the convergence point. It also captures the basal position of PIN proteins in the vein precursor cells, the gradual narrowing of vein-defining canals, and the towards-the-vein orientation of PINs in the cells adjacent to these canals. The model predicted a transient polarization of PIN1 proteins in the subepidermis towards the epidermis at the onset of the primordium formation. This phenomenon was subsequently observed microscopically.

While formulating their model, Bayer et al. (2009) observed that canalizing strands cannot easily find sinks representing previously formed veins. To overcome this problem, they introduced a hypothetical diffusing substance that was produced

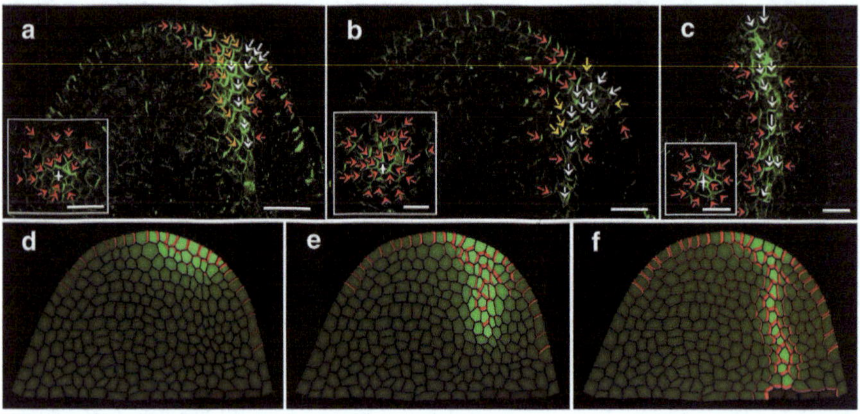


**Fig. 15.12** Comparison of experimental observations with the dual model for PIN1 polarization. (**a–c**) Three stages of midvein initiation in a tomato shoot apex. Scale bars: 20 μm. *White stars* in the insets indicate the PIN1 convergence point in the L1. PIN1 immunolocalization (*green*) suggests that PINs are oriented up the gradient of auxin concentration both in the L1 and in the subepidermal tissues that surround the initiating vein (*red arrows*). In contrast, PINs at the center of the convergence point and along the midvein appear to be oriented with the auxin flux (*white arrows*). Intermediate polarities are observed at the boundary between both regions (*yellow arrows*). (**d–f**) Successive stages of a simulation of PIN polarization using the dual-polarization model. The simulation employs a cellular template approximating microscopic image (**a**). Auxin concentration is shown in *green* and PIN localization in *red*. (**d**) PINs in the inner tissue are polarized towards the convergence point forming in the epidermis. (**e**) PINs near the convergence point switch polarity as the auxin flow extends into the subepidermis. (**f**) Auxin flux reaches the sink that represents preexisting vasculature (dark cells at the bottom) and becomes refined into a narrow strand. Figure adapted from Bayer et al. (2009)

in the vasculature and polarized cells towards existing veins. The problem of finding the sink was revisited in the recent work by O'Connor et al. (2014), in the context of phyllotactic and vascular patterning in the shoot apical meristem of the grass *Brachypodium distachyon*. Confocal microscopy observations indicated that three different proteins (termed PIN1a, PIN1b, and SoPIN1—sister of PIN1) were involved in Brachypodium, in contrast to the single PIN1 in Arabidopsis. These PINs have distinct expression and cellular localization patterns, which points to differences in the mechanisms determining their polarization. Assuming that SoPIN1 proteins polarize up the gradient and the remaining two PINs polarize with the flux at different rates (linear and quadratic, respectively), O'Connor et al. modeled the observed patterns of DR5 expression and PIN localization. The model suggests that each PIN plays a distinct role. SoPIN1 generates convergence points in the epidermis. PIN1b broadly polarizes cells towards nearby vasculature, which provides a sink-finding mechanism similar to that described by Feugier et al. (2005) and Stoma et al. (2008). Finally, PIN1a canalizes broad auxin flow towards the sink into a narrow high-flux path. These results show that phyllotaxis and vascular patterning in Brachypodium can be explained by concurrent up-the-gradient and with-the-flux polarization. Their coordinated action is consistent with the dual-polarization model. The progression from broad to canalizing auxin-driven PIN polarization suggests a mechanism for directing the emerging vein to the sink, alternative to the hypothetical factor introduced by Bayer et al. (2009).

Another model integrating up-the-gradient polarization (which leads to the emergence of convergence points) and with-the-flux polarization (which leads to the production of canals) captures the formation of the midvein and first-order laterals veins in open venation patterns, i.e., patterns without loops (Smith and Bayer 2009; Smith 2011). The model is driven by growth of the leaf blade, approximated as a single cellular layer. Questions related to the coupling of canalization and growth in the context of vein pattern formation have been highlighted and analyzed in a preliminary model study by Lee et al. (2014).

Observations by Scarpella et al. (2006) indicate that loops are formed by anastomosis, i.e., connection of canals. PIN proteins in these canals have opposite orientations, pointing away from a bipolar cell at which both canals meet. Mitchison's 1980 model and its recreation by Rolland-Lagan and Prusinkiewicz (2005) show that such a scenario of loop creation is possible if the bipolar cell is a source of auxin, turned on at a precisely defined time. A separate model of vein patterning in areoles (Dimitrov and Zucker 2006) also relies on elevated auxin concentration to localize the meeting point. However, the data of Scarpella et al. (2006) do not show an elevated auxin concentration at the meeting points. It is possible that bipolar cells are located at weak maxima of auxin concentration, not detected using experimental techniques of Scarpella et al. (2006). Another possibility, investigated using a computational model by Feugier and Iwasa (2006), is that proposed anastomosing canals are guided towards each other by a hypothetical diffusing substance. The existence of such a substance has not been experimentally confirmed. Vein pattern formation beyond the formation of the midvein and first-order lateral branches thus remains unclear.

## 5.4 Apical Dominance and Bud Activation

From leaves, auxin flows to the stem. There, auxin not only patterns the stem vascular system in a manner similar to the patterning of leaf veins, but also coordinates the activation of lateral buds, and thus the development of the branching structure as a whole. This coordination includes the phenomenon of apical dominance: a strong inhibitory influence of the shoot apical meristem in the vegetative state on the lateral buds below. Apical dominance is lifted upon the transition of the apex to the flowering state, resulting in the activation of one or more lateral buds in a basipetal sequence. Thimann and Skoog (1933) suggested that the inhibitory signal is auxin, produced by the shoot apex and actively transported down the plant. The use of computational models in the study of apical dominance has a particularly long history, rivaled only by Mitchison's (1980) models of auxin canalization.

The first family of models of apical dominance was created by Lindenmayer and his collaborators (Lindenmayer 1984; Janssen and Lindenmayer 1987; Prusinkiewicz et al. 1988; Prusinkiewicz and Lindenmayer 1990). The models aimed at elucidating the dynamics of branch initiation and flowering in compound inflorescences, using the herbaceous plant *Mycelis muralis* as a case study. To switch apical meristems in the main and lateral branches from the vegetative to flowering state, the models incorporated an additional long-distance signal, representing a then hypothetical flower-inducing substance, "florigen." The nature of florigen has since then been established (Lifschitz et al. 2006; Lifschitz and Eshed 2006; Shalit et al. 2009; Zeevaart 2008), opening the door for future models that may lead to a deeper understanding of inflorescence development.

Lindenmayer and his collaborators hypothesized that the timing of activation of successive buds reflects the speed with which the wave of auxin depletion propagates down the stem after the transition of the apical meristem to flowering. This hypothesis put in focus several questions. First, it is not clear how the resulting models could account for the activation sequences of buds within rosettes. There, extremely short internodes should lead to almost simultaneous activation of lateral buds, yet in Arabidopsis, for example, a basipetal sequence is observed in the rosette in spite of the short internodes (Stirnberg et al. 1999). Second, the depletion-wave models do not take into consideration contributions of the lateral branches to the auxin flow in the stem, contrary to experimental data (Morris 1977). Third, experiments with radiolabeled auxin show that auxin transported from the main apex through the stem does not pass in the vicinity of the dormant buds and does not enter them (Morris 1977). It is thus not clear how the auxin signal is conveyed to the bud. One possibility is that auxin acts on the lateral bud indirectly, through the intermediacy of one or more other hormones that move freely between the stem and the bud and act as secondary messengers. Candidate hormones are cytokinin (Muüller and Leyser 2011; Shimizu-Sato et al. 2008) and strigolactone (Gomez-Roldan et al. 2008; Dun et al. 2009; Brewer et al. 2009; Agusti et al. 2011), possibly acting jointly (Dun et al. 2012). Modeling shows, however, that this intermediacy is not necessary, and all three shortcomings of the auxin-depletion model can be addressed with a "relay" model using only auxin (Prusinkiewicz et al. 2009; Leyser 2009; Shinohara et al. 2013).

The relay model is based on the assumption, most recently supported by Furutani et al. (2014), that a lateral bud remains dormant until it can effectively export the auxin it produces through a polar transport mechanism. Bifurcation analysis of the with-the-flux model of auxin transport (Prusinkiewicz et al. 2009) shows that such export can be triggered by a temporary decrease of auxin concentration in the stem segment (metamer) supporting the bud, which the bud senses through an increase in the background auxin flow from the bud. Once triggered, the polar auxin transport is maintained even after the high level of auxin concentration in the metamer is restored by the auxin efflux from the activated bud.

In the context of a branched shoot, the temporary decrease of auxin concentration in a metamer results from a decrease in auxin supply from the shoot apical meristem and/or lateral meristems positioned further up the stem. A more basipetal

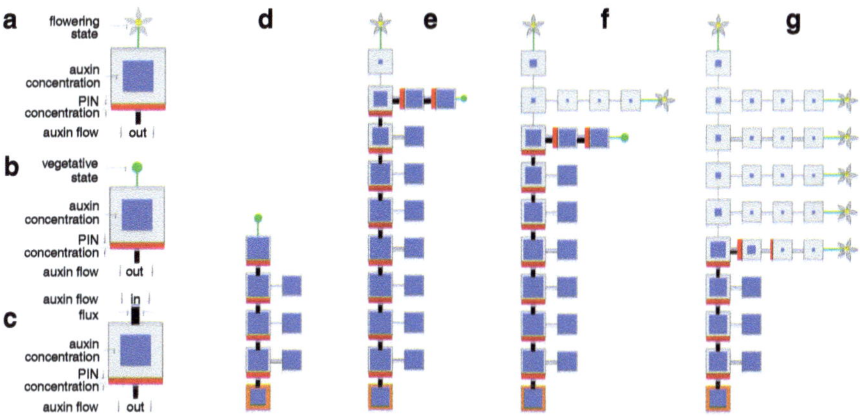


**Fig. 15.13** The relay model of bud activation at the metamer level. (**a**, **b**) Schematic representations of an apex in the flowering and vegetative state. (**c**) Representation of a metamer. (**d**–**g**) Selected stages of the simulation. At the beginning of the simulation, the main apex creates a sequence of metamers with the associated lateral buds. Due to the high supply of auxin from the apex, the concentration of auxin in the metamers is high (**d**). Upon transition to flowering, production of auxin in the main apex decreases, causing a decrease in auxin concentration in the stem. This decrease is the strongest in the topmost metamer, triggering polar auxin efflux from the associated lateral bud that activates it. Auxin produced by this bud re-saturates the stem (**e**). After transition of the topmost bud to the flowering state, next lateral bud becomes activated (**f**). The resulting relay process continues (**g**) until all buds have become activated. Figure based on Prusinkiewicz et al. (2009)

bud is thus activated when the bud higher up switches from the vegetative to the flowering state. By the same mechanism, the subsequent switch to flowering of the most recently activated bud triggers activation of the next one, and the relay progresses (Fig. 15.13). In contrast to the depletion model, the timing of this progression is determined by the delay between the activation of a lateral bud and its switch to flowering. Auxin propagation times thus play a secondary role.

The relay model extrapolates the with-the-flux auxin polarization model from the level of individual cells to the level of architectural modules of a plant: apices, buds, and metamers. An important aspect of with-the-flux polarization is its ability to canalize auxin flow into narrow streams, precursors of vascular strands (Fig. 15.7c). In the case of lateral buds, vascular connections may be formed concurrently with, and indeed as an integral part of, increased auxin outflow from the buds (Grbic and Bleecker 2000). A cellular-level version of the relay model (Fig. 15.14) shows that it is compatible with such a behavior.

The secondary messenger model and the model explaining apical dominance and bud activation in terms of the properties of with-the-flux polarization are not mutually exclusive. As the mechanisms of apical control continue to be actively studied, the use of models elucidating logical consequences of different assumptions fulfills one of key roles of modeling: to sharpen the questions.

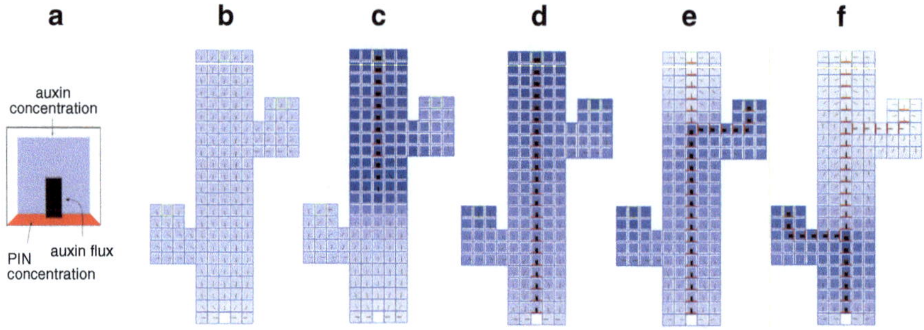


**Fig. 15.14** The relay model of bud activation at the cellular level. The cellular grid represents a longitudinal section of a stem with two buds. (**a**) Iconic representation of a cell. (**b–f**) Selected stages of the simulation. Following the placement of an auxin source at the top of the main segment, a vascular strand running through the segment emerges (**c**). Subsequent placement of auxin sources in the two buds (**d**) does not trigger formation of lateral vasculature until the auxin source at the top of the main stem is deactivated. The resulting decrease of auxin concentration in the main vasculature then triggers the formation of a vein connecting the higher bud to the central vasculature (**e**). When the source of auxin associated with this bud is deactivated, a similar process occurs in the lower bud (**f**). Figure adapted from Prusinkiewicz et al. (2009)

## *5.5 Root Development*

As described by the reverse/inverse fountain model (Fig. 15.1), auxin from the stem flows into the root. There, PINs are localized towards the root apex in the vasculature and away from it in the epidermis and cortex. Consistent with these localizations, auxin flows towards the root apex in the subepidermal layers and away from it in the epidermis. In the outer layers PIN proteins are also partially polarized towards the central axis of the root. As a result, auxin from the outer layers reenters the inner layers and is recycled towards the root tip. This recycling underlies the maintenance of an auxin maximum at the root apex. Grieneisen et al. (2007) capture this phenomenon using a model operating on a static grid and a model incorporating growth and division of approximately rectangular cells. In both cases PIN polarities were predefined. Similar spatial patterns of auxin concentration were subsequently obtained by Santuari et al. (2011), who used static cellular templates based on digitized microscopy images (Fig. 15.15). Cellular templates were also used by Stoma et al. (2008), who assumed that PINs are polarized according to the with-the-flux model. They showed that auxin maxima are maintained in this case as well.

In contrast to the above models, which were focused on the maintenance of the auxin maximum at the root tip, Mironova et al. (2010) addressed the problem of the emergence of this maximum and its regeneration after the removal of the root tip. They modeled these phenomena by assuming that PINs in different root layers have predefined polarizations, but their concentrations depend on the concentration of auxin. Mironova et al. (2012) extended that model by incorporating three different PIN proteins, PIN1, PIN2, and PIN3, and assuming that their expression and turnover rates respond differently to auxin concentration levels.

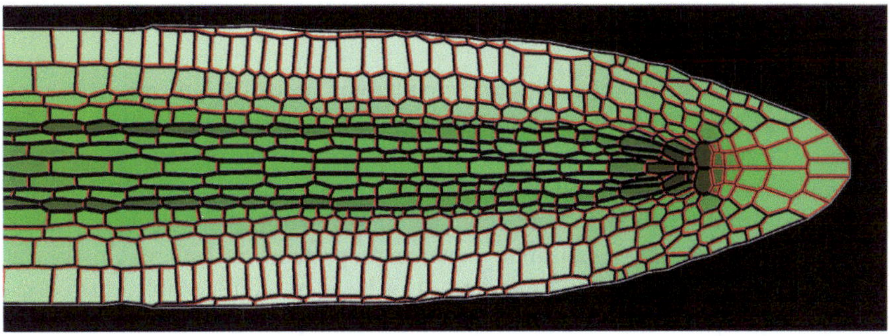

**Fig. 15.15** A model of auxin fluxes in the root. PINs (*red*) polarize towards the root tip in the central vascular system and away from the tip in the epidermis and cortex. Note the presence of PINs directing auxin back into the vasculature in the outer cell layers. This causes a reflux of auxin (*green*) back to the tip, allowing the system to store auxin even after the shoot is removed (Grieneisen et al. 2007). Simulation adapted from Santuari et al. (2011)

Grieneisen et al. (2007) used their model to propose that the recycling ("reflux") of auxin at the root tip produces an "auxin capacitor," where auxin is gradually accumulated. An extension of this idea underlies the mechanistic models of lateral root initiation proposed by Lucas et al. (2008a, b). In these models, the auxin capacitor at the root tip is charged by the basipetal flux of auxin. The capacitor is periodically discharged when the auxin level exceeds a threshold. The discharge triggers the formation of a lateral root. The models explain the timing of the initiation of lateral roots. Although they do not have a spatial character, they yield a spatial distribution of lateral roots when a rate of main root growth is assumed.

Fortin et al. (1989) observed that the sites of lateral root initiation are primed by root geometry, as lateral roots tend to form on the convex side of a curved main root. Investigating this phenomenon with computational models, Laskowski et al. (2008) found that longer cells on the convex side accumulate more auxin than shorter cells on the concave side. These differences are amplified by the auxin-dependent upregulation of auxin transport to cells by the AUX1 proteins. The higher auxin concentration on the convex side prompts preferential establishment of auxin maxima on the same side. These maxima induce lateral roots (Benková et al. 2003).

The model by Laskowski et al. (2008) showed that a feedback between auxin and its importers may play a role in auxin-driven patterning. As shown in Sect. 4.4, such a feedback can theoretically create a pattern of approximately equidistant auxin maxima even in the absence of the more widely considered feedback between auxin and its exporters.

# 6 Conclusions

Computational modeling of auxin-driven patterning got off to an early start with Mitchison's (1980) exploration of Sachs's canalization theory (Sachs 1969), but for the next 25 years the area remained dormant. The situation changed in the early 2000s with the explosion of new experimental techniques. Immunological detection (Sauer and Friml 2010) and fluorescent tagging (Millar et al. 2009) have made it possible to display the localization patterns of proteins in different tissues at subcellular resolution. In vivo techniques (Heisler et al. 2005) enabled the observation of these localizations over time. Genetic manipulations led to remarkable advancements in the study of protein functions. The experimental results prompted by these techniques have led to new hypotheses regarding mechanisms of pattern formation in plants. Computational models turned out to be useful in testing whether these hypotheses are plausible. They also put into focus many crucial questions, in particular regarding the biological mechanisms of PIN polarization (Bennett et al. 2013), which are subject of current research.

In contrast to the experimental systems, where causal relations are not directly observed, in computational models such relations are explicitly assumed. This makes models indispensable, as they provide the only rigorous means to examine whether proposed mechanisms can indeed yield the observed patterns and forms. The use of computational models for this purpose began with the examination (Jönsson et al. 2006; Smith et al. 2006a; Barbier de Reuille et al. 2006) of the conceptual model of phyllotaxis introduced by Reinhardt et al. (2003). Models examining further auxin-driven processes in plants, including apical control and the development of leaves and roots, quickly followed (Sect. 5). Extensions to other processes and systems (e.g., regulation of embryonic development and mediation of tropic responses in plants) are forthcoming.

In addition to explaining phenomena based on experimental data, computational models highlight areas where experimental data are insufficient. At present, one such area encompasses molecular mechanisms of PIN polarization. Models examining different hypothetical mechanisms have been proposed (Sect. 4.5), but data that would put them on a solid experimental foundation are still not available. Their theoretical implications and relations to the higher-level up-the-gradient, with-the-flux, and dual-polarization models also require a better understanding. This is needed to connect models operating at different scales and levels of abstraction (Fig. 15.2). An intriguing element of the puzzle is the extent to which the interplay between mechanical stresses and the distribution of auxin transporters (Hamant et al. 2008; Hamant and Traas 2010; Boudaoud 2010; Heisler et al. 2010), as well as growth, may play a role in PIN polarization and the resulting regulation of development. Furthermore, production of leaves and vascular systems in *pin* mutants (Barkoulas et al. 2008; Bilsborough et al. 2011) indicate that the feedback between PIN proteins and polar auxin transport represents only one facet of the relevant patterning processes (Guenot et al. 2012; Kierzkowski et al. 2013). Problems of current interest also concern the interplay between auxin and other substances (e.g., cytokinin, strigolactone) and the role of auxin synthesis in patterning. Furthermore, mechanistic links between auxin-related patterns and the resulting macroscopic forms, for example, the diverse forms of leaves, remain an area of active study.

From the methodological perspective, most models of auxin-driven patterning and growth devised to date operate on surfaces. In some cases, however, the use of three-dimensional models appears to be essential. One example is the modeling of vascular pattern development in stems. The methodology for creating and visualizing three-dimensional models, especially those operating on growing tissues as opposed to static templates, is yet to be fully developed.

**Acknowledgment** The authors' research is supported by the Swiss National Science Foundation (RSS), the Max Planck Society (RSS), the Natural Sciences and Engineering Research Council of Canada (PP), and the Human Frontier Science Program (PP).

## References

Abley K, Barbier de Reuille P, Strutt D, Bangham A, Prusinkiewicz P, Marée A, Grieneisen V, Coen E (2013) An intracellular partitioning-based framework for tissue cell polarity in plants and animals. Development 140(10):2061–2074

Agusti J, Herold S, Schwarz M, Sanchez P, Ljung K, Dun E, Brewer P, Beveridge C, Sieberer T, Sehr E, Greb T (2011) Strigolactone signaling is required for auxin-dependent stimulation of secondary growth in plants. Proc Natl Acad Sci U S A 108(50):20242–20247

Arber A (1950) The natural philosophy of plant form. University Press, Cambridge

Bainbridge K, Guyomarc'h S, Bayer E, Swarup R, Bennett M, Mandel T, Kuhlemeier C (2008) Auxin influx carriers stabilize phyllotactic patterning. Genes Dev 22(6):810–823

Banasiak A (2011) Putative dual pathway of auxin transport in organogenesis of *Arabidopsis*. Planta 233(1):49–61

Barbier de Reuille P, Bohn-Courseau I, Ljung K, Morin H, Carraro N, Godin C, Traas J (2006) Computer simulations reveal properties of the cell-cell signaling network at the shoot apex in *Arabidopsis*. Proc Natl Acad Sci U S A 103(5):1627–1632

Barkoulas M, Hay A, Kougioumoutzi E, Tsiantis M (2008) A developmental framework for dissected leaf formation in the *Arabidopsis* relative *Cardamine hirsuta*. Nat Genet 40 (9):1136–1141

Bayer E, Smith R, Mandel T, Nakayama N, Sauer M, Prusinkiewicz P, Kuhlemeier C (2009) Integration of transport-based models for phyllotaxis and midvein formation. Genes Dev 23 (3):373–384

Ben-Gera H, Shwartz I, Shao M, Shani E, Estelle M, Ori N (2012) ENTIRE and GOBLET promote leaflet development in tomato by modulating auxin response. Plant J 70(6):903–915

Benková E, Michniewicz M, Sauer M, Teichmann T, Seifertova D, Jürgens G, Friml J (2003) Local, efflux-dependent auxin gradients as a common module for plant organ formation. Cell 115(5):591–602

Bennett M, Marchant A, Green H, May S, Ward S, Millner P, Walker A, Schulz B, Feldmann K (1996) Arabidopsis AUX1 gene: a permease-like regulator of root gravitropism. Science 273 (5277):948–950

Bennett T, Sieberer T, Willett B, Booker J, Luschnig C, Leyser O (2006) The Arabidopsis MAX pathway controls shoot branching by regulating auxin transport. Curr Biol 16(6):553–563

Bennett T, Hines G, Leyser O (2013) Canalization: what the flux? Trends Genet 30(2):41–48

Berger Y, Harpaz-Saad S, Brand A, Melnik H, Sirding N, Alvarez J, Zinder M, Samach A, Eshed Y, Ori N (2009) The NAC-domain transcription factor GOBLET specifies leaflet boundaries in compound tomato leaves. Development 136(5):823–832

Berleth T, Scarpella E, Prusinkiewicz P (2007) Towards the systems biology of auxin-transport-mediated patterning. Trends Plant Sci 12(14):151–159

Besnard F, Refahi Y, Morin V, Marteaux B, Brunoud G, Chambrier P, Rozier F, Mirabet V, Legrand J, Lainé S, Thévenon E, Farcot E, Cellier C, Das P, Bishopp A, Dumas R, Parcy F, Helariutta Y, Boudaoud A, Godin C, Traas J, Guédon Y, Vernoux T (2014) Cytokinin signalling inhibitory fields provide robustness to phyllotaxis. Nature 505:417–421

Besson S, Dumais J (2011) Universal rule for the symmetric division of plant cells. Proc Natl Acad Sci U S A 108(15):6294–6299

Bilsborough G, Runions A, Barkoulas M, Jenkins H, Hasson A, Galinha C, Laufs P, Hay A, Prusinkiewicz P, Tsiantis M (2011) Model for the regulation of *Arabidopsis thaliana* leaf margin development. Proc Natl Acad Sci U S A 108(8):3424–3429

Boudaoud A (2010) An introduction to the mechanics of morphogenesis for plant biologists. Trends Plant Sci 15(6):353–360

Brewer P, Dun E, Ferguson B, Rameau C, Beveridge C (2009) Strigolactone acts downstream of auxin to regulate bud outgrowth in pea and *Arabidopsis*. Plant Physiol 150(1):482–493

Camazine S, Deneubourg J, Franks N, Sneyd J, Theraulaz G, Bonabeau E (2001) Self-organization in biological systems. Princeton University Press, Princeton, NJ

Champagne C, Sinha N (2004) Compound leaves: equal to the sum of their parts? Development 131(18):4401–4412

Chitwood D, Headland L, Ranjan A, Martinez C, Braybrook S, Koenig D, Kuhlemeier C, Smith R, Sinha N (2012) Leaf asymmetry as a developmental constraint imposed by auxin-dependent phyllotactic patterning. Plant Cell 24(16):2318–2327

Corson F, Adda-Bedia M, Boudaoud A (2009) In silico leaf venation networks: growth and reorganization driven by mechanical forces. J Theor Biol 259(3):440–448

Couder Y, Pauchard L, Allain C, Adda-Bedia M, Douady S (2002) The leaf venation as formed in a tensorial field. Eur Phys J B Condens Matter Complex Syst 28(2):135–138

Crawford S, Shinohara N, Sieberer T, Williamson L, George G, Hepworth J, Muller D, Domagalska M, Leyser O (2010) Strigolactones enhance competition between shoot branches by dampening auxin transport. Development 137(17):2905–2913

Dimitrov P, Zucker S (2006) A constant production hypothesis guides leaf venation patterning. Proc Natl Acad Sci U S A 103(24):9363–9368

Draelants D, Broeckhove J, Beemster G, Vanroose W (2012) Numerical bifurcation analysis of the pattern formation in a cell based auxin transport model. J Math Biol 67(5):1279–1305

Dun E, Brewer P, Beveridge C (2009) Strigolactones: discovery of the elusive shoot branching hormone. Trends Plant Sci 14(7):364–372

Dun E, de Saint Germain A, Rameau C, Beveridge C (2012) Antagonistic action of strigolactone and cytokinin in bud outgrowth control. Plant Physiol 158(1):487–498

Edmonds J (1960) A combinatorial representation for polyhedral surfaces. Notices Amer Math Soc 7:646

Errera L (1886) Sur une condition fondamentale d'équilibre des cellules vivantes. Comptes Rendus Hebdomadaires des Seances de l'Academie des Sciences 103:822–824

Fan Y, Litven J, Levin D, Pai D (2013) Eulerian-on-Lagrangian simulation. ACM Trans Graph 32 (3):22:1–22:9

Farcot E, Yuan Y (2013) Homogenous auxin steady states and spontaneous oscillations in flux-based auxin transport models. SIAM J Appl Dyn Syst 12(3):1330–1353

Feugier F, Iwasa Y (2006) How canalization can make loops: a new model of reticulated leaf vascular pattern formation. J Theor Biol 243(2):235–244

Feugier F, Mochizuki A, Iwasa Y (2005) Self-organization of the vascular system in plant leaves: inter-dependent dynamics of auxin flux and carrier proteins. J Theor Biol 236(4):366–375

Fortin M, Pierce F, Poff K (1989) The pattern of secondary root formation in curving roots of *Arabidopsis thaliana* (L.) Heynh. Plant Cell Environ 12(3):337–339

Fozard J, King J, Bennett M (2013) Modelling auxin efflux carrier phosphorylation and localization. J Theor Biol 319:34–49

Friml J, Yang X, Michniewicz M, Weijers D, Quint A, Tietz O, Benjamins R, Ouwerkerk P, Ljung K, Sandberg G, Hooykaas P, Palme K, Offringa R (2004) A PINOID-dependent binary switch in apical-basal PIN polar targeting directs auxin efflux. Science 306(5697):862–865

Fujita H, Mochizuki A (2006) Pattern formation of leaf veins by the positive feedback regulation between auxin flow and auxin efflux carrier. J Theor Biol 241(3):541–551

Furutani M, Nakano Y, Tasaka M (2014) MAB4-induced auxin sink generates local auxin gradients in *Arabidopsis* organ formation. Proc Natl Acad Sci U S A 111(3):1198–1203

Garnett P, Stepney S, Leyser O (2008) Towards an executable model of auxin transport canalisation. In: Stepney S, Polack F, Welch P (eds) Proceedings of the 2008 workshop on complex systems modelling and simulation. Luniver Press, Frome, pp 63–91

Geldner N, Friml J, Stierhof Y, Jürgens G, Palme K (2001) Auxin transport inhibitors block PIN1 cycling and vesicle trafficking. Nature 413(6854):425–428

Gillespie D (1976) A general method for numerically simulating the stochastic time evolution of coupled chemical reactions. J Comput Phys 22(4):403–434

Gillespie D (1977) Exact stochastic simulation of coupled chemical reactions. J Phys Chem 81 (25):2340–2361

Goldsmith M (1977) The polar transport of auxin. Annu Rev Plant Physiol 28:439–478

Gomez-Roldan V, Fermas S, Brewer P, Puech-Pages V, Dun E, Pillot J, Letisse F, Matusova R, Danoun S, Portais J, Bouwmeester H, Bécard G, Beveridge C, Rameau C, Rochange S (2008) Strigolactone inhibition of shoot branching. Nature 455(7210):189–194

Grbic V, Bleecker AB (2000) Axillary meristem development in *Arabidopsis thaliana*. Plant J 21 (2):215–223

Grieneisen V, Xu J, Marée A, Hogeweg P, Scheres B (2007) Auxin transport is sufficient to generate a maximum and gradient guiding root growth. Nature 449(7165):1008–1013

Guenot B, Bayer E, Kierzkowski D, Smith R, Mandel T, Zadnikova P, Benková E, Kuhlemeier C (2012) PIN1-independent leaf initiation in Arabidopsis. Plant Physiol 159(4):1501–1510

Hamant O, Traas J (2010) The mechanics behind plant development. New Phytol 185(2):369–385

Hamant O, Heisler M, Jönsson H, Krupinski P, Uyttewaal M, Bokov P, Corson F, Sahlin P, Boudaoud A, Meyerowitz E, Couder Y, Traas J (2008) Developmental patterning by mechanical signals in *Arabidopsis*. Science 322(5908):1650–1655

Hay A, Barkoulas M, Tsiantis M (2006) ASYMMETRIC LEAVES1 and auxin activities converge to repress *BREVIPEDICELLUS* expression and promote leaf development in *Arabidopsis*. Development 133(20):3955–3961

Heisler M, Jönsson H (2006) Modeling auxin transport and plant development. J Plant Growth Regul 25(4):302–312

Heisler M, Ohno C, Das P, Sieber P, Reddy G, Long J, Meyerowitz E (2005) Patterns of auxin transport and gene expression during primordium development revealed by live imaging of the Arabidopsis inflorescence meristem. Curr Biol 15(21):1899–1911

Heisler M, Hamant O, Krupinski P, Uyttewaal M, Ohno C, Jönsson H, Traas J, Meyerowitz E (2010) Alignment between PIN1 polarity and microtubule orientation in the shoot apical meristem reveals a tight coupling between morphogenesis and auxin transport. PLoS Biol 8 (10):e1000516

Hofmeister W (1868) Allgemeine Morphologie der Gewachse. Engelmann, Leipzig, pp 405–664

Hošek P, Kubes M, Lankova M, Dobrev P, Klima P, Kohoutova M, Petrasek J, Hoyerova K, Jirina M, Zažímalová E (2012) Auxin transport at cellular level: new insights supported by mathematical modelling. J Exp Bot 63(10):3815–3827

Janssen J, Lindenmayer A (1987) Models for the control of branch positions and flowering sequences of capitula in *Mycelis muralis* (L.) Dumont (Compositae). New Phytol 105 (2):191–220

Jönsson H, Heisler M, Shapiro B, Meyerowitz E, Mjolsness E (2006) An auxin-driven polarized transport model for phyllotaxis. Proc Natl Acad Sci U S A 103(5):1633–1638

Kawamura E, Horiguchi G, Tsukaya H (2010) Mechanisms of leaf tooth formation in *Arabidopsis*. Plant J 62(3):429–441

Kierzkowski D, Lenhard M, Smith R, Kuhlemeier C (2013) Interaction between meristem tissue layers controls phyllotaxis. Dev Cell 26(6):616–628

Koenig D, Bayer E, Kang J, Kuhlemeier C, Sinha N (2009) Auxin patterns *Solanum lycopersicua*m leaf morphogenesis. Development 136(17):2997–3006

Kramer E (2004) PIN and AUX/LAX proteins: their role in auxin accumulation. Trends Plant Sci 9 (12):578–582

Kramer E (2009) Auxin-regulated cell polarity: an inside job? Trends Plant Sci 14(5):242–247

Laguna M, Bohn S, Jagla E (2008) The role of elastic stresses on leaf venation morphogenesis. PLoS Comput Biol 4(4):e1000055

Larson P (1975) Development and organization of the primary vascular system in *Populus deltoides* according to phyllotaxy. Am J Bot 62(10):1084–1099

Laskowski M, Biller S, Stanley K, Kajstura T, Prusty R (2006) Expression profiling of auxin-treated *Arabidopsis* roots: toward a molecular analysis of lateral root emergence. Plant Cell Physiol 47(6):788–792

Laskowski M, Grieneisen V, Hofhuis H, ten Hove C, Hogeweg P, Marée A, Scheres B (2008) Root system architecture from coupling cell shape to auxin transport. PLoS Biol 6(12):e307

Lee S-W, Feugier F, Morishita Y (2014) Canalization-based vein formation in a growing leaf. J Theor Biol 353C:104–120

Leyser O (2009) The control of shoot branching: an example of plant information processing. Plant Cell Environ 32(6):694–703

Leyser O (2011) Auxin, self-organization, and the colonial nature of plants. Curr Biol 21(9):R331–R337

Lifschitz E, Eshed Y (2006) Universal florigenic signals triggered by FT homologues regulate growth and flowering cycles in perennial day-neutral tomato. J Exp Bot 57(13):3405–3414

Lifschitz E, Eviatar T, Rozman A, Shalit A, Goldshmidt A, Amsellem Z, Alvarez J, Eshed Y (2006) The tomato FT ortholog triggers systemic signals that regulate growth and flowering and substitute for diverse environmental stimuli. Proc Natl Acad Sci U S A 103(16):6398–6403

Lindenmayer A (1984) Positional and temporal control mechanisms in inflorescence development. In: Barlow PW, Carr DJ (eds) Positional controls in plant development. University Press, Cambridge, pp 461–486

Lockhart J (1965) An analysis of irreversible plant cell elongation. J Theor Biol 8(2):264–275

Lucas M, Godin C, Jay-Allemand C, Laplaze L (2008a) Auxin fluxes in the root apex coregulate gravitropism and lateral root initiation. J Exp Bot 59(1):55–66

Lucas M, Guédon Y, Jay-Allemand C, Godin C, Laplaze L (2008b) An auxin transport-based model of root branching in *Arabidopsis thaliana*. PLoS One 3(11):e3673

Marhavý P, Bielach A, Abas L, Abuzeineh A, Duclercq J, Tanaka H, Pařezová M, Petrášek J, Friml J, Kleine-Vehn J, Benková E (2011) Cytokinin modulates endocytic trafficking of PIN1 auxin efflux carrier to control plant organogenesis. Dev Cell 21(4):796–804

Meinhardt H (1982) Models of biological pattern formation. Academic, London

Merks R, Glazier J (2005) A cell-centered approach to developmental biology. Phys A 352:113–130

Merks R, Van de Peer Y, Inzé D, Beemster G (2007) Canalization without flux sensors: a traveling-wave hypothesis. Trends Plant Sci 12(9):384–390

Merks R, Guravage M, Inzé D, Beemster G (2011) VirtualLeaf: an open-source framework for cell-based modeling of plant tissue growth and development. Plant Physiol 155(2):656–666

Millar A, Carrie C, Pogson B, Whelan J (2009) Exploring the function-location nexus: using multiple lines of evidence in defining the subcellular location of plant proteins. Plant Cell 21 (6):1625–1631

Mirabet V, Besnard F, Vernoux T, Boudaoud A (2012) Noise and robustness in phyllotaxis. PLoS Comput Biol 8(2):e1002389
Mironova V, Omelyanchuk N, Yosiphon G, Fadeev S, Kolchanov N, Mjolsness E, Likhoshvai V (2010) A plausible mechanism for auxin patterning along the developing root. BMC Syst Biol 4(98):1–19
Mironova V, Omelyanchuk N, Novoselova E, Doroshkov A, Kazantsev F, Kochetov A, Kolchanov N, Mjolsness E, Likhoshvai V (2012) Combined in silico/in vivo analysis of mechanisms providing for root apical meristem self-organization and maintenance. Ann Bot 110(2):349–360
Mitchison G (1980) A model for vein formation in higher plants. Proc R Soc B 207:79–109
Mitchison G (1981) The polar transport of auxin and vein patterns in plants. Philos Trans Roy Soc Lond B 295:461–470
Moreno-Risueno M, Van Norman J, Moreno A, Zhang J, Ahnert S, Benfey P (2010) Oscillating gene expression determines competence for periodic Arabidopsis root branching. Science 329 (5997):1306–1311
Morris D (1977) Transport of exogenous auxin in two-branched dwarf pea seedlings (*Pisum sativum* L.). Planta 136(1):91–96
Muller D, Leyser O (2011) Auxin, cytokinin and the control of shoot branching. Ann Bot 107 (7):1203–1212
Napier R, David K, Perrot-Rechenmann C (2002) A short history of auxin-binding proteins. Plant Mol Biol 49(3–4):339–348
Nikovics K, Blein T, Peaucelle A, Ishida T, Morin H, Aida M, Laufs P (2006) The balance between the MIR164A and CUC2 genes controls leaf margin serration in *Arabidopsis*. Plant Cell 18(11):2929–2945
O'Connor D, Runions A, Sluis A, Bragg J, Vogel J, Prusinkiewicz P, Hake S (2014) A division in PIN-mediated patterning during lateral organ initiation in the grasses. PLoS Comput Biol 10 (1):e1003447
Paciorek T, Zažímalová E, Ruthardt N, Petrasek J, Stierhof Y, Kleine-Vehn J, Morris D, Emans N, Jürgens G, Geldner N, Friml J (2005) Auxin inhibits endocytosis and promotes its own efflux from cells. Nature 435(7046):1251–1256
Paponov I, Paponov M, Teale W, Menges M, Chakrabortee S, Murray J, Palme K (2008) Comprehensive transcriptome analysis of auxin responses in *Arabidopsis*. Mol Plant 1 (2):321–337
Parry G, Marchant A, May S, Swarup R, Swarup K, James N, Graham N, Allen T, Martucci T, Yemm A, Napier R, Manning K, King G, Bennett M (2001) Quick on the uptake: characterization of a family of plant auxin influx carriers. J Plant Growth Regul 20(30):217–225
Pinon V, Prasad K, Grigg S, Sanchez-Perez G, Scheres B (2013) Local auxin biosynthesis regulation by PLETHORA transcription factors controls phyllotaxis in *Arabidopsis*. Proc Natl Acad Sci U S A 110(3):1107–1112
Prusinkiewicz P, Lindenmayer A (1990) The algorithmic beauty of plants. Springer, New York, NY. With JS Hanan, FD Fracchia, DR Fowler, MJM de Boer, and L Mercer
Prusinkiewicz P, Runions A (2012) Computational models of plant form and development. New Phytol 193(3):549–569
Prusinkiewicz P, Lindenmayer A, Hanan J (1988) Developmental models of herbaceous plants for computer imagery purposes. Comput Graph 22:141–150
Prusinkiewicz P, Crawford S, Smith R, Ljung K, Bennett T, Ongaro V, Leyser O (2009) Control of bud activation by an auxin transport switch. Proc Natl Acad Sci U S A 106(41):17431–17436
Raven J (1975) Transport of indoleacetic acid in plant cells in relation to pH and electrical potential gradients, and its significance for polar IAA transport. New Phytol 74(2):163–172
Reinhardt D, Pesce E, Stieger P, Mandel T, Baltensperger K, Bennett M, Traas J, Friml J, Kuhlemeier C (2003) Regulation of phyllotaxis by polar auxin transport. Nature 426 (6964):255–260

Renton M, Hanan J, Ferguson B, Beveridge C (2012) Models of long-distance transport: how is carrier-dependent auxin transport regulated in the stem? New Phytol 194(3):704–715

Robert S, Kleine-Vehn J, Barbez E, Sauer M, Paciorek T, Baster P, Vanneste S, Zhang J, Simon S, Covanova M, Hayashi K, Dhonukshe P, Yang Z, Bednarek S, Jones A, Luschnig C, Aniento F, Zažímalová E, Friml J (2010) ABP1 mediates auxin inhibition of clathrin-dependent endocytosis in *Arabidopsis*. Cell 143(1):111–121

Rolland-Lagan A, Prusinkiewicz P (2005) Reviewing models of auxin canalization in the context of leaf vein pattern formation in *Arabidopsis*. Plant J 44(5):854–865

Roussel M, Slingerland M (2012) A biochemically semi-detailed model of auxin-mediated vein formation in plant leaves. BioSyst 109(3):475–487

Rubery P, Sheldrake R (1974) Carrier-mediated auxin transport. Planta 118:101–121

Rutschow H, Baskin T, Kramer E (2011) Regulation of solute flux through plasmodesmata in the root meristem. Plant Physiol 155(4):1817–1826

Sachs T (1969) Polarity and the induction of organized vascular tissues. Ann Bot 33(2):263–275

Sachs T (1981) The control of patterned differentiation of vascular tissues. Adv Bot Res 9:151–262

Sachs T (1991) Pattern formation in plant tissues. Cambridge University Press, Cambridge

Sachs T (2003) Collective specification of cellular development. Bioessays 25(9):897–903

Sahlin P, Söderberg B, Jönsson H (2009) Regulated transport as a mechanism for pattern generation: capabilities for phyllotaxis and beyond. J Theor Biol 258(1):60–70

Santuari L, Scacchi E, Rodriguez-Villalon A, Salinas P, Dohmann E, Brunoud G, Vernoux T, Smith R, Hardtke C (2011) Positional information by differential endocytosis splits auxin response to drive Arabidopsis root meristem growth. Curr Biol 21(22):1918–1923

Sauer M, Friml J (2010) Immunolocalization of proteins in plants. In: Hennig L, Köhler C (eds) Plant developmental biology: methods ad protocols. Humana, New York, NY, pp 253–263

Scarpella E, Marcos D, Friml J, Berleth T (2006) Control of leaf vascular patterning by polar auxin transport. Genes Dev 20(8):1015–1027

Shalit A, Rozman A, Goldshmidt A, Alvarez J, Bowman J, Eshed Y, Lifschitz E (2009) The flowering hormone florigen functions as a general systemic regulator of growth and termination. Proc Natl Acad Sci U S A 106(20):8392–8397

Shapiro B, Meyerowitz E, Mjolsness E (2013) Using Cellzilla for plant growth simulations at the cellular level. Front Plant Sci 4(408):1–9

Shimizu-Sato S, Tanaka M, Mori H (2008) Auxin–cytokinin interactions in the control of shoot branching. Plant Mol Biol 69(4):429–435

Shinohara N, Taylor C, Leyser O (2013) Strigolactone can promote or inhibit shoot branching by triggering rapid depletion of the auxin efflux protein PIN1 from the plasma membrane. PLoS Biol 11(1):e1001474

Smith C (2006) On vertex-vertex systems and their use in geometric and biological modelling. PhD Thesis, University of Calgary, Calgary

Smith R (2011) Modeling plant morphogenesis and growth. In: Amar M, Goriely A, Mueller M, Cugliandolo L (eds) New trends in the physics and mechanics of biological systems, vol 92, Lecture notes of the Les Houches summer school, July 2009. Oxford University Press, Oxford, pp 301–336

Smith R, Bayer E (2009) Auxin transport-feedback models of patterning in plants. Plant Cell Environ 32(5):1258–1271

Smith C, Prusinkiewicz P, Samavati F (2004) Local specification of subdivision algorithms. In: Pfaltz J (ed) Applications of graph transformations with industrial relevance, vol 3062, Lecture notes in computer science. Springer, Berlin, pp 313–327

Smith R, Guyomarc'h S, Mandel T, Reinhardt D, Kuhlemeier C, Prusinkiewicz P (2006a) A plausible model of phyllotaxis. Proc Natl Acad Sci U S A 103:1301–1306

Smith R, Kuhlemeier C, Prusinkiewicz P (2006b) Inhibition fields for phyllotactic pattern formation: a simulation study. Can J Bot 84:1635–1649

Steinacher A, Leyser O, Clayton R (2012) A computational model of auxin and pH dynamics in a single plant cell. J Theor Biol 296:84–94

Stirnberg P, Chatfield S, Leyser O (1999) AXR1 acts after lateral bud formation to inhibit lateral bud growth in *Arabidopsis*. Plant Physiol 121(3):839–847

Stoma S, Lucas M, Chopard J, Schaedel M, Traas J, Godin C (2008) Flux-based transport enhancement as a plausible unifying mechanism for auxin transport in meristem development. PLoS Comput Biol 4(10):e1000207

Swarup R, Péret B (2012) AUX/LAX family of auxin influx carriers — an overview. Front Plant Sci 3(225):1–11

Swarup R, Friml J, Marchant A, Ljung K, Sandberg G, Palme K, Bennett M (2001) Localization of the auxin permease AUX1 suggests two functionally distinct hormone transport pathways operate in the *Arabidopsis* root apex. Genes Dev 15(20):2648–2653

Swarup R, Kramer E, Perry P, Knox K, Leyser O, Haseloff J, Beemster G, Bhalerao R, Bennett M (2005) Root gravitropism requires lateral root cap and epidermal cells for transport and response to a mobile auxin signal. Nat Cell Biol 7(11):1057–1065

Thimann K, Skoog F (1933) Studies on the growth hormone of plants III. The inhibitory action of the growth substance on bud development. Proc Natl Acad Sci U S A 19:714–716

Townsley B, Sinha N (2012) A new development: evolving concepts in leaf ontogeny. Annu Rev Plant Biol 63(1):535–562

van Berkel K, de Boer R, Scheres B, ten Tusscher K (2013) Polar auxin transport: models and mechanisms. Development 140(11):2253–2268

van Mourik S, Kaufmann K, van Dijk A, Angenent G, Merks R, Molenaar J (2012) Simulation of organ patterning on the floral meristem using a polar auxin transport model. PLoS One 7(1): e28762

Vieten A, Vanneste S, Wisniewska J, Benková E, Benjamins R, Beeckman T, Luschnig C, Friml J (2005) Functional redundancy of PIN proteins is accompanied by auxin-dependent cross-regulation of PIN expression. Development 132(20):4521–4531

Wabnik K, Kleine-Vehn J, Balla J, Sauer M, Naramoto S, Reinöhl V, Merks R, Govaerts W, Friml J (2010) Emergence of tissue polarization from synergy of intracellular and extracellular auxin signaling. Mol Syst Biol 6:447

Weijers D, Jürgens G (2005) Auxin and embryo axis formation: the ends in sight? Curr Opin Plant Biol 8(1):32–37

Weiss T (1996) Cellular biophysics: transport, vol 1. Bradford Books, Cambridge, MA

White A (1973) Graphs, groups and surfaces. North-Holland, Amsterdam

Xu T, Wen M, Nagawa S, Fu Y, Chen J, Wu M, Perrot-Rechenmann C, Friml J, Jones A, Yang Z (2010) Cell surface- and Rho GTPase-based auxin signaling controls cellular interdigitation in *Arabidopsis*. Cell 143(1):99–110

Zažímalová E, Murphy A, Yang H, Hoyerová K, Hošek P (2010) Auxin transporters–why so many? Cold Spring Harb Perspect Biol 2(3):a001552

Zeevaart J (2008) Leaf-produced floral signals. Curr Opin Plant Biol 11(5):541–547

# Part III
# Auxin versus Environment

# Chapter 16
# Auxin and Tropisms

**Katarzyna Retzer, Barbara Korbei, and Christian Luschnig**

**Abstract** From the very beginnings, attempts to identify mechanisms underlying polar auxin transport in higher plants have been intimately linked to studies on the regulation of plant tropisms. Already in the nineteenth century Charles Darwin came up with a concept, suggesting that a transmissible signal might be involved in controlling directional plant growth in response to an environmental stimulus. Much later, plant physiologists identified auxin as a candidate molecule that could mediate tropic growth responses. However, it was not until establishment of *Arabidopsis* genetics and novel molecular techniques at the end of the twentieth century that enabled the characterization of auxin-signaling pathways and resulted in mechanistic insights into control of polar auxin transport and its significance for plant tropisms. In this chapter, essential aspects of the current framework of molecular events are presented, highlighting the role of auxin in directional plant growth.

## 1 A Moving Story

When taking basic courses in plant physiology at the—then brand new—first Biozentrum building in Vienna, Professor Karl Burian entertained us with a conversation he had with one of the construction site engineers. They walked through one of the greenhouses, and by accident the engineer touched one of the mimosas that were cultivated for student courses. The plant responded to this stimulus, and after a few seconds of silence, the stunned engineer exclaimed: "For a moment I thought that it's alive!"

K. Retzer • B. Korbei • C. Luschnig (✉)
Department of Applied Genetics and Cell Biology, University of Natural Resources and Life Sciences, BOKU, Wien, Austria
e-mail: christian.luschnig@boku.ac.at

E. Zažímalová et al. (eds.), *Auxin and Its Role in Plant Development*,
DOI 10.1007/978-3-7091-1526-8_16, © Springer-Verlag Wien 2014

The experienced professor used this anecdote as a starter for his lecture on plant movements, one of the most fascinating aspects of plant physiology. When going through the old notes on this lecture, it seems that not much has changed in conceptual terms: Plants sense an environmental stimulus that triggers transmission of endogenous signals, ultimately resulting in, sometimes quite fast responses like the nastic movements of mimosa leaves, or, comparably slow growth responses like organ movement in accordance with the gravity vector or gravitropism. Some of these concepts are ancient, dating back long before Darwin's influential book on the movement of plants (Darwin and Darwin 1881). However, what has changed in recent years is the availability of powerful molecular techniques, producing amazing insights into principles of signal perception and transduction that guide plant movements.

## 2 Plant Movements: Overlapping and Distinct Concepts

Plants are continuously exposed to variations in their respective environment, which are perceived by a sophisticated sensorium of receptors and signaling pathways in order to adjust growth parameters accordingly. Given that higher plants are sessile, they have evolved mechanisms that allow them to modify their organ growth in response to environmental stimuli. These activities differ from preprogrammed autonomous plant movements, as they depend on perception of an external stimulus. Two major types of such responses, tropism and nastic movements have been described, where tropic responses depends on the direction of the stimulus, like a chemical gradient or local variations in illumination, whereas nastic movements summarize nondirectional responses. A large variety of different stimuli cause tropic responses in plants, and a range of different tropisms have been described. Moreover, they come in different flavors. Depending on the direction of movements, either toward or away from the external stimulus, resulting movements are referred to as positive or negative tropisms. In special cases, even transversal tropisms have been observed, like, for instance, strawberry runners, defined by a growth direction on the soil's surface, i.e., perpendicular to the main growth axis.

Empirical records describing *phototropism* (or *heliotropism*), i.e., directional movement of plant organs in response to a light stimulus, date back to Greek philosophers. For the longest time though, Aristotelian views, that discriminated plants from animals based on their supposed "insensitivity," hampered proper interpretation of this growth response (for an excellent summary on the history of phototropism research, see Whippo and Hangarter 2006). It was not until the seventeenth century, that researchers began to hypothesize that tropic curvature involves an actual growth response and it took until the late nineteenth century, to show that phototropism is an inductive response, triggered by a directional light stimulus (Sharrock 1672; Darwin and Darwin 1881; Sachs 1882). Darwin's interpretation of growth experiments suggesting that organ circumnutation functions as a uniform mechanism that drives directional plant growth responses could not be verified in subsequent experiments (Darwin and Darwin 1881). However, he was the first to come up with a concept, predicting that a transmissible signal would control events, leading from perception of a stimulus to the actual growth response. This concept still holds true and represents one of the cornerstones of contemporary models for plant tropisms.

The term *geotropism* (nowadays most often replaced by the term *gravitropism*) was introduced by Albert B. Frank (1868) and by Julius von Sachs (1868) and defines directional growth movements of plants in response to gravity. Early experiments by Andrew Knight and Henry Johnson (1829) proposed that downward growth of roots might be controlled by gravity. But it was the work especially by Sachs, who introduced the clinostat and initiated a systematic analysis of plant organ bending in response to external stimuli, that led to our current views on mechanisms of plant tropisms (Sachs 1879, 1892).

*Thigmotropism* involves perception of and response to mechanical stimulation (Migliaccio et al. 2009), whereas *chemotropism* describes directional movement of plant organs in response to chemical gradients. A prominent example is pollen tube chemotropism that describes directed pollen tube growth from the stigma to ovules in a process that seemingly involves pollen tube attraction by chemotropic factors (Chae and Lord 2011). *Hydrotropism* finally could be considered a special case of chemotropism, in which directional growth of roots is influenced by the availability of water, a vital growth response that has been described already by early plant physiologists like Sachs (1872).

## 3 Plant Movements and the Influence of Auxin: A Historical Perspective

Plant tropisms depend on spatiotemporal control of asymmetric organ growth. This typically involves differential cell expansion in distinct portions of an organ, like in the apical region of hypocotyls bending toward light. In some instances, differential cell elongation appears to be intimately linked to the control of cell proliferation, as in graviresponding root tips. Regardless of mechanisms involved, it appears that activity of the phytohormone auxin is crucial for all tropisms, and there is now convincing evidence that this growth regulator represents Darwin's transmissible signal.

A number of plant researchers deserve credit for initial characterization of the hormone and for putting forward working models that explain the contribution of auxin to the control of organ bending, thereby generating a solid foundation for our current understanding of polar auxin transport (PAT). Based on Darwin's experiments, which show that removal of the very tip of canary grass coleoptiles will block organ bending toward a light source (Darwin and Darwin 1881), researchers tested the concept of a transmissible signal causing this growth response. It was Peter Boysen Jensen (1910), who demonstrated the existence of such a transmissible signal that will penetrate through a layer of gelatine, introduced subapically into oat coleoptiles. Furthermore, by inserting impermeable mica sheets into cuts made into coleoptiles before phototropic stimulation, Boysen Jensen was able to conclude that signal transmission at the shaded side of illuminated coleoptiles is essential for their bending, an elegant, first demonstration of signal gradient formation taking place in plants (Boysen Jensen 1913). Discovery and initial characterization of auxin, as a mediator of tropic growth responses, came from work by Frits Went (1926) and Nikolai Cholodny (1927), whereas the chemical characterization of the compound was achieved by Kögl and Haagen-Smits (1931). Went and Cholodny, independently, suggested quite similar mechanisms by which auxin could influence organ bending, now known as the Cholodny–Went theory (Went and Thimann 1937). In simple terms, this theory proposes that lateral, unequal auxin distribution within organs, induces differential growth, ultimately causing organ curvature (Cholodny 1928; Went 1928; Fig. 16.1a).

Whereas the Cholodny–Went theory offers a simple explanation for organ bending, it is not fully resolved, if lateral auxin transport to the shaded side of organs is the only cause for phototropic curvature. Additional mechanisms, such as light-induced growth inhibition at the illuminated side of the organ (Overbeek 1932; Baskin 1986), or variable photo-inactivation of auxin (Kögl and Schuringa 1944), could contribute to the growth response as well (Fig. 16.1a). In addition, spatiotemporal variations in cellular auxin sensitivity could lead to differential auxin responsiveness and resulting growth adjustments (Kang and Burg 1974; Fig. 16.1a). Experimental evidence in support of the Cholodny–Went theory was provided by Winslow Briggs and coworkers, demonstrating light-controlled lateral auxin redistribution in phototropically responding corn coleoptiles (Briggs et al. 1957). Some years later, Briggs showed a direct correlation between phototropic coleoptile curvature and auxin found on the shaded side of the organ, suggesting that it is the steepness of lateral auxin concentration gradients that defines phototropic organ bending (Briggs 1963). Related studies demonstrated that unilateral illumination by blue light is sufficient for establishment of a lateral auxin gradient in coleoptiles, thus predicting the involvement of blue-light responsive photoreceptors in gradient establishment (Pickard and Thimann 1964; Gardner et al. 1974).

Alternatives to upregulation of lateral auxin transport in response to unilateral illumination have been proposed. For example, work by Shen-Miller and colleagues (1969) suggested light-controlled asymmetric inhibition of PAT in phototropically stimulated oat coleoptiles. Recent, molecular characterization of auxin transport activities in Arabidopsis hypocotyls undergoing phototropic curvature provided evidence for both scenarios; induction and downregulation of PAT (Christie et al. 2011; Ding et al. 2011), pointing toward a more complex array of events involved in light-mediated control of auxin distribution.

Similar to the situation in above-ground organs, control of root tropisms appears to involve establishment of a lateral auxin gradient. That is, an environmental stimulus like gravity would promote auxin accumulation at the "lower" side of a horizontally positioned root, where it would cause growth inhibition, resulting in a

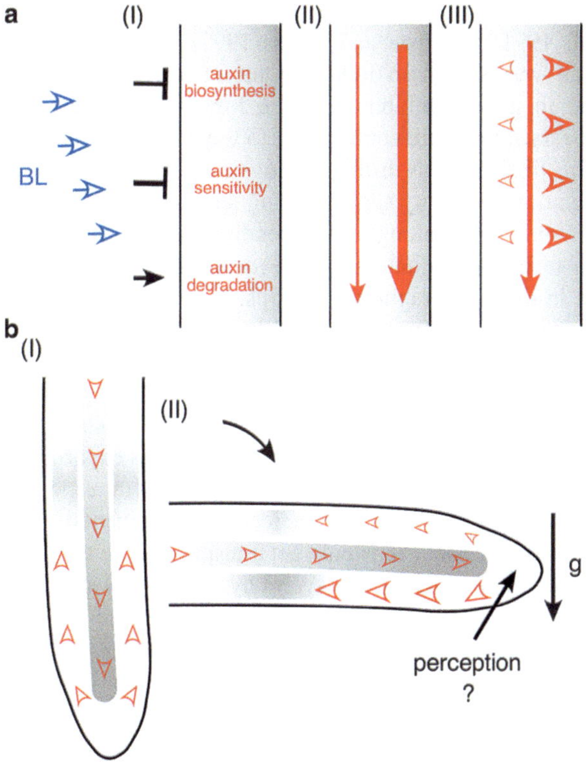


**Fig. 16.1** Suggested modes of auxin action upon tropic organ growth. (**a**) Differential auxin responses upon phototropic stimulation by unilateral illumination with blue light (BL) were suggested to result from local variations in auxin biosynthesis, sensitivity, or degradation (I). In alternative models, asymmetric auxin flow from the apical portion of the stimulated organ (*red arrows*) was suggested to induce differential organ growth (II). This might result from local inhibition or induction of PAT. Another model suggested establishment of a lateral auxin concentration gradient, via differential lateral auxin transport (*red arrowheads*), with more auxin transferred into the shaded, elongating portion of the organ (III). (**b**) Fountain model describing auxin transport routes in root meristems. Rootward auxin transport in the stele mediates auxin accumulation in the root tip (*red arrowheads*). From there it is redistributed for further shootward transport via outer cell layers into the root meristem elongation zone (I). Upon gravistimulation (*black arrow* indicates direction of gravity vector), more auxin is transported to the lower side of the root tip, resulting in asymmetric, shootward auxin flow into the elongation zone (*red arrowheads*), ultimately causing organ bending (II). *Shaded areas* indicate zones accumulating auxin and/or exhibiting increased auxin responsiveness

downward bending of the root tip (Kaufman et al. 1988). Differences in cell expansion rates in gravistimulated roots have been described, specifically in the Distal Elongation Zone (DEZ) a region proximal to the mitotically active zone in root meristems (Ishikawa et al. 1991). In numerous studies, a complex interplay of spatiotemporal variations in cell elongation has been described, and variations have been observed when comparing responses under different conditions or in different species (Pilet and Ney 1981; Nelson and Evans 1986; Firn and Myers 1989; Ishikawa et al. 1991; Ishikawa and Evans 1997). In Arabidopsis for example, gravistimulation was shown to induce transient, differential elongation in the DEZ, reflected in growth inhibition at the lower side, whereas growth at the upper side appears stimulated, contributing to the control of root bending (Mullen et al. 1998). This likely reflects distinct and overlapping growth responses, ensuring proper interpretation of gravity signals during the continuous growth of roots.

Linking differential root elongation in bending roots to auxin has been difficult, mainly due to technical limitations in the determination of variations in auxin concentrations in tissue as fragile as root tips. Moreover, in those cases in which researchers managed to determine auxin distribution in graviresponding roots, differences in auxin concentration were rather small, which made it difficult to link root gravity responses to lateral auxin concentration gradients in the DEZ (Mertens and Weiler 1983; Young et al. 1990). Perception and transduction of the gravity signal requires the root cap, verified by numerous experiments demonstrating that root decapitation blocks any further tropic growth responses (Darwin and Darwin 1881; Juniper et al. 1966). This indicates involvement of a signal that, after gravity perception in the root cap, would get transmitted into the root bending zone. In a modified version of the Cholodny–Went hypothesis it was proposed that auxin itself would function as that signal, first being translocated via the stele into the columella root cap, from where it would get redistributed into the root elongation zone, with asymmetries in auxin flow, resulting in directional root growth (Konings 1968; Hasenstein and Evans 1986; Boonsirichai et al. 2002). This fountain model for PAT in root meristems accounts for variations in auxin flow rates and consequences for directional root growth and is meanwhile supported by several studies, describing molecular determinants of PAT (Fig. 16.1b).

# 4 Molecular Determinants of Auxin Transport and Signaling in the Regulation of Tropisms

It was not until the advent of *Arabidopsis thaliana* as a model system for plant molecular genetics that led to identification of molecular determinants of signaling cascades involved in plant tropisms. In this section, selected aspects of stimulus perception as well as transmission and translation of auxin signals into tropic growth responses are summarized.

## 4.1 Phototropism

### Stimulus Perception

In the nineteenth century Julius von Sachs demonstrated that blue light is most efficient for induction of phototropic responses (Sachs 1882), but it took another century and mutant screens performed in Arabidopsis to characterize the genes involved. In such screens, hypocotyl phototropism mutants that no longer bend toward a light source were identified (Khurana and Poff 1989; Liscum and Briggs 1995), which eventually led to characterization of *NON PHOTOTROPIC RESPONSE1* (*NPH1*) an AGC-type kinase protein (Huala et al. 1997; Galvan-Ampudia and Offringa 2007). This protein, together with its close relative *NPH1-LIKE 1* (*NPL1*; Jarillo et al. 2001), represents key regulators of blue light-mediated tropic responses and hence were renamed *phototropin1* (*phot1*) and *phototropin2* (*phot2*; Briggs et al. 2001). Both proteins are characterized by two Flavin Mononucleotide (FMN)-binding sites, so-called LOV (light-, oxygen- or voltage-sensing) domains that bind FMN chromophores noncovalently in the dark. Blue light causes covalent binding of FMN to the LOV domains, which results in conformational changes that expose the C-terminal phototropin kinase domain (Salomon et al. 2000; Christie et al. 2002). Blue light-induced kinase activation then appears to cause phototropin autophosphorylation at conserved Serine residues, a modification indispensable for phototropic responses (Christie et al. 1998, 2002; Sakai et al. 2001; Inoue et al. 2008). Although phototropins seem to represent the only class of receptors capable of sensing directionality of a blue light stimulus (Sakai et al. 2001), additional light receptors have been implicated in the regulation of tropic growth responses, acting via, e.g., modulation of transcriptional control and intracellular protein distribution (Lariguet and Fankhauser 2004; Han et al. 2008; Wu et al. 2010; Kami et al. 2012). This apparent involvement of additional light receptors might reflect parts of a complex network of interactions among distinct classes of light receptors, essential for fine-tuning of tropic growth responses (Hohm et al. 2013).

Identification of phototropin-interacting proteins elucidated potential links to auxin transport. Cloning of *NPH3*, which was originally identified as another hypocotyl phototropism mutant, revealed a protein featuring conserved domains that function in protein–protein interaction (Motchoulski and Liscum 1999): A coiled-coil domain, in the C-terminal portion of NPH3 is required for its interaction with phot1 (Motchoulski and Liscum 1999), whereas a BTB/POZ (broad-complex, tramtrack, bric à brac/Pox virus and zinc finger) domain mediates interaction with ROOT PHOTOTROPISM2 (RPT2), a protein similar to NPH3 and belonging to the NRL (NPH3/RPT2-Like) protein family (Inada et al. 2004; Sakai 2005). NPH3 has been found to localize to specific domains at the plasma membrane, similar to its interaction partner phot1 (Motchoulski and Liscum 1999). In addition its rice ortholog has been implicated in auxin redistribution in phototropically responding coleoptiles (Haga et al. 2005); however, functions of NPH3 and related proteins in mechanistic terms remained elusive.

Work by Roberts and others (2011) might provide answers to questions regarding the function of NRL proteins. The authors found that NPH3 acts as adaptor protein that binds to CUL3, an SCF-type E3 ubiquitin ligase complex subunit, apparently facilitating phot1 ubiquitylation in a light-dependent manner (Roberts et al. 2011). Under conditions of low blue light intensity, phot1 is mono(multi) ubiquitylated, whereas higher irradiation intensities correlated with phot1 polyubiquitylation. Such different ubiquitylation modes could have a strong impact on the fate of phot1, with monoubiquitylation affecting intracellular sorting, and polyubiquitylation triggering proteasome-dependent phot1 degradation (Roberts et al. 2011). Resulting variations in phot1 distribution and/or abundance were suggested to feed back on PAT in Arabidopsis hypocotyls (Roberts et al. 2011).

Additional members of the NRL family have been linked to the regulation of directional growth responses. Next to *RPT2,* required for root phototropism and demonstrated to interact with phot1 (Sakai et al. 2000), *NAKED PINS IN YUCCA/ ENHANCER OF PINOID/MACHI-BOU* (*NPY/ENP/MAB*) genes were recently shown to function in auxin-controlled processes involving root gravitropism (Treml et al. 2005; Cheng et al. 2007, 2008; Furutani et al. 2007, 2011; Li et al. 2011). By analogy to interaction of NPH3 and phot1, there is evidence for similar crosstalk between NPYs and additional AGC-kinases involving PINOID (PID) and some of its close homologs (Cheng et al. 2007, 2008). Given the function of PID in adjusting directionality of PAT by controlling phosphorylation status and localization of PIN-FORMED (PIN) auxin transport proteins (Friml et al. 2004; Michniewicz et al. 2007), it is tempting to speculate about a function for NPY proteins, similar to the one shown for NPH3. In such a scenario, NPYs might function as adaptors, recruiting E3 ubiquitin ligases for ubiquitylation of PID or related AGC kinases, in order to modulate phosphorylation and intracellular sorting of auxin transport proteins (Michniewicz et al. 2007). Indirect evidence supporting such a scenario comes from a recent study, demonstrating mislocalization of PINs auxin transport proteins in *npy* mutant background, presumably causing pronounced aberrations in PAT (Furutani et al. 2011). Moreover, a function in the regulation of PIN sorting was recently suggested for NPH3, which together with phot1 appears to modulate PIN2 localization in root meristem cells (Wan et al. 2012). We still need to await further experimental proof for such a concept, but a regulatory module consisting of SCF-ubiquitin E3 ligases, NRL adaptor proteins, and AGC-type protein kinases could represent a highly versatile regulatory hub, integrating environmental and intrinsic cues and their various impacts on PAT.

## Transmission of the Light Stimulus

After perception of a tropic stimulus, it is necessary to adjust auxin signaling and responses in respective tissues, in order to orchestrate growth events. Transmission of the initial light signal to the auxin transport/signaling machinery has remained a black box for quite some years, until recently, two reports provided mechanistic insights (Christie et al. 2011; Ding et al. 2011). Christie and others demonstrate interaction between phot1 and auxin efflux carrier ABCB19 (ATP-BINDING CASSETTE B19), which appears to result in ABCB19 phosphorylation and inhibition of its transport activities (Christie et al. 2011). The authors suggested that such inhibition of auxin transport, preferentially in the illuminated portions of hypocotyls, could establish an ABCB19 activity gradient, with more hormone being transported at the shaded side from the hypocotyl apex into the bending zone (Christie et al. 2011). In this model, an auxin concentration gradient would be established already in the hypocotyl apex (Fig. 16.2a, b). In another model suggested by Ding and coworkers establishment of lateral auxin gradients takes place directly in the bending zone of hypocotyls (Ding et al. 2011). Together with additional plasma membrane-localized PIN-type auxin transport proteins, PIN3 has been proposed to act in hypocotyl phototropism (Friml et al. 2002; Haga et al. 2005), which is expressed at the plasma membrane of hypocotyl stele and endodermis cells. The authors found that unilateral blue-light treatment resulted in a relocation of endodermal PIN3 reporter signals, giving rise to polar distribution at the plasma membrane that would favor lateral auxin transport from the illuminated towards the shaded side of hypocotyls (Ding et al. 2011; Fig. 16.2c and 16.3a). Such control of PIN3 distribution apparently depends on intracellular protein sorting via ADP ribosylation factor guanine nucleotide exchange factor (ARF-GEF) GNOM-pathways and is controlled by PID. According to this model, blue-light-mediated gradual downregulation of *PID* transcription enforces PIN3 polarization due to resulting variations in PID kinase activities (Ding et al. 2011; Fig. 16.2d).

Both mechanisms offer plausible scenarios, connecting blue-light signaling and auxin gradient establishment in the control of hypocotyl phototropism. It remains to be resolved, however, how these distinct mechanisms converge to control lateral auxin gradient establishment (Haga and Sakai 2012; Sakai and Haga 2012; Hohm et al. 2013).

Regardless of mechanisms ultimately causing auxin gradient establishment in illuminated hypocotyls, variations in auxin concentration need to be perceived by hormone receptors and translated further into local adjustments of gene expression programs that drive differential growth. Some lines of evidence suggest involvement of $SCF^{TIR1/AFB}$-type E3 ubiquitin ligases that function as auxin receptors and are essential for transcriptional responses induced by the growth regulator (see Chap. 6 and Dharmasiri et al. 2005). Möller and others (2010) demonstrated phototropism defects in *tir1 afb1 afb2 afb3*, deficient in four F-box proteins, each of which constituting a subunit of SCF auxin receptor complexes. Additional evidence for involvement of the SCF-auxin receptor pathway comes from analysis of *MASSUGU2(MSG2)/IAA19* one of the likely substrates for $SCF^{TIR1/AFB}$ complex-mediated polyubiquitylation (Tatematsu et al. 2004). *MSG2* represents one out of 29 Aux/IAA proteins in Arabidopsis (Liscum and Reed 2002), several of which have been demonstrated to be short lived and degraded in response to auxin by means of $SCF^{TIR1/AFB}$ E3 ubiquitin ligase activity (Gray et al. 2001). Analysis of

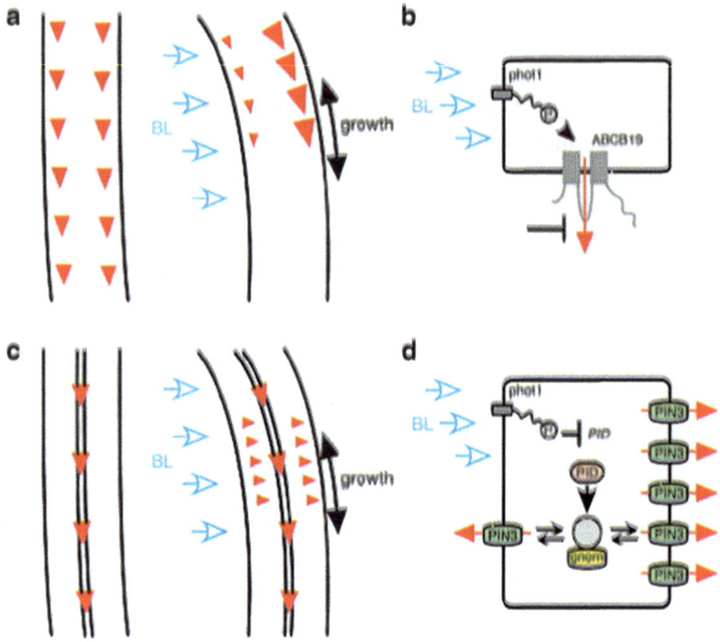


**Fig. 16.2** Current models describing establishment of lateral auxin gradients upon phototropic stimulation of Arabidopsis hypocotyls. (**a**) According to Christie and colleagues (Christie et al. 2011), differential auxin flow is established in the most apical portion of unilaterally illuminated hypocotyls (*blue arrows*), with more auxin transported at the shaded side to induce asymmetric growth in the hypocotyl bending zone. (**b**) This response is proposed to involve phot1-mediated phosphorylation of ABCB19 resulting in local inhibition of ABCB19-mediated rootward auxin transport (*red arrow*). (**c**) Ding and colleagues (2011) proposed PIN3-mediated lateral auxin transport in the hypocotyl bending zone toward the shaded side (*small red arrowheads*). (**d**) Mechanistically, this response is suggested to involve phot1-controlled inhibition of *PID* transcription, resulting in local adjustments in PIN3 distribution at the plasma membrane controlled by PID- and GNOM-dependent sorting pathways

dominant, presumably stabilized, *msg2* alleles revealed severe defects in hypocotyl tropisms and alterations in auxin-controlled gene expression, suggesting involvement of *MSG2* in the control of differential hypocotyl growth responses (Tatematsu et al. 2004). Aux/IAA proteins form heterodimers with AUXIN RESPONSE FACTOR (ARF) transcriptional regulators, and these dimers are suggested to attenuate *ARF* function in transcriptional control (Guilfoyle and Hagen 2007). Intriguingly, Yeast-2-Hybrid assays demonstrated interaction between *MSG2* and *NPH4*/*ARF7* (Tatematsu et al. 2004), representing an *ARF* gene with a nonredundant function in hypocotyl phototropism (Harper et al. 2000). This led to models in which variations in $SCF^{TIR1/AFB}$-mediated MSG2 turnover cause adjustments in *NPH4*/*ARF7*-controlled transcriptional programs, essential for phototropic organ bending (Tatematsu et al. 2004).

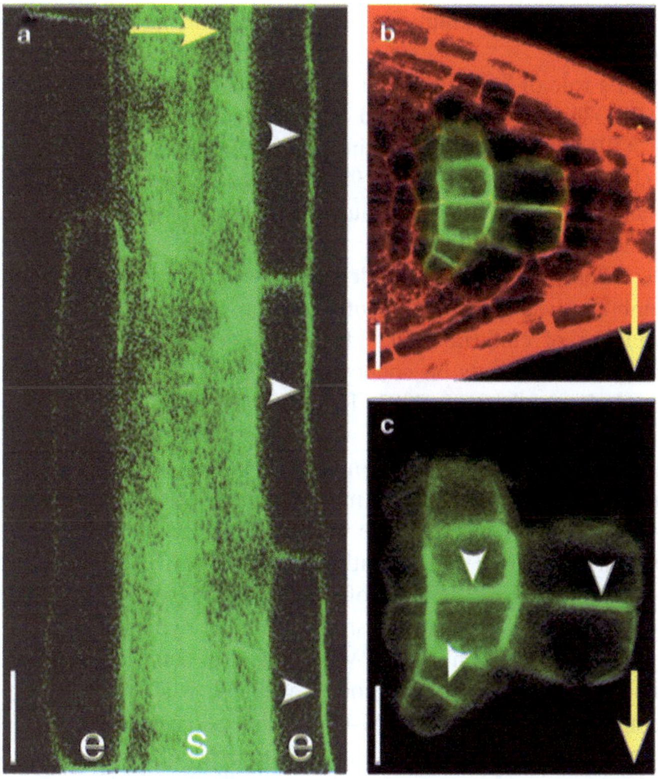


**Fig. 16.3** Relocation of PIN auxin carriers as potential determinants of directionality of auxin transport in *Arabidopsis thaliana*. (**a**) Distribution of PIN3:YFP signals (*green*) in unilaterally illuminated *PIN3::PIN3:YFP* hypocotyl cells (*yellow arrow*). *White arrowheads* indicate enrichment of fluorescent signals at the lateral, outer plasma membrane domain of endodermis cells at the shaded side. This might result in increased, lateral auxin flow. (**b**, **c**) Accumulation of PIN3: YFP signals (*green*) at the *bottom side* of columella root cap cells gravistimulated for 10 min (*yellow arrow* indicates direction of gravity vector). *White arrowheads* in (**c**) indicate signal accumulation at the *lower side* of these cells. *Red signals*: Propidium iodide cell wall staining. Scale bars: $a = 50$ μm; $b$, $c = 10$ μm

Only little is known about downstream targets of the *MSG2/NPH4* regulatory module that could bring about physiological adjustments essential for differential cell elongation. A function in differential growth has been attributed to expansins, a group of proteins implicated in controlled cell wall loosening, particularly under conditions of acidic pH, thereby allowing for turgor-driven cell expansion (Rayle and Cleland 1970; Hager et al. 1971; Cosgrove 2005). Whilst it is not fully resolved, how expansins function in mechanistic terms, expression of some members of this gene family has been associated with hypocotyl elongation (Caderas et al. 2000). Furthermore, by analyzing unilaterally illuminated *B. oleracea* seedlings, Esmon and colleagues demonstrated differential expression of presumptive *NPH4/ARF7* target genes, including expansins *EXPA1* and *EXPA8* in hypocotyls, with more transcripts accumulating at the elongating, shaded side (Esmon et al. 2006).

Apart from a role of $SCF^{TIR1/AFB}$ in phototropism, auxin signaling via AUXIN BINDING PROTEIN1 (ABP1) could be involved as well. Evidence for auxin binding by ABP1 was first provided in maize coleoptile membrane preparations (Hertel et al. 1972; Batt and Venis 1976). Subsequently, molecular cloning and functional characterization suggested auxin perception by ABP1 at the periphery of the plasma membrane (Barbier-Brygoo et al. 1989; Inohara et al. 1989; Tillmann et al. 1989). Clearcut molecular evidence for ABP1-controlled signaling at the plasma membrane was provided recently, demonstrating that ABP1 functions as positive regulator of clathrin-mediated endocytosis (Robert et al. 2010). Upon auxin binding, ABP1 effects on endocytic sorting of plasma membrane proteins are attenuated (Paciorek et al. 2005; Robert et al. 2010). This might affect the equilibrium between intracellular and plasma membrane localized pools of auxin carrier proteins, which in turn might impact on PAT activities essential for phototropic bending. Consistent with this model, defects in phototropism have been described for *abp1/ABP1* heterozygote seedlings (Effendi et al. 2011).

In addition to a function in modulating auxin flow, ABP1 appears relevant for apoplastic acidification and control of cell expansion (Cosgrove 2005). Auxin binding by ABP1 correlates with plasma membrane hyperpolarization, resulting from elevated plasma membrane-(PM)-$H^+$ATPase activity that is detectable already within minutes (Barbier-Brygoo et al. 1989; Rück et al. 1993). This has been attributed to auxin-induced PM-$H^+$ATPase phosphorylation as well as to increased abundance at the plasma membrane (Hager et al. 1991; Takahashi et al. 2012). Together with comparably slow $SCF^{TIR1/AFB}$-mediated transcriptional responses, such rapid ABP1-induced proton extrusion might constitute a regulatory module for transmission of auxin signals, controlling phototropic growth.

## Root Phototropism

Compared to the wealth of information, addressing regulation of phototropism in above-ground organs, substantially less is known about the situation in roots. The more so, as root phototropism appears to be a less uniform response than light-regulated growth of aerial organs (Kutschera and Briggs 2012) and as the ecological significance of soil-borne organs responding to light remains unclear (Kiss et al. 2003; Galen et al. 2007). Roots of several plant species, including those of *Arabidopsis* exhibit negative phototropism (Wiesner 1884; Kutschera and Briggs 2012), which allowed the characterization of mutants with deficiencies in light-controlled directional root growth (Okada and Shimura 1992; Liscum and Briggs 1995). Some of these mutant alleles turned out to be allelic to *phot1*, suggesting participation of the blue light receptor in modulating light-responsive root growth. Notably, *phot1* appears to be expressed in the root meristem (Sakamoto and Briggs 2002), and analysis of PHOT1-GFP reporter expression in roots demonstrated a positive correlation between root growth efficiency and phot1 abundance in cells close to the soil surface, suggesting that light perception by root-expressed phot1 is involved in directional root growth (Galen et al. 2007).

In a study by Wan and colleagues (2012), the authors provide indirect evidence for asymmetric distribution of auxin in unilaterally illuminated roots, with more auxin accumulating at the shaded side. Control of such asymmetric auxin distribution might involve *PIN2* activity, indicated by reduced responsiveness of a *pin2* null allele, exposed to blue light. Notably, blue light appears to modulate intracellular distribution of a PIN2 reporter protein by mechanisms that seemingly require NPH3 (Wan et al. 2012). It remains to be determined, as to how such regulation of PIN2 localization might affect PAT in phototropically responding roots.

## 4.2 Gravitropism

### Gravity Perception and Signaling

Quite similar to analysis of phototropism, early experiments addressing root gravitropism identified sites of gravity perception as well as zones that respond with differential growth to the stimulus (reviewed in, e.g., Boonsirichai et al. 2002; Arnaud et al. 2010). However, unlike phototropism, a gravity receptor, acting analogous to light receptors has not been identified.

Physical ablation experiments in which the root cap, the very tip of the root, was removed demonstrated its significance for gravity responsiveness (Darwin and Darwin 1881; Juniper et al. 1966). In more recent years, manipulation of Arabidopsis root meristems by genetic or laser-mediated removal of the root cap confirmed earlier findings and characterized starch-accumulating columella root cap cells as potential gravity sensors (Blancaflor et al. 1998; Tsugeki and Fedoroff 1999). These and numerous further studies are in agreement with the starch-statolith hypothesis, originally coined by Gottlieb Haberlandt, suggesting that sedimentation of starch-filled amyloplasts (statoliths) within specialized statocytes is essential for perception of variations in the direction of the gravity vector (Haberlandt 1900). Analogous to statolith sedimentation in roots, relocation of plastids in endodermal cells was suggested to function in gravity perception in above-ground organs (Kiss et al. 1997; Morita 2010).

A role for starch-accumulating plastids in gravity perception is supported by analysis of starch metabolism mutants, like *pgm1* alleles, deficient in *PHOSPHO-GLUCOMUTASE1*, which exhibit a reduced, but still significant gravity responsiveness (Caspar and Pickard 1989; Sack 1991). This led to modified concepts, suggesting that starch is important but not absolutely required for initial gravity perception by plastids in gravity responsive cells (Morita 2010). However, to this day, mechanisms that would translate gravity-induced plastid relocation into a cellular response remain elusive, both, in roots and stems. Mutual interactions between the actin cytoskeleton and plastids have been implicated in early events of gravity signal perception (Hou et al. 2003; Nakamura et al. 2011), which might affect activity of mechano-sensitive ion channels (Sievers et al. 1989; Perbal and Driss-Ecole 2003). This in turn might cause variations in secondary messenger signaling and/or in a shift in cytoplasmic pH, but the physiological significance of these responses is not fully resolved (Daye et al. 1984; Sievers et al. 1984; Gehring et al. 1990; Fasano et al. 2001; Plieth and Trewavas 2002; Boonsirichai et al. 2003; Perera et al. 2006). Intriguingly, some of these responses, like $Ca^{2+}$ and inositol trisphosphate signaling have been linked to control of auxin transport, which might reflect a function in controlling auxin distribution in gravistimulated organs (Lee et al. 1984; Ettlinger and Lehle 1988; Hasenstein and Evans 1988; Zhang et al. 2011; Cho et al. 2012).

## The Role of Auxin Transport

A combination of genetics, cell biology, and biochemistry identified the molecular machinery orchestrating PAT in higher plants (see Chaps. 5 and 8) and established its function in control of tropisms. Identification of auxin response mutants provided genetic evidence, as these mutants exhibit often quite pronounced defects in directional growth responses (Lincoln et al. 1990; Pickett et al. 1990; Timpte et al. 1995). Mutations in the AUXIN TRANSPORTER PROTEIN1 (AUX1) permease-like protein result in pronounced resistance to externally applied IAA and cause strong defects in tropisms, which led to the suggestion that AUX1 functions in cellular uptake of auxin thereby modulating directional growth (Pickett et al. 1990; Bennett et al. 1996; Stone et al. 2008). Indeed, functional analysis in heterologous systems demonstrated auxin transport activity, capable of translocating IAA across membrane boundaries (Yang et al. 2006). Characterization of additional mutants deficient in root gravitropism led to identification of proteins required for cellular auxin efflux. Mutants deficient in *AGRAVITROPIC ROOT/ETHYLENE INSENSITIVE ROOT1/PIN-FORMED2/WAVY GROWTH6* (*AGR/EIR1/PIN2/WAV6*) turned out to be defective in a plasma membrane protein expressed in the outer cell layers of root meristems and became the founding member of the plant-specific family of PIN-type auxin efflux carrier proteins (Chen et al. 1998; Luschnig et al. 1998; Müller et al. 1998; Utsuno et al. 1998). Furthermore, mutations affecting members of the p-glycoprotein PGP/ABCB family exhibit alterations in auxin distribution and in control of tropisms, arguing for an involvement of plant ABC-transporters in directional growth responses (Noh et al. 2001; Terasaka et al. 2005; Nagashima et al. 2008). For both, plasma membrane-localized PINs and ABCBs, there is now experimental evidence for their activity as auxin carrier proteins (Petrášek et al. 2006; Yang and Murphy 2009), but it took extensive analysis of auxin carrier expression, subcellular localization, and its dynamics to obtain a comprehensive picture of events, orchestrating auxin distribution in graviresponding organs.

According to the fountain model, acropetal/rootward PAT in the stele would deliver auxin from the shoot into the root tip. PIN1 auxin efflux carrier expressed in the stele appears to be involved in this process, indicated by defects in rootward

PAT in *pin1* loss-of-function alleles, and by PIN1 localization at the lower, basal end of stele cells, which would favor transport toward the root tip (Okada et al. 1991; Gälweiler et al. 1998; Geldner et al. 2001). Redistribution of auxin for further basipetal/shootward transport toward the elongation zone is suggested to take place in the root cap (Hasenstein and Evans 1988), and might involve activity of PIN3 and PIN7 efflux carriers that localize to the plasma membrane of root cap columella cells (Friml et al. 2002; Kleine-Vehn et al. 2010). Upon gravistimulation, both PINs rapidly accumulate at the plasma membrane domain at the lower side of columella cells (Friml et al. 2002; Kleine-Vehn et al. 2010; Fig. 16.3b, c). As a result, auxin would be relocated predominantly to the lower side of the root tip ready for its passage to the DEZ, where it would induce differential cell elongation and root bending (Ottenschlager et al. 2003; Band et al. 2012). Such a PIN-dependent redistribution mechanism is also utilized in lateral roots, allowing defined auxin flux and lateral root gravitropic set point regulation (Rosquete et al. 2013). PIN relocation in columella cells is detectable within a few minutes after gravistimulation and presumably occurs via ARF-GEF GNOM-dependent protein transcytosis (Friml et al. 2002; Kleine-Vehn et al. 2010; Fig. 16.4a, b). In addition, mutants affected in *ALTERED RESPONSE TO GRAVITY1/ROOT AND HYPOCOTYL GRAVITROPISM* (*ARG1/RHG*) or *ARG1-LIKE2* (*ARL2*) were shown to exhibit defects in gravistimulated relocation of PIN3 (Harrison and Masson 2008). ARG1 and ARL2 represent DnaJ-domain containing peripheral membrane proteins, implicated in regulation of vesicular transport that might link early events in gravity perception to the control of directional auxin transport (Boonsirichai et al. 2003; Harrison and Masson 2008; Fig. 16.4a, b).

Somewhat surprisingly, *pin3* or *pin7* single mutants and its combinations exhibit only subtle defects in root gravitropism, suggesting redundant activities of additional auxin transport proteins (Kleine-Vehn et al. 2010). This differs from mutants deficient in AUX1 and PIN2, which exhibit pronounced agravitropic root growth and defects in PAT, indicating rate-limiting activities in the control of root gravitropism (Bennett et al. 1996; Chen et al. 1998; Luschnig et al. 1998; Utsuno et al. 1998; Rashotte et al. 2000). Indepth analysis of AUX1 function revealed a significant role for AUX1 expression in lateral root cap cells proximal to the central root cap, as conditional *aux1* mutants lacking *AUX1* expression in these cells fail to exhibit gravitropic root growth (Swarup et al. 2005). Further auxin transport via lateral root cap and epidermis cells into the DEZ requires PIN2 (Müller et al. 1998; Blilou et al. 2005; Vieten et al. 2005), with a tight regulation of its localization and turnover. PIN2 protein stability regulation is seemingly essential for establishment of a lateral auxin gradient in graviresponding roots (Paciorek et al. 2005; Abas et al. 2006; Fig. 16.4c, d). Specifically, analysis of PIN2 signals in gravity-responding roots revealed establishment of a transient expression gradient, with more PIN2 accumulating at the lower side vs. the upper side of horizontally positioned roots (Paciorek et al. 2005; Abas et al. 2006). This involves antagonistic processes, with PIN2 retention at the plasma membrane of epidermis cells at the lower side, and enhanced PIN2 endocytosis and vacuolar targeting at the upper side, altogether resulting in differential auxin flow toward the DEZ, and -ultimately-

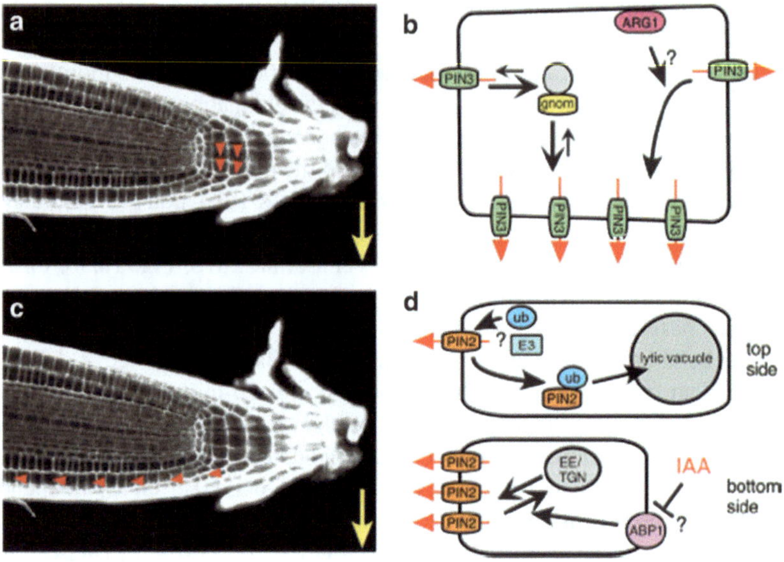


**Fig. 16.4** Mechanisms potentially involved in auxin relocation to the lower portion of gravistimulated Arabidopsis root meristems. (**a**) Auxin relocation to the lower side of the root tip is suggested to initiate within minutes after gravistimulation (*yellow arrow* indicates direction of the gravity vector; *red arrowheads* highlight directionality of auxin flow). (**b**) Rapid relocation of PIN3 auxin carrier to the plasma membrane domain at the lower side of gravistimulated root cap cells involves GNOM ARF-GEF regulated sorting and transcytosis. In addition, a function for peripheral membrane protein ARG1 in PIN3 relocation has been shown but remains to be characterized in mechanistic terms. (**c**) At later stages, auxin flow is increased at the lower side of root meristems, when compared to the upper side (*red arrowheads*), eventually causing downward bending of the root tip. (**d**) Control of differential PIN2 expression has been implicated as an essential determinant for this process. At the upper side vacuolar PIN2 targeting is suggested to cause a reduction in plasma membrane-localized PIN2 levels. Vacuolar PIN2 targeting has been suggested to require its (poly)-ubiquitylation, however, E3 ubiquitin ligase controlling this step is still unknown. At the lower side, PIN2 appears to be stabilized at the plasma membrane promoting auxin transport into the root elongation zone. This response is suggested to involve inhibition of ABP1-mediated clathrin-dependent PIN2 endocytosis into Early Endosome/Trans Golgi Network (EE/TGN) compartments. Inhibition of ABP1 activity might arise as a consequence of its binding to auxin. Whether such ABP1–auxin interaction occurs at the cell's periphery, or intracellularly, is not known

downward bending of the root tip (Paciorek et al. 2005; Abas et al. 2006; Kleine-Vehn et al. 2008; Robert et al. 2010; Leitner et al. 2012). It is not fully resolved, how such differential protein sorting might be regulated, but apparently PIN2 ubiquitylation signals variations in its sorting and proteolytic turnover in an auxin-dependent feedback regulatory loop (Sieberer et al. 2000; Leitner et al. 2012; Baster et al. 2013). This regulatory switch allows for continuous adjustments in auxin flow rates in the coordination of directional root growth (Band et al. 2012), but molecular determinants, sensing, and controlling the interplay between auxin levels and PIN2 sorting remain to be determined in more detail (Fig. 16.4d).

PIN expression gradient formation could as well be involved in control of hypocotyl gravitropism. Rakusova and others observed asymmetric abundance of PIN3-GFP reporter signals in endodermis cells of gravistimulated hypocotyls that would favor auxin flow to the organ's lower, elongating side (Rakusova et al. 2011). Such gravity-induced PIN3 relocation in hypocotyls was found to require GNOM ARF-GEF and PID activity, which appears comparable to PIN3 relocation upon phototropic stimulation and further highlights the critical role of intracellular protein sorting in the control of tropic growth responses (Ding et al. 2011; Rakusova et al. 2011).

Variations in auxin distribution, resulting from differential auxin transport activities in graviresponding organs, appear to cause induction of distinct transcriptional programs that would induce differential cell expansion. By analogy to the transcriptional output resulting from phototropic signaling, transcriptional responses to gravistimulation could involve activity of Aux/IAA and ARF proteins, as indicated by agravitropic growth phenotypes associated with corresponding mutant and misexpression lines (Okushima et al. 2005; Wang et al. 2005; Weijers et al. 2005; Yan et al. 2013). However, detailed insights into phenotypic consequences of variable transcriptional control of downstream target loci are still limited. Perhaps, transcriptome analyses based on cell-sorting approaches will help to elucidate differences in gene expression at a sufficiently high resolution, both, over time, and in different portions of graviresponding organs (Birnbaum et al. 2005).

## 4.3 Hydrotropism

Molecular analysis of physiological events controlling directional plant growth in response to humidity gradients has been problematic for the longest time, due to difficulties in separating gravitropic and hydrotropic responses (Sachs 1872). Notably, decapitation and ablation experiments revealed a role for the columella root cap in moisture gradient perception (Jaffe et al. 1985; Miyazawa et al. 2008), thus, overlapping with sites of gravity perception and raising questions as to how root tip cells perceive and transmit these different environmental stimuli (Takahashi et al. 2009). However, in the 1980s, an agravitropic pea mutant was found to exhibit positive hydrotropism, suggesting that control of these tropic responses requires distinct regulatory switches (Jaffe et al. 1985). There is only limited information available on early events of hydrotropic signaling, which was suggested to involve $Ca^{2+}$ signaling and was found to coincide with degradation of starch granules in columella root cap cells, the significance of which remains elusive (Takano et al. 1997; Takahashi et al. 2003).

Conflicting evidence suggests involvement of auxin in regulation of hydrotropism. Auxin transport inhibitors, like 2,3,5-triiodobenzoic acid (TIBA) were found to block hydrotropism in some plant species, whereas in Arabidopsis no comparable effect was observed (Mizuno et al. 2002; Kaneyasu et al. 2007). Likewise, while some reports suggest local differences in auxin signaling and/or concentration in roots exposed to moisture gradients, other reports failed to verify such variations (Mizuno et al. 2002; Takahashi et al. 2009). Moreover, Arabidopsis loss-of-function mutants deficient in *PIN2* and *AUX1* exhibit normal root hydrotropism, which uncouples Arabidopsis hydrotropic responses from these key mediators of PAT in root meristems (Takahashi et al. 2002).

Mutant screens performed in Arabidopsis allowed for characterization of determinants involved in root hydrotropism (Eapen et al. 2003; Kobayashi et al. 2007). Cloning of *MIZU-KUSSEI1* (*MIZ1*) revealed a gene of unknown function that is strongly expressed in root columella cells; however, its function in root hydrotropism remains unanswered (Kobayashi et al. 2007). Notably, when analyzing lateral root formation in *MIZ1* loss-of-function and overexpression lines, it turned out that *MIZ1* acts as a negative regulator of lateral root formation in an auxin-dependent manner (Moriwaki et al. 2011). This could be linked to variations in free IAA content, with *miz1* roots accumulating more auxin than wild-type controls and might suggest that *MIZ1* function in root hydrotropism relates to modulation of auxin levels (Moriwaki et al. 2011).

Cloning of *MIZ2* provided additional indirect evidence for an involvement of auxin in regulation of hydrotropism. This mutant turned out to represent a novel allele of ARF-GEF *GNOM*; however unlike *gnom* alleles described earlier, $gnom^{miz2}$ does not exhibit comparable defects in overall plant morphology and in gravitropic growth responses (Geldner et al. 2004; Miyazawa et al. 2009). These findings indicated a role for GNOM-dependent vesicle sorting in the regulation of hydrotropism, and, given the function of GNOM in controlling localization of auxin transport components (Geldner et al. 2003), it is tempting to speculate about similar, potentially subtler defects in $gnom^{miz2}$. Limited or delayed GNOM-dependent rerouting of auxin carrier proteins might account for moderate variations in auxin distribution or signaling, causing growth defects restricted to hydrotropism. However, no variations in auxin carrier protein localization or abundance have so far been described for *miz2* (Moriwaki et al. 2013). Notably, MIZ1 dosage effects on lateral root formation are blocked by $gnom^{miz2}$, suggesting indirect effects of GNOM on MIZ1-mediated variations in auxin homeostasis (Moriwaki, et al. 2011). Mechanisms controlling such crosstalk remain to be determined.

## 5 Concluding Remarks

More than a century after pioneering experimental work by early plant physiologists, a decent framework of molecular events controlling directional plant growth responses, is finally available. Even though current models still suffer from major shortcomings, like the vaguely understood mechanism of gravity-perception, we experienced the evolution of a once uncertain idea of transmissible signals being involved in tropism to sophisticated molecular models, underpinning various functions of auxin in these growth processes. With all these concepts available, attracting students probably would have been easier for our professor.

## References


Abas L, Benjamins R, Malenica N, Paciorek T, Wisniewska J, Moulinier-Anzola JC, Sieberer T, Friml J, Luschnig C (2006) Intracellular trafficking and proteolysis of the Arabidopsis auxin-efflux facilitator PIN2 are involved in root gravitropism. Nat Cell Biol 8(3):249–256

Arnaud C, Bonnot C, Desnos T, Nussaume L (2010) The root cap at the forefront. C R Biol 333 (4):335–343

Band LR, Wells DM, Larrieu A, Sun JY, Middleton AM, French AP, Brunoud G, Sato EM, Wilson MH, Peret B, Oliva M, Swarup R, Sairanen I, Parry G, Ljung K, Beeckman T, Garibaldi JM, Estelle M, Owen MR, Vissenberg K, Hodgman TC, Pridmore TP, King JR, Vernoux T, Bennett MJ (2012) Root gravitropism is regulated by a transient lateral auxin gradient controlled by a tipping-point mechanism. Proc Natl Acad Sci USA 109(12):4668–4673

Barbier-Brygoo H, Ephritikhine G, Klambt D, Ghislain M, Guern J (1989) Functional evidence for an auxin receptor at the plasmalemma of tobacco mesophyll protoplasts. Proc Natl Acad Sci USA 86(3):891–895

Baskin T (1986) Redistribution of growth during phototropism and nutation in the pea epicotyl. Planta 169:406–414

Baster P, Robert S, Kleine-Vehn J, Vanneste S, Kania U, Grunewald W, De Rybel B, Beeckman T, Friml J (2013) SCF(TIR1/AFB)-auxin signalling regulates PIN vacuolar trafficking and auxin fluxes during root gravitropism. EMBO J 32(2):260–274

Batt S, Venis MA (1976) Separation and localization of two classes of auxin binding sites in corn coleoptile membranes. Planta 130:15–21

Bennett MJ, Marchant A, Green HG, May ST, Ward SP, Millner PA, Walker AR, Schulz B, Feldmann KA (1996) Arabidopsis AUX1 gene: a permease-like regulator of root gravitropism. Science 273(5277):948–950

Birnbaum K, Jung JW, Wang JY, Lambert GM, Hirst JA, Galbraith DW, Benfey PN (2005) Cell type-specific expression profiling in plants via cell sorting of protoplasts from fluorescent reporter lines. Nat Methods 2(8):615–619

Blancaflor EB, Fasano JM, Gilroy S (1998) Mapping the functional roles of cap cells in the response of Arabidopsis primary roots to gravity. Plant Physiol 116(1):213–222

Blilou I, Xu J, Wildwater M, Willemsen V, Paponov I, Friml J, Heidstra R, Aida M, Palme K, Scheres B (2005) The PIN auxin efflux facilitator network controls growth and patterning in Arabidopsis roots. Nature 433(7021):39–44

Boonsirichai K, Guan C, Chen R, Masson PH (2002) Root gravitropism: an experimental tool to investigate basic cellular and molecular processes underlying mechanosensing and signal transmission in plants. Annu Rev Plant Biol 53:421–447

Boonsirichai K, Sedbrook JC, Chen R, Gilroy S, Masson PH (2003) ALTERED RESPONSE TO GRAVITY is a peripheral membrane protein that modulates gravity-induced cytoplasmic alkalinization and lateral auxin transport in plant statocytes. Plant Cell 15(11):2612–2625

Boysen Jensen P (1910) Über die Leitung des phototropischen Reizes in Avenakeimpflanzen. Ber Deutsch Bot Ges 28:118–120

Boysen Jensen P (1913) Über die Leitung des phototropischen Reizes in der Avenakoleoptile. Ber Deutsch Bot Ges 31:559–566

Briggs WR (1963) Mediation of phototropic responses of corn coleoptiles by lateral transport of auxin. Plant Physiol 38:237–247

Briggs WR, Tochter RD, Wilson JF (1957) Phototropic auxin redistribution in corn coleoptiles. Science 126:210–212
Briggs WR, Beck CF, Cashmore AR, Christie JM, Hughes J, Jarillo JA, Kagawa T, Kanegae H, Liscum E, Nagatani A, Okada K, Salomon M, Rudiger W, Sakai T, Takano M, Wada M, Watson JC (2001) The phototropin family of photoreceptors. Plant Cell 13(5):993–997
Caderas D, Muster M, Vogler H, Mandel T, Rose JK, McQueen-Mason S, Kuhlemeier C (2000) Limited correlation between expansin gene expression and elongation growth rate. Plant Physiol 123(4):1399–1414
Caspar T, Pickard BG (1989) Gravitropism in a starchless mutant of Arabidopsis: implications for the starch-statolith theory of gravity sensing. Planta 177:185–197
Chae K, Lord EM (2011) Pollen tube growth and guidance: roles of small, secreted proteins. Ann Bot 108(4):627–636
Chen R, Hilson P, Sedbrook J, Rosen E, Caspar T, Masson PH (1998) The arabidopsis thaliana AGRAVITROPIC 1 gene encodes a component of the polar-auxin-transport efflux carrier. Proc Natl Acad Sci USA 95(25):15112–15117
Cheng Y, Qin G, Dai X, Zhao Y (2007) NPY1, a BTB-NPH3-like protein, plays a critical role in auxin-regulated organogenesis in Arabidopsis. Proc Natl Acad Sci USA 104(47):18825–18829
Cheng Y, Qin G, Dai X, Zhao Y (2008) NPY genes and AGC kinases define two key steps in auxin-mediated organogenesis in Arabidopsis. Proc Natl Acad Sci USA 105(52):21017–21022
Cho D, Villiers F, Kroniewicz L, Lee S, Seo YJ, Hirschi KD, Leonhardt N, Kwak JM (2012) Vacuolar CAX1 and CAX3 influence auxin transport in guard cells via regulation of apoplastic pH. Plant Physiol 160(3):1293–1302
Cholodny N (1927) Wuchshormone und Tropismem bei den Pflanzen. Biol Zent Bl 47:604–626
Cholodny N (1928) Beiträge zur hormonalen Theorie von Tropismen. Planta 6:118–134
Christie JM, Reymond P, Powell GK, Bernasconi P, Raibekas AA, Liscum E, Briggs WR (1998) Arabidopsis NPH1: a flavoprotein with the properties of a photoreceptor for phototropism. Science 282(5394):1698–1701
Christie JM, Swartz TE, Bogomolni RA, Briggs WR (2002) Phototropin LOV domains exhibit distinct roles in regulating photoreceptor function. Plant J 32(2):205–219
Christie JM, Yang H, Richter GL, Sullivan S, Thomson CE, Lin J, Titapiwatanakun B, Ennis M, Kaiserli E, Lee OR, Adamec J, Peer WA, Murphy AS (2011) Phot1 inhibition of ABCB19 primes lateral auxin fluxes in the shoot apex required for phototropism. PLoS Biol 9(6): e1001076
Cosgrove DJ (2005) Growth of the plant cell wall. Nat Rev Mol Cell Biol 6(11):850–861
Darwin C, Darwin FE (1881) The power of movement in plants. Appleton, New York
Daye S, Biro RL, Roux SJ (1984) Inhibition of gravitropism in oat coleoptiles by the calcium chelator, ethyleneglycol-bis-(beta-aminoethyl ether)-N, N'-tetraacetic acid. Physiol Plant 61 (3):449–454
Dharmasiri N, Dharmasiri S, Weijers D, Lechner E, Yamada M, Hobbie L, Ehrismann JS, Jurgens G, Estelle M (2005) Plant development is regulated by a family of auxin receptor F box proteins. Dev Cell 9(1):109–119
Ding Z, Galvan-Ampudia CS, Demarsy E, Langowski L, Kleine-Vehn J, Fan Y, Morita MT, Tasaka M, Fankhauser C, Offringa R, Friml J (2011) Light-mediated polarization of the PIN3 auxin transporter for the phototropic response in Arabidopsis. Nat Cell Biol 13(4):447–452
Eapen D, Barroso ML, Campos ME, Ponce G, Corkidi G, Dubrovsky JG, Cassab GI (2003) A no hydrotropic response root mutant that responds positively to gravitropism in Arabidopsis. Plant Physiol 131(2):536–546
Effendi Y, Rietz S, Fischer U, Scherer GF (2011) The heterozygous abp1/ABP1 insertional mutant has defects in functions requiring polar auxin transport and in regulation of early auxin-regulated genes. Plant J 65(2):282–294
Esmon CA, Tinsley AG, Ljung K, Sandberg G, Hearne LB, Liscum E (2006) A gradient of auxin and auxin-dependent transcription precedes tropic growth responses. Proc Natl Acad Sci USA 103(1):236–241

Ettlinger C, Lehle L (1988) Auxin induces rapid changes in phosphatidylinositol metabolites. Nature 331(6152):176–178

Fasano JM, Swanson SJ, Blancaflor EB, Dowd PE, Kao TH, Gilroy S (2001) Changes in root cap pH are required for the gravity response of the Arabidopsis root. Plant Cell 13(4):907–921

Firn RD, Myers AB (1989) Plant movements caused by differential growth–unity or diversity of mechanisms? Environ Exp Bot 29(1):47–55

Frank AB (1868) Über die durch Schwerkraft verursachten Bewegungen von Pflanzenteilen. Beitrage zur Pflanzenphysiologie 8:1–99

Friml J, Wisniewska J, Benkova E, Mendgen K, Palme K (2002) Lateral relocation of auxin efflux regulator PIN3 mediates tropism in Arabidopsis. Nature 415(6873):806–809

Friml J, Yang X, Michniewicz M, Weijers D, Quint A, Tietz O, Benjamins R, Ouwerkerk PB, Ljung K, Sandberg G, Hooykaas PJ, Palme K, Offringa R (2004) A PINOID-dependent binary switch in apical-basal PIN polar targeting directs auxin efflux. Science 306(5697):862–865

Furutani M, Kajiwara T, Kato T, Treml BS, Stockum C, Torres-Ruiz RA, Tasaka M (2007) The gene MACCHI-BOU 4/ENHANCER OF PINOID encodes a NPH3-like protein and reveals similarities between organogenesis and phototropism at the molecular level. Development 134 (21):3849–3859

Furutani M, Sakamoto N, Yoshida S, Kajiwara T, Robert HS, Friml J, Tasaka M (2011) Polar-localized NPH3-like proteins regulate polarity and endocytosis of PIN-FORMED auxin efflux carriers. Development 138(10):2069–2078

Galen C, Rabenold JJ, Liscum E (2007) Functional ecology of a blue light photoreceptor: effects of phototropin-1 on root growth enhance drought tolerance in Arabidopsis thaliana. New Phytol 173(1):91–99

Galvan-Ampudia CS, Offringa R (2007) Plant evolution: AGC kinases tell the auxin tale. Trends Plant Sci 12(12):541–547

Gälweiler L, Guan C, Müller A, Wisman E, Mendgen K, Yephremov A, Palme K (1998) Regulation of polar auxin transport by AtPIN1 in Arabidopsis vascular tissue. Science 282 (5397):2226–2230

Gardner G, Shaw S, Wilkins MB (1974) IAA transport during the phototropic responses of intact Zea and Avena coleoptiles. Planta 28:439–478

Gehring CA, Williams DA, Cody SH, Parish RW (1990) Phototropism and geotropism in maize coleoptiles are spatially correlated with increases in cytosolic free calcium. Nature 345:528–530

Geldner N, Friml J, Stierhof YD, Jurgens G, Palme K (2001) Auxin transport inhibitors block PIN1 cycling and vesicle trafficking. Nature 413(6854):425–428

Geldner N, Anders N, Wolters H, Keicher J, Kornberger W, Muller P, Delbarre A, Ueda T, Nakano A, Jurgens G (2003) The Arabidopsis GNOM ARF-GEF mediates endosomal recycling, auxin transport, and auxin-dependent plant growth. Cell 112(2):219–230

Geldner N, Richter S, Vieten A, Marquardt S, Torres-Ruiz RA, Mayer U, Jurgens G (2004) Partial loss-of-function alleles reveal a role for GNOM in auxin transport-related, post-embryonic development of Arabidopsis. Development 131(2):389–400

Gray WM, Kepinski S, Rouse D, Leyser O, Estelle M (2001) Auxin regulates SCF(TIR1)-dependent degradation of AUX/IAA proteins. Nature 414(6861):271–276

Guilfoyle TJ, Hagen G (2007) Auxin response factors. Curr Opin Plant Biol 10(5):453–460

Haberlandt G (1900) Über die Perzeption des geotropischen Reizes. Ber Deutsch Bot Ges 18:261–272

Haga K, Sakai T (2012) PIN auxin efflux carriers are necessary for pulse-induced but not continuous light-induced phototropism in Arabidopsis. Plant Physiol 160(2):763–776

Haga K, Takano M, Neumann R, Iino M (2005) The Rice COLEOPTILE PHOTOTROPISM1 gene encoding an ortholog of Arabidopsis NPH3 is required for phototropism of coleoptiles and lateral translocation of auxin. Plant Cell 17(1):103–115

Hager A, Menzel H, Krauss A (1971) Versuche und Hypothese zur Primärwirkung des Auxins beim Streckungswachstum. Planta 100:47–75

Hager A, Debus G, Edel H-G, Stransky H, Serrano R (1991) Auxin induces exocytosis and the rapid synthesis of a high-turnover pool of plasma-membrane H+-ATPase. Planta 185:527–537
Han IS, Tseng TS, Eisinger W, Briggs WR (2008) Phytochrome A regulates the intracellular distribution of phototropin 1-green fluorescent protein in Arabidopsis thaliana. Plant Cell 20 (10):2835–2847
Harper RM, Stowe-Evans EL, Luesse DR, Muto H, Tatematsu K, Watahiki MK, Yamamoto K, Liscum E (2000) The NPH4 locus encodes the auxin response factor ARF7, a conditional regulator of differential growth in aerial Arabidopsis tissue. Plant Cell 12(5):757–770
Harrison BR, Masson PH (2008) ARL2, ARG1 and PIN3 define a gravity signal transduction pathway in root statocytes. Plant J 53(2):380–392
Hasenstein KH, Evans ML (1986) Calcium dependence of rapid auxin action in maize roots. Plant Physiol 81:439–443
Hasenstein KH, Evans ML (1988) Effects of cations on hormone transport in primary roots of Zea mays. Plant Physiol 86:890–894
Hertel R, Thomson K-S, Russo VEA (1972) In-vitro auxin binding to particulate cell fractions from corn coleoptiles. Planta 107:325–340
Hohm T, Preuten T, Fankhauser C (2013) Phototropism: translating light into directional growth. Am J Bot 100(1):47–59
Hou G, Mohamalawari DR, Blancaflor EB (2003) Enhanced gravitropism of roots with a disrupted cap actin cytoskeleton. Plant Physiol 131(3):1360–1373
Huala E, Oeller PW, Liscum E, Han IS, Larsen E, Briggs WR (1997) Arabidopsis NPH1: a protein kinase with a putative redox-sensing domain. Science 278(5346):2120–2123
Inada S, Ohgishi M, Mayama T, Okada K, Sakai T (2004) RPT2 is a signal transducer involved in phototropic response and stomatal opening by association with phototropin 1 in Arabidopsis thaliana. Plant Cell 16(4):887–896
Inohara N, Shimomura S, Fukui T, Futai M (1989) Auxin-binding protein located in the endoplasmic reticulum of maize shoots: molecular cloning and complete primary structure. Proc Natl Acad Sci USA 86(10):3564–3568
Inoue S, Kinoshita T, Matsumoto M, Nakayama KI, Doi M, Shimazaki K (2008) Blue light-induced autophosphorylation of phototropin is a primary step for signaling. Proc Natl Acad Sci USA 105(14):5626–5631
Ishikawa H, Evans ML (1997) Novel software for analysis of root gravitropism: comparative response patterns of Arabidopsis wild-type and axr1 seedlings. Plant Cell Environ 20(7):919–928
Ishikawa H, Hasenstein KH, Evans ML (1991) Computer-based video digitizer analysis of surface extension in maize roots: kinetics of growth rate changes during gravitropism. Planta 183 (3):381–390
Jaffe MJ, Takahashi H, Biro RL (1985) A pea mutant for the study of hydrotropism in roots. Science 230:445–447
Jarillo JA, Gabrys H, Capel J, Alonso JM, Ecker JR, Cashmore AR (2001) Phototropin-related NPL1 controls chloroplast relocation induced by blue light. Nature 410(6831):952–954
Johnson H (1829) The unsatisfactory nature of the theories proposed to account for the descent of the radicles in the germination of seeds, shown by experiments. Edinburgh New Pil J 312–317
Juniper B, Groves S, Landau-Schchar B, Audus L (1966) Root cap and the perception of gravity. Nature 209:93–94
Kami C, Hersch M, Trevisan M, Genoud T, Hiltbrunner A, Bergmann S, Fankhauser C (2012) Nuclear phytochrome A signaling promotes phototropism in Arabidopsis. Plant Cell 24 (2):566–576
Kaneyasu T, Kobayashi A, Nakayama M, Fujii N, Takahashi H, Miyazawa Y (2007) Auxin response, but not its polar transport, plays a role in hydrotropism of Arabidopsis roots. J Exp Bot 58:1143–1150
Kang BG, Burg SP (1974) Red Light enhancement of the phototropic response in pea stems. Plant Physiol 53:445–448

Kaufman PB, Wu L-L, Brock TG, Kim D (1988) Hormones and the orientation of growth. In: Davies PJ (ed) Plant hormones physiology, biochemistry and molecular biology. Kluwer Academic, Dordrecht

Khurana JP, Poff KL (1989) Mutants of Arabidopsis thaliana with altered phototropism. Planta 178:400–406

Kiss JZ, Guisinger MM, Miller AJ, Stackhouse KS (1997) Reduced gravitropism in hypocotyls of starch-deficient mutants of Arabidopsis. Plant Cell Physiol 38(5):518–525

Kiss JZ, Mullen JL, Correll MJ, Hangarter RP (2003) Phytochromes A and B mediate red-light-induced positive phototropism in roots. Plant Physiol 131(3):1411–1417

Kleine-Vehn J, Leitner J, Zwiewka M, Sauer M, Abas L, Luschnig C, Friml J (2008) Differential degradation of PIN2 auxin efflux carrier by retromer-dependent vacuolar targeting. Proc Natl Acad Sci USA 105(46):17812–17817

Kleine-Vehn J, Ding Z, Jones AR, Tasaka M, Morita MT, Friml J (2010) Gravity-induced PIN transcytosis for polarization of auxin fluxes in gravity-sensing root cells. Proc Natl Acad Sci USA 107(51):22344–22349

Kobayashi A, Takahashi A, Kakimoto Y, Miyazawa Y, Fujii N, Higashitani A, Takahashi H (2007) A gene essential for hydrotropism in roots. Proc Natl Acad Sci USA 104(11):4724–4729

Kögl F, Haagen-Smits AJ (1931) Über die Chemie des Wuchsstoffs. Proc Koninklijke Nederlandse Akademie Wetenschappen 34:1411–1416

Kögl F, Schuringa GJ (1944) Über die Inaktivierung von Auxin-a-lacton bei verschiedenen Wellenlängen und den Einfluß von Carotinoiden auf die Lichtreaktion. Hoppe Seylers Z Physiol Chem 280:148–161

Konings H (1968) The significance of the root cap for geotropism. Acta Bot Neerl 17:203–211

Kutschera U, Briggs WR (2012) Root phototropism: from dogma to the mechanism of blue light perception. Planta 235(3):443–452

Lariguet P, Fankhauser C (2004) Hypocotyl growth orientation in blue light is determined by phytochrome A inhibition of gravitropism and phototropin promotion of phototropism. Plant J 40(5):826–834

Lee JS, Mulkey TJ, Evans ML (1984) Inhibition of polar calcium movement and gravitropism in roots treated with auxin-transport inhibitors. Planta 160:536–543

Leitner J, Petrasek J, Tomanov K, Retzer K, Parezova M, Korbei B, Bachmair A, Zazimalova E, Luschnig C (2012) Lysine63-linked ubiquitylation of PIN2 auxin carrier protein governs hormonally controlled adaptation of Arabidopsis root growth. Proc Natl Acad Sci USA 109 (21):8322–8327

Li Y, Dai X, Cheng Y, Zhao Y (2011) NPY genes play an essential role in root gravitropic responses in Arabidopsis. Mol Plant 4(1):171–179

Lincoln C, Britton JH, Estelle M (1990) Growth and development of the axr1 mutants of Arabidopsis. Plant Cell 2(11):1071–1080

Liscum E, Briggs WR (1995) Mutations in the NPH1 locus of Arabidopsis disrupt the perception of phototropic stimuli. Plant Cell 7(4):473–485

Liscum E, Reed JW (2002) Genetics of Aux/IAA and ARF action in plant growth and development. Plant Mol Biol 49(3–4):387–400

Luschnig C, Gaxiola RA, Grisafi P, Fink GR (1998) EIR1, a root-specific protein involved in auxin transport, is required for gravitropism in Arabidopsis thaliana. Genes Dev 12(14):2175–2187

Mertens R, Weiler EW (1983) Kinetic studies on the redistribution of endogenous growth regulators in gravireacting plant organs. Planta 158:339–348

Michniewicz M, Zago MK, Abas L, Weijers D, Schweighofer A, Meskiene I, Heisler MG, Ohno C, Zhang J, Huang F, Schwab R, Weigel D, Meyerowitz EM, Luschnig C, Offringa R, Friml J (2007) Antagonistic regulation of PIN phosphorylation by PP2A and PINOID directs auxin flux. Cell 130(6):1044–1056

Migliaccio F, Fortunati A, Tassone P (2009) Arabidopsis root growth movements and their symmetry: progress and problems arising from recent work. Plant Signal Behav 4(3):183–190

Miyazawa Y, Sakashita T, Funayama T, Hamada N, Negishi H, Kobayashi A, Kaneyasu T, Ooba A, Morohashi K, Kakizaki T, Wada S, Kobayashi Y, Fujii N, Takahashi H (2008) Effects of locally targeted heavy-ion and laser microbeam on root hydrotropism in Arabidopsis thaliana. J Radiat Res 49(4):373–379

Miyazawa Y, Takahashi A, Kobayashi A, Kaneyasu T, Fujii N, Takahashi H (2009) GNOM-mediated vesicular trafficking plays an essential role in hydrotropism of Arabidopsis roots. Plant Physiol 149(2):835–840

Mizuno H, Kobayashi A, Fujii N, Yamashita M, Takahashi H (2002) Hydrotropic response and expression pattern of auxin-inducible gene, CS-IAA1, in the primary roots of clinorotated cucumber seedlings. Plant Cell Physiol 43(7):793–801

Möller B, Schenck D, Luthen H (2010) Exploring the link between auxin receptors, rapid cell elongation and organ tropisms. Plant Signal Behav 5(5)

Morita MT (2010) Directional gravity sensing in gravitropism. Annu Rev Plant Biol 61:705–720

Moriwaki T, Miyazawa Y, Kobayashi A, Uchida M, Watanabe C, Fujii N, Takahashi H (2011) Hormonal regulation of lateral root development in Arabidopsis modulated by MIZ1 and requirement of GNOM activity for MIZ1 function. Plant Physiol 157(3):1209–1220

Moriwaki T, Miyazawa Y, Kobayashi A, Takahashi H (2013) Molecular mechanisms of hydrotropism in seedling roots of Arabidopsis thaliana (Brassicaceae). Am J Bot 100(1):25–34

Motchoulski A, Liscum E (1999) Arabidopsis NPH3: a NPH1 photoreceptor-interacting protein essential for phototropism. Science 286(5441):961–964

Mullen JL, Ishikawa H, Evans ML (1998) Analysis of changes in relative elemental growth rate patterns in the elongation zone of Arabidopsis roots upon gravistimulation. Planta 206(4):598–603

Müller A, Guan C, Gälweiler L, Tanzler P, Huijser P, Marchant A, Parry G, Bennett M, Wisman E, Palme K (1998) AtPIN2 defines a locus of Arabidopsis for root gravitropism control. EMBO J 17(23):6903–6911

Nagashima A, Uehara Y, Sakai T (2008) The ABC subfamily B auxin transporter AtABCB19 is involved in the inhibitory effects of N-1-naphthyphthalamic acid on the phototropic and gravitropic responses of Arabidopsis hypocotyls. Plant Cell Physiol 49(8):1250–1255

Nakamura M, Toyota M, Tasaka M, Morita MT (2011) An Arabidopsis E3 ligase, SHOOT GRAVITROPISM9, modulates the interaction between statoliths and F-actin in gravity sensing. Plant Cell 23(5):1830–1848

Nelson AJ, Evans ML (1986) Analysis of growth patterns during gravitropic curvature in roots of Zea mays by use of a computer-based video digitizer. J Plant Growth Regul 5:73–83

Noh B, Murphy AS, Spalding EP (2001) Multidrug resistance-like genes of Arabidopsis required for auxin transport and auxin-mediated development. Plant Cell 13(11):2441–2454

Okada K, Shimura Y (1992) Mutational analysis of root gravitropism and phototropism of Arabidopsis thaliana seedlings. Aust J Plant Physiol 19:439–448

Okada K, Ueda J, Komaki MK, Bell CJ, Shimura Y (1991) Requirement of the Auxin Polar Transport System in Early Stages of Arabidopsis Floral Bud Formation. Plant cell 3(7):677–684

Okushima Y, Overvoorde PJ, Arima K, Alonso JM, Chan A, Chang C, Ecker JR, Hughes B, Lui A, Nguyen D, Onodera C, Quach H, Smith A, Yu G, Theologis A (2005) Functional genomic analysis of the AUXIN RESPONSE FACTOR gene family members in Arabidopsis thaliana: unique and overlapping functions of ARF7 and ARF19. Plant Cell 17(2):444–463

Ottenschlager I, Wolff P, Wolverton C, Bhalerao RP, Sandberg G, Ishikawa H, Evans M, Palme K (2003) Gravity-regulated differential auxin transport from columella to lateral root cap cells. Proc Natl Acad Sci USA 100(5):2987–2991

Overbeek JV (1932) An analysis of phototropism in dicotyledons. Proc Koninklijke Nederlandse Akademie Wetenschappen 35:1325–1335

Paciorek T, Zažímalová E, Ruthardt N, Petrášek J, Stierhof YD, Kleine-Vehn J, Morris DA, Emans N, Jurgens G, Geldner N, Friml J (2005) Auxin inhibits endocytosis and promotes its own efflux from cells. Nature 435(7046):1251–1256

Perbal G, Driss-Ecole D (2003) Mechanotransduction in gravisensing cells. Trends Plant Sci 8 (10):498–504
Perera IY, Hung CY, Brady S, Muday GK, Boss WF (2006) A universal role for inositol 1,4,5-trisphosphate-mediated signaling in plant gravitropism. Plant Physiol 140(2):746–760
Petrášek J, Mravec J, Bouchard R, Blakeslee JJ, Abas M, Seifertová D, Wisniewska J, Tadele Z, Kubeš M, Čovanová M, Dhonukshe P, Skůpa P, Benková E, Perry L, Křeček P, Lee OR, Fink GR, Geisler M, Murphy AS, Luschnig C, Zažímalová E, Friml J (2006) PIN proteins perform a rate-limiting function in cellular auxin efflux. Science 312(5775):914–918
Pickard BG, Thimann KV (1964) Transport and distribution of auxin during tropistic response. II. The lateral migration of auxin in phototropism of coleoptiles. Plant Physiol 39:341–350
Pickett FB, Wilson AK, Estelle M (1990) The aux1 Mutation of Arabidopsis Confers Both Auxin and Ethylene Resistance. Plant Physiol 94(3):1462–1466
Pilet P-E, Ney D (1981) Differential growth of georeacting maize roots. Planta 151:146–150
Plieth C, Trewavas AJ (2002) Reorientation of seedlings in the earth's gravitational field induces cytosolic calcium transients. Plant Physiol 129(2):786–796
Rakusova H, Gallego-Bartolome J, Vanstraelen M, Robert HS, Alabadi D, Blazquez MA, Benkova E, Friml J (2011) Polarization of PIN3-dependent auxin transport for hypocotyl gravitropic response in Arabidopsis thaliana. Plant J 67(5):817–826
Rashotte AM, Brady SR, Reed RC, Ante SJ, Muday GK (2000) Basipetal auxin transport is required for gravitropism in roots of arabidopsis. Plant Physiol 122(2):481–490
Rayle DL, Cleland RE (1970) Enhancement of wall loosening and elongation by acid solutions. Plant Physiol 46:250–253
Robert S, Kleine-Vehn J, Barbez E, Sauer M, Paciorek T, Baster P, Vanneste S, Zhang J, Simon S, Covanova M, Hayashi K, Dhonukshe P, Yang Z, Bednarek SY, Jones AM, Luschnig C, Aniento F, Zazimalova E, Friml J (2010) ABP1 mediates auxin inhibition of clathrin-dependent endocytosis in Arabidopsis. Cell 143(1):111–121
Roberts D, Pedmale UV, Morrow J, Sachdev S, Lechner E, Tang X, Zheng N, Hannink M, Genschik P, Liscum E (2011) Modulation of phototropic responsiveness in Arabidopsis through ubiquitination of phototropin 1 by the CUL3-Ring E3 ubiquitin ligase CRL3 (NPH3). Plant Cell 23(10):3627–3640
Rosquete MR, von Wangenheim D, Marhavy P, Barbez E, Stelzer EH, Benkova E, Maizel A, Kleine-Vehn J (2013) An auxin transport mechanism restricts positive orthogravitropism in lateral roots. Curr Biol 23(9):817–822
Rück A, Palme K, Venis MA, Napier RM, Felle HH (1993) Patch-clamp analysis establishes a role for an auxin binding protein in the auxin stimulation of plasma membrane current in Zea mays protoplasts. Plant J 4:41–46
Sachs J (1868) Lehrbuch der Botanik. Engelmann, Leipzig
Sachs J (1872) Ablenkung der Wurzel von ihrer normalen Wachstumsrichtung durch feuchte Köper. Arb D Bot Inst Würzburg 1:209–222
Sachs J (1879) Über Ausschließung der geotropischen und heliotropischen Krümmungen während des Wachsthums. Würzburger Arbeiten 2:209–225
Sachs J (1882) Vorlesungen über Pflanzen-Physiologie. Engelmann, Leipzig
Sachs J (1892) Gesammelte Abhandlungen über Pflanzen-Physiologie, vol Erster Band. Wilhem Engelmann, Leipzig
Sack FD (1991) Plant gravity sensing. Int Rev Cytol 127:193–252
Sakai T (2005) NPH3 and RPT2: signal transducers in phototropin signaling pathways. In: Wada M, Shimazaki K, Iino M (eds) Light sensing in plants. Springer, Tokyo
Sakai T, Haga K (2012) Molecular genetic analysis of phototropism in Arabidopsis. Plant Cell Physiol 53(9):1517–1534
Sakai T, Wada T, Ishiguro S, Okada K (2000) RPT2. A signal transducer of the phototropic response in Arabidopsis. Plant Cell 12(2):225–236

Sakai T, Kagawa T, Kasahara M, Swartz TE, Christie JM, Briggs WR, Wada M, Okada K (2001) Arabidopsis nph1 and npl1: blue light receptors that mediate both phototropism and chloroplast relocation. Proc Natl Acad Sci USA 98(12):6969–6974

Sakamoto K, Briggs WR (2002) Cellular and subcellular localization of phototropin 1. Plant Cell 14(8):1723–1735

Salomon M, Christie JM, Knieb E, Lempert U, Briggs WR (2000) Photochemical and mutational analysis of the FMN-binding domains of the plant blue light receptor, phototropin. Biochemistry 39(31):9401–9410

Sharrock R (1672) The history of the propagation and improvement of vegetables. W. Hall, Oxford

Shen-Miller J, Cooper P, Gordon SA (1969) Phototropism and photoinhibiiton of basipolar transport of auxin in oat coleoptiles. Plant Physiol 44:491–496

Sieberer T, Seifert GJ, Hauser MT, Grisafi P, Fink GR, Luschnig C (2000) Post-transcriptional control of the Arabidopsis auxin efflux carrier EIR1 requires AXR1. Curr Biol 10(24):1595–1598

Sievers A, Behrens HM, Buckhout TJ, Gradmann D (1984) Can a Ca2+ pump in the endoplasmic reticulum of the Lepidium root be the trigger for rapid changes in membrane potential after gravistimulation? J Plant Physiol 114(3):195–200

Sievers A, Kruse S, Kuo-Huang LL, Wendt M (1989) Statoliths and microfilaments in plant cells. Planta 179(2):275–278

Stone BB, Stowe-Evans EL, Harper RM, Celaya RB, Ljung K, Sandberg G, Liscum E (2008) Disruptions in AUX1-dependent auxin influx alter hypocotyl phototropism in Arabidopsis. Mol Plant 1(1):129–144

Swarup R, Kramer EM, Perry P, Knox K, Leyser HM, Haseloff J, Beemster GT, Bhalerao R, Bennett MJ (2005) Root gravitropism requires lateral root cap and epidermal cells for transport and response to a mobile auxin signal. Nat Cell Biol 7(11):1057–1065

Takahashi N, Goto N, Okada K, Takahashi H (2002) Hydrotropism in abscisic acid, wavy, and gravitropic mutants of Arabidopsis thaliana. Planta 216(2):203–211

Takahashi N, Yamazaki Y, Kobayashi A, Higashitani A, Takahashi H (2003) Hydrotropism interacts with gravitropism by degrading amyloplasts in seedling roots of Arabidopsis and radish. Plant Physiol 132(2):805–810

Takahashi H, Miyazawa Y, Fujii N (2009) Hormonal interactions during root tropic growth: hydrotropism versus gravitropism. Plant Mol Biol 69(4):489–502

Takahashi K, Hayashi K, Kinoshita T (2012) Auxin activates the plasma membrane H+-ATPase by phosphorylation during hypocotyl elongation in Arabidopsis. Plant Physiol 159(2):632–641

Takano M, Takahashi H, Suge H (1997) Calcium requirement for the induction of hydrotropism and enhancement of calcium-induced curvature by water stress in primary roots of pea, Pisum sativum L. Plant Cell Physiol 38:385–391

Tatematsu K, Kumagai S, Muto H, Sato A, Watahiki MK, Harper RM, Liscum E, Yamamoto KT (2004) MASSUGU2 encodes Aux/IAA19, an auxin-regulated protein that functions together with the transcriptional activator NPH4/ARF7 to regulate differential growth responses of hypocotyl and formation of lateral roots in Arabidopsis thaliana. Plant Cell 16(2):379–393

Terasaka K, Blakeslee JJ, Titapiwatanakun B, Peer WA, Bandyopadhyay A, Makam SN, Lee OR, Richards EL, Murphy AS, Sato F, Yazaki K (2005) PGP4, an ATP binding cassette P-glycoprotein, catalyzes auxin transport in Arabidopsis thaliana roots. Plant Cell 17 (11):2922–2939

Tillmann U, Viola G, Kayser B, Siemeister G, Hesse T, Palme K, Lobler M, Klambt D (1989) cDNA clones of the auxin-binding protein from corn coleoptiles (Zea mays L.): isolation and characterization by immunological methods. EMBO J 8(9):2463–2467

Timpte C, Lincoln C, Pickett FB, Turner J, Estelle M (1995) The AXR1 and AUX1 genes of Arabidopsis function in separate auxin-response pathways. Plant J 8(4):561–569

Treml BS, Winderl S, Radykewicz R, Herz M, Schweizer G, Hutzler P, Glawischnig E, Ruiz RA (2005) The gene ENHANCER OF PINOID controls cotyledon development in the Arabidopsis embryo. Development 132(18):4063–4074

Tsugeki R, Fedoroff NV (1999) Genetic ablation of root cap cells in Arabidopsis. Proc Natl Acad Sci USA 96(22):12941–12946

Utsuno K, Shikanai T, Yamada Y, Hashimoto T (1998) Agr, an Agravitropic locus of Arabidopsis thaliana, encodes a novel membrane-protein family member. Plant Cell Physiol 39(10):1111–1118

Vieten A, Vanneste S, Wisniewska J, Benkova E, Benjamins R, Beeckman T, Luschnig C, Friml J (2005) Functional redundancy of PIN proteins is accompanied by auxin-dependent cross-regulation of PIN expression. Development 132(20):4521–4531

Wan Y, Jasik J, Wang L, Hao H, Volkmann D, Menzel D, Mancuso S, Baluska F, Lin J (2012) The signal transducer NPH3 integrates the phototropin1 photosensor with PIN2-based polar auxin transport in Arabidopsis root phototropism. Plant Cell 24(2):551–565

Wang JW, Wang LJ, Mao YB, Cai WJ, Xue HW, Chen XY (2005) Control of root cap formation by MicroRNA-targeted auxin response factors in Arabidopsis. Plant Cell 17(8):2204–2216

Weijers D, Benkova E, Jager KE, Schlereth A, Hamann T, Kientz M, Wilmoth JC, Reed JW, Jurgens G (2005) Developmental specificity of auxin response by pairs of ARF and Aux/IAA transcriptional regulators. EMBO J 24(10):1874–1885

Went FW (1926) On growth accelerating substances in the coleoptile of Avena sativa. Proc Koninklijke Nederlandse Akademie Wetenschappen 30:10–19

Went FW (1928) Wuchsstoff und Wachstum. Receuil des Travaux Botaniques Neerlandais 25:1–116

Went FW, Thimann KV (1937) Phytohormones. Macmillan, New York

Whippo CW, Hangarter RP (2006) Phototropism: bending towards enlightenment. Plant Cell 18 (5):1110–1119

Wiesner J (1884) Untersuchungen über die Wachstumsbewegungen der Wurzeln. Sitzungsber Akad Wiss Wien Math-Naturwiss Kl I 89:223–302

Wu G, Cameron JN, Ljung K, Spalding EP (2010) A role for ABCB19-mediated polar auxin transport in seedling photomorphogenesis mediated by cryptochrome 1 and phytochrome B. Plant J 62(2):179–191

Yan DW, Wang J, Yuan TT, Hong LW, Gao X, Lu YT (2013) Perturbation of auxin homeostasis by overexpression of wild-type IAA15 results in impaired stem cell differentiation and gravitropism in roots. PLoS One 8(3):e58103

Yang H, Murphy AS (2009) Functional expression and characterization of Arabidopsis ABCB, AUX 1 and PIN auxin transporters in Schizosaccharomyces pombe. Plant J 59(1):179–191

Yang Y, Hammes UZ, Taylor CG, Schachtman DP, Nielsen E (2006) High-affinity auxin transport by the AUX1 influx carrier protein. Curr Biol 16(11):1123–1127

Young LM, Evans ML, Hertel R (1990) Correlations between gravitropic curvature and auxin movement across gravistimulated roots of Zea mays. Plant Physiol 92:792–796

Zhang J, Vanneste S, Brewer PB, Michniewicz M, Grones P, Kleine-Vehn J, Lofke C, Teichmann T, Bielach A, Cannoot B, Hoyerova K, Chen X, Xue HW, Benkova E, Zazimalova E, Friml J (2011) Inositol trisphosphate-induced $Ca^{2+}$ signaling modulates auxin transport and PIN polarity. Dev Cell 20(6):855–866

# Chapter 17
# Auxin Coordinates Shoot and Root Development During Shade Avoidance Response

**Valentino Ruzza, Giovanna Sessa, Massimiliano Sassi, Giorgio Morelli, and Ida Ruberti**

**Abstract** Plants have evolved sophisticated mechanisms to sense the presence of other plants growing nearby and adjust their growth rate accordingly. The early perception of neighbor proximity depends on the detection of light quality changes. Within a vegetation community, the ratio of red (R) to far-red (FR) light is lowered by the absorption of R light by photosynthetic pigments. This light quality change is perceived through phytochrome (phyB, phyD, and phyE in Arabidopsis) as a signal of the proximity of neighbors and induces a suite of developmental responses (termed the shade avoidance response). In Arabidopsis shade avoidance is regulated by a balance of positive (PIF) and negative (HFR1/SICS1) regulators of gene expression which ensures a fast reshaping of the plant body toward an environment optimal for growth while at the same time avoiding an exaggerated reaction to low R/FR. Persistency of a low R/FR signal enhances the activity of phyA and, in turn, of HY5, a master regulator of seedling de-etiolation. Several hormones, such as gibberellins and brassinosteroids, have been implicated in shade-induced elongation. However, a compelling amount of evidence indicates that low R/FR-induced changes in auxin homeostasis and auxin transport are central in the shade avoidance response. This chapter describes the recent advances in understanding how auxin coordinates plant growth in a low R/FR light environment.

V. Ruzza • G. Sessa • I. Ruberti (✉)
Institute of Biology, Molecular Medicine and NanoBiotechnology, National Research Council, P.le A. Moro 5, 00185, Rome, Italy
e-mail: ida.ruberti@uniroma1.it

M. Sassi
Institute of Biology, Molecular Medicine and NanoBiotechnology, National Research Council, P.le A. Moro 5, 00185, Rome, Italy

Laboratoire de Reproduction et Développement des Plantes, École Normale Supérieure de Lyon 46, Allée d'Italie, 69364, Lyon, CEDEX 07, France

G. Morelli
Food and Nutrition Research Centre, Agricultural Research Council, Via Ardeatina 546, 00178, Rome, Italy

E. Zažímalová et al. (eds.), *Auxin and Its Role in Plant Development*,
DOI 10.1007/978-3-7091-1526-8_17, © Springer-Verlag Wien 2014

# 1 Plant Responses to Light

To grow and develop properly, all organisms need to perceive and process information from their environment. As sessile organisms, plants have evolved sophisticated mechanisms to sense, respond, and adapt to fluctuating environmental cues. Light is likely the most important environmental factor influencing plant growth and development. It provides energy for photosynthesis, but also circadian, seasonal, and positional information critical for plant survival and reproduction.

Developmental responses to light, although especially evident at the seedling stage, occur throughout the life of the plant. Depending on whether germination occurs in darkness or in light, angiosperms choose between two distinct developmental pathways: skotomorphogenesis, also known as etiolation, and photomorphogenesis (Arsovski et al. 2012). When germination occurs underground, in the absence of light, seedling development is characterized by rapid hypocotyl elongation, slow root growth, and unexpanded cotyledons which enclose an inactive shoot apical meristem (SAM). The seedling utilizes its energy resources to rapidly emerge from below the soil surface and reach the light. Under sunlight, the photomorphogenic pattern is followed rapidly establishing the seedling as a photoautotrophic organism. The plant's energy is essentially used for leaf and root development, while longitudinal extension growth is minimized. However, in shade-avoiding plants the photoautotrophic seedling rapidly gains the capacity to perceive the presence of neighboring vegetation and compete for the light resource by stimulation of elongation growth (Casal 2013; Franklin 2008; Ruberti et al. 2012).

## 1.1 *Shade Avoidance Response*

Plants have evolved two opposing strategies in response to competition for light: shade tolerance and shade avoidance. Angiosperms have an impressive capacity to avoid shade. Daylight contains roughly equal proportions of red (R) and far-red (FR), but within vegetation that ratio is lowered as a result of R absorption by photosynthetic pigments. The reduction in the R/FR ratio is perceived through the phytochrome photoreceptors as an early signal of neighbor proximity resulting in a suite of developmental responses known as shade avoidance. The most dramatic response to low R/FR light is the stimulation of elongation growth. This response is remarkably rapid, with a lag phase of a few minutes, and its magnitude inversely relates to the R/FR ratio. In dicotyledonous plants, elongation growth induced by low R/FR is often associated with a reduction of leaf development. Root growth is also often impaired in low R/FR light environments. In the long term, low R/FR exposure leads to early flowering with a reduced seed set, which is considered an escape mechanism because it shortens generation time. All of these responses occur both in natural dense communities and in shade simulations (low R/FR).

Furthermore, similar responses are induced by exposing plants to horizontal FR radiation with white light from above. This is because shade-avoiding plants are able to perceive light reflected by neighboring plants as partially depleted of the R wavelengths, and they can activate responses to avoid shade even before canopy closure and actual shading occurs (Casal 2013; Franklin 2008; Ruberti et al. 2012). However, at high canopy density multiple light signals control shade avoidance response (Ballaré 1999). There is evidence that both low R/FR and reduced blue (B) light are required for full expression of shade avoidance in plant canopies. Interestingly, the B light responses seem to be mediated through pathways that showed only limited overlap with those activated by low R/FR (Keller et al. 2011; Keuskamp et al. 2011).

## 2 Perception of Shade Light Signals by Photoreceptors

An ever-increasing body of evidence highlights the significance of the reduction in the R/FR ratio as a signal that triggers shade avoidance response. However, the reduced irradiance and the blue/green ratio of shade also provide signals which are important for plant responses to canopy light, and phoreceptors other than phytochromes—cryptochromes, phototropins, UV RESISTANCE LOCUS 8—are increasingly being implicated to play a role in perceiving differences between light and shade (Casal 2013).

### *2.1 Phytochromes*

Phytochromes are photochromic biliproteins that exist in two photo-convertible isoforms: a R-light-absorbing form (Pr) and a FR-light-absorbing form (Pfr). They exist as dimers with each monomer consisting of an apoprotein covalently attached to a tetrapyrrole chromophore, phytochromobilin. Phytochromes are synthesized in the dark in their inactive Pr form. Upon absorption of red light, Pr is converted into the biological active Pfr form which can absorb FR light and switch back to Pr, resulting in a dynamic photoequilibrium between the two forms of phytochrome. Following conversion to the Pfr form, phytochromes translocate to the nucleus (Bae and Choi 2008; Casal 2013).

The phytochrome apoproteins are encoded by a small gene family in most plant species. Three phytochrome encoding genes, *PHYA*, *PHYB*, and *PHYC*, are conserved in angiosperms (Mathews 2006). Additional *PHY* genes are found in dicots, perhaps the products of more recent duplications within the *PHYB* lineage (Mathews and Sharrock 1997). In Arabidopsis, the phytochrome apoproteins are encoded by five genes, *PHYA-PHYE*. *PHYE* is thought to have originated from a duplication within the *PHYB* lineage early on in the evolution of dicots. *PHYD*, which encodes a protein that shares approximately 80 % amino acid sequence identity with PHYB, apparently arose from a relatively recent gene duplication within the *Brassicaceae* (Mathews and Sharrock 1997). *PHYC* is thought to have originated from a duplication within the *PHYA* lineage (Mathews and Sharrock 1997). However, phyA is the only phytochrome that is rapidly degraded in its Pfr form and can signal during rapid photoconversion between Pr and Pfr form. It is the predominant phytochrome in etiolated seedlings, and it plays a major role in the rapid promotion of de-etiolation upon emergence from the soil. All the other phytochromes are relatively stable in the Pfr form, and they control several aspects of plant growth and development (Bae and Choi 2008; Casal 2013; Franklin and Quail 2010).

A major function of phytochromes in photoautotrophic seedlings and in adult plants involves the perception of changes in the R/FR ratio of the light environment. Among the light-stable phytochromes, phyB plays a central role in shade avoidance. Arabidopsis *phyB* loss-of-function mutants constitutively display shade avoidance traits such as elongated hypocotyl, stem, petioles, and leaves, acceleration of flowering, and higher apical dominance under high R/FR light (Reed et al. 1993). However, *phyB* mutants also show typical shade avoidance responses under low R/FR light, indicating that other phytochromes contribute to these responses (Franklin and Quail 2010; Smith and Whitelam 1997). *phyD* and *phyE* single loss-of-function mutants are essentially indistinguishable from wild-type plants. However, *phyB phyE*, and, to a lesser extent, *phyB phyD* double mutants had longer petioles and flower earlier than *phyB* mutants in high R/FR light. This led to the proposal that in conjunction with phyB, phyD and phyE function in the regulation of shade avoidance responses (Aukerman et al. 1997; Devlin et al. 1998, 1999). In agreement, *phyB phyD phyE* triple mutants show no response to low R/FR light (Franklin et al. 2003). On the other hand, phyA seems to attenuate the elongation response induced by low R/FR light (Devlin et al. 2003; Johnson et al. 1994; Wang et al. 2011; Ciolfi et al. 2013).

## 3 PIF Proteins in the Control of the Shade Avoidance Response

In the nucleus, phytochromes physically interact with a subfamily of basic helix-loop-helix (bHLH) proteins, the PHYTOCHROME INTERACTING FACTORS (PIFs), controlling several aspects of photomorphogenesis (Castillon et al. 2007; Jiao et al. 2007; Leivar and Quail 2011). This interaction in turn leads to PIF's phosphorylation, ubiquitination, and degradation via the 26S proteasome, providing an elegant mechanism for rapid regulation of gene expression in response to changes in the light environment (Al-Sady et al. 2006; Bauer et al. 2004; Nozue et al. 2007; Park et al. 2004; Shen et al. 2005, 2007).

### 3.1 PIF Proteins Promote the Shade Avoidance Response

Signaling downstream of the photoreceptors involves two main pathways: CONSTITUTIVE PHOTOMORPHOGENIC 1 (COP1)-ELONGATED HYPOCOTYL 5 (HY5) and PIFs. COP/DE-ETIOLATED(DET)/FUSCA(FUS) are central repressors of photomorphogenesis, which, in the dark, function in concert to target positive regulators of photomorphogenesis (i.e., HY5) for degradation through the 26S proteasome, thus preventing de-etiolation. In daylight, the activity of COP/DET/FUS proteins is reduced resulting in the accumulation of transcription factors required for photomorphogenesis. COP1, one of the COP/DET/FUS proteins, is an E3 ligase that interacts with several transcription factors and promotes their ubiquitination together with the SUPPRESSOR OF PHYA-105 (SPA) proteins (Lau and Deng 2010). On the other hand, a *pif1 pif3 pif4 pif5* quadruple (*pifq*) mutant displays a *cop*-like phenotype in darkness, demonstrating that these PIF transcription factors function in the dark to promote skotomorphogenesis (Leivar et al. 2008a; Shin et al. 2009). Upon light exposure, photoactivated phytochromes interact with PIF proteins and promote their degradation via the ubiquitin-proteasome pathway (Leivar and Quail 2011). Interestingly, the stability of PIF3 is dependent on COP1 and SPA proteins, suggesting that these molecules, besides targeting photomorphogenesis-promoting transcription factors for degradation, may positively affect PIF protein levels in the dark (Bauer et al. 2004; Leivar et al. 2008a). The rapid, light-induced degradation of PIF transcription factors does not lead to their disappearance, but rather it results in a lower steady-state level of these proteins in daylight (Leivar and Quail 2011).

PIF1, PIF3, PIF4, PIF5, and PIF7 have been demonstrated to directly contribute to shade avoidance (Hornitschek et al. 2012; Leivar et al. 2012a; Li et al. 2012; Lorrain et al. 2008). They all interact physically with phyB through the conserved N-terminal sequence, called the active phyB-binding motif; PIF1 and PIF3 also interact with phyA through a distinct motif (Leivar and Quail 2011). As a result, PIF1, PIF3, PIF4, and PIF5 become phosphorylated and degraded via the ubiquitin-proteasome system, with degradation half-times in the range of 5–20 min (Leivar and Quail 2011). PIF3, PIF4, and PIF5 protein levels increase rapidly in photoautothrophic seedlings upon exposure to low R/FR light (Leivar et al. 2012a; Lorrain et al. 2008). Unlike its close relatives, PIF7 is not rapidly degraded in high R/FR light (Leivar et al. 2008b). However, this PIF protein accumulates in its dephosphorylated form in shade, suggesting the existence of a protein phosphatase and a protein kinase whose activities or availability is regulated by light quality changes (Li et al. 2012). Shade-induced elongation response is significantly attenuated in *pif4 pif5* and, to an even greater extent, in *pifq* and *pif7* mutants (Leivar et al. 2012a; Li et al. 2012; Lorrain et al. 2008). Conversely, PIF4 and PIF5 overexpressing seedlings have constitutively long hypocotyls and petioles (Lorrain et al. 2008).

In high R/FR light, PIF proteins negatively regulate phyB levels promoting the polyubiquitination of active phyB by COP1, which in turn leads to degradation of phyB through the proteasome (Jang et al. 2010; Leivar et al. 2008b). In agreement, *pifq* mutants contain significantly higher levels of PHYB than wild-type seedlings in high R/FR light (Leivar et al. 2012b). However, when seedlings grown under high R/FR are exposed to simulated shade, the PHYB levels increase and become largely independent of PIFs (within 12 h), thus suggesting that PIFs contribute directly to trigger shade avoidance through their transcriptional regulatory activity rather than via feedback regulation of phyB abundance (Leivar et al. 2012b).

### 3.2 PIF Proteins Directly Regulate Transcription Factor Genes Promoting Shade Avoidance

Consistent with the rapidity of the elongation growth response to low R/FR and its reversibility upon perception of high R/FR, changes in gene expression are very rapid and reversible (Carabelli et al. 1996; Salter et al. 2003). The transcript levels of the Homeodomain-Leucine Zipper (HD-Zip) *ARABIDOPSIS THALIANA HOMEOBOX2* (*ATHB2*) and bHLH *PIF3-LIKE1* (*PIL1*) transcription factor genes, functionally implicated in the elongation response provoked by light quality changes (Salter et al. 2003; Steindler et al. 1999), increase within a few minutes of low R/FR exposure (Carabelli et al. 1996; Salter et al. 2003). Significantly, *ATHB2* and *PIL1* transcript levels fall very rapidly after transfer from low R/FR to high R/FR (Carabelli et al. 1996; Salter et al. 2003). phyB, phyD, and phyE are all involved in the regulation of both *ATHB2* and *PIL1* by light quality changes, further indicating the redundant action of these photoreceptors in regulation of shade avoidance (Franklin et al. 2003; Salter et al. 2003).

*ATHB2* and *PIL1* induction by low R/FR does not require de novo protein synthesis (Roig-Villanova et al. 2006) and is significantly reduced in loss-of-function *pif* mutants (*pif1 pif3*; *pif4 pif5; pif7*; Hornitschek et al. 2009; Leivar et al. 2012a; Li et al. 2012; Lorrain et al. 2008). There is evidence that *PIL1* and *ATHB2* are recognized in vivo by PIF4 and PIF5 (de Lucas et al. 2008; Hornitschek et al. 2009, 2012), and physical interaction between *PIL1* promoter and PIF7 has also been reported (Li et al. 2012).

*ATHB2* is a member of the HD-Zip II family consisting of 10 genes. Phylogeny reconstruction revealed that almost all of the HD-Zip II genes can be subdivided into 4 clades (α–δ), each clade comprising 2–3 paralogs (Ciarbelli et al. 2008). All the γ [*ATHB2*, *HOMEOBOX ARABIDOPSIS THALIANA 1* (*HAT1*), *HAT2*] and δ genes (*ATHB4*, *HAT3*) are regulated by light quality changes that induce shade avoidance, and kinetics of induction and low R/FR/high R/FR reversibility strongly suggest that *HAT1*, *HAT3*, and *ATHB4*, as *ATHB2*, are under the control of the phytochrome system (Ciarbelli et al. 2008).

Transgenic plants bearing constructs that alter *ATHB2* expression display a series of interesting developmental phenotypes (Schena et al. 1993; Steindler et al. 1999). For example, seedlings overproducing ATHB2 have longer hypocotyls and petioles and smaller and fewer leaves. Moreover, these seedlings also have a thinner root mass; that is, they produce less lateral roots than wild-type controls. The phenotypes of adult transgenic plants are similar to those of seedlings but more exaggerated. Altogether the phenotypes of plants overexpressing ATHB2 are reminiscent of those displayed by wild-type plants grown in low R/FR light, further suggesting a role for this HD-Zip protein in the regulation of shade avoidance response (Morelli and Ruberti 2000, 2002; Steindler et al. 1999). Overexpression of *HAT1*, *HAT2*, initially isolated as an auxin-inducible gene by DNA microarray (Sawa et al. 2002), *HAT3*, and *ATHB4* results in phenotypes similar to those caused by elevated levels of ATHB2 (Ciarbelli et al. 2008; Ruberti et al. 2012; Sawa et al. 2002; Sorin et al. 2009), further suggesting a redundant function of these transcription factors in shade avoidance.

Interestingly, members of the HD-Zip II $\gamma$ and $\delta$ subfamilies also seem to play a central role in several aspects of plant development in a high R/FR environment. *ATHB2, ATHB4, HAT1* (also known as *JAIBA*; Zúñiga-Mayo et al. 2012), and HAT3 were indeed recently identified as genes positively regulated by SPATULA, a bHLH protein related to PIFs but lacking an active phytochrome binding domain involved in their negative regulation by the phytochrome in high R/FR, and proposed to be involved in carpel margin development (Reymond et al. 2012). In addition, there is evidence that ATHB2, ATHB4, and HAT3 control embryonic apical patterning and SAM function at least in part through interaction with HD-Zip III proteins (Turchi et al. 2013). Importantly, they seem to play a critical role in promoting auxin transport and auxin responses during embryogenesis (Turchi et al. 2013). Finally, simultaneous lack of ATHB4 and HAT3 is associated with the loss of adaxial identity of lateral organs (Bou-Torrent et al. 2012; Turchi et al. 2013).

The dual role of HD-Zip II proteins in development and in shade avoidance suggests that the spatial expression pattern of these transcription factors may change when plants perceive a low R/FR light signal. In agreement, ATHB2 is indeed induced by low R/FR in cell types that do not normally express this protein. Under a high R/FR light environment, ATHB2 expression is mainly localized in provascular cells in either the embryo or leaf primordia (Turchi et al. 2013). Low R/FR light rapidly induces ATHB2:GUS expression in all cell layers of the elongating portion of the hypocotyl and cotyledon petioles (Fig. 17.1a), thus suggesting that ATHB2 acts, at least in part, in these organs to control shade avoidance. Consistent with the transient induction of *ATHB2* by light quality changes (Sessa et al. 2005), both ATHB2:GUS and ATHB2:GFP protein levels decrease upon prolonged exposure to low R/FR (Fig. 17.1a, b), implying that ATHB2 may also be regulated at the level of protein stability.

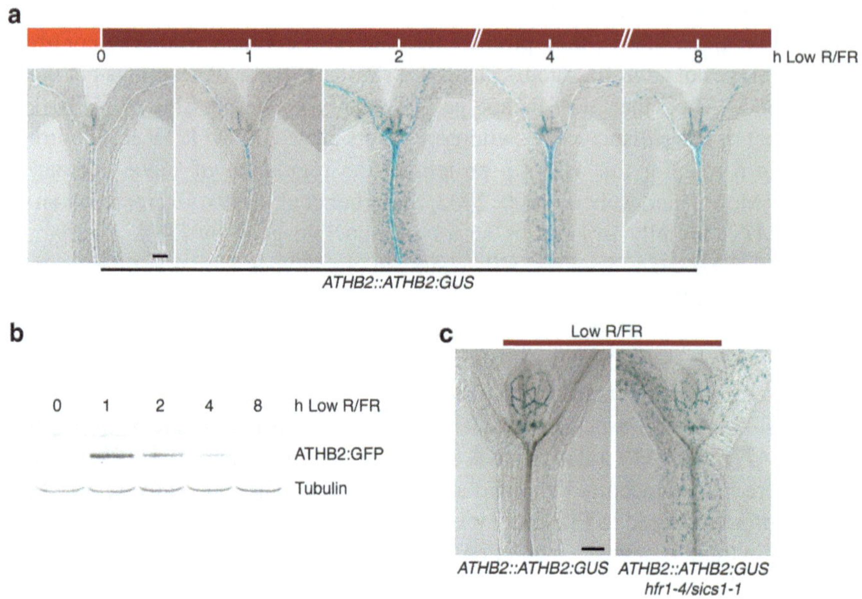


**Fig. 17.1** ATHB2 is rapidly and transiently induced by low R/FR. (**a**) Time-course analysis of the histochemical localization of GUS activity in ATHB2::ATHB2:GUS seedlings grown for 4 days in a light (L)/dark (D)cycle (16/8 h) in high R/FR and then exposed to low R/FR under the same L/D regimen for the indicated times. Bar, 0.1 mm. (**b**) Time-course analysis of ATHB2:GFP protein levels in response to low R/FR. ATHB2::ATHB2:GFP seedlings were grown for 4 days in a light (L)/dark (D)cycle (16/8 h) in high R/FR and then exposed to low R/FR for the indicated times before immunoblot analysis with GFP antibody (sc-9996 HRP, Santa Cruz Biotech). Tubulin was used as a loading control. (**c**) Histochemical localization of GUS activity in ATHB2::ATHB2:GUS and ATHB2::ATHB2:GUS *hfr1-4/sics1-1* seedlings grown for 5 days in a L/D cycle (16/8 h) in high R/FR and then exposed to low R/FR under the same L/D regimen for 1 day. Bar, 0.1 mm. Growth conditions, light settings, and GUS staining procedure were as previously reported (Carabelli et al. 2007; Sessa et al. 2005). Immunoblot analysis was performed as described by Jang et al. (2005)

## *3.3 PIF Proteins Directly Regulate Atypical bHLH Factor Genes Attenuating Shade Avoidance*

By exploiting mutant analysis in combination with genome-wide expression profiling, Sessa et al. (2005) uncovered a negative regulatory mechanism active in low R/FR that involves *HYPOCOTYL FAR RED1/SLENDER IN CANOPY SHADE1* (*HFR1/SICS1*). HFR1/SICS1 acting as a negative controller of the shade avoidance response ensures that in the presence of a persistent low R/FR light signal an exaggerated plant response does not occur (Hornitschek et al. 2009; Sessa et al. 2005). *HFR1/SICS1* was originally described as a downstream component of phyA and cryptochrome 1 in the de-etiolation process (Duek and Fankhauser 2003; Fairchild et al. 2000; Fankhauser and Chory 2000; Soh et al. 2000). However, there is evidence that in photoautotrophic seedlings *HFR1/SICS1* is largely regulated through phyB and that its promoter is directly recognized by PIF5 (Hornitschek et al. 2009; Ciolfi et al. 2013). The *HFR1/SICS1* transcript increases significantly upon exposure to low R/FR (15–30 fold; Sessa et al. 2005). HFR1/SICS1 is a short-lived protein in darkness and is degraded through a 26S proteasome-dependent pathway and this process requires COP1. Light, however, irrespective of its quality, enhances HFR1/SICS1 protein stability (Duek et al. 2004; Jang et al. 2005; Yang et al. 2005).

HFR1/SICS1 encodes an atypical bHLH protein and acts as a HLH inhibitor. Upon prolonged exposure to low R/FR, HFR1/SICS1 accumulates and interacts with PIF4 and PIF5 forming non-DNA-binding heterodimers, thus limiting PIF-mediated gene expression (Hornitschek et al. 2009). Consistent with this, several genes rapidly and transiently induced by low R/FR are significantly upregulated in *hfr1/sics1* loss-of-function mutants upon prolonged exposure to simulated shade (Sessa et al. 2005; Fig. 17.1c).

Another atypical bHLH protein gene, *HELIX LOOP HELIX1/PHYTOCHROME RAPIDLY REGULATED1* (*HLH1/PAR1*; Roig-Villanova et al. 2006; Sessa et al. 2005), is also rapidly regulated by low R/FR, and its induction does not require de novo protein synthesis. HLH1/PAR1 has also been involved in negative regulation of shade-induced elongation and proposed to act as a dominant-negative antagonist of conventional bHLH transcription factors (Galstyan et al. 2011; Hao et al. 2012; Roig-Villanova et al. 2007).

# 4 COP1 in the Control of the Shade Avoidance Response

*cop1* mutants lack the low R/FR-induced elongation response in young seedlings, thus implying a central role for COP1 in shade avoidance (McNellis et al. 1994). In agreement, the low R/FR induction of several transcription factor genes (i.e., *ATHB2*, *PIL1*) rapidly regulated by light quality changes is reduced in weak *cop1* alleles (Roig-Villanova et al. 2006).

More recently, Rolauffs et al. (2012) provided evidence that both *COP1* and the four *SPA* genes (*SPA1-SPA4*) are essential for hypocotyl and leaf petiole elongation in response to low R/FR, in a fashion that seems to involve the COP1/SPA ubiquitination target HFR1/SICS1 but not HY5. This led to the proposal that COP1/SPA activity may be important in modulating HFR1/SICS1 protein levels in low R/FR and thus in turn PIF activity (Rolauffs et al. 2012).

The *cop1* mutant does respond to shade in the *b-box domain protein 21* (*bbx21*) *bbx22* double mutant background (Crocco et al. 2010). The *BBX21* (also known as *SALT TOLERANCE HOMOLOG 2*) gene, encoding a B-box-containing zinc-finger transcription factor, plays a central role during seedling de-etiolation and functions as a negative regulator of the shade avoidance response (Crocco et al. 2010; Datta et al. 2007; Khanna et al. 2009). BBX22, a homolog of BBX21, operates additively with BBX21 in early seedling development, inducing hypocotyl inhibition, anthocyanin accumulation, and chloroplast biogenesis (Chang et al. 2008; Datta et al. 2008). BBX22 is degraded through the proteasome; this reaction is faster in darkness than in light and requires COP1 (Chang et al. 2011). This further suggests that COP1 may be required in low R/FR to control protein levels of negative regulators of shade avoidance.

# 5 Hormones in the Control of the Shade Avoidance Response

An ever-increasing body of evidence shows that plant responses to low R/FR involve changes in hormone signaling. Several hormones, such as auxin, gibberellins (GA), and brassinosteroids (BR), have been functionally involved in the promotion of elongation growth by low R/FR (Casal 2013; Ruberti et al. 2012; Stamm and Kumar 2010), and links between these hormones and PIF proteins have been established (Casal 2013). There is also evidence of a high degree of interaction between the different hormonal pathways (Depuydt and Hardtke 2011). The interactions may involve the regulation of the homeostasis of another hormone and/or the shared participation of signaling factors in more than one pathway (Depuydt and Hardtke 2011; Jaillais and Chory 2010; Kuppusamy et al. 2009; see Chap. 12). For example, auxin regulates GA biosynthesis (Frigerio et al. 2006), and DELLA proteins negatively regulate both GA and BR signaling. GA promotes plant growth by removing DELLA proteins. Binding of GA to its nuclear receptor GA-INSENSITIVE DWARF 1 (GID1) enhances GID1–DELLA interaction and association with the E3 ubiquitin ligase $SCF^{SLY1/GID2}$, leading to polyubiquitylation and degradation of DELLAs (Murase et al. 2008; Sun 2011). When GA levels are low, DELLAs accumulate and directly inactivate several transcription factors, including PIF3 and PIF4 (de Lucas et al. 2008; Feng et al. 2008; Harberd et al. 2009; Sun 2011). Low R/FR reduces the stability of DELLA proteins, likely as a consequence of increased GA levels (Djakovic-Petrovic et al. 2007). Furthermore, there is evidence that enhanced DELLA stability [i.e., *gibberellic acid-insensitive* dominant mutant] inhibits shade-induced elongation, suggesting that DELLAs constrain shade avoidance (Djakovic-Petrovic et al. 2007). BR is perceived by the receptor kinase Brassinosteroid-Insensitive 1 (BRI1), and downstream signal transduction leads to activation of the BRASSINAZOLE-RESISTANT 1 (BZR1) family transcription factors, which control BR-responsive gene expression (Kim and Wang 2010). BZR1 and PIF4 physically interact and synergistically regulate common target genes, including the PACLOBUTRAZOL RESISTANCE family HLH factors required for promoting cell elongation (Oh et al. 2012). DELLAs negatively regulate BR signaling by interacting with BZR1 and inhibiting its ability to bind to target genes (Bai et al. 2012; Gallego-Bartolomé et al. 2012). These findings lead to the suggestion that DELLAs, BZR1/2, and PIFs form the central command system in the control of the elongation growth processes (Bai et al. 2012), potentially including those occurring during shade avoidance (Casal 2013).

Auxin however is unique among plant hormones for exhibiting polar transport, and a key role of auxin in triggering shade avoidance is clearly emerging. Auxin-related genes are overrepresented among those rapidly and transiently induced by low R/FR (Devlin et al. 2003; Hornitschek et al. 2012; Sessa et al. 2005; Tao et al. 2008; Ciolfi et al. 2013). The expression of the synthetic auxin-inducible marker *DR5::GUS* (Ulmasov et al. 1997) is promoted in the cotyledons and hypocotyl by low R/FR (Carabelli et al. 2007; Morelli and Ruberti 2002; Salisbury et al. 2007; Tao et al. 2008). However, the expression of auxin-responsive genes seems to be enhanced by low R/FR in a tissue-specific manner. For example, *AUXIN/INDOLE-3-ACETIC ACID* (AUX/IAA) *19*, a gene implicated in phototropic growth (Tatematsu et al. 2004), is strongly induced in the vasculature of cotyledon petioles and hypocotyl upon low R/FR exposure (Pierik et al. 2009; Ruberti et al. 2012). Significantly, auxin-related genes are upregulated in *hfr1/sics1* loss-of-function mutants upon prolonged exposure to low R/FR (Sessa et al. 2005; Ciolfi et al. 2013; Fig. 17.2). Importantly, besides promoting cell elongation, auxin controls other aspects of shade avoidance, such as reduced leaf and root growth, thus coordinating plant growth in low R/FR (Carabelli et al. 2007, 2008; Morelli and Ruberti 2000, 2002; Sassi et al. 2012, 2013).

# 6 Auxin Has a Central Role in Many Aspects of the Shade Avoidance Response

Several features of plants grown in low R/FR light, such as increased elongation of stem-like organs (including hypocotyl and petioles), reduced leaf growth, leaf hyponasty, and apical dominance, are characteristic of high auxin levels. Conversely, other aspects of the shade avoidance response, such as reduced root growth and development, are typical of plants with low auxin content.

## *6.1 Auxin and Hypocotyl Growth*

Low R/FR light produces distinct but coordinated effects on different cell types within an organ. The increase in extension growth of a seedling in low R/FR is indeed the consequence of two events: a change in the orientation of cell expansion toward elongation in cells that do not divide, as the epidermal and cortical cells in the hypocotyl, and the inhibition of cambial cell proliferation that contributes to radial growth. Auxin has a central role in cell division and elongation, and therefore it was suggested to act as a coordinator of organ growth in low R/FR. Consistent with this suggestion, *auxin resistant 1* mutant seedlings do not elongate

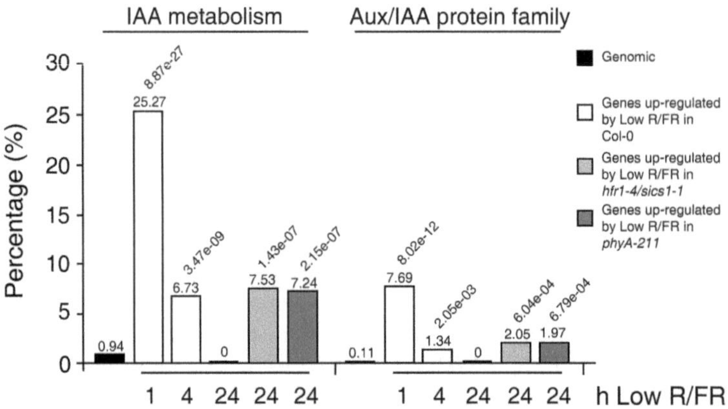


**Fig. 17.2** Dynamic regulation of auxin genes during shade avoidance. Histograms show percentage and statistical overrepresentation of auxin metabolism ("IAA metabolism") and auxin signaling ("Aux/IAA protein family") functional classes among the genes upregulated after 1 h (ArrayExpress, E-MEXP-443), 4 h (ArrayExpress, E-MEXP-3266), and 24 h (ArrayExpress, E-MEXP-3266) in Low R/FR in Col-0 and after 24 h in the same light environment in *hfr1-4/sics1-1* (ArrayExpress, E-MEXP-444) and *phyA-211* (ArrayExpress, E-MEXP-3267) mutants. All Affymetrix GeneChips were normalized together in a single normalization step by mean of GeneSpring implementation of RMA algorithm. Statistically significantly regulated genes were defined by fold change >2 on expression level and FDR < 0.05 (one-way ANOVA, Student's *t*-test with Benjamini and Hochberg correction). For each group, statistical overrepresentation analyses of functional classes were carried out using the ORA of MapMan/PageMan webtool. The percentage of each class is compared with that obtained considering all genes present on the ATH1 Chip (indicated as genome) in terms of Fisher's Exact Test. P-values are shown above the bars

significantly in low R/FR, and the hypocotyl response to simulated shade is blocked by the addition of the auxin transport inhibitor 1-N-naphthylphthalamic acid (NPA). These results lead to a model for Arabidopsis shade-induced responses. The model postulated that low R/FR light may produce a reorientation of the auxin transport stream through a spatial redistribution of a specific auxin efflux carrier protein or the activation of regulatory protein(s) controlling specific auxin efflux carrier protein(s) or both (see Chaps. 5 and 8). A higher lateral transport of auxin in the hypocotyl may promote the elongation of epidermal and cortical cells; on the other hand, a reduction of polar auxin transport through the central cylinder is likely to cause a let up in vascular differentiation and a decrement in auxin concentration reaching the root. This, in turn, may result in a reduction of lateral root formation and, eventually, primary root growth. In support of this hypothesis is the root phenotype of seedlings overexpressing ATHB2 in high R/FR. Primary root growth and lateral root formation are both inhibited by elevated levels of ATHB2, and at least the lateral root phenotype of ATHB2 seedlings is rescued by exogenous IAA (Morelli and Ruberti 2000, 2002; Steindler et al. 1999).

## 6.2 Auxin and Leaf Growth

In dicotyledonous plants, elongation growth induced by low R/FR is often associated with a reduction of leaf development. In Arabidopsis, there is evidence that cell number contributes to the reduced leaf size of plants grown under simulated shade (Carabelli et al. 2007). Low R/FR rapidly and transiently reduces the frequency of cell division in young leaf primordia through a non-cell-autonomous mechanism that requires the action of the auxin receptor TRANSPORT INHIBITOR RESISTANT 1 (TIR1). The auxin increase perceived through TIR1 induces *CYTOKININ OXIDASE/DEHYDROGENASE 6* (*CKX6*), a gene encoding an enzyme involved in cytokinin degradation (Redman et al. 2004; Werner et al. 2003), which in turn promoting cytokinin breakdown diminishes cell proliferation in developing leaf primordia. Interestingly, the upregulation of *DR5::GUS* as well as that of *CKX6::GUS* by low R/FR occurs in pre-provascular cells of young leaf primordia, suggesting that induction of cytokinin degradation in the developing vasculature may be sufficient to arrest leaf primordium growth in low R/FR (Carabelli et al. 2007, 2008). However, the mechanism through which the incipient vein cells might signal to all the other cells of the young primordium to arrest their division upon perception of neighbors remains to be elucidated.

## 6.3 Auxin and Root Growth

Low R/FR severely reduces cell proliferation in the root apical meristem (RAM) of young seedlings as evidenced by diminished RAM size and reduced number of RAM cells expressing the *CYCB1;1:GUS* cell proliferation marker (Colón-Carmona et al. 1999). This inhibition is fully reversible as RAM cell proliferation is recovered upon subsequent exposure of the young seedlings to high R/FR light (Sassi and Ruberti, unpublished data). The model proposed for shade-induced responses described in Sect. 6.1 postulated that a reduction of polar auxin transport through the central cylinder is likely to cause a decrement in auxin concentration reaching the root. The RAM phenotype of seedlings exposed to low R/FR indeed resembles that of seedlings in which shoot-to-root polar auxin transport is impaired (Sassi et al. 2012). In agreement, the expression of *DR5::GUS*, which is strongly enhanced in the aerial tissues of seedlings exposed to low R/FR, is markedly reduced in the RAM under simulated shade. DR5::GUS expression is restored by reexposure to high R/FR light (Sassi et al. 2013).

# 7 Auxin Homeostasis and Auxin Transport in the Control of the Shade Avoidance Response

A growing body of evidence indicates that auxin homeostasis and auxin transport are central to many aspects of shade avoidance response.

## 7.1 *Auxin Homeostasis*

During the first 1–2 h of low R/FR, the levels of auxin in the Arabidopsis shoot increase (Hornitschek et al. 2012; Li et al. 2012; Tao et al. 2008). New auxin is synthesized from tryptophan (Trp) through TRYPTOPHAN AMINOTRANSFERASE OF ARABIDOPSIS 1 (TAA1), an enzyme encoded by the *SHADE AVOIDANCE 3* (*SAV3*) gene. Loss-of-function *sav3* mutants have reduced auxin in high R/FR light and are impaired in de novo auxin synthesis under simulated shade (Tao et al. 2008, see Chap. 2). In agreement with the central role of auxin in shade avoidance, *sav3* mutants are defective in several plant responses to low R/FR, including reduced hypocotyl elongation and leaf hyponasty (Tao et al. 2008). *SAV3* is highly expressed in cotyledons and leaves and not in the hypocotyl, further confirming that shade-induced elongation requires auxin transport (Steindler et al. 1999; Tao et al. 2008). Low R/FR however has no effect on *TAA1* transcript levels at the early stages of shade avoidance response (Tao et al. 2008).

A revised pathway for IAA biosynthesis has recently been proposed (Mashiguchi et al. 2011; Stepanova et al. 2011; Won et al. 2011, see Chap. 2) in which Trp is converted to indole-3-pyruvic acid (IPA) through TAA1, and IPA is converted to IAA through the action of a family of flavin monooxygenases encoded by the *YUCCA* (*YUC*) gene family (Zhao et al. 2001). *YUC2*, *YUC5*, *YUC8*, and *YUC9* are induced by low R/FR (Li et al. 2012; Tao et al. 2008). *YUCs* appear to encode a rate-limiting step in auxin biosynthesis (Mashiguchi et al. 2011; Won et al. 2011; Zhao et al. 2001), and thus the increase of *YUC2*, *YUC5*, *YUC8*, and *YUC9* RNA in low R/FR is likely to contribute to increased auxin synthesis during early stages of shade avoidance (Li et al. 2012). *yuc1-163 yuc4* double mutant, and to a lesser extent *yuc3 yuc5 yuc7 yuc8 yuc9* quintuple mutant, displays reduced hypocotyl elongation in response to low R/FR (Li et al. 2012; Won et al. 2011). It is possible that the weak phenotype of *yuc* quintuple mutant in low R/FR is due to redundancy among YUC family members (Li et al. 2012).

The identification of genome-wide PIF5-binding sites during shade avoidance revealed that this bHLH transcription factor physically interacts not only with the promoters of transcritption factor genes early induced by low R/FR (i.e., *ATHB2*, *HFR1*/*SICS1*) but also with those of auxin biosynthesis (*YUC8*) and auxin signaling (*AUX*/*IAA29*) genes. However, whereas *ATHB2* and *HFR1*/*SICS1* show reduced induction by low R/FR in *pif4 pif5* double mutant, the expression of both *YUC8* and *IAA29* is only slightly diminished in seedlings lacking PIF4 and PIF5 exposed to simulated shade (Hornitschek et al. 2012). More significantly, PIF7 in its dephosphorylated form binds G-boxes of the auxin biosynthetic genes *YUC5*, *YUC8*, and *YUC9* and increases their expression, thus directly linking the perception of a low R/FR signal to changes in free IAA required for shade-induced growth (Li et al. 2012). In agreement, *pif7* loss-of-function mutants show impaired auxin level increase in response to low R/FR (Li et al. 2012).

Consistent with the proposal that the COP1/SPA complex may influence HFR1/SICS1 protein levels in low R/FR, and in turn PIF protein activity, *spa1 spa2 spa3 spa4* quadruple (*spaq*) mutant seedlings fail to exhibit an increase in the transcript levels of the auxin biosynthesis genes *YUC2*, *YUC8*, and *YUC9* in response to simulated shade. This suggests that the increase in auxin biosynthesis promoted by low R/FR requires the activity of the SPA proteins. In agreement, the expression of the auxin-inducible marker *DR5::GUS* does not increase in *spaq* mutant seedlings upon exposure to simulated shade (Rolauffs et al. 2012).

It is worth to point out that low R/FR might further control auxin levels by regulating auxin inactivation as well. Several auxin-inducible genes of the GH3 family have been shown to be early upregulated by low R/FR (Devlin et al. 2003; Carabelli et al. 2007). GH3 proteins are known to promote the conjugation of free IAA to different amino acids, likely reducing the pool of free IAA (Staswick et al. 2005, see Chap. 2). Relevantly, plants with altered levels of GH3 proteins display defects in light-mediated hypocotyl elongation responses (Nakazawa et al. 2001; Takase et al. 2004). Thus the concurrent induction of auxin biosynthesis and inactivation by low R/FR might be required to fine-tune auxin levels during the shade avoidance response.

### *7.2 Auxin Transport*

Cells with the highest expression levels of *SAV3* are distinct from the ones that elongate in response to low R/FR (Tao et al. 2008). It was thus proposed that the low R/FR signal is perceived by phytochrome in cotyledons or leaves where *SAV3* is highly expressed. SAV3 then mediates an increase in free IAA, which is transported to hypocotyls, leading to promotion of cell elongation. In agreement, it was found that NPA blocks the shade-induced increase of *DR5::GUS* in hypocotyls but not in other aerial organs indicating that shade-induced increase in auxin synthesis occurs in the upper part of the shoot, and auxin is transported to the hypocotyls (Tao et al. 2008).

An increasing amount of evidence indicates that polar auxin transport is actively regulated during the shade avoidance response. Low R/FR has been shown to regulate the expression of PIN-FORMED (PIN) auxin efflux carriers PIN1, PIN3 and PIN7 (Devlin et al. 2003; Carabelli et al. 2007; Keuskamp et al. 2010; Sassi et al. 2013). Relevantly, among all the members of the PIN family, only PIN1, PIN3, and PIN7 are expressed in the vasculature of the hypocotyl (Blakeslee et al. 2007; Sassi et al. 2012), suggesting that low R/FR specifically regulates auxin transport in this tissue.

In the hypocotyls low R/FR light also regulates the localization of the auxin efflux carrier PIN-FORMED (PIN) 3 (Keuskamp et al. 2010), known to play a central role in tropic responses (Friml et al. 2002, see Chap. 16). Recent work revealed how light perception initiates auxin redistribution that leads to directional growth during phototropic responses (Ding et al. 2011). In the dark, high activity of the PINOID (PID) kinase correlates with apolar localization of PIN3 in the endodermal cells of the hypocotyls. Light represses *PID* transcription and PIN3 is polarized specifically to the inner cell sides by GNOM ARF GTPase GEF (guanine nucleotide exchange factor)-dependent trafficking. Upon exposure to unilateral blue light, the differential recruitment of PIN3 into different trafficking pathways at the two sides of the hypocotyl is followed by a redirection of auxin flow toward the shaded part where auxin promotes cell elongation and thus bending (Ding et al. 2011). By analogy to tropic responses, it was proposed more than a decade ago that elongation growth induced by neighbor detection and shade is the result of a laterally symmetric redistribution of auxin (see Sect. 6.1; Steindler et al. 1999; Morelli and Ruberti 2000, 2002). In agreement, it was recently found that low R/FR promotes PIN3 lateral localization in the endodermal cells toward the outer cells of the hypocotyl (Keuskamp et al. 2010). This shade-induced relocalization of PIN3 redirects auxin efflux toward the cortical and epidermal cells of the hypocotyl, promoting the cell elongation response (Keuskamp et al. 2010). Notably, the fitness of seedlings lacking PIN3 (*pin3-3* mutants), which are impaired in hypocotyl elongation in low R/FR, is suppressed by 40 % when competing with wild-type neighbors (Keuskamp et al. 2010).

Recent work also demonstrates that the regulation of auxin fluxes plays a central role in coordinating shoot and root growth in response to changes in the light environment (Sassi et al. 2012, 2013). *PIN1* is expressed at relatively low levels in etiolated hypocotyls, and it is induced upon exposure to light. This suggested that light may control shoot-to-root polar auxin transport in the hypocotyl primarily by regulating *PIN1* transcription. In accordance, *pin1* loss-of-function mutants display reduced root length and RAM defects identical to those of plants with inhibited shoot-to-root polar auxin transport. Relevantly, the light-mediated regulation of *PIN1* expression in the hypocotyl depends on the action of COP1, which can therefore fine-tune shoot-derived auxin levels in the root. This in turn influences auxin transport and cell proliferation in the RAM by modulating PIN1 and PIN2 intracellular distribution in the root in a COP1-dependent fashion (Sassi et al. 2012). Notably, shade induces a strong downregulation of *PIN1* in the hypocotyls, along with a concurrent decrease in auxin levels in the RAM, suggesting that low R/FR may activate a PIN1-dependent mechanism, analogous to that observed in etiolated seedlings, to partition auxin levels between shoot and root (Sassi et al. 2012, 2013). Similarly to what occurs during etiolation, this regulatory mechanism might inhibit root growth, allowing to tune the development of the whole plant to the light environment (Sassi et al. 2012, 2013).

# 8 Plant Adaptation to Low R/FR Environment

Plant responses to light quality changes are regulated by a balance of positive (PIFs) and negative (HFR1/SICS1) regulators of gene expression which ensures a fast reshaping of the plant body toward an environment optimal for growth while at the same time avoiding an exaggerated reaction to low R/FR (Hornitschek et al. 2009; Lorrain et al. 2008; Sessa et al. 2005; Fig. 17.3). Recent work demonstrated that downregulation of genes early induced by light quality changes upon prolonged exposure to low R/FR depends not only on HFR1/SICS1 (Sessa et al. 2005; Fig. 17.2), which interacts with PIF transcription factors forming non-DNA-binding heterodimers, thus limiting PIF-mediated gene expression (Hornitschek et al. 2009), but also on phyA (Ciolfi et al. 2013; Fig. 17.2). Furthermore, phyA and not HFR1/SICS1 is required for upregulation of several genes late induced by low R/FR. Remarkably, among them is the HY5 transcription factor gene (Ciolfi et al. 2013; Fig. 17.3). Expression studies in young seedlings exposed to simulated shade for different times have shown that *HY5-HOMOLOG* (*HYH*), a gene functionally involved in inhibition of hypocotyl elongation (Holm et al. 2002) and known to be a direct target of HY5 (Lee et al. 2007), is late induced by low R/FR, and its upregulation depends on the action of HY5 since it does not occur in *hy5* mutant (Ciolfi et al. 2013).

The role of HY5 has been mostly studied at the early stages of seedling development. Initially identified as a negative regulator of cell elongation functioning downstream of multiple families of the photoreceptors (Oyama et al. 1997; Osterlund et al. 2000), it has been then shown to act as a key controller of the transcriptional cascades promoting seedling de-etiolation (Lau and Deng 2010). More recently, HY5 has also been implicated in inhibition of hypocotyl elongation induced in shaded plants by brief exposure to direct sunlight perceived primarily by phyB (Sellaro et al. 2011). In agreement with its prominent role in de-etiolation, HY5 protein reaches its highest level 2–3 days after germination and then dramatically decreases at later times of seedling development (Hardtke et al. 2000). Notably, HY5 abundance directly correlates with the degree of photomorphogenic development (Osterlund et al. 2000). Thus, it seems likely that increased HY5 expression upon prolonged exposure to low R/FR may be a mechanism through which phyA exerts its regulatory role in shade avoidance response (Ciolfi et al. 2013).

Relevantly, several links between auxin and HY5/HYH have been established. On the basis of the misexpression of key components of the auxin signaling observed in the *hy5* and *hy5 hyh* mutants, it has been proposed that HY5 and HYH act as negative regulators of auxin response (Cluis et al. 2004; Sibout et al. 2006). Moreover, HY5 and HYH may also play a role in the regulation of polar auxin transport. Indeed, *hy5 hyh* double mutants displayed altered auxin fluxes in roots, as well as shoot phenotypes reminiscent of defective auxin transport (Sibout et al. 2006). In agreement with this, HY5-binding sites in *PIN1* and *PIN3* loci have also been found (Lee et al. 2007), further suggesting a direct regulation on

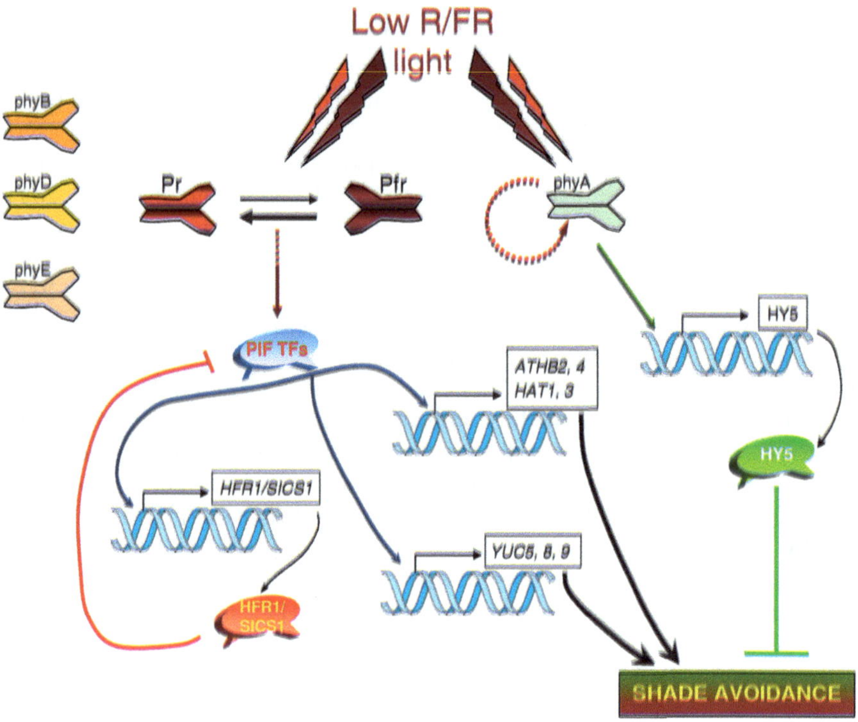


**Fig. 17.3** Transcriptional networks in shade avoidance response. Changes in the R/FR ratio of the light environment are perceived by multiple phytochromes (phyB, phyD, and phyE) and result in a shift of the photoequilibrium between the Pr and Pfr forms toward Pr which, in turn, leads to an increase in the level or activity of several PIF proteins (PIF1, 3, 4, 5, 7). These phy-regulated changes in PIFs rapidly induce the transcription of genes encoding positive (HD-Zips II and YUCs) and negative (HFR1/SICS1) regulators of shade avoidance response. By forming non-DNA-binding heterodimers, HFR1/SICS1 inhibits PIF activity ensuring that an exaggerated response to low R/FR light does not occur when the plant is unsuccessful in escaping canopy shade. Shade avoidance is further attenuated by phyA which positively regulates *HY5*, a master regulator of seedling de-etiolation

the expression of auxin efflux carriers. More recently, Sassi et al. (2012) have also suggested that HY5 might be involved in the COP1-mediated regulation of *PIN1* transcription in the hypocotyl in response to light. However, although this is an attractive possibility, whether HY5 regulates the expression of *PIN* genes in response to changes in the light environment remains to be determined.

# 9 Conclusions

The shade avoidance signaling network involves the reduction of active phyB, phyD, and phyE by low R/FR, the subsequent increase in PIF activity, and the induction of transcription factor genes that function as positive and negative regulators of plant responses to light quality changes. Among the positive regulators of shade avoidance are members of the HD-Zip II protein family (Ciarbelli et al. 2008; Sawa et al. 2002; Sorin et al. 2009; Steindler et al. 1999). Several links between these transcription factors and auxin have been established (Sorin et al. 2009; Steindler et al. 1999; Turchi et al. 2013); however the mechanisms through which ATHB2, ATHB4, and HAT3 influence auxin transport and response in low R/FR remain to be investigated. Among the genes directly regulated by PIF proteins there are also several auxin biosynthetic genes, thus directly linking the perception of a low R/FR signal to changes in free IAA required for shade-induced growth (Li et al. 2012). Shade-induced increase in auxin synthesis occurs in the upper part of the shoot, and auxin is transported to the hypocotyls (Tao et al. 2008). Polar auxin transport is actively regulated during the shade avoidance response. In the hypocotyl, low R/FR promotes PIN3 lateral localization in the endodermal cells toward the outer cell layers (Keuskamp et al. 2010) and strongly downregulates *PIN1* in the vasculature (Sassi et al. 2012, 2013). In agreement with the model for Arabidopsis shade-induced responses (Steindler et al. 1999; Morelli and Ruberti 2000, 2002), these changes in PIN expression and localization, on the one hand, enhance lateral transport of auxin in the hypocotyl promoting elongation and, on the other hand, reduce polar auxin transport through the central cylinder and, in turn, shoot-derived auxin levels in the root (Keuskamp et al. 2010; Sassi et al. 2012, 2013). This in turn is likely to influence auxin transport and cell proliferation in the RAM by modulating PIN1 and PIN2 intracellular distribution in the root in a COP1-dependent fashion (Sassi et al. 2012, 2013).

**Acknowledgments** We thank all our collaborators who made the work on shade avoidance a rewarding experience. Our apologies to the many researchers whose work or original publications could not be cited here because of space constraints. Authors work was funded by grants from the Italian Ministry of Education, University and Research, FIRB-ERA-PG Program, and from the Italian Ministry of Agricultural, Food and Forestry Policies, AGRONANOTECH and NUTRIGEA Programs. This work was also supported by a grant from the Italian Ministry of Economy and Finance to the CNR for the Project FaReBio di Qualità.

# References

Al-Sady B, Ni W, Kircher S et al (2006) Photoactivated phytochrome induces rapid PIF3 phosphorylation prior to proteasome-mediated degradation. Mol Cell 23:439–446

Arsovski AA, Galstyan A, Guseman JM et al (2012) Photomorphogenesis. Arabidopsis Book 10: e0147

Aukerman MJ, Hirschfeld M, Wester L et al (1997) A deletion in the *PHYD* gene of the Arabidopsis Wassilewskija ecotype defines a role for phytochrome D in red/far-red light sensing. Plant Cell 9:1317–1326

Bae G, Choi G (2008) Decoding of light signals by plant phytochromes and their interacting proteins. Annu Rev Plant Biol 59:281–311

Bai MY, Shang JX, Oh E et al (2012) Brassinosteroid, gibberellin and phytochrome impinge on a common transcription module in Arabidopsis. Nat Cell Biol 14:810–817

Ballaré CL (1999) Keeping up with the neighbours: phytochrome sensing and other signaling mechanisms. Trends Plant Sci 4:97–102

Bauer D, Viczián A, Kircher S et al (2004) Constitutive photomorphogenesis 1 and multiple photoreceptors control degradation of phytochrome interacting factor 3, a transcription factor required for light signaling in Arabidopsis. Plant Cell 16:1433–1445

Blakeslee JJ, Bandyopadhyay A, Lee OR et al (2007) Interactions among PIN-FORMED and P-glycoprotein auxin transporters in *Arabidopsis*. Plant Cell 19:131–147

Bou-Torrent J, Salla-Martret M, Brandt R et al (2012) ATHB4 and HAT3, two class II HD-ZIP transcription factors, control leaf development in Arabidopsis. Plant Signal Behav 7:1382–1387

Carabelli M, Morelli G, Whitelam G et al (1996) Twilight-zone and canopy shade induction of the *ATHB-2* homeobox gene in green plants. Proc Natl Acad Sci USA 93:3530–3535

Carabelli M, Possenti M, Sessa G et al (2007) Canopy shade causes a rapid and transient arrest in leaf development through auxin-induced cytokinin oxidase activity. Genes Dev 21:1863–1868

Carabelli M, Possenti M, Sessa G et al (2008) A novel regulatory circuit underlying plant response to canopy shade. Plant Signal Behav 3:137–139

Casal JJ (2013) Photoreceptor signaling networks in plant responses to shade. Annu Rev Plant Biol 64:403–427

Castillon A, Shen H, Huq E (2007) Phytochrome interacting factors: central players in phytochrome-mediated light signaling networks. Trends Plant Sci 12:514–521

Chang CS, Li YH, Chen LT et al (2008) LZF1, a HY5-regulated transcriptional factor, functions in Arabidopsis de-etiolation. Plant J 54:205–219

Chang CSJ, Maloof JN, Wu SH (2011) COP1 mediated degradation of BBX22/LZF1 optimizes seedling development in Arabidopsis. Plant Physiol 156:228–239

Ciarbelli AR, Ciolfi A, Salvucci S et al (2008) The Arabidopsis homeodomain-leucine zipper II gene family: diversity and redundancy. Plant Mol Biol 68:465–478

Ciolfi A, Sessa G, Sassi M et al (2013) Dynamics of the shade-avoidance response in Arabidopsis. Plant Physiol 163:331–353

Cluis CP, Mouchel CF, Hardtke CS (2004) The Arabidopsis transcription factor HY5 integrates light and hormone signaling pathways. Plant J 38:332–347

Colón-Carmona A, You R, Haimovitch-Gal T et al (1999) Spatio-temporal analysis of mitotic activity with a labile cyclin-GUS fusion protein. Plant J 20:503–508

Crocco CD, Holm M, Yanovsky MJ et al (2010) AtBBX21 and COP1 genetically interact in the regulation of shade avoidance. Plant J 64:551–562

Datta S, Hettiarachchi C, Johansson H et al (2007) SALT TOLERANCE HOMOLOG2, a B-box protein in Arabidopsis that activates transcription and positively regulates light-mediated development. Plant Cell 19:3242–3255

Datta S, Johansson H, Hettiarachchi C et al (2008) LZF1/SALT TOLERANCE HOMOLOG3, an Arabidopsis B-box protein involved in light-dependent development and gene expression, undergoes COP1-mediated ubiquitination. Plant Cell 20:2324–2338

de Lucas M, Daviere J-M, Rodriguez-Falcon M et al (2008) A molecular framework for light and gibberellin control of cell elongation. Nature 451:480–484

Depuydt S, Hardtke CS (2011) Hormone signalling crosstalk in plant growth regulation. Curr Biol 21:R365–R373

Devlin PF, Patel SR, Whitelam GC (1998) Phytochrome E influences internode elongation and flowering time in Arabidopsis. Plant Cell 10:1479–1487

Devlin PF, Robson PRH, Patel SR et al (1999) Phytochrome D acts in the shade-avoidance syndrome in Arabidopsis by controlling elongation and flowering time. Plant Physiol 119:909–915

Devlin PF, Yanovsky MJ, Kay SA (2003) A genomic analysis of the shade avoidance response in Arabidopsis. Plant Physiol 133:1617–1629

Ding Z, Galván-Ampudia CS, Demarsy E et al (2011) Light-mediated polarization of the PIN3 auxin transporter for the phototropic response in Arabidopsis. Nat Cell Biol 13:447–452

Djakovic-Petrovic T, de Wit M, Voesenek LACJ et al (2007) DELLA protein function in growth responses to canopy signals. Plant J 51:117–126

Duek PD, Fankhauser C (2003) HFR1, a putative bHLH transcription factor, mediates both phytochrome A and cryptochrome signaling. Plant J 34:827–836

Duek PD, Elmer MV, van Oosten VR et al (2004) The degradation of HFR1, a putative bHLH class transcription factor involved in light signaling, is regulated by phosphorylation and requires COP1. Curr Biol 14:2296–2301

Fairchild CD, Schumaker MA, Quail PH (2000) *HFR1* encodes an atypical bHLH protein that acts in phytochrome A signal transduction. Genes Dev 14:2377–2391

Fankhauser C, Chory J (2000) *RSF1*, an Arabidopsis locus implicated in phytochrome A signaling. Plant Physiol 124:39–45

Feng S, Martinez C, Gusmaroli G et al (2008) Coordinated regulation of *Arabidopsis thaliana* development by light and gibberellins. Nature 451:475–479

Franklin KA (2008) Shade avoidance. New Phytol 179:930–944

Franklin KA, Quail PH (2010) Phytochrome functions in Arabidopsis development. J Exp Bot 61:11–24

Franklin KA, Praekelt U, Stoddart WM et al (2003) Phytochromes B, D, and E act redundantly to control multiple physiological responses in Arabidopsis. Plant Physiol 131:1340–1346

Frigerio M, Alabadi D, Perez-Gomez J et al (2006) Transcriptional regulation of gibberellin metabolism genes by auxin signaling in Arabidopsis. Plant Physiol 142:553–563

Friml J, Wiśniewska J, Benková E et al (2002) Lateral relocation of auxin efflux regulator PIN3 mediates tropism in Arabidopsis. Nature 415:806–809

Gallego-Bartolomé J, Minguet EG, Grau-Enguix F et al (2012) Molecular mechanism for the interaction between gibberellin and brassinosteroid signaling pathways in Arabidopsis. Proc Natl Acad Sci USA 109:13446–13451

Galstyan A, Cifuentes-Esquivel N, Bou-Torrent J et al (2011) The shade avoidance syndrome in Arabidopsis: a fundamental role for atypical basic helix-loop-helix proteins as transcriptional cofactors. Plant J 66:258–267

Hao Y, Oh E, Choi G, Liang Z et al (2012) Interactions between HLH and bHLH factors modulate light-regulated plant development. Mol Plant 5:688–697

Harberd NP, Belfield E, Yasumura Y (2009) The angiosperm gibberellin-GID1–DELLA growth regulatory mechanism: how an 'inhibitor of an inhibitor' enables flexible response to fluctuating environments. Plant Cell 21:1328–1339

Hardtke CS, Gohda K, Osterlund MT et al (2000) HY5 stability and activity in *Arabidopsis* is regulated by phosphorylation in its COP1 binding domain. EMBO J 19:4997–5006

Holm M, Ma L-G, Qu L-J et al (2002) Two interacting bZIP proteins are direct targets of COP1-mediated control of light-dependent gene expression in Arabidopsis. Genes Dev 16:1247–1259

Hornitschek P, Lorrain S, Zoete V et al (2009) Inhibition of the shade avoidance response by formation of non-DNA binding bHLH heterodimers. EMBO J 28:3893–3902

Hornitschek P, Kohnen MV, Lorrain S et al (2012) Phytochrome interactin factors 4 and 5 control seedling growth in changing light conditions by directly controlling auxin signaling. Plant J 71:699–711

Jaillais Y, Chory J (2010) Unraveling the paradoxes of plant hormone signaling in- tegration. Nat Struct Mol Biol 17:642–645

Jang I-C, Yang J-Y, Soo Seo H et al (2005) HFR1 is targeted by COP1 E3 ligase for post-translational proteolysis during phytochrome A signaling. Genes Dev 19:593–602

Jang IC, Henriques R, Seo HS et al (2010) Arabidopsis PHYTOCHROME INTERACTING FACTOR proteins promote phytochrome B polyubiquitination by COP1 E3 ligase in the nucleus. Plant Cell 22:2370–2383

Jiao Y, Lau OS, Deng XW (2007) Light-regulated transcriptional networks in higher plants. Nat Rev Genet 8:217–230

Johnson E, Bradley JM, Harberd NP et al (1994) Photoresponses of light-grown *phyA* mutants of Arabidopsis: phytochrome A is required for the perception of daylength extensions. Plant Physiol 105:141–149

Keller MM, Jaillais Y, Pedmale UV et al (2011) Cryptochrome 1 and phytochrome B control shade-avoidance responses in Arabidopsis via partially independent hormonal cascades. Plant J 67:195–207

Keuskamp DH, Pollmann S, Voesenek LACJ et al (2010) Auxin transport through PIN-FORMED 3 (PIN3) controls shade avoidance and fitness during competition. Proc Natl Acad Sci USA 107:22740–22744

Keuskamp DH, Sasidharan R, Vos I et al (2011) Blue-light-mediated shade avoidance requires combined auxin and brassinosteroid action in Arabidopsis seedlings. Plant J 67:208–217

Khanna R, Kronmiller B, Maszle DR et al (2009) The Arabidopsis B-box zinc finger family. Plant Cell 21:3416–3420

Kim TW, Wang ZY (2010) Brassinosteroid signal transduction from receptor kinases to transcription factors. Annu Rev Plant Biol 61:681–704

Kuppusamy KT, Walcher CL, Nemhauser JL (2009) Cross-regulatory mechanisms in hormone signaling. Plant Mol Biol 69:375–381

Lau OS, Deng XW (2010) Plant hormone signaling lightens up: integrators of light and hormones. Curr Opin Plant Biol 13:571–577

Lee J, He K, Stolc V et al (2007) Analysis of transcription factor HY5 genomic binding sites revealed its hierarchical role in light regulation of development. Plant Cell 19:731–749

Leivar P, Quail PH (2011) PIFs: pivotal components in a cellular signaling hub. Trends Plant Sci 16:19–28

Leivar P, Monte E, Oka Y et al (2008a) Multiple phytochrome-interacting bHLH transcription factors repress premature seedling photomorphogenesis in darkness. Curr Biol 18:1815–1823

Leivar P, Monte E, Al-Sady B et al (2008b) The Arabidopsis phytochrome-interacting factor PIF7, together with PIF3 and PIF4, regulates responses to prolonged red light by modulating phyB levels. Plant Cell 20:337–352

Leivar P, Tepperman JM, Cohn MM et al (2012a) Dynamic antagonism between phytochromes and PIF family basic helix-loop-helix factors induces selective reciprocal responses to light and shade in a rapidly responsive transcriptional network in Arabidopsis. Plant Cell 24:1398–1419

Leivar P, Monte E, Cohn MM et al (2012b) Phytochrome signaling in green Arabidopsis seedlings: impact assessment of a mutually negative phyB–PIF feedback loop. Mol Plant 5:734–749

Li L, Ljung K, Breton G et al (2012) Linking photoreceptor excitation to changes in plant architecture. Genes Dev 26:785–790

Lorrain S, Allen T, Duek PD et al (2008) Phytochrome-mediated inhibition of shade avoidance involves degradation of growth-promoting bHLH transcription factors. Plant J 53:312–323

Mashiguchi K, Tanaka K, Sakai T et al (2011) The main auxin biosynthesis pathway in Arabidopsis. Proc Natl Acad Sci USA 108:18512–18517

Mathews S (2006) Phytochrome-mediated development in land plants: red light sensing evolves to meet the challenges of changing light environments. Mol Ecol 15:3483–3503

Mathews S, Sharrock RA (1997) Phytochrome gene diversity. Plant Cell Environ 20:666–671

McNellis TW, von Arnim AG, Araki T et al (1994) Genetic and molecular analysis of an allelic series of *cop1* mutants suggests functional roles for the multiple protein domains. Plant Cell 6:487–500

Morelli G, Ruberti I (2000) Shade avoidance responses. Driving auxin along lateral routes. Plant Physiol 122:621–626

Morelli G, Ruberti I (2002) Light and shade in the photocontrol of Arabidopsis growth. Trends Plant Sci 7:399–404

Murase K, Hirano Y, Sun TP et al (2008) Gibberellin-induced DELLA recognition by the gibberellin receptor GID1. Nature 456:459–463

Nakazawa M, Yabe N, Ichikawa T et al (2001) *DFL1*, an auxin-responsive *GH3* gene homologue, negatively regulates shoot cell elongation and lateral root formation, and positively regulates the light response of hypocotyl length. Plant J 25:213–221

Nozue K, Covington MF, Duek PD et al (2007) Rhythmic growth explained by coincidence between internal and external cues. Nature 448:358–361

Oh E, Zhu JY, Wang ZY (2012) Interaction between BZR1 and PIF4 integrates brassinosteroid and environmental responses. Nat Cell Biol 14:802–809

Osterlund MT, Hardtke CS, Wei N et al (2000) Targeted destabilization of HY5 during light-regulated development of Arabidopsis. Nature 405:462–466

Oyama T, Shimura Y, Okada K (1997) The Arabidopsis *HY5* gene encodes a bZIP protein that regulates stimulus-induced development of root and hypocotyl. Genes Dev 11:2983–2995

Park E, Kim J, Lee Y et al (2004) Degradation of phytochrome interacting factor 3 in phytochrome-mediated light signaling. Plant Cell Physiol 45:968–975

Pierik R, Djakovic-Petrovic T, Keuskamp DH et al (2009) Auxin and ethylene regulate elongation responses to neighbor proximity signals independent of gibberellin and DELLA proteins in Arabidopsis. Plant Physiol 149:1701–1712

Redman JC, Haas BJ, Tanimoto G et al (2004) Development and evaluation of an Arabidopsis whole genome Affymetrix probe array. Plant J 38:545–561

Reed JW, Nagpal P, Poole DS et al (1993) Mutations in the gene for the red/far-red light receptor phytochrome B alter cell elongation and physiological responses throughout Arabidopsis development. Plant Cell 5:147–157

Reymond MC, Brunoud G, Chauvet A et al (2012) A light-regulated genetic module was recruited to carpel development in Arabidopsis following a structural change to SPATULA. Plant Cell 24:2812–2825

Roig-Villanova I, Bou J, Sorin C et al (2006) Identification of primary target genes of phytochrome signaling. Early transcriptional control during shade avoidance responses in Arabidopsis. Plant Physiol 141:85–96

Roig-Villanova I, Bou-Torrent J, Galstyan A et al (2007) Interaction of shade avoidance and auxin responses: a role for two novel atypical bHLH proteins. EMBO J 26:4756–4767

Rolauffs S, Fackendah P, Sahm J (2012) Arabidopsis *COP1* and *SPA* genes are essential for plant elongation but not for acceleration of flowering time in response to a low red light to far-red light ratio. Plant Physiol 160:2015–2027

Ruberti I, Sessa G, Ciolfi A et al (2012) Plant adaptation to dynamically changing environment: the shade avoidance response. Biotechnol Adv 30:1047–1058

Salisbury FJ, Hall A, Grierson CS et al (2007) Phytochrome coordinates Arabidopsis shoot and root development. Plant J 50:429–438

Salter MG, Franklin KA, Whitelam GC (2003) Gating of the rapid shade avoidance response by the circadian clock in plants. Nature 426:680–683

Sassi M, Lu Y, Zhang Y et al (2012) COP1 mediates the coordination of root and shoot growth by light through modulation of PIN1- and PIN2-dependent auxin transport in Arabidopsis. Development 139:3402–3412

Sassi M, Wang J, Ruberti I et al (2013) Shedding light on auxin movement: light-regulation of polar auxin transport in the photocontrol of plant development. Plant Signal Behav 8:e23355

Sawa S, Ohgishi M, Goda H et al (2002) The *HAT2* gene, a member of the HD-Zip gene family, isolated as an auxin inducible gene by DNA microarray screening, affects auxin response in Arabidopsis. Plant J 32:1011–1022

Schena M, Lloyd AM, Davis RW (1993) The *HAT4* gene of Arabidopsis encodes a developmental regulator. Genes Dev 7:367–379

Sellaro R, Yanovsky MJ, Casal JJ (2011) Repression of shade-avoidance reactions by sunfleck-induction of *HY5* expression in Arabidopsis. Plant J 68:919–928

Sessa G, Carabelli M, Sassi M et al (2005) A dynamic balance between gene activation and repression regulates the shade avoidance response in Arabidopsis. Genes Dev 19:2811–2815

Shen H, Moon J, Huq E (2005) PIF1 is regulated by light-mediated degradation through the ubiquitin-26S proteasome pathway to optimize photomorphogenesis of seedlings in Arabidopsis. Plant J 44:1023–1035

Shen Y, Khanna R, Carle CM et al (2007) Phytochrome induces rapid PIF5 phosphorylation and degradation in response to red-light activation. Plant Physiol 145:1043–1051

Shin J, Kim K, Kang H et al (2009) Phytochromes promote seedling light responses by inhibiting four negatively-acting phytochrome-interacting factors. Proc Natl Acad Sci USA 106:7660–7665

Sibout R, Sukumar P, Hettiarachchi C et al (2006) Opposite root growth phenotypes of *hy5* versus *hy5 hyh* mutants correlate with increased constitutive auxin signaling. PLoS Genet 11:e202

Smith H, Whitelam GC (1997) The shade avoidance syndrome: multiple responses mediated by multiple phytochromes. Plant Cell Environ 20:840–844

Soh M-S, Kim Y-M, Han S-J et al (2000) *REP1*, a basic helix-loop-helix protein, is required for a branch pathway of phytochrome A signaling in Arabidopsis. Plant Cell 12:2061–2073

Sorin C, Salla-Martret M, Bou-Torrent J et al (2009) ATHB4, a regulator of shade avoidance, modulates hormone response in Arabidopsis seedlings. Plant J 59:266–277

Stamm P, Kumar PP (2010) The phytohormone signal network regulating elongation growth during shade avoidance. J Exp Bot 61:2889–2903

Staswick PE, Serban B, Rowe M et al (2005) Characterization of an Arabidopsis enzyme family that conjugates amino acids to indole-3-acetic acid. Plant Cell 17:616–627

Steindler C, Matteucci A, Sessa G et al (1999) Shade avoidance responses are mediated by the ATHB-2 HD-Zip protein, a negative regulator of gene expression. Development 125:4235–4245

Stepanova AN, Yun J, Robles LM et al (2011) The Arabidopsis YUCCA1 flavin monooxygenase functions in the indole-3-pyruvic acid branch of auxin biosynthesis. Plant Cell 23:3961–3973

Sun T-P (2011) The molecular mechanism and evolution of the GA-GID1-DELLA signaling module in plants. Curr Biol 21:R338–R345

Takase T, Nakazawa M, Ishikawa A et al (2004) *ydk1-D*, an auxin-responsive *GH3* mutant that is involved in hypocotyl and root elongation. Plant J 37:471–483

Tao Y, Ferrer JL, Ljung K et al (2008) Rapid synthesis of auxin via a new tryptophan-dependent pathway is required for shade avoidance in plants. Cell 133:164–176

Tatematsu K, Kumagai S, Muto H et al (2004) *MASSUGU2* encodes Aux/IAA19, an auxin-regulated protein that functions together with the transcriptional activator NPH4/ARF7 to regulate differential growth responses of hypocotyl and formation of lateral roots in *Arabidopsis thaliana*. Plant Cell 16:379–393

Turchi L, Carabelli M, Ruzza V et al (2013) Arabidopsis HD-Zip II transcription factors control embryo development and meristem function. Development 140:2118–2129

Ulmasov T, Murfett J, Hagen G et al (1997) Aux/IAA proteins repress expression of reporter genes containing natural and highly active synthetic auxin response elements. Plant Cell 9:1963–1971

Wang X, Roig-Villanova I, Khan S et al (2011) A novel high-throughput in vivo molecular screen for shade avoidance mutants identifies a novel phyA mutation. J Exp Bot 62:2973–2987

Werner T, Motyka V, Laucou V et al (2003) Cytokinin-deficient transgenic Arabidopsis plants show multiple developmental alterations indicating opposite functions of cytokinins in the regulation of shoot and root meristem activity. Plant Cell 11:2532–2550

Won C, Shen X, Mashiguchi K et al (2011) Conversion of tryptophan to indole- 3-acetic acid by TRYPTOPHAN AMINOTRANSFERASES OF ARABIDOPSIS and YUCCAs in Arabidopsis. Proc Natl Acad Sci USA 108:18518–18523

Yang J, Lin R, Sullivan J et al (2005) Light regulates COP1-mediated degradation of HFR1, a transcription factor essential for light signaling in Arabidopsis. Plant Cell 17:804–821

Zhao Y, Christensen SK, Fankhauser C et al (2001) A role for flavin monooxygenase-like enzymes in auxin biosynthesis. Science 291:306–309

Zúñiga-Mayo VM, Marsch-Martínez N, de Folter S (2012) JAIBA, a class-II HD-ZIP transcription factor involved in the regulation of meristematic activity, and important for correct gynoecium and fruit development in Arabidopsis. Plant J 71:314–326

# Chapter 18
# Auxin and the Interaction Between Plants and Microorganisms

**Jutta Ludwig-Müller**

**Abstract** While auxin is involved in virtually every process in plant development and orientation in the environment, plant pathogens have exploited the auxin machinery of the plant to alter plant growth and development in their favor. On the opposite, the plant is able to turn this against invaders and uses the same pathways for defense reactions. Also, plant beneficial microbes can interfere with the auxin metabolism of the host plant to induce growth of the plant for mutual benefits. Here, the role of auxin in disease symptom development will be reviewed, where either the plant or the pathogen contributes to alterations in host auxin synthesis and metabolism. Due to the many interactions known, the focus here will be on bacteria, protists, and fungi. On the one hand, auxin can be rated as pathogenicity factor, but also on the other hand alterations in auxin levels can result in changes of target genes, which then lead to changes in plant defense. In addition, the auxin-signaling pathway is directly utilized in the defense reaction against some pathogens.

## 1 Introduction

Plant diseases are economical factors influencing agricultural and horticultural applications. Plant diseases cause huge losses by changing the quality and quantity of harvested crops. Other costs are created by the chemical defense using pesticides in general, which is ecologically not desirable. To understand how disease symptoms develop and how the plant's defense machinery can be employed for breeding resistant varieties is of importance. Many disease symptoms caused by bacteria or

J. Ludwig-Müller (✉)
Technische Universität Dresden, 01062 Dresden, Germany
e-mail: jutta.ludwig-mueller@tu-dresden.de

E. Zažímalová et al. (eds.), *Auxin and Its Role in Plant Development*,
DOI 10.1007/978-3-7091-1526-8_18, © Springer-Verlag Wien 2014

fungi rely on the involvement of plant hormones, while other plant hormones act as defense signals in the plant. The latter encompass ethylene, jasmonic acid, and salicylic acid. In some cases these defense compounds show also cross talk with auxin. The growth promoting plant hormones causing disease symptoms are gibberellins, factors for the "foolish seedling" (bakanae) disease in rice (Mitchell and Angel 1951); cytokinins, which stimulate cell division and thereby resulting in tumorous growth of infected tissues (Zupan and Zambryski 1995) or growth anomalies such as fasciation (Crespi et al. 1992); and auxins, which also stimulate cell division, but also cell elongation, resulting in hypertrophied cells (Ludwig-Müller et al. 2009). In many cases auxins and cytokinins are inseparable in causing disease symptoms, such as tumor formation. Finally, brassinosteroids have dual roles, first in some cases they are involved in defense (Krishna 2003), but they could also act together with auxins on cell elongation and thus tissue growth. During an infection cycle auxin can play a role at different points in the plant being involved in either pathogenicity or defense responses (Fig. 18.1). Auxin is involved in symptom development after colonization and thus important for pathogenesis. It can also direct plant defense signaling or act directly as defense molecule with antimicrobial activity. Alterations in auxin levels result in changes of target genes of the auxin response. For example, a reduction of auxin could alter cell wall functionality by decreasing cell wall loosening enzymes. In this case the auxin reduction would increase plant defense. However, auxin can also be a factor used by beneficial microbes to alter plant growth and development.

## 2 Auxin in Symbiotic Interactions

Soil microbes are known to alter plant growth for example by alteration of hormone levels, either by themselves or by induction of auxin synthesis in the host plant. Beneficial soil bacteria produce indole-3-acetic acid (IAA), which induces the plant's root growth (Patten and Glick 1996, 2002). Whether there is a beneficial effect for the bacteria through an enlarged root system of the host plants has not directly been shown, but possibly, the larger root system secretes more nutrients, which is of advantage for the bacteria. In agreement with this hypothesis, maize roots inoculated with bacteria from the genus *Azospirillum* had higher levels of indole-3-butyric acid (IBA) (Fallik et al. 1989). Also, there are soil bacteria, which hydrolyze with high specificity different IAA conjugates with amino acids, thus transforming inactive auxins into active IAA (Chou et al. 1998; Chou and Huang 2005).

Rhizobia are able to induce root nodules on their hosts from the family of Leguminoseae (reviewed in Long 2001). They were shown to synthesize IAA by either the indole-3-acetamide or the indole-3-pyruvic acid pathways (reviewed in Spaepen et al. 2007; see also Chap. 2), but also the host IAA homeostasis plays a role. For the recognition, the symbiotic bacteria use host plant compounds, the flavonoids. These compounds are secreted into the rhizosphere to attract the rhizobia. They are recognized by specific bacterial receptors, which lead to

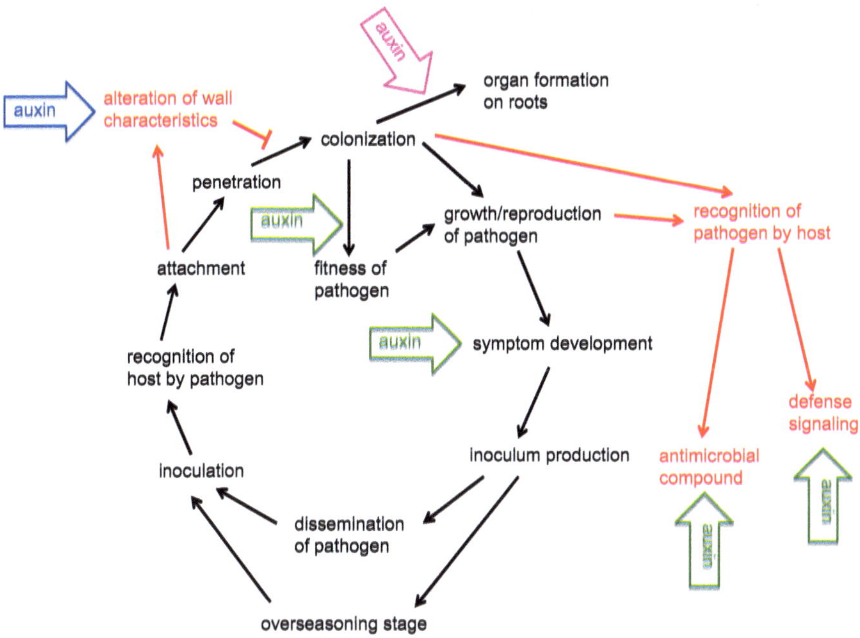


**Fig. 18.1** A model of a disease cycle and the possible auxin functions in pathogenicity or defense reactions. In *black* a basic infection cycle is shown and in *red* defense responses. The *green arrows* show where auxin has a positive effect and the *blue* where auxin is negatively influencing plant defense. In addition, the entry points for auxin in beneficial interactions are shown in *magenta*, such as organ formation as an example

flavonoid-dependent gene expression in the free-living bacteria. Later, in the plant the first response is a curvature of the root hairs, which is prerequisite for the colonization process. The symbiosis is fully established upon symbiosome formation, which terminates into the macroscopically visible nodules. However, flavonoids are not only recognition signals, but also discussed to be involved in the formation of the root nodule structure itself. Flavonoids accumulate at the entrance site for the rhizobia. Since they inhibit auxin efflux from cells (Peer et al. 2004; see also Chap. 5), it is hypothesized that high levels of flavonoids would lead to a local increase in IAA which might trigger the formation of the nodules, in turn leading to increased cell division rates. Root nodules are discussed to be organs, such as lateral roots, so their initiation needs a plant-derived signal, which could be auxin (Hirsch 1992; Mathesius 2008). While auxin is not essential for the initial stages in colonization, it might be important for nodule development. Auxin-responsive promoter::reporter constructs were used as tool to get an indication for increased auxin levels in root nodules (e.g., Huo et al. 2006). Consequently, mutants with altered flavonoid synthesis did not form nodules, even though normal root hair curling occurred after inoculation with rhizobia (Wasson et al. 2006). The inhibition of chalcone synthase, a protein essential for flavonoid biosynthesis, in

*Medicago truncatula* results in reduced flavonoid accumulation and reduced auxin transport, which in turn reduces nodulation (Wasson et al. 2006). More directly, the reduction of a PIN transporter gene expression involved in IAA transport in *M. truncatula* also results in reduced nodulation (Huo et al. 2006). Rhizobia have been shown to synthesize auxin, probably from tryptophan exuded by the host plant (Kefford et al. 1960). A role for the hydrolysis of auxin conjugates in nodule formation was also suggested (Campanella et al. 2008). Since the transcript levels of several auxin amido hydrolases from *Medicago truncatula* were dramatically increased during early stages of nodule formation, the authors hypothesized that the conjugate hydrolysis could play a role to increase free auxin levels.

Auxins can be produced by several fungi, which form ectomycorrhizal associations and thus influence the host plant to increase the root system to form more interaction sites (Splivallo et al. 2009). Also, arbuscular mycorrhizal (AM) fungi colonize preferentially young host roots. They form appressoria, which develop into intraradical hyphae and then arbuscules. The latter are the exchange places for nutrients from plant to fungus and vice versa. Several publications have shown that auxins are increased in roots colonized by AM fungi (e.g., Meixner et al. 2005). Besides IAA other auxin compounds such as IBA occur naturally in plants (Epstein and Ludwig-Müller 1993; Ludwig-Müller 2000). In maize roots the accumulation of IBA but not IAA might trigger the enlargement of the root system, where especially lateral roots are newly induced (Kaldorf and Ludwig-Müller 2000). These are thought to be the preferential entry points for the fungi. In *Tropaeolum majus* also an increase in several molecules with auxin activity (Ludwig-Müller and Cohen 2002) was found after mycorrhization (Jentschel et al. 2007). In addition, Fitze et al. (2005) reported an increase in IAA amino acid conjugates in AM-inoculated roots, but not ester conjugates. Also, a systemic increase of IBA and IBA conjugates was noted. In *Medicago truncatula* roots the increase in auxin was accompanied by differential upregulation of several transcripts belonging to a family of auxin amino acid conjugate hydrolases (Campanella et al. 2008). This could be an indication for the involvement of auxin conjugate hydrolysis to contribute to high auxin levels.

The endophytic basidiomycete *Piriformospora indica* (Sebacinaceae) has various beneficial effects on plants (Oelmüller et al. 2009). It was shown that the major growth promoting effect derives from cytokinins synthesized by the fungus and perceived by the plant (Vadassery et al. 2008). However, the fungus is also able to synthesize auxins (Sirrenberg et al. 2007; Vadassery et al. 2008), which might, similarly to arbuscular mycorrhiza, result in alterations of the root system of host plants. Using auxin and cytokinin mutants, the role of each hormone for the observed growth promotion was investigated, indicating that only cytokinin, but not auxin, was responsible for induction of shoot growth (Vadassery et al. 2008), despite the auxin production of the fungus.

# 3 Auxin as Pathogenicity Factor

## 3.1 Symptoms Likely Dependent on Auxin

Many symptoms of plant diseases displaying changes of organs or turning the plant tissue into tumorous swellings are dependent on alterations in auxin metabolism. Such tumors are dependent on the activation of the cell cycle, in which auxin is also involved (see also Chap. 7). The focus of plant–microbe interactions will lie on bacteria, protists, and fungi, even though for other pests such as nematodes and some insects an involvement of auxins has also been hypothesized and in some cases experimentally verified. Among the disease symptoms caused by bacteria, protists, and fungi are changes in organ structure, for example, leaf and stem fasciation, other leaf anomalies, i.e., leaf curling or the witches' broom symptoms, and the hairy root symptom, but also tumor formation. The latter can be divided into two groups, the first being the tumors of *Agrobacterium tumefaciens*, which result from stable transformation of the tissue (Zupan and Zambryski 1995), and the second comprises undirected growth of host tissues caused by infection, but not transformation. Among the latter are the clubroot symptoms caused by *Plasmodiophora brassicae* (Ludwig-Müller et al. 2009). Not in all cases it is clear that the cause of the anomaly is dependent on auxin, but due to the type of alterations observed, an involvement can be postulated. Typical symptoms thought to be associated with auxin are connected with changes in phenotypes, such as tumors, galls, and root or shoot like structures (Fig. 18.2).

Among the disease symptoms changes in organ structure, for example, leaf and stem fasciation, caused by the bacterium *Rhodococcus fascians*, sometimes also called "leafy galls" (Goethals et al. 2001), can be found. Fasciation is an unnatural status of growth originating from the meristem producing flattened, ribbon-like, crested, or contorted tissue. This might be due to hormonal imbalances in the meristematic cells of plants or induced by phytopathogens. Also, the witches' broom disease caused by phytoplasma results in alterations of host tissue. The latter appear as brushlike clusters of dwarfed and weak shoots emerging at the same site of the original stem. They may be also due to changes in auxin levels (Hoshi et al. 2009).

Alternatively, the tissue is transformed into indeterminate growing cell clumps (tumor) or organs (hairy roots), which grow in theory without additional plant hormones (Georgiev et al. 2010). The hairy root disease caused by *Agrobacterium rhizogenes* (Altamura 2004) is also brought into the context of changes in auxin levels. Infection triggers organogenesis and transforms (in theory) any plant tissue to the "hairy root" phenotype, consisting of strongly branching roots with many lateral emergences. Here, in contrast to the other bacterial pathogens mentioned, the plant is stably transformed by the *rol* genes (Altamura 2004), but auxin synthesis is caused by two additional loci of the bacteria, namely, the *aux1* and *aux2* genes (Nemoto et al. 2009). Plant tumors are caused by stable transformation with *Agrobacterium tumefaciens* causing an imbalance in the hormone homeostasis of

**Fig. 18.2** Some typical disease symptoms associated with auxin. (**a**) Bacterial tumor (unknown cause), (**b**) *Agrobacterium rhizogenes* hairy root culture (picture courtesy of Annemarie Lippert, Technische Universität Dresden), (**c**) *Plasmodiophora brassicae* root gall on oilseed rape seedlings, (**d**) Witches' broom, unknown cause, (**e**) *Ustilago maydis* tumor (*left* corn cob; *right* male flower)

auxin and cytokinin, resulting in tumorous tissues (Zupan et al. 2000). However, free-living bacteria can also synthesize IAA (Lee et al. 2009a). While *Pseudomonas syringae* infection is associated with leaf necroses or partial tissue death (Canfield et al. 1986), another species *Pseudomonas savastanoi* is the causal agent of olive (*Olea europaea*) knot disease and an unorthodox member of the *P. syringae* complex, causing tumorous structures (Ramos et al. 2012). Symptoms of infected trees include hypertrophy formation on the stems and branches and occasionally on the leaves and fruits. Here, in addition to a pathway for IAA biosynthesis, the bacteria also possess an enzyme for the conjugation of IAA with the amino acid lysine (Matas et al. 2009). Since IAA lysine seems to be a conjugate which cannot be hydrolyzed by the plant, the formation of this protein could lead to alterations in the hormone homeostasis, thus inducing the disease phenotype. The tumors formed by the phytopathogenic bacterium *Pantoea agglomerans* (formerly *Enterobacter agglomerans*) are different from those induced by *A. tumefaciens*, because the presence of *Pantoea agglomerans* is essential for tumor development via synthesis of IAA by different pathways as the tissue is not stably transformed (Clark et al. 1989).

Growth anomalies are also induced by a group of phytopathogenic protists of the Plasmodiophoromycota. Among these are *Plasmodiophora brassicae* (Ludwig-Müller et al. 2009), the causal agent of clubroot disease, *Polymyxa betae* transmitting a virus and thereby causing the rhizomania disease on host roots (Ciafardini 1991), and *Spongospora subterranea*, the causal agent of potato powdery scab, inducer of the disease itself, but also virus transmitter (Merz and Falloon 2009). *P. brassicae* infection is characterized by changing the complete root tissue into large undifferentiated galls or tumors. Contrary to the tumors induced by *A. tumefaciens*, the *P. brassicae*-induced tumorous tissue is not stably transformed, but relies on continuous synthesis of auxins and cytokinins, the former only by the plant and the latter by the plant and the protist (Ludwig-Müller et al. 2009). Not much is known about the involvement of hormones in disease development caused by the other two protists.

Many other growth anomalies are known to be induced by fungi. The causal agent of black wart on potatos, *Synchytrium endobioticum*, induces host cells to proliferate into a warty gall containing sporangia (Laidlaw 1985). Very often whole plant organs are transformed into the tumorous structure, such as in the case of the conversion of plum fruit into the pocket plum galls after *Taphrina pruni* infection (Ogawa et al. 1995). Some pathogens induce formation of tissue alterations only on small parts of an organ, for example, the leaf curling ("peach leaf curl") induced by *Taphrina deformans* (Bassi et al. 1984). The transformation can also occur at small sites, as is the case in the formation of witches' brooms by phytoplasma or various *Taphrina* species among them *T. betulina* or *T. wiesneri* (Spanos and Woodward 1994; Komatsu et al. 2010). In several of these *Taphrina* species the IAA synthesis could be demonstrated in vitro, so it was assumed that they alter plant growth by their own hormone synthesis (Yamada et al. 1990). However, without direct genetic proof it cannot be verified that IAA is indeed a pathogenicity factor. Investigations on auxin synthesis in the smut fungus *Ustilago maydis*, infecting maize kernels and turning these into large gall structures, have shown that the fungus is indeed able to produce IAA and that this IAA is responsible for the increase of auxin in infected kernels, but the tumor formation was not affected in IAA biosynthesis mutants (Basse et al. 1996; Reineke et al. 2008).

## *3.2 Auxin Biosynthesis*

There are different possibilities how pathogens can alter the level of plant hormones, here auxins (Fig. 18.3). First, they can synthesize the hormones by themselves; these are secreted into the host plant and change the growth responses. Second, the plant is taking part in this process. Here, two main possibilities exist: one involves transformation of the plant by genes of the pathogen, but the plant cell itself is making the hormone. The other involves partial synthesis of the hormone by

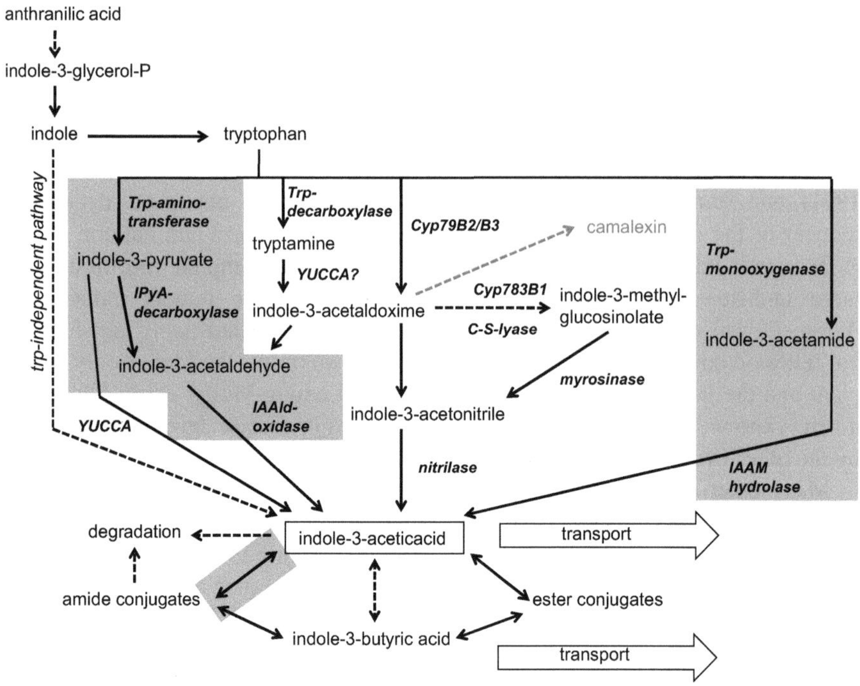


**Fig. 18.3** Pathways of auxin synthesis and metabolism, including those routes that might be utilized by pathogens shaded in *gray*

the pathogen, so that the plant is delivering precursors for the biosynthesis or the pathogen is delivering intermediates in auxin synthesis to the plant. Also, the hydrolysis of inactive auxin conjugates by bacteria, albeit not phytopathogenic ones, has been described (Chou et al. 1998; Chou and Huang 2005), which would lead to an increase in the levels of active plant hormones. Third, the plant is alone responsible for the synthesis of auxin, but the pathogen is somehow stimulating the plant to do so. In addition, the hydrolysis from inactive auxin conjugates can increase the levels of free auxins, which is used by some soil bacteria or symbiotic fungi to stimulate plant root growth for their own benefit. Vice versa, the synthesis of auxin conjugates decreases free active auxin, which results in altered auxin responses, i.e., lower levels of cell wall loosening proteins, which reduces possibilities to penetrate (Ding et al. 2008). Of course, not for all plant–pathogen interactions where auxin might play a role the precise mechanisms have yet been understood.

### Auxin Synthesized by Phytopathogens

Even though the metabolism of auxin is described in Chap. 2, a short overview on biosynthesis and metabolism will be given, including the routes that might be used by the pathogens (Fig. 18.3). There are several possible pathways for auxin biosynthesis, which exist parallel to each other in plants. Some are specific for the family Brassicaceae, including the indole glucosinolates as possible intermediates. The indole phytoalexins, in *Arabidopsis thaliana* it is camalexin (Glawischnig 2007), are also deriving from the indole pathway, but will not be discussed further. Anthranilic acid is converted to indole-3-glycerol-phosphate (IGP) and then to indole, which is then turned into the aromatic amino acid tryptophan (Trp). From either IGP or indole the Trp-independent pathway is leading directly to IAA. Other routes all depend on Trp as precursor. Several of these pathways are used by microbes. First, the one via indole-3-acetamide (IAM) is used by phytopathogenic bacteria, which are able to transform the plant with the two genes for IAM and IAA synthesis, Trp-monooxygenase and IAM-hydrolase (amidase) (Patten and Glick 1996). The second major pathway is that leading via indole-3-pyruvic acid (IPyA) to indole-3-acetaldehyde (IAAld) and IAA. However, microbes also possess nitrilases capable of transforming indole-3-acetonitrile (IAN) to IAA (Yamada et al. 1990). In addition, IAA can be stored as inactive ester or amide conjugates, which can release by hydrolysis the active auxin moiety (reviewed in Ludwig-Müller 2011). These reactions also play a role for the levels of IAA and thus the auxin status of the plant (Fig. 18.3).

Many phytopathogenic and beneficial bacteria can synthesize auxin (for review, see Spaepen et al. 2007). Maybe the most prominent example is the induction of tumors by the soil bacterium *Agrobacterium tumefaciens*. Here, the host is directly genetically transformed with genes involved in auxin and cytokinin synthesis (Zupan et al. 2000). The two genes involved in auxin synthesis encoded on the T-DNA of the Ti-plasmid are known and convert tryptophan to IAM, which in turn yields IAA. Several other phytopathogenic and also beneficial bacteria use this pathway via IAM for IAA synthesis, so it was concluded that these routes might have a common evolutionary origin (Morris 1995). Among the phytopathogens possessing the genes *iaaM* and *iaaH* are *Agrobacterium rhizogenes*, *Pseudomonas savastanoi*, and *Pantoea agglomerans*. The latter also possesses a second pathway via IPyA, encoded on the bacterial chromosome. Only those isolates with the plasmid-encoded IAM pathway were virulent, whereas deletion of the IPyA pathway did not result in the reduction of gall size (Clark et al. 1993). In some bacteria also nitrilases were detected, which converted IAN to IAA (see Spaepen et al. 2007) (Fig. 18.3). However, nitrilases might also be involved in the detoxification of nitriles other than IAN. For *Rhodococcus fascians* it was shown that it can synthesize and secrete IAA (Vandeputte et al. 2005). Interestingly, the synthesis of IAA was only induced when a compound from infected plants was added. In addition, it was shown that the plant symptoms are a result of cytokinins from the bacteria and induction of the auxin biosynthetic pathway via IPyA in the plant resulting in enhanced auxin signaling in infected tissues (Stes et al. 2012).

If a pathogen secretes IAA in vitro, it can be concluded that the complete pathway has to be present in the organism. For some *Taphrina* species it was shown that they possess nitrilase activity (Yamada et al. 1990). In addition, they were able to synthesize IAA from Trp via the IPyA and IAAld pathway in cultures. Interestingly, nitrilase activity was inducible by the substrate, while the other enzymatic reactions were not, implying that maybe the nitrilase pathway might only be used in the plant and the fungus would thus contribute only part of the biosynthetic route (Fig. 18.3). Hence, if the organism is contributing to the plant's IAA synthesis, it cannot be excluded that only part of the biosynthetic route is derived from the pathogen.

The smut fungus *Ustilago maydis* has stimulated many investigations on its possible production of IAA, likely due to the galls formed on the host tissues. Selection of *U. maydis* strains for different IAA levels indicated a correlation between high pathogenicity of the different strains and high IAA content and vice versa, so the authors assumed that IAA is necessary for tumor formation (Guevara-Lara et al. 2000). Genetic evidence showed that *U. maydis* possesses genes for the biosynthesis of IAA via the IPyA pathway, aromatic amino acid aminotransferase, and IAAld dehydrogenase genes and the respective enzymatic activities were confirmed as well as IAA production (Basse et al. 1996; Reineke et al. 2008). Mutants were generated in different combinations, up to quadruple mutant strains, showing that indeed IPyA is a precursor for increased IAA formation in *U. maydis*-induced tumors. However, tumor induction itself was not compromised in these strains (Reineke et al. 2008). Another gall-inducing *Ustilago* species also showed IAA production. *Ustilago esculenta* is the cause of galls in the aquatic perennial grass, *Zizania latifolia* (Chung and Tzeng 2004). The galls had higher IAA levels than control tissues and *Ustilago esculenta* was shown to produce IAA from Trp via IPyA and IAAld similar to *U. maydis*, but the fungus did not convert IAM. In contrast, the sugarcane smut fungus *U. scitaminea* produced less amount of IAA in comparison to the other two species (Chung and Tzeng 2004).

IAA is also produced by phytopathogenic fungi not altering plant growth or development. The biosynthesis of IAA has been detected in the rice blast fungus *Magnaporthe oryzae* during the biotrophic growth phase (Tanaka et al. 2011). Immunolocalization of IAA in infected tissue has localized IAA to infection hyphae of the fungus. Host plants expressing the auxin-responsive DR5::reporter construct showed activation of the GUS gene locally restricted to those cells in contact with the apex of the infection hyphae (Tanaka et al. 2011), indicating that the host can respond to the auxin secreted by the fungus. In two other hemibiotrophic fungal species *Colletotrichum gloeosporioides* and *Colletotrichum acutatum*, also IAA synthesis was observed (Chung et al. 2003; Maor et al. 2004). There are more examples of auxin producing phytopathogenic fungi, which cannot all be mentioned here, but they all highlight the importance of auxin for various diseases.

**Auxin Involved in Pathogenesis and Synthesized by Plants**

It cannot be excluded that in the examples described above the plant is also contributing to the biosynthesis of auxin. Sometimes the involvement of the pathogen can be ruled out and thus the increased levels of IAA are solely the result of altered auxin biosynthesis of the host plant. This is the case for the clubroot disease, caused by *Plasmodiophora brassicae*, a plant pathogenic obligate biotrophic protist on host plants of the Brassicales. Auxin and cytokinin play major roles in this root gall (=club) formation, where auxins are only formed by the host (Ludwig-Müller et al. 2009). For the interaction of *P. brassicae* with its host plants from the family Brassicaceae, indole glucosinolates (Fig. 18.2) have been proposed to play a role in symptom development and the nitrilase pathway of IAA biosynthesis is thought to be the major source of free IAA (reviewed in Ludwig-Müller 2009a). In Arabidopsis nitrilase transcript levels and also IAN as well as indole glucosinolates are increased during pathogenesis (Ludwig-Müller et al. 1999). The nitrilase protein was found to be elevated in cells harboring plasmodia of the pathogen (Grsic-Rausch et al. 2000) and nitrilase mutants displayed smaller gall phenotypes (Neuhaus et al. 2000). In addition, nitrilase transcription was also increased in pathogen-containing cells (Päsold et al. 2010). However, blocking the initial pathway for indole glucosinolate biosynthesis completely did not result in the expected reduction of club size and IAA levels, suggesting that alternative routes also contribute to IAA (Siemens et al. 2008). For different *Brassica* species the role for nitrilase in clubroot formation was confirmed. Ishikawa et al. (2007a) reported an increase in nitrilase transcripts in clubroots compared to controls and Ando et al. (2008) observed an alterative splicing for nitrilase transcripts only in infected roots. In addition, evidence for the involvement of the IAM pathway in *Brassica* root galls was provided (Ishikawa et al. 2007b). As a third pathway in *Brassica* an increase in aldehyde oxidase transcript level accompanied with increased enzymatic activities suggests the activation of the IPyA pathway (Ando et al. 2006). Additionally, the hydrolysis of auxin conjugates was differentially regulated in *B. rapa* clubroots (Schuller and Ludwig-Müller 2006).

## 3.3 *Modulation of Auxin Signaling and Transport of the Plant in Pathogenesis*

Another strategy would be to interfere with the auxin transport or signaling pathway. In some examples it is not easy to distinguish between the involvement of auxin in symptom development or defense. Alterations of the auxin pathway could also lead to the reduction of disease symptoms and could thus be considered as mechanism involved in resistance. By interference with auxin signaling, the plant pathogenic bacterium phytoplasma was able to induce witches' broom symptoms (Hoshi et al. 2009). This was demonstrated by overexpressing a single virulence factor called TENGU, a small secreted protein of this bacterium, in *Nicotiana benthamiana* and Arabidopsis plants. These transgenic plants showed symptoms of witches' broom and dwarfism, the typical symptoms of phytoplasma infection, without being infected. Furthermore, microarray analyses showed that auxin-responsive genes were significantly downregulated in the transgenic plants expressing *TENGU* compared with control plants. The results presented in this work suggest that TENGU inhibits auxin-related signaling and subsequent gene expression pathways, thereby affecting plant development.

IAA can also act as signaling molecule to induce specific genes of the pathogen, which are in turn able to increase the fitness of the colonizing organism (see review by Spaepen et al. 2007). Ideally, this leads then to better colonization. This strategy is obviously of advantage, if the pathogen is the cause of increased IAA production but can also be exploited, if the plant can be triggered to synthesize more auxin (Fig. 18.1).

The auxin transport rate also influences the levels of auxin at specific tissue sites (see also Chap. 5). This is exploited by rhizobia and also plant pathogens. For example, an increase in local auxin levels can be achieved by inhibition of auxin efflux. In the case of phytoplasma infection a reduced auxin efflux has been observed in samples showing strong witches' broom symptoms on apple trees (Aldaghi et al. 2009). In the case of the clubroot disease several lines of evidence point to an altered auxin transport through club development. Some evidence has accumulated that flavonoids might act as auxin efflux modulators in clubroots, thereby contributing to increased auxin levels in galls (Päsold et al. 2010). Indeed, blocking IAA transport during early stages of infection disturbs clubroot development (Devos and Prinsen 2006). In addition, an Arabidopsis mutant *alh1* was more tolerant to clubroot (Devos et al. 2006). This mutant has a defect in the cross talk between ethylene and auxins, probably at the level of auxin transport (Vandenbussche et al. 2003). Consequently, it was proposed that the mutant was resistant because host IAA transport was hampered.

## 3.4 Auxin as Indirect Factor to Influence Pathogenicity

Since auxins are key regulators of many developmental processes, alterations of auxin levels could result in phenotypical changes associated with colonization events by pathogens. The cell wall is one example for this observation. It is well known that auxins can induce loosening of plant cell walls (Cosgrove 1993). Consequently, the stimulation of the expression of genes encoding proteins involved in cell wall loosening are upregulated by auxin, for example, Xyloglucan endoTransglycosylase/Hydrolase (XTH) (Yokoyama and Nishitani 2001) and expansins (Cosgrove et al. 2002). For the hypertrophied root galls induced by *P. brassicae* a loosening of cell wall by expansins has been postulated based on microarray experiments (Siemens et al. 2006; Ludwig-Müller 2009b). Also, the higher activity of a XTH was shown early in club development (Devos et al. 2005). The advantage for the protist is that more resting spores can be formed in larger cells of the host. Reduction of these protein activities should therefore result in reduced disease symptoms.

Contrary, strengthening the cell wall could be beneficial in the case of penetration by hemibiotrophic fungi. Since expansins are controlled partially by auxin, the reduction of active IAA from the total auxin pool should reduce expansin levels and thus render the cell wall more rigid. Overexpression of a gene encoding an IAA amino acid conjugate synthetase *GH3.8* in rice resulted in smaller cells and lesser disease symptoms after infection with bacteria presumably based on reduced expansin production (Ding et al. 2008), because it was shown that overexpression of expansin genes renders the plant more susceptible. Similarly, two additional rice *GH3* genes *GH3.1* and *GH3.2* confer broad-spectrum resistance against *Xanthomonas* species and *Magnaporthe grisea* to rice plants by suppressing pathogen-induced accumulation of IAA (Domingo et al. 2009; Fu et al. 2011). The conjugate of IAA with the amino acid aspartate also plays a role in disease development (González-Lamothe et al. 2012). Contrary to the results obtained from rice, the conjugation to the amino acid renders the Arabidopsis plant more susceptible to infection with *Botrytis cinerea* and *Pseudomonas syringae*. This work has shown that infection leads to transcriptional activation of the Arabidopsis *GH3.2* gene and consequently to accumulation of IAA-Asp. It is interesting to note that IAA-Asp was categorized as auxin conjugate, which is rather degraded than hydrolyzed (Fig. 18.3) (reviewed in Ludwig-Müller 2011). IAA-Asp was able to promote disease development as a compound in several plant species, indicating that the conjugate itself and not low levels of IAA might be responsible for disease severity (González-Lamothe et al. 2012). For the induction of GH3.5 in Arabidopsis a cross talk between auxin and salicylic acid (SA) is discussed for this protein, because it can convert both IAA and SA to amino acid conjugates (Zhang et al. 2007). The authors hypothesized that GH3.5 might play dual roles in disease resistance and susceptibility against avirulent or virulent *Pseudomonas* races, respectively. The dual enzymatic function could be the consequence of the evolution of an auxin-conjugating enzyme, which promotes susceptibility to the bacteria, but also capable of regulating the SA pathway to trigger plant defense responses (Fig. 18.4).

## 4 Auxin in Plant Defense Reactions

### *4.1 Auxin as a Signal Triggering Plant Defense*

Since auxins have been recognized to be important for plant defense (Kazan and Manners 2009), many investigations have been conducted to elucidate the pathways involved (Fu and Wang 2011). Some studies have demonstrated that exogenous

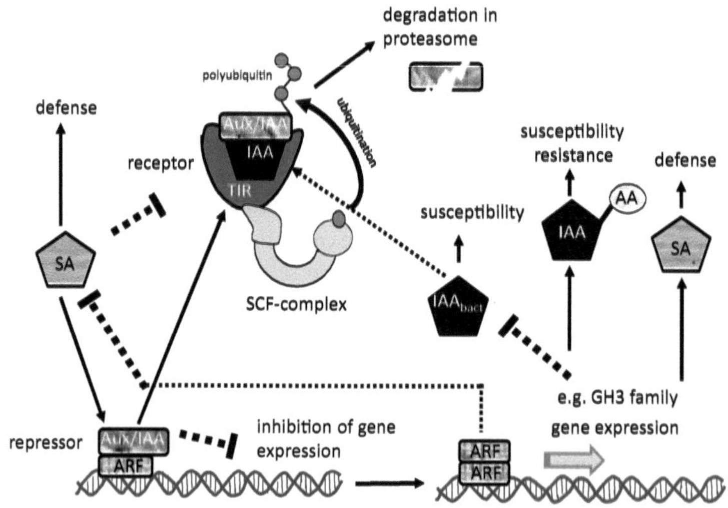


**Fig. 18.4** Regulation of auxin signaling with respect to defense mechanisms and also salicylic acid. Auxin is perceived by the TIR family of receptors, leading to the degradation of transcriptional inhibitors, the Aux/IAA proteins. That in turn allows auxin-dependent gene expression to occur. Among the auxin-inducible genes are the genes of the GH3 family involved in the conjugation of IAA, jasmonic acid, and salicylic acid (SA). Also there is a cross talk between auxin and SA at different levels. ARFs negatively regulate SA, but SA induces Aux/IAA function and inhibits TIR1 signaling

IAA induces a resistance response (for example Ueno et al. 2004; Terrile et al. 2006). Over the recent years it was shown that auxin influences resistance against pathogens, probably because many plant pathogenic microorganisms produce auxins during their interactions with plants. An infection with *Pseudomonas syringae* pv. *tomato* (Pst) DC3000, but also the expression of *P. syringae* type III effector AvrRpt2 in plants increased the level of IAA, leading to enhanced disease development (Chen et al. 2007). Here, a fast inactivation of excess IAA would help the plant in the resistance response, suggesting that auxin homeostasis plays an important role in direct or indirect defense responses. Downregulation of auxin signaling by bacterial PAMPs (pathogen-associated molecular patterns), specifically by the flagellin *flg22*-dependent recognition in Arabidopsis, which play an important role in the first basal defense reaction of a plant, enhances bacterial disease resistance in Arabidopsis (Navarro et al. 2006; Wang et al. 2007). The repression results from stabilization of Aux/IAA proteins via either microRNA- (Navarro et al. 2006) or SA-mediated (Wang et al. 2007) negative regulation of F-box auxin receptors. Downregulation of TIR and stabilization of Aux/IAA proteins occur simultaneously (see also Chap. 6). Thus, repression of auxin signaling by the SA pathway contributes to antibacterial resistance. By showing that a knockout mutant in a GDSL lipase has a more susceptible phenotype concerning some pathogens, it could be established that auxin signaling is also involved in this defense response. In comparison to wild-type plants the mutants exhibited enhanced auxin responses and elevated Aux/IAA gene expression (Lee et al. 2009b). When the plants were inoculated with the necrotropic bacteria *Erwinia carotovora*, *glip2* mutants were more susceptible than wild-type plants. Thus, GLIP2 is an additional factor, which plays a role in plant defense by negatively regulating the auxin-signaling pathway.

Another plant defense strategy, occurring after the PAMP-induced immunity, is the hypersensitive response (HR), characterized by necrosis of plant cells in the inoculated area. By this mechanism mainly biotrophic pathogens are excluded from living tissue and limited to the inoculation site. It was shown that auxin produced by the bacterium *Pseudomonas savastanoi* is required to block the HR in tobacco leaves induced by a second phytopathogenic bacterium *Pseudomonas syringae* pv. *phaseolicola* (Robinette and Matthysse 1990). The results were extended by using a bacterial elicitor, harpin, from the phytopathogen *Erwinia amylovora* (Gopalan 2008). Like the bacterium itself, the elicitor can induce cell death on plant leaves. Auxin was able to inhibit this HR over a wide period of time during the programmed cell death response, but gene expression for local and systemic resistance responses were not affected by the treatment. Interestingly, plants overproducing auxin did not show this reversal of HR (Gopalan 2008), indicating that exogenous auxin is perceived differently than the endogenously produced hormone.

In the case of necrotrophic fungi the functional auxin-signaling pathway is needed for the resistance response. The strategy of these organisms is therefore to suppress auxin signaling. For the resistance against the fungi *Plectosphaerella cucumerina* and *Botrytis cinerea* it was shown that the components of the auxin signaling pathway need to be intact (Llorente et al. 2008), while repression of the auxin-response pathway, either in mutants or by a pharmacological approach blocking proteasome function, increases the pathogenicity in Arabidopsis. It has to be assumed that the fungi might possess mechanisms to suppress this signaling pathway. Similarly, the fungus *Macrophomina phaseolina*, causing the charcoal rot disease on many plant species suppresses the auxin signaling response of host plants, here *Medicago truncatula* (Mah et al. 2012). Transcriptome analysis revealed the downregulation of genes involved in auxin homeostasis, transport, and signal transduction during the disease progression, suggesting that plant susceptibility is closely connected to this downregulation. Hence, treatment with auxin reduced disease symptoms. On the contrary, resistance to the obligate biotrophic oomycete *Hyaloperonospora arabidopsidis* (formerly *H. paràsitica*) was not dependent on the TIR-pathway of auxin signal transduction (Llorente et al. 2008), suggesting that an adaptation has occurred for different pathogens to the existing auxin pathways.

## 4.2 Auxin as Direct Defense Molecule

There are some indications that the different auxins naturally occurring in plant species might be directly toxic to pathogenic microbes. However, other organisms might have developed a tolerance for auxin, because they themselves synthesize IAA like *Agrobacterium tumefaciens*. Despite this assumption, one investigation has shown that eight out of ten plant-associated bacteria, including one *A. tumefaciens* strain, were inhibited in their growth by IAA, while this was not the case using the same IAA concentration on bacteria not in nature associated with plants (Liu and Nester 2006). The authors concluded that, even if the number of organisms tested was small, a bias was found toward growth inhibition by IAA of plant-associated bacteria. A decrease of phytoplasma cells in infected *Catharanthus roseus* treated with high concentrations of auxins were reported (Pertot et al. 1998). This result supports the idea that auxins may interfere with the phytoplasma growth in the plant host. Certain phytoplasma strains were completely reduced in *C. roseus* after treatment with IBA (Curkovic-Perica et al. 2007).

It was reported by Walker et al. (2003) that indole-3-propionic acid (IPA) was one of the metabolites found in the root exudate of *Arabidopsis* upon treatment with SA, which is usually produced by plants as a response to microbial attack. In *Brassica rapa* an auxin conjugate hydrolase was identified with high preference for conjugates of IPA and the amino acid alanine (Savić et al. 2009). A function for this hydrolase could not be determined, but it was suggested that maybe IPA could act as antimicrobial compound, because it was shown that IPA and some derivatives are partially toxic for phytopathogenic bacteria, i.e., *Pseudomonas solanacearum* and *Ralstonia solanacearum* (Matsuda et al. 1993, 1998). This observation makes those indoles possibly interesting as antimicrobial agents. Corroborating the possibility of antimicrobial compounds is the developed of resistance to IPA through conjugation with amino acids and sugars by some bacteria like *Bacillus megatherium* (Tabone and Tabone 1953; Tabone 1958).

The common scab disease of potato is caused by *Streptomyces* spp. It was shown that the foliar application of auxin-like compounds such as 2,4-dichlorophenoxyacetic acid (2,4-D) and others somehow systemically reduced severity and occurrence of common scab in subsequently produced tubers (Tegg et al. 2008). Virulence was dependent on the production of a toxin by the pathogen and the treatment with 2,4-D led to increased tolerance to the pathogen-derived toxin in tubers. Confirmatory evidence came from studies with *Arabidopsis thaliana* seedlings demonstrating that 2,4-D and IAA reduced toxicity of the virulence factor secreted by the pathogen (Tegg et al. 2008).

The amount of IAA synthesized by the pathogen might also influence the balance of the interaction toward susceptibility or tolerance in the case of the host plant. While low auxin synthesis levels of the oomycete *Pythium ultimum* were increasing disease symptoms on tomato, higher levels induced plant defense reactions (Gravel et al. 2007). This shows that the pathogen itself can trigger the outcome of the interaction with its host. The small amounts of IAA might mimic the host levels and are thus not detected by the plant, while larger amounts of IAA trigger the defense responses, because they have been sensed as unusual. This is an attractive hypothesis that of course needs experimental support.

In experiments controlling head blight of barley by *Fusarium culmorum* it could be shown that IAA treatment effectively reduced the disease, while treatment with abscisic acid did not (Petti et al. 2012). The study was conducted to find out whether IAA is involved in biocontrol of the disease by a biocontrol agent *Pseudomonas fluorescence*. The bacterium is contributing to the higher IAA levels in co-inoculated plants; thus IAA synthesis by beneficial bacteria could be one of the puzzle stones to achieve control of plant diseases.

## 5 Concluding Remarks

As much as auxin regulates a plethora of plant developmental processes, its role in the interaction with other organisms is likewise of plentiful nature. Some examples were given how auxin is involved in the interaction between plants and beneficial organisms as well as pathogens. The auxin response of the plant can be used to transform the host tissue into a paradise for the parasites by either synthesizing auxin themselves or inducing the plant's auxin metabolism. On the other hand, the plant is using this molecule as signal in its defense responses against the pathogens. These are co-evolutionary processes, which result in complicated networks, still needed to be completely unraveled in the future.

## References

Aldaghi M, Massart S, Bertaccini A, Jijakli MH, Lepoivre P (2009) Identification of host genes potentially implicated in the *Malus pumila* and '*Candidatus Phytoplasma mali*' interactions. In: Proceedings of the 21st ICVF conference, Berichte aus dem Julius Kühn-Institut, Neustadt, Germany, pp 43–44

Altamura MM (2004) Agrobacterium rhizogenes rolB and rolD genes: regulation and involvement in plant development. Plant Cell Tissue Organ Cult 77:89–101

Ando S, Tsushima S, Tagiri A, Kamachi S, Konagaya K-I, Hagio T, Tabei Y (2006) Increase in *BrAO1* gene expression and aldehyde oxidase activity during clubroot development in Chinese cabbage (*Brassica rapa* L.). Mol. Plant Pathol 7:223–234

Ando S, Tsushima S, Kamachi S, Konagaya K, Tabei Y (2008) Alternative transcription initiation of the nitrilase gene (*BrNIT2*) caused by infection with *Plasmodiophora brassicae* Woron. in Chinese cabbage (*Brassica rapa* L.). Plant Mol Biol 68:557–569

Basse CW, Lottspeich F, Steglich W, Kahmann R (1996) Two potential indole-3-acetaldehyde dehydrogenases in the phytopathogenic fungus *Ustilago maydis*. Eur J Biochem 242:648–656

Bassi M, Conti GG, Barbieri N (1984) Cell wall degradation by *Taphrina deformans* in host leaf cells. Mycopathologia 88:115–125

Campanella JJ, Smith SM, Leibu D, Wexler S, Ludwig-Müller J (2008) The auxin conjugate hydrolase family of *Medicago truncatula* and their expression during the interaction with two symbionts. J Plant Growth Regul 27:26–38

Canfield ML, Baca S, Moore LW (1986) Isolation of *Pseudomonas syringae* from 40 cultivars of diseased woody plants with tip dieback in Pacific Northwest nurseries. Plant Dis 70:647–650

Chen Z, Agnew JL, Cohen JD, He P, Shan L, Sheen J, Kunkel BN (2007) *Pseudomonas syringae* type III effector AvrRpt2 alters *Arabidopsis thaliana* auxin physiology. Proc Natl Acad Sci USA 104:20131–20136

Chou J-C, Huang Y-B (2005) Induction and characterization of an indole-3-acetyl-L-alanine hydrolase from *Arthrobacter ilicis*. J Plant Growth Regul 24:11–18

Chou J-C, Mulbry WW, Cohen JD (1998) The gene for indole-3-acetyl-L-aspartic acid hydrolase from *Enterobacter agglomerans*: molecular cloning, nucleotide sequence, and expression in *Escherichia coli*. Mol Gen Genet 259:172–178

Chung KR, Tzeng DD (2004) Biosynthesis of indole-3-acetic acid by the gall-inducing fungus *Ustilago esculenta*. J Biol Sci 4:744–750

Chung K-R, Shilts T, Ertürk Ü, Timmer LW, Ueng PP (2003) Indole derivatives produced by the fungus *Colletotrichum acutatum* causing lime anthracnose and postbloom fruit drop of citrus. FEMS Microbiol Lett 226:23–30

Ciafardini G (1991) Evaluation of *Polymyxa betae* Keskin contaminated by Beet Necrotic Yellow Vein Virus in soil. Appl Environ Microbiol 57:1817–1821

Clark E, Vigodskyhaas H, Gafni Y (1989) Characteristics in tissue-culture of hyperplasias induced by *Erwinia herbicola* pathovar *gypsophilae*. Physiol Mol Plant Pathol 35:383–390

Clark E, Manulis S, Ophir Y, Barash I, Gafni Y (1993) Cloning and characterization of *iaaM* and *iaaH* from *Erwinia herbicola* pathovar *gypsophilae*. Phytopathology 83:234–240

Cosgrove D (1993) How do plant cell walls extend? Plant Physiol 102:1–6

Cosgrove D, Li L, Cho H, Hoffmann-Benning S, Moore R, Blecker D (2002) The growing world of expansins. Plant Cell Physiol 43:1436–1444

Crespi M, Messens E, Caplan AB, van Montagu M, Desomer J (1992) Fasciation induction by the phytopathogen *Rhodococcus fascians* depends upon a linear plasmid encoding a cytokinin synthase gene. EMBO J 11:795–804

Curkovic-Perica M, Lepedus H, Music MS (2007) Effect of indole-3-butyric acid on phytoplasmas in infected *Catharanthus roseus* shoots grown in vitro. FEMS Microbiol Lett 268:171–177

Devos S, Prinsen E (2006) Plant hormones: a key in clubroot development. Commun Agric Appl Biol Sci 71((3 Pt B)):869–872

Devos S, Vissenberg K, Verbelen J-P, Prinsen E (2005) Infection of Chinese cabbage by *Plasmodiophora brassicae* leads to a stimulation of plant growth: inpacts on cell wall metabolism and hormone balance. New Phytol 166:241–250

Devos S, Laukens K, Deckers P, Van Der Straeten D, Beeckman T, Inze D, van Onckelen H, Witters E, Prinsen E (2006) A hormone and proteome approach to picturing the initial metabolic events during *Plasmodiophora brassicae* infection on Arabidopsis. Mol Plant Microbe Interact 19:1431–1433

Ding X, Cao Y, Huang L, Zhao J, Xu C, Li X, Wang S (2008) Activation of the indole-3-acetic acid–amido synthetase GH3-8 suppresses expansin expression and promotes salicylate- and jasmonate-independent basal immunity in rice. Plant Cell 20:228–240

Domingo C, Andres F, Tharreau D, Iglesias DJ, Talon M (2009) Constitutive expression of *OsGH3.1* reduces auxin content and enhances defense response and resistance to a fungal pathogen in rice. Mol Plant-Microbe Interact 22:201–210

Epstein E, Ludwig-Müller J (1993) Indole-3-butyric acid in plants: occurrence, synthesis, metabolism and transport. Physiol Plant 88:382–389

Fallik E, Okon Y, Epstein E, Goldman A, Fischer M (1989) Identification and quantification of IAA and IBA in *Azospirillum brasilense*-inoculated maize roots. Soil Biol Biochem 21:147–153

Fitze D, Wiepning A, Kaldorf M, Ludwig-Müller J (2005) Auxins in the development of an arbuscular mycorrhizal symbiosis in maize. J Plant Physiol 162:1210–1219

Fu J, Wang S (2011) Insights into auxin signaling in plant-pathogen interactions. Front Plant Sci 2:74. doi:10.3389/fpls.2011.00074

Fu J, Liu H, Li Y, Yu H, Li X, Xiao J, Wang S (2011) Manipulating broad-spectrum disease resistance by suppressing pathogen-induced auxin accumulation in rice. Plant Physiol 155:589–602

Georgiev MI, Ludwig-Müller J, Bley T (2010) Hairy root culture: copying nature in new bioprocesses. In: Arora R (ed) Medicinal plant biotechnology. CAB International, Oxon, pp 156–175

Glawischnig E (2007) Camalexin. Phytochemistry 68:401–406

Goethals K, Vereecke D, Jaziri M, Van Montagu M, Holsters M (2001) Leafy gall formation by *Rhodococcus fascians*. Annu Rev Phytopathol 39:27–52

González-Lamothe R, El Oirdi M, Brisson N, Bouarab K (2012) The conjugated auxin indole-3-acetic acid–aspartic acid promotes plant disease development. Plant Cell 24:762–777

Gopalan S (2008) Reversal of an immunity associated plant cell death program by the growth regulator auxin. BMC Res Notes 1:126

Gravel V, Antoun H, Tweddell R (2007) Effect of indole-acetic acid (IAA) on the development of symptoms caused by *Pythium ultimum* on tomato plants. Eur J Plant Pathol 119:457–462

Grsic-Rausch S, Kobelt P, Siemens J, Bischoff M, Ludwig-Müller J (2000) Expression and localization of nitrilase during symptom development of the clubroot disease in *Arabidopsis thaliana*. Plant Physiol 122:369–378

Guevara-Lara F, Valverde ME, Paredes-Lopez O (2000) Is pathogenicity of *Ustilago maydis* (huitlacoche) strains on maize related to in vitro production of indole-3-acetic acid? World J Microbiol Biotechnol 16:481–490

Hirsch A (1992) Developmental biology of legume nodulation. New Phytol 122:211–237

Hoshi A, Oshima K, Kakizawa S, Ishii Y, Ozeki J, Hashimoto M, Komatsu K, Kagiwada S, Yamaji Y, Namba S (2009) A unique virulence factor for proliferation and dwarfism in plants identified from a phytopathogenic bacterium. Proc Natl Acad Sci USA 106:6416–6421

Huo X, Schnabel E, Hughes K, Frugoli J (2006) RNAi phenotypes and the localization of a protein:GUS fusion imply a role for Medicago truncatula PIN genes in nodulation. J Plant Growth Regul 25:156–165

Ishikawa T, Okazaki K, Kuroda H, Itoh K, Mitsui T, Hori H (2007a) Molecular cloning of Brassica rapa nitrilases and their expression during clubroot development. Mol Plant Pathol 8:623–637

Ishikawa T, Kuroda H, Okazaki K, Itoh K, Mitsui T, Hori H (2007b) Evaluation of roles of amidase which converts indole-3-acetamide to indole-3-acetic acid, in formation of clubroot in turnip. Bull Facul Agric Niigata Univ 60:53–60

Jentschel K, Thiel D, Rehn F, Ludwig-Müller J (2007) Arbuscular mycorrhiza enhances auxin levels and alters auxin biosynthesis in *Tropaeolum majus* during early stages of colonization. Physiol Plant 129:320–333

Kaldorf M, Ludwig-Müller J (2000) AM fungi might affect the root morphology of maize by increasing indole-3-butyric acid biosynthesis. Physiol Plant 109:58–67

Kazan K, Manners J (2009) Linking development to defense: auxin in plant-pathogen interactions. Trends Plant Sci 14:373–382

Kefford NP, Brockwell J, Zwar JA (1960) The symbiotic synthesis of auxin by legumes and nodule bacteria and its role in nodule development. Aust J Biol Sci 13:456–467

Komatsu M, Taniguchi M, Matsushita N, Takahashi Y, Hogetsu T (2010) Overwintering of Taphrina wiesneri within cherry shoots monitored with species-specific PCR. J Gen Plant Pathol 76:363–369

Krishna P (2003) Brassinosteroid-mediated stress responses. J Plant Growth Regul 22:289–297

Laidlaw WMR (1985) A method for the detection of resting sporangia of the potato wart disease (*Synchytrium endobioticum*) in the soil of old outbreak sites. Potato Res 28:223–232

Lee C-W, Efetova M, Engelmann JC, Kramell R, Wasternack C, Ludwig-Müller J, Hedrich R, Deeken R (2009a) *Agrobacterium tumefaciens* promotes tumor induction by modulating pathogen defense in *Arabidopsis thaliana*. Plant Cell 21:2948–2962

Lee DS, Kim BK, Kwon SJ, Jin HC, Park OK (2009b) Arabidopsis GDSL lipase 2 plays a role in pathogen defense via negative regulation of auxin signaling. Biochem Biophys Res Commun 379:1038–1042

Liu P, Nester EW (2006) Indoleacetic acid, a product of transferred DNA, inhibits *vir* gene expression and growth of *Agrobacterium tumefaciens* C58. Proc Natl Acad Sci USA 103:4658–4662

Llorente F, Muskett P, Sanchez-Vallet A, Lopez G, Ramos B, Sanchez-Rodriguez C, Jorda L, Parker J, Molina A (2008) Repression of the auxin response pathway increases Arabidopsis susceptibility to necrotrophic fungi. Mol Plant 1:496–509

Long SR (2001) Genes and signals in the *Rhizobium*-legume symbiosis. Plant Physiol 125:69–72

Ludwig-Müller J (2000) Indole-3-butyric acid in plant growth and development. Plant Growth Regul 32:219–230

Ludwig-Müller J (2009a) Glucosinolates and the clubroot disease: defense compounds or auxin precursors? Phytochem Rev 8:135–148

Ludwig-Müller J (2009b) Plant defense – what can we learn from clubroots? Aust Plant Pathol 38:318–324

Ludwig-Müller J (2011) Auxin conjugates: their role for plant development and in the evolution of land plants. J Exp Bot 62:1757–1773

Ludwig-Müller J, Cohen JD (2002) Identification and quantification of three active auxins in different tissues of *Tropaeolum majus*. Physiol Plant 115:320–329

Ludwig-Müller J, Pieper K, Ruppel M, Cohen JD, Epstein E, Kiddle G, Bennett R (1999) Indole glucosinolate and auxin biosynthesis in *Arabidopsis thaliana* L. glucosinolate mutants and the development of the clubroot disease. Planta 208:409–419

Ludwig-Müller J, Prinsen E, Rolfe S, Scholes J (2009) Metabolism and plant hormone action during the clubroot disease. J Plant Growth Regul 28:229–244

Mah KM, Uppalapati SR, Tang Y, Allen S, Shuai B (2012) Gene expression profiling of *Macrophomina phaseolina* infected *Medicago truncatula* roots reveals a role for auxin in plant tolerance against the charcoal rot pathogen. Physiol Mol Plant Pathol 79:21–30

Maor R, Haskin S, Levi-Kedmi H, Sharon A (2004) *In planta* production of indole-3-acetic acid by *Colletotrichum gloeosporioides* f. sp. *aeschynomene*. Appl Environ Microbiol 70:1852–1854

Matas IM, Perez-Martinez I, Quesada JM, Rodriguez-Herva JJ, Penyalver R, Ramos C (2009) *Pseudomonas savastanoi* pv. *savastanoi* contains two *iaaL* paralogs, one of which exhibits a variable number of a trinucleotide (TAC) tandem repeat. Appl Environ Microbiol 75:1030–1035

Mathesius U (2008) Auxin: at the root of nodule development? Funct Plant Biol 35:651–668

Matsuda K, Toyoda H, Yokoyama K, Wakita K, Nishio H, Nishida T et al (1993) Growth inhibition of *Pseudomonas solanacearum* by substituted 3-indolepropionic acids and related compounds. Biosci Biotechnol Biochem 57:1766–1767

Matsuda K, Toyoda H, Nishio H, Nishida T, Bingo M, Matsuda Y et al (1998) Control of the bacterial wilt of tomato plants by a derivative of 3-indolepropionic acid based on selective actions on *Ralstonia solanacearum*. J Agric Food Chem 46:4416–4419

Meixner C, Ludwig-Müller J, Miersch O, Gresshoff P, Staehelin C, Vierheilig H (2005) Lack of mycorrhizal autoregulation and phytohormonal changes in the supernodulating soybean mutant nts1007. Planta 222:709–715

Merz U, Falloon RE (2009) Review: powdery scab of potato - increased knowledge of pathogen biology and disease epidemiology for effective disease management. Potato Res 52:17–37

Mitchell JE, Angel CR (1951) The growth-stimulating properties of a metabolic product of *Fusarium moniliforme*. Phytopathology 41:26–27

Morris RO (1995) Genes specifying auxin and cytokinin biosynthesis in prokaryotes. In: Davies PJ (ed) Plant hormones. Kluwer Academic, Dordrecht, pp 318–339

Navarro L, Dunoyer P, Jay F, Arnold B, Dharmasiri N, Estelle M, Voinnet O, Jones JDG (2006) A plant miRNA contributes to antibacterial resistance by repressing auxin signaling. Science 312:436–439

Nemoto K, Hara M, Goto S, Kasai K, Seki H, Suzuki M, Oka A, Muranaka T, Mano Y (2009) The *aux1* gene of the Ri plasmid is sufficient to confer auxin autotrophy in tobacco BY-2 cells. J Plant Physiol 166:729–738

Neuhaus K, Grsic-Rausch S, Sauerteig S, Ludwig-Müller J (2000) *Arabidopsis* plants transformed with nitrilase 1 or 2 in antisense direction are delayed in clubroot development. J Plant Physiol 156:756–761

Oelmüller R, Sherameti I, Tripathi S, Varma A (2009) *Piriformospora indica*, a cultivable root endophyte with multiple biotechnological applications. Symbiosis 49:1–17

Ogawa JM, Zehr EI, Bird GW, Ritchie DF, Uriu K, Uyemoto JK (1995) Compendium of stone fruit diseases. APS Press, St. Paul, MN

Päsold S, Siegel I, Seidel C, Ludwig-Müller J (2010) Flavonoid accumulation in *Arabidopsis thaliana* root galls caused by the obligate biotrophic pathogen *Plasmodiophora brassicae*. Mol Plant Pathol 11:545–562

Patten CL, Glick BR (1996) Bacterial biosynthesis of indole-3-acetic acid. Can J Microbiol 42:207–220

Patten CL, Glick BR (2002) Role of *Pseudomonas putida* indole-acetic acid in development of the host plant root system. Appl Environ Microbiol 68:3795–3801

Peer WA, Bandyopadhyay A, Blakeslee JJ, Makam SN, Chen RJ, Masson PH, Murphy AS (2004) Variation in expression and protein localization of the PIN family of auxin efflux facilitator proteins in flavonoid mutants with altered auxin transport in *Arabidopsis thaliana*. Plant Cell 16:1898–1911

Pertot I, Musetti R, Pressacco L, Osler R (1998) Changes in Indole-3-acetic acid level in micropropagated tissues of *Catharanthus roseus* L. infected by the agent of the clover phyllody and effect of exogenous auxins on phytoplasma morphology. Cytobios 95:13–23

Petti C, Reiber K, Ali SS, Berney M, Doohan FM (2012) Auxin as a player in the biocontrol of Fusarium head blight disease of barley and its potential as a disease control agent. BMC Plant Biol 12:224

Ramos C, Matas IM, Bardaji L, Aragon IM, Murillo J (2012) *Pseudomonas savastanoi* pv. *savastanoi*: some like it knot. Mol. Plant Pathol 13:998–1009

Reineke G, Heinze B, Schirawaski J, Buettner H, Kahmann R, Basse CW (2008) Indole-3-acetic acid (IAA) biosynthesis in the smut fungus *Ustilago maydis* and its relevance for increased IAA levels in infected tissue and host tumour formation. Mol Plant Pathol 9:339–355

Robinette D, Matthysse AG (1990) Inhibition by *Agrobacterium tumefaciens* and *Pseudomonas savastanoi* of development of the hypersensitive response elicited by *Pseudomonas syringae* pv *phaseolicola*. J Bacteriol 172:5742–5749

Savić B, Tomić S, Magnus V, Gruden K, Barle K, Grenković R, Ludwig-Müller J, Salopek-Sondi B (2009) Auxin amidohydrolases from *Brassica rapa* cleave the alanine conjugate of indolepropionic acid as a preferable substrate: a biochemical and modeling approach. Plant Cell Physiol 50:1587–1599

Schuller A, Ludwig-Müller J (2006) A family of auxin conjugate hydrolases from *Brassica rapa*: characterization and expression during clubroot disease. New Phytol 171:145–158

Siemens J, Keller I, Sarx J, Kunz S, Schuller A, Nagel W, Schmülling T, Parniske M, Ludwig-Müller J (2006) Transcriptome analysis of *Arabidopsis* clubroots indicate a key role for cytokinins in disease development. Mol Plant-Microbe Interact 19:480–494

Siemens J, Glawischnig E, Ludwig-Müller J (2008) Indole glucosinolates and camalexin do not influence the development of the clubroot disease in *Arabidopsis thaliana*. J Phytopathol 156:332–337

Sirrenberg A, Göbel C, Grond S, Czempinski N, Ratzinger A, Karlovsky P, Santos P, Feussner I, Pawlowski K (2007) *Piriformospora indica* affects plant growth by auxin production. Physiol Plant 13:581–589

Spaepen S, Vanderleyden J, Remans R (2007) Indole-3-acetic acid in microbial and microorganism-plant signaling. FEMS Microbiol Rev 31:425–448

Spanos YA, Woodward S (1994) The effects of *Taphrina betulina* infection on growth of *Betula pubescens*. Eur J Forest Pathol 24:277–286

Splivallo R, Fischer U, Göbel C, Feussner I, Karlovsky P (2009) Truffles regulate plant root morphogenesis *via* the production of auxin and ethylene. Plant Physiol 150:2018–2029

Stes E, Prinsen E, Holsters M, Vereecke D (2012) Plant-derived auxin plays an accessory role in symptom development upon *Rhodococcus fascians* infection. Plant J 70:513–527

Tabone D (1958) Biosynthèse par *B. megatherium* de combinaisons de l'acide indol propionique avec certains acides aminés. Bull Soc Chim Biol 40:5–6

Tabone J, Tabone D (1953) Bio-esterification du glucose. V. Bio-synthèse par Bacillus megatherium de l'ester β glucosidique de l'acide indolpropionique. C R Hebd Seances Acad Sci 237:943–944

Tanaka E, Koga H, Mori M, Mori M (2011) Auxin production by the rice blast fungus and its localization in host tissue. J Phytopathol 159:522–530

Tegg RS, Gill WM, Thompson HK, Davies NW, Ross JJ, WIlson CR (2008) Auxin-induced resistance to common scab disease of potato linked to inhibition of thaxtomin A toxicity. Plant Dis 92:1321–1328

Terrile M, Olivieri F, Bottini R, Casalongue C (2006) Indole-3-acetic acid attenuates the fungal lesions in infected potato tubers. Physiol Plant 127:205–211

Ueno M, Kihara J, Honda Y, Arase S (2004) Indole-related compounds induce the resistance to rice blast fungus, *Magnaporthe grisea* in barley. J Phytopathol 152:606–612

Vadassery J, Ritter C, Venus Y, Camehl I, Varma A, Shahollari B, Novák O, Strnad M, Ludwig-Müller J, Oelmüller R (2008) The role of auxins and cytokinins in the mutualistic interaction between *Arabidopsis* and *Piriformospora indica*. Mol Plant-Microbe Interact 21:1371–1383

Vandenbussche F, Smalle J, Le J, Saibo NJM, De Paepe A, Chaerle L, Tietz O, Smets R, Laarhoven LJJ, Harren FJM, Van Onckelen H, Palme K, Verbelen J-P, Van Der Straeten D (2003) The Arabidopsis mutant *alh1* illustrates a cross talk between ethylene and auxin. Plant Physiol 131:1228–1238

Vandeputte O, Öden S, Mol A, Vereecke D, Goethals K, El Jaziri M, Prinsen E (2005) Biosynthesis of auxin by the gram-positive phytopathogen *Rhodococcus fascians* is controlled by compounds specific to infected plant tissues. Appl Environ Microbiol 71:1169–1177

Walker TS, Bais HP, Halligan KM, Stermitz FR, Vivanco JM (2003) Metabolic profiling of root exudates of *Arabidopsis thaliana*. J Agric Food Chem 51:2548–2554

Wang D, Pajerowska-Mukhtar K, Culler A, Dong X (2007) Salicylic acid inhibits pathogen growth in plants through repression of the auxin signaling pathway. Curr Biol 17:1784–1790

Wasson AP, Pellerone FI, Mathesius U (2006) Silencing the flavonoid pathway in *Medicago truncatula* inhibits root nodule formation and prevents auxin transport regulation by rhizobia. Plant Cell 18:1617–1629

Yamada T, Tsukamoto H, Shirashi T, Nomura T, Oku H (1990) Detection of indoleacetic acid biosynthesis in some species of *Taphrina* causing hyperplastic diseases in plants. Ann Phytopath Soc Jpn 56:532–540

Yokoyama R, Nishitani K (2001) A comprehensive expression analysis of all members of a gene family encoding cell-wall enzymes allowed us to predict cis-regulatory regions involved in cell-wall construction in specific organs of Arabidopsis. Plant Cell Physiol 42:1025–1033

Zhang Z, Li Q, Li Z, Staswick PE, Wang M, Zhu Y, He Z (2007) Dual regulation role of *GH3.5* in salicylic acid and auxin signaling during Arabidopsis-*Pseudomonas syringae* interaction. Plant Physiol 145:450–464

Zupan JR, Zambryski P (1995) Transfer of T-DNA from Agrobacterium to the plant cell. Plant Physiol 107:1041–1047

Zupan J, Muth TR, Draper O, Zambryski P (2000) The transfer of DNA from *Agrobacterium tumefaciens* into plants: a feast of fundamental insights. Plant J 23:11–28

# Index




DOI 10.1007/978-3-7091-1526-8, © Springer-Verlag Wien 2014

## Q

MIX
Papier aus verantwortungsvollen Quellen
Paper from responsible sources
FSC® C105338

If you have any concerns about our products,
you can contact us on
**ProductSafety@springernature.com**

In case Publisher is established outside the EU,
the EU authorized representative is:
**Springer Nature Customer Service Center GmbH**
**Europaplatz 3, 69115 Heidelberg, Germany**

Printed by Libri Plureos GmbH
in Hamburg, Germany